# Field Guide to the Flower Flies of Northeastern North America

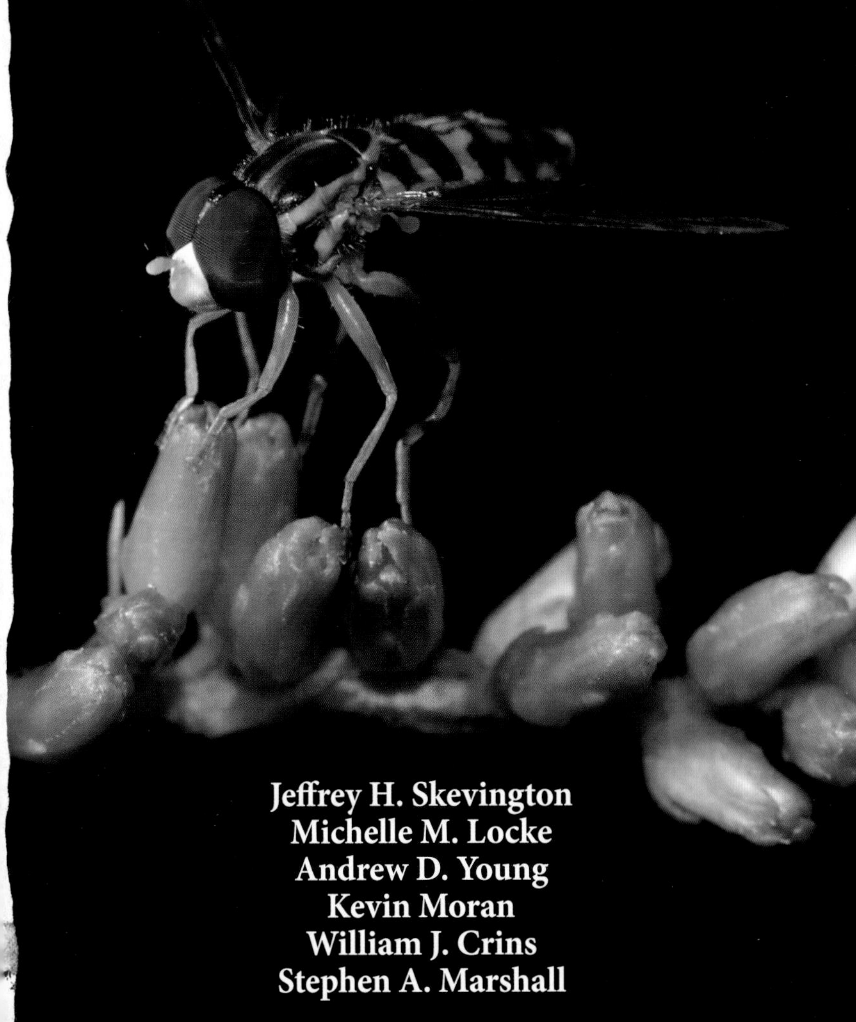

Jeffrey H. Skevington
Michelle M. Locke
Andrew D. Young
Kevin Moran
William J. Crins
Stephen A. Marshall

PRINCETON UNIVERSITY PRESS
PRINCETON AND OXFORD

Published by Princeton University Press
41 William Street, Princeton, New Jersey 08540
6 Oxford Street, Woodstock, Oxfordshire OX20 1TR

press.princeton.edu



Library of Congress Control Number: 2018953762
ISBN (pbk.) 9780691189406

British Library Cataloging-in-Publication Data is available

Editorial: Robert Kirk and Kristin Zodrow
Production Editorial: Mark Bellis
Text Design: Rob Still
Cover Design: Carmina Alvarez
Front Cover Credit: *Doros aequalis* (photo by Steve Marshall)
Page 1 Credit: *Toxomerus geminatus* (photo by Andrew Young)
Production: Steven Sears
Publicity: Sara Henning-Stout
Copyeditor: Laurel Anderton

This book has been composed in Minion Pro and Myriad Pro

Printed on acid-free paper. ∞

Printed in China

10 9 8 7 6 5 4 3 2 1

# Contents

# Dedication

This book is dedicated to our colleagues and syrphidologist mentors J. Richard (Dick) Vockeroth and F. Christian (Chris) Thompson. Dick provided the incredible taxonomic framework for Nearctic Syrphidae that we have been building on. He freely exchanged ideas with us, made his unpublished keys available for general use, and left a remarkable legacy of specimens supporting this research in the Canadian National Collection of Insects, Arachnids and Nematodes (CNC). Dick passed away in 2012 but before that worked with Chris for many years to combine their immense knowledge. In addition to dozens of publications, Chris has always made his unpublished work freely available and encouraged a cadre of new syrphid students to pick up the torch and continue his work. Without Chris making his "Conspectus of Nearctic Syrphidae" available to us years ago, this field guide could never have been envisioned. We have relied heavily on discussions with Chris and the use of his conspectus to build our knowledge and the content of this book.

Dick Vockeroth on a field trip near Kyushu, Japan, after the International Congress of Dipterology (September 2006).

Dick Vockeroth (left) and Chris Thompson (right) in Ottawa in February 2010. Chris was in Ottawa to coteach a syrphid course.

From left: Jeff Skevington, Andrew Young, Chris Thompson, and Kevin Moran at Chris's home in Florida in February 2016.

Chris Thompson and Michelle Locke at the syrphid course in Ottawa in 2010.

# Foreword

The first thing that delighted me about this book was the cover, especially because it did not feature a photograph of a bee. After all, many books and articles about bees have been illustrated with hover fly photos by mistake, eliciting groans from hover fly enthusiasts. I guess it turns out that hover flies are more familiar and beloved than most people realize. We simply mistake them for bees and, in fact, we seem to prefer them to bees. Let's face it: hover flies are vastly more photogenic. And yes, they do pollinate flowers, as you will learn in the pages that follow.

My own interest in the Syrphidae began back in my amateur entomology days, as a teenage insect collector, when I was deeply in love with beetles and butterflies. Flies in general failed to capture my attention, but I realized that at least some groups of flies (hover flies, bee flies, and robber flies especially) contained many good-looking, sizable, and ecologically interesting species. It was difficult back then to find identification resources for any of these families of flies, so my interest failed to blossom. I did devote one episode of my former television series, *Acorn the Nature Nut*, to the subject of flies, but it took until my fifth decade before I began exploring the marvels of two-winged insect life, helped along by some dipterist friends, improvements in my ability to work through the technical literature, and some excellent online identification resources put together by many of the same specialists who have now produced this book.

I don't live in northeastern North America; I live in Alberta, but that does not diminish my enthusiasm for this volume. I grew up using eastern guides to butterflies and beetles (in the days when there were no western guides), and I found that they worked quite well east of the foothills of the Rocky Mountains. As a field guide author myself, I am fully aware of the need to focus on the geographic area that you know well, rather than attempt to market your book more widely than the region for which it was intended. However, because this book contains so much excellent information and so many clear, instructive photographs of syrphid structures, it will be an essential component of my dipterist's library, and those of hover fly enthusiasts worldwide. And who knows, perhaps the existence of a *Field Guide to the Flower Flies of Northeastern North America* will inspire others to produce an equivalent guide for the west, or perhaps the prairie region of Canada. If so, I hope that some of these same authors are involved.

I now have a modest collection of hover fly specimens (it currently fills four Cornell-style insect collection drawers), and I genuinely enjoy the process of examining each under the microscope (good stereomicroscopes are more accessible and affordable than ever before, as are gooseneck LED lamps, of the kind sold at IKEA). I realize, however, that not every naturalist is inclined toward collecting specimens. In my opinion, it is unfortunate that the ethical concerns that typically underlie environmentalism ("what, you mean you kill those pollinators!?") make it difficult for so many of us to recognize the value of specimens, and the fact that this value vastly outweighs any perceived damage caused by collecting. Trust me, insect collecting is not a threat to insect populations, especially the populations of such fecund creatures as hover flies. Other, real threats to pollinators, including habitat destruction and pesticides, can be identified and reversed only if we know exactly what effect they are having, and the only way to obtain these data is through specimen collecting. If you are just beginning to appreciate the Syrphidae, you may be skeptical at this point, but once you have gotten your feet wet, you will surely agree that attempting to identify most syrphids in the field (especially any that look like they might be a member of the genus *Syrphus*), or from photographs, is more or less impossible.

Having said that, I do enjoy insect photography, and I find hover flies remarkably photogenic. While bees engage in hyperactive, hunched-over, vibratory flower visits, making decent photographs frustratingly difficult (most of my bee photos are of fuzzy butts protruding from flowers), hover flies are insect supermodels, posing majestically, head up, eyes shining, colors glinting in the sunlight. I find that having a combination of specimens I can examine under the scope, and photographs taken in the same area at the same time, allows me to identify a majority of my images with much more confidence. Last month, for example, I went through all my marsh flies (*Helophilus* spp.), first the specimens, then the photos. What was originally a group of nearly identical flies has now coalesced into five quite recognizable species, and I'm feeling pretty good about that group, at least. On the other hand, in my photographs from before I became enamored with syrphids, I discovered a shot of the rare Vockeroth's Pond Fly, *Sericomyia vockerothi*. Wow! With that photo as incontrovertible evidence, I can hardly wait to go back to that spot to see if I can find another.

I anticipate keeping my copy of the field guide right beside my microscope, along with Dick Vockeroth's original monograph on the Canadian Syrphinae. On my computer, I have PDF files for the *Manual of Nearctic Diptera*, and a few key papers on particular groups of syrphids. Anyone who has ascended to the level of entomological awareness that inspires them to obtain this book will also quickly discover the standard online resources for hover fly identification, including BugGuide.net, and a number of sites for the fauna of the United Kingdom. On my main bookshelf, I might have another copy of this field guide (I'm that kind of person), right beside Steve Marshall's *Flies: The Natural History and Diversity of Diptera*, Swedish author Fredrik Sjöberg's *The Fly Trap* (my main source of inspiration and wisdom when it comes to the pursuit of syrphids), and Francis Gilbert's *Hoverflies*, from the Naturalists' Handbook series for Britain.

I purchased the latter volume at the gift shop of the Natural History Museum in London, England, in the late 1980s. The clerk, an older woman with half glasses perched at the tip of her nose and a sophisticated British accent, looked up at me and asked, "Are you a specialist in this area?" I replied no, but I was willing to learn, and then I asked why she wanted to know. Was it rare for her to sell a copy? "I don't think we've ever sold a single one" was her reply. Fortunately, things have now changed, and hover flies are all the rage with entomologically inclined naturalists. I still treasure that small book, with its excellent color illustrations.

Let's face it, the state of pollinators is now a major conservation concern, and there are plenty of signs that enthusiasm for the Syrphidae is spreading as a consequence. The other day, a colleague asked me if I had any promising undergraduate students with an interest in syrphids or bees, who might be interested in taking on a master's project involving wild pollinators. I did indeed and introduced my colleague to a young woman just now completing a study of hover flies at our university's botanical garden, under my guidance. Perhaps this will be the beginning of a productive study, and if so, I'm sure she will be delighted at the timing of the publication of this field guide.

Either way, pollination may be timely, but good natural history and taxonomy are timeless. A book of this nature will retain its usefulness as long as books exist, and I am delighted to have had the opportunity to participate in the arrival of this marvelous resource. I congratulate the authors, encourage the readers, and applaud the whole shebang. As I write, spring is on the way, and one of the great things about hover fly studies is that they begin as soon as insect activity begins in springtime and persist to the last warm days of autumn. I can hardly wait.

*John Acorn*

# Acknowledgments

It's hard to know where to start, as so many people have contributed. Although our field guide has liberally borrowed ideas from many of the leading vertebrate field guides, this book is very unlike these guides in many aspects. Vertebrate taxonomy (and that of the prominent insect groups like butterflies and dragonflies) is robust, and preparation of a guide to these groups requires relatively little primary research. Despite the fact that Syrphidae taxonomy is more robust than that of many fly groups, a huge amount of research was required to complete this guide. We examined and compiled data from 147,971 specimens for this project. Two MSc theses, one undergraduate thesis, parts of other student theses, and several scientific papers were published in order to further our understanding of some of the groups involved. DNA sequencing of most species was completed to test existing species concepts. Many new species were discovered during this process, and many existing species concepts were revised. This book thus serves as more than just a field guide. It is a snapshot of our current understanding of the taxonomy of eastern Nearctic Syrphidae. It is clear from this work that a second edition will be needed for refined species concepts, but we hope this will be on the back of a swell of knowledge that would have been impossible without the guide.

With this context in mind, we owe a huge debt of gratitude to the 3,835 collectors who contributed syrphid research specimens to collections that we examined. Our national, provincial/state, and university collections are pivotal to research on taxonomy of syrphids and are a treasure trove of historical data. Many scientists contributed to the solid foundation from which we worked, most notably Chris Thompson and Dick Vockeroth (to whom we have dedicated this book). The origin of the project can be traced back to discussions with Ian Carmichael about creating some quality Diptera field guides. Ian worked tirelessly with Bill and Eileene Stewart to document the fauna of Elgin County, Ontario and Ian's series of guides to Elgin County nature served as a catalyst to this field guide. In fact, the first version of this guide was started as a local guide to Ontario species following Ian's model. It morphed from there to what you now see.

Many students helped to database material that we examined and thus made this material available for analysis and inclusion in the guide. Thanks very much to Sheri Albers, Anna Baillie, Lisa Bartels, Meagan Blair, Nadia Boukina, Abigail Boursiquot, Veronica Bura, Rene Chabot, Anita Cheng, Kevin Colmenares, Jessica Diguer, Bridget Dueck, Mariah Fleck, Ela Godbole, Claire Hobden, Adam Jewiss-Gaines, Bronwyn Kelly, Scott Kelso, Sarah King, Melissa Kohlman, Sarah Matheson, James McConnell, Meredith Miller, Tejal Mistry, Ombor Mitra, Sebastian Namek, Nestor Nebesio, Victoria Nowell, Purvasha Patnaik, Emily Sayadi, Jenny Smith, Andrea Sugarman, and Lyn Vakulenko for their professional work on this material and their dedication to ensuring that the label data were interpreted correctly. Everyone was fabulous to work with, but in particular, we want to single out Victoria Nowell. who provided leadership to many in this group, setting the bar very high in terms of productivity, accuracy, and attitude. Victoria also prepared all of the maps in this guide and we owe her a great debt for this.

Scott Kelso did an enormous amount of molecular work to support species concepts used in the guide. His data, together with data collected at the Biodiversity Institute of Ontario, constitute an impressive 13,263 specimens with DNA barcode data that were evaluated as part of this project. All of these data are stored in the online BOLD database, and most are publicly accessible.

Gil Miranda took the lead on a key to the genera of Nearctic Syrphidae, which stands as a superb supplement to this field guide (Miranda *et al.* 2013). Martin Speight's work to compile information on European Syrphidae (Speight 2016) and

Graham Rotheray and Francis Gilbert's (Rotheray and Gilbert 2011) work on the natural history of Syrphidae were also very important and key to our being able to present so much natural history information on our flower flies. Very little ecological work has been conducted on North American syrphids, so these European data give us background we would not otherwise have. Finally, the review of Syrphinae prey by Rojo and his colleagues was critical in providing insight into the natural history of some of our North American species (Rojo *et al.* 2003).

Funding for the project was provided by grants from Agriculture and Agri-Food Canada (AAFC) to Skevington, and from the Natural Sciences and Engineering Research Council of Canada (NSERC) to Marshall, Skevington, and Peter Kevan. The latter grant (NSERC-CANPOLIN project) was particularly critical, as it supported several graduate student projects on syrphids as well as part of the databasing initiative. Laurence Packer played a key role in the CANPOLIN grant and ensured that funding was provided to support fly research. Skevington's colleagues at the CNC were all very supportive of the project. Andy Bennett, Jeff Cumming, and Jim O'Hara supplemented Skevington's AAFC funding to help reach some of the goals. Scott Brooks and O'Hara sorted material out of trap residues that were used to support the project. Cumming, O'Hara, Brooks, and Brad Sinclair all contributed specimens to the project and supported the rapidly expanding infrastructure and new material for the project.

Loans of specimens from several collections were necessary to complete this work. Thanks go to the following for coordinating these loans: Jim Boone and Crystal Maier (Field Museum of Natural History), Stephanie Boucher and Terry Wheeler (Lyman Entomological Museum, McGill University), Don Bright and Boris Kondratieff (C. P. Gillette Museum of Arthropod Diversity, Colorado State University), Brian Brown (Los Angeles County Museum), Rob Cannings and Joel Gibson (Royal British Columbia Museum), Doug Currie and Brad Hubley (Royal Ontario Museum), Zack Falin, Michael Engel, and Jennifer Thomas (Snow Entomological Museum, University of Kansas), Terry Galloway, Rob Roughley, and Jason Gibbs (Wallis-Roughley Museum of Entomology, University of Manitoba), David Grimaldi (American Museum of Natural History), Chris Grinter (Illinois Natural History Survey), Martin Hauser and Steve Gaimari (California State Collection of Arthropods), Steve Heydon and Lynn Kimsey (Richard M. Bohart Museum of Entomology, University of California, Davis), Pam Horsely (Albert J. Cook Arthropod Research Collection, Michigan State University), Norm Johnson and Luciana Musetti (C. A. Triplehorn Insect Collection, Ohio State University), Valerie Levesque-Beaudin (Biodiversity Institute of Ontario, University of Guelph), Karen Needham (Spencer Entomological Museum, University of British Columbia), Peter Oboyski (Essig Museum of Entomology, University of California, Berkley), Steve Paiero (University of Guelph Insect Collection), Norm Penny and Michelle Trautwein (California Academy of Sciences), Phil Perkins (Museum of Comparative Zoology, Harvard University), James Pitts and Wilford Hansen (Utah State University), Terry Schiefer (Mississippi State University), Virginia Scott (University of Colorado), Derek Sikes (University of Alaska), Zoë Simmons (Oxford University Museum of Natural History), Felix Sperling (E. H. Strickland Entomological Museum, University of Alberta), Chris Thompson and Torsten Dikow (National Museum of Natural History), Jason Weintraub (Academy of Natural Sciences of Drexel University), and Doug Yanega (Entomology Research Museum, University of California, Riverside). Mike Irwin, Martin Hauser, John Klymko, and Eleanor Proctor went to extra lengths to ensure that their personal collections were available for the project. Irwin and Proctor donated substantial collections to the CNC.

Kathryn Vezsenyi and David Beresford's work on northern Ontario and Nunavut Syrphidae contributed greatly to our understanding of syrphid distributions in the north. Beresford collected specimens from Akimiski Island with support provided by the Ontario Ministry of Natural Resources Wildlife Research and Development Section (in collaboration with Ken Abraham and Rod Brook). Far north samples from Ontario were provided by the Ontario Ministry of Natural Resources and Forestry and the Far North Biodiversity Project (FNBP). Significant support from local communities allowed this work to be completed (including Peawanuck, Marten Falls, Webequie, Nibinamik, Kitchenuhmaykoosib Inninuwug, Keewaywin, Fort Albany, and Fort Severn). Dean Phoenix, Dave Etheridge, and John Ringrose were key contributors to this work.

Wing drawings and drawings for the morphology section of the guide were made by Jessica Hsiung. Genitalia drawings that were previously published were provided with permission from AAFC.

Many people contributed comments on the guide, but John Klymko, Martin Hauser, Bill Dean, and Barry Cottam completed the Herculean task of reviewing the book. Jeff Skevington's wife, Angela, and son Alexander were a great help on the many field trips collecting specimens for the project and also provided advice and feedback on design of the book.

# General Introduction

Although bees get most of the press when it comes to pollination, recent research has shown that flies are also extremely important and carry out about one-third of our pollination services. Among the flies, syrphids are usually the most important pollinators. Together, insect pollinators provide over 500 billion dollars' worth of direct benefit to agriculture annually. The actual value when pollination of native plant species is factored in is incalculable.

In addition to the value of pollination services, the importance of immature hover flies must be considered. About one-third of all syrphid larvae (mostly Syrphinae and Pipizinae) are predators of soft-bodied insects like aphids and scales and as such bring huge benefits in natural and biological control of pests. Other syrphid larvae are extremely important in the recycling chain, with species of drone flies (*Eristalis*) and lagoon flies (*Eristalinus*) being the major players in sewage recycling, and jewel flies (*Ornidia*) playing a big role in tropical compost turnover.

Many syrphid larvae in the subfamily Eristalinae are bacterial filter feeders in sap runs or under bark. Many are old-growth forest specialists and are important indicators of overall habitat quality, and some are among the most threatened insects in degraded habitats.

Ant flies (Microdontinae) are among the most ecologically amazing groups of syrphids. Larvae of species with known biologies are found in ant nests, where they feed on ant eggs and larvae while mimicking their host's pheromones to avoid detection. Most known ant flies are predators, but recently the life history of the first parasitoid in the family, *Hypselosyrphus trigonus*, was discovered. This recently discovered Central American species develops externally on ant prepupae, thus avoiding the need to overcome the immune system of the host, as in internal parasitoids.

If you are starting to get the impression that flower flies do almost

*Cheilosia* covered with pollen. Recent research has shown that nonbees perform 39% of our agricultural pollination services.

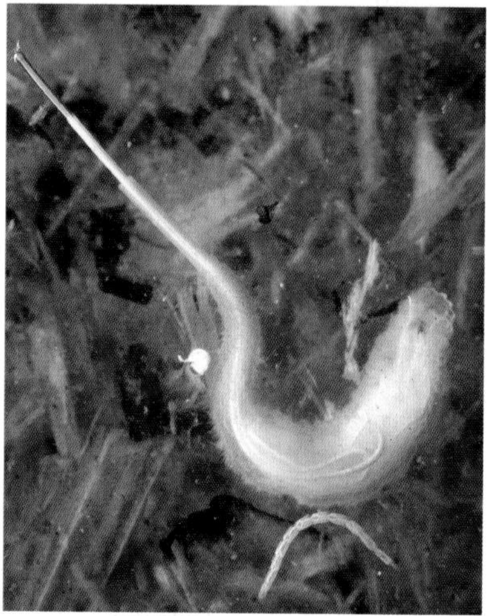

Many rat-tailed maggot larvae (*Eristalis* and relatives) are important recyclers in ponds and lagoons, while others live in tree holes, bromeliads, and other standing water. This one was in a tree hole in Backus Woods, Ontario.

*Callicera erratica*, a presumed old-growth forest specialist.

everything as larvae, you are correct. In addition to the larval life histories mentioned above, there are also plant feeders (including root, stem, and leaf feeders, even leaf miners), fungal feeders, specialized pollen feeders, predators in bee nests, aquatic filter feeders in ponds, puddles, and bromeliads, heartwood borers in old trees, and much more. Few families of insects have such a diverse range of feeding styles, and because of this, syrphids are a model system for the evolution of feeding strategies.

All of our *Microdon* larvae with known biologies are predators in ant nests. At least some mimic the pheromones of their host ants to avoid detection.

Syrphids are also a model system for research into mimicry. Most adult syrphids mimic wasps or bees in some way. There are some spectacular cases of "perfect mimicry" where syrphids look and behave almost exactly the same as their models. Have a look through the wasp flies (*Ceriana*), swiftwings (*Volucella*), yellowjacket flies (*Sphecomyia*), and mimics (*Mallota*) to get an appreciation for these remarkable feats of evolution. If you think that the physical resemblance is impressive, wait until you see some syrphids in the field. Some falsehorns (*Temnostoma*), pond flies (*Sericomyia*), and hornet flies (*Spilomyia*) waggle their prolegs to mimic wasp antennae. Their legs are patterned just like their wasp models'. Catch a swiftwing in your hand and you will likely release it on impulse when it buzzes just like a bumblebee. These mimics obtain considerable protection by looking and behaving like stinging wasps and bees. No syrphids bite or sting, but they certainly advertise that they do. While the larger syrphids are often perfect mimics, most of the smaller species (most of the Syrphinae, for example) are known to be "imperfect" mimics. These species have converged on general wasp models (i.e., black and yellow stripes/spots) and are thought to gain protection by looking more or less like bees and wasps and being small enough not to merit a second look by vertebrate predators.

*Xylota* larva under wet bark.

*Criorhina nigriventris*, a perfect mimic of an eastern *Bombus* (bumblebee).

We may be a bit biased, but on top of all of these marvelous natural history traits, we think that syrphids are among the most attractive of all flies. Most are field identifiable and the diversity is comparable to that of other "accessible" animal groups such as birds. There are over 6,300 described species of hover flies in the world, but only 413 species in the area covered by this book. With a bit of time and dedication, it is just as easy to learn to recognize the majority of adult hover flies as it is to learn to identify most birds.

## How to Use This Book

Our goal was to make everything in this book as self-evident as possible, but below are some notes to support the field guide pages.

## Guide Coverage

This book covers all 413 known syrphid species that occur in or north of Virginia, Kentucky, and Missouri, west to include Iowa, Minnesota, Ontario, and Nunavut, and east to the Atlantic Ocean, including Greenland (the area in green on the map below).

*Range Maps*
Specimen data were accumulated in the Canadian National Collection of Insects, Arachnids and Nematodes (CNC) database (available from http://www.cnc-ottawa.ca) based largely on specimens in the CNC as well as the American Museum of Natural History (AMNH), the University of Guelph insect collection (DEBU), and the Smithsonian National Museum of Natural History (USNM). BugGuide (http://bugguide.net/) data were incorporated from states not found in collections examined. Dots on the maps represent specimen records. Colored areas on the maps were inferred using these point data in combination with vegetation, elevation, and climate layers (using the program MaxEnt and compiled in QGIS). Darker shading on the maps indicates areas where the species is more likely to occur. The entire Nearctic range of included species is shown.

*Illustrations*
Field photos were used when available and when we were able to definitively identify species in the photos. Specimen vouchers support many of these field photos and are available in the CNC and DEBU. Museum photos are linked to specimens in the AMNH, CNC, DEBU, and USNM. If there are ever questions about the identity of specimens, we can trace the photos back to these specimen vouchers to check. Lab photos were mostly taken with a Canon EOS 50D

Imperfect mimics like this *Eupeodes* all converge on a similar gestalt, making them very difficult to identify.

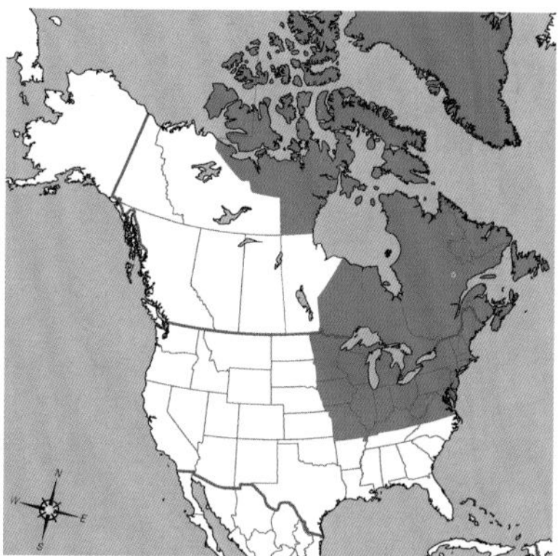

All syrphid species known from the green-shaded area are included in the guide.

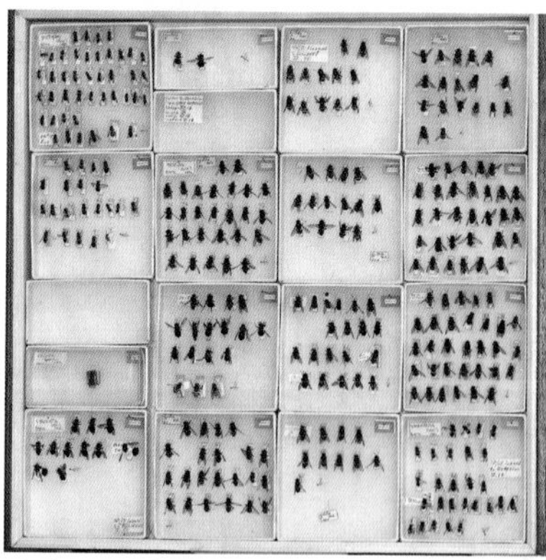

A drawer of syrphid specimens (*Blera*) from the CNC, showing regional colored labels and a type specimen.

and 100 mm or 65 mm macro lens on a StackShot rail. Multiple photos were usually montaged into single images using Zerene Stacker to provide complete depth of focus. Other specimen photos were taken with a DFC495 five-megapixel camera attached to a Leica M205C stereomicroscope. Image stacks were acquired with the Leica Application Suite software and stacked using Zerene Stacker. Images were processed using Adobe Photoshop.

*Toxomerus marginatus*, arguably the most abundant syrphid in northeastern North America.

*Estimates of Abundance*

We apply qualitative terms in the text for each species to give a rough idea of how common the species are in the region of greatest abundance and during their flight periods. Our definitions for these terms are the following: **Abundant** – can be found in large numbers daily, **Common** – found daily without much effort, **Uncommon** – either low density or local and require special effort to be found most days, **Rare** – very low density or very local species, likely to be found annually, **Very Rare** – very low density or very local species unlikely to be found annually, **Vagrant** – presumed not to be able to survive within the region of the guide, occurring only rarely as a migrant.

## It's All in a Name

Flower flies are also known as hover flies (and incorrectly as hoverflies). We prefer the name flower fly, as it best captures the true essence of most adult flies in this family. However, there is no dispute that some syrphids are among the best fliers in the animal kingdom, capable of hovering motionless for minutes at a time, backing up out of flowers, and zooming off in an instant. We concede that most people will still connect with the name hover fly and have thus included it here. However, we use flower fly interchangeably with hover fly throughout the text, as we find the former more descriptive.

Although lots of syrphids hover, many more, like this *Temnostoma*, visit flowers; thus our preference for the common name flower flies.

Ant flies, such as this *Microdon*, are in the least-encountered subfamily of syrphids (Microdontinae).

Many of the small black syrphids are in the subfamily Pipizinae.

Eristalinae, the most varied subfamily, includes many bee-like members such as this *Mallota posticata*.

Syrphinae mostly have a generalized wasp-like gestalt, similar to this *Syrphus*.

### *Subfamilies and Tribes of Flower Flies*

There are currently four subfamilies of flower flies recognized: **Microdontinae**, **Eristalinae**, **Pipizinae**, and **Syrphinae**. Each subfamily is color-coded throughout the book. Microdontines (ant flies) are ant associates. Eristalines make up most of the family's diversity, varying widely in morphology and natural history, and possibly not forming a natural group. Pipizines are small black syrphids that specialize on root aphids. Syrphines are mostly imperfect mimics as adults, and predators of soft-bodied arthropods as larvae. The term "tribe" is often used to circumscribe smaller and more manageable groups of species. Phylogenetic evidence suggests that many tribes are not natural groups, and most are of little value in a field guide. We therefore avoid the use of tribal names.

### *Scientific Names for Species*

The best current database of scientific names of syrphids can be found at http://www.cnc-ottawa.ca (adapted from http://www.diptera.dk). Based on these databases and new research, we provide a list of valid Nearctic Syrphidae names at http://www.canacoll.org/Diptera/Staff/Skevington/Syrphidae/Syrphidae_Nearctic_Checklist.htm. Flower fly taxonomy is more in flux than taxonomy of vertebrate groups but is among the best available for any group of flies. There are still new flower fly species to be discovered and described (in fact, we have discovered many in the lead-up to publishing this book), and taxonomic hypotheses are typically based only on a combination of adult morphology and DNA. Larvae of very few Nearctic Syrphidae have been described, and behavioral observations are primitive compared to those for vertebrates. Research on interbreeding and knowledge of hybrid zones are still largely lacking. We hope that this field guide will help fill some of these gaps and produce more rigorous species hypotheses as field identification of syrphids becomes attainable.

*Common Names for Species*

Very few syrphids have been given common names, and many websites are now starting to offer common names for species with no attempt to standardize. We have proposed a set of common names for all Nearctic syrphids (http://www.canacoll.org/Diptera/Staff/Skevington/Syrphidae/Syrphidae_Nearctic_Checklist.htm) and provide them here with the species accounts. Our goal was to provide a single name for each group of species with descriptors added to each. This gives an idea of how species are related, much as generic names do. We tried to create these names with the global fauna in mind and thus hope that syrphidologists in other regions will follow our lead.

This is an undescribed species of *Hammerschmidtia*. Some undescribed species like this have been known for years, while others were discovered while producing this guide.

We created common names that we hope are memorable and descriptive. The sedgesitters (*Platycheirus*) are often found on sedge and grass flowers, where they presumably feed on pollen.

Almost all adult flies have two wings (most insects have four).

Flower flies are variable in appearance, but most of our species have a spurious vein. Only a few Conopidae share this trait.

## Identifying Flower Flies

Is it a fly? Diptera is the scientific name for flies and derives from the Greek words *di* (two) and *ptera* (wings). Whereas most insects have four wings, flies have only two. The hind pair of wings are reduced to small clubs, called halteres, for orientation.

### *Is It a Flower Fly?*

Syrphids usually have large heads, large eyes, and short antennae; many are mimics of wasps or bees. If in doubt, check the wing venation. All of the syrphids in our region except *Psilota* and *Syritta flaviventris* have a spurious vein (a "false" vein, meaning that it is not joined to any others).

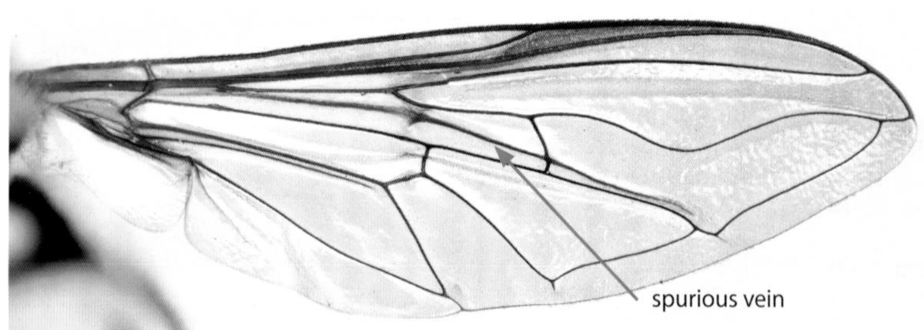

The online key to Nearctic syrphid genera will help narrow things down if you can't find your species by flipping pages.

Identifiable without having to catch it if you get a good view

Identification requires careful examination with a hand lens

Identification requires microscopic examination

Indicates species that hilltop (*see* Hilltopping *page 22*)

Shows the middle of the typical size range for each species (*see* Measurements *page 22*)

Icons used in the book.

*Parts of a Flower Fly?*

As with all groups, you have to learn a specific terminology in order to converse and use the literature. We cover all of the terms used in this book in the illustrations and the morphology sections in the glossary.

*Identification to Genus and Species*

We have laid out the book so that similar species appear together to facilitate flipping through the images in search of a match. If you cannot match the species you see to photos, figuring out what genus you are dealing with is critical. Learning to recognize the genera of syrphids in the field will help immensely with species identification. A free online illustrated key to Nearctic syrphid genera is available at http://cjai.biologicalsurvey.ca/mylmst_23/mylmst_23.html. Each species in the guide has a symbol associated with it to indicate the detail required for identification. Some species are quite easy to recognize (the eye symbol indicates you should not have to catch it if you get a good view), while others will require careful examination with a hand lens in the field (hand lens symbol) or even a microscope in the lab (microscope symbol). Species that require you to see the microtrichia on the wings require either a high-powered hand lens or microscope. Don't expect to be able to immediately field identify all of the species marked with the eye (👁). You may need to capture and examine many of these "easier" species to first become familiar with them.

*Life Cycle*

Syrphids are typical of all flies (and in fact all higher insects) in that they go through a complete metamorphosis. Eggs hatch into larvae, larvae transform into pupae, and pupae transform into adults. Syrphid larvae go through three "molts" before forming a pupa within a puparium, which is essentially a cocoon made from the skin of the last larval stage. In contrast, lower flies (including groups like mosquitoes and horse flies) usually have an exposed pupa.

*Sexual Dimorphism*

Males and females of most syrphids look the same except for their genitalia and the relative size of their eyes. In most species, the eyes of males are larger and touch above

Most males can be distinguished from females by looking at the eyes. In these mating *Toxomerus*, the male's eyes (top) touch in front but the female's (bottom) do not.

Both photos are of *Volucella facialis*. As with their model bumblebees, they have both orange and yellow marked color morphs.

their antennae (holoptic), whereas most females have smaller, completely separated eyes (dichoptic). Males also typically have one extra visible abdominal segment and often appear to have a swollen tip to their abdomen because of their genital capsule. Some syrphid species are sexually dimorphic, with males and females looking quite different. This gives us a bit more to learn and also means that the dimorphic species pages will be more crowded with illustrations of the characters of both sexes (see the sickleleg [*Polydontomyia*] page 84 for an example of this).

### *Variation within Species*

Some flower flies have different color morphs (for example, the bumbleflies [*Criorhina*] often have red and yellow morphs like the bumblebees they are mimicking). Other syrphids look generally the same but vary in more subtle details such as leg color. This type of variation often confounds species hypotheses, and some variable species have been described under many different names. Molecular data are often useful in distinguishing intraspecific variation (variation within a species) from interspecific variation (variation between species).

*Measurements*
Size variation within syrphid species is often considerable and presumably relates to larval diet. Sometimes, when different generations of larvae feed on different things at different times of the year, size varies seasonally. Measurements of the typical size range of each species are presented in millimeters. A life-size gray silhouette in the middle of the typical size range is provided for each species. Holding a specimen up to these shapes should give a quick idea of whether or not you are in the correct ballpark.

*DNA*
We have been building a DNA database using several genes. Our most extensive dataset uses part of the mitochondrial cytochrome c oxidase 1 gene (the "DNA barcoding gene" region). Although this is not a panacea for all of our taxonomic problems, it has proven to be useful in delimiting about 80% of Nearctic syrphid taxa at the species level. The most important direct use of this database (accessible at http://www.boldsystems.org/) is in the identification of immature stages. Very few larvae of Nearctic Syrphidae have been studied or are even known. If you find a larva, you have the option of trying to rear it (time consuming but possible) or putting it in 95% alcohol and sending it to someone who can sequence it (Jeff Skevington [JHS] is often available to help with this). It is important to photograph the larva and take good notes on the habitat in which it was found.

Unlike most syrphids, *Platycheirus* species can often be found in cloudy or even rainy weather.

## Observing Flower Flies

*When to Find Flower Flies*
Flower flies are most common in the shoulder seasons (spring and fall), but species vary markedly in time of emergence from the southern to northern ends of their range. Small regional field guides often show flight periods for each species, but given the broad latitudinal range covered by this guide, this is not attempted here and we provide only general estimates of flight periods for most species. In the southern part of the area covered by the guide, some syrphids may be found on warm days in any month of the year. Around the latitude of southern Ontario, the first syrphids start to appear in late March and the peak flight period is from late May to mid-June. Different habitats also see different peak periods, with flower flies in cool environments such as bogs flying up to a month later than those in surrounding habitats. Search the CNC database (http://www.cnc-ottawa.ca/) to find dates of occurrence for particular syrphid species within your area.

Focus your observational efforts from early morning to early afternoon. Flies are morning creatures and although you may find them all day, most flies are not very active from mid- to late afternoon. The best temperature for syrphid activity is typically between about 15 and 25 degrees Celsius. If it is too hot, there is a risk of desiccation, so they are not active. Cool, sunny spring days with low wind encourage activity at lower temperatures. Sun is key for finding most species. Even a passing cloud will radically reduce flower fly activity. Don't entirely give up on overcast or rainy days, though. Some syrphids, including many of the sedgesitters (*Platycheirus*), appear to be most active when it is overcast and even raining lightly.

*Flower Fly Habitats*

Flower flies can occur in any terrestrial habitat but are least diverse in arid and semiarid environments. They lack water retention systems such as those found in bee flies (Bombyliidae) and many beetles, so most flower flies do poorly in dry environments. The best environments for finding a diversity of syrphids are forest openings, natural meadows, and riparian areas. Flowers can be very productive; flat-topped white compound flowers are particularly attractive to flies. Slowly walking back and forth through a productive area of flowers is a good strategy, and when one plant turns out to be attractive, sitting and watching it can produce a great variety of species. Remember to watch flowers on trees and shrubs as well as herbaceous plants. When not watching flowers, try looking for syrphids around freshly fallen trees or on tree wounds. Other species will be found sitting on low vegetation in sunny patches along forest edges/paths. Patiently watching at these sites will often yield some of the least common species. Some species (typically males) hover over woodland paths, while others hover over shrubs or on hilltops.

*Contrast between Europe and North America*

Disturbed areas (weedy meadows) in North America can be rich in flowers but almost devoid of syrphids. This is in stark contrast to Europe, where similar-looking meadows are literally teeming with syrphids. Flower flies are often the most abundant, conspicuous, and diverse inhabitants of meadows in Europe. People used to hunting for syrphids in Europe will be shocked at how outwardly similar habitats here contain only a few common flower fly species. We assume that this is because disturbed meadows here are dominated by nonnative plants (in fact, they look so much like European meadows because the floral diversity is so similar). To find a good diversity of syrphids, you thus need to find meadows dominated by native plants. Wet meadows, prairies, and savannas typically fill this role. Unfortunately, the latter two habitats have been decimated in North America and are now among the most endangered ecosystems that we have. For examples of these systems, look at Cook County Forest Preserves (Illinois), Indiana Dunes National Lakeshore, Neal Smith National Wildlife Refuge (Iowa), Oak Openings Preserve Metropark (Ohio), or Pinery Provincial Park (Ontario).

*Hilltopping*

One of the best ways to find the rarest syrphids is to go to hilltops. Males of rare or unpredictably distributed species cannot easily find a mate and use hilltops as landmark mating sites. Hilltops do not have to be high; they just need to be prominent to attract a hilltopping fauna. In flat areas, a low rise can function as a hilltop, whereas in rolling areas you should look for the most prominent hilltop to start your search. Ridges or escarpments tend to spread out the hilltoppers, so in these situations look for spots that are slightly higher or jut out. Hilltopping is very much a morning phenomenon, starting early and winding down quickly after noon. Ensure you look everywhere, for flies are very specific as to where they occur on hilltops (hovering in the open or under cover, sitting on shrubs, on rocks, on human-made vertical or horizontal structures, on tree trunks, twigs, etc.). Every year we visit the same hilltops and find new syrphids that we overlooked on past visits. In most cases, we can find

*Ceriana willistoni* is a rare syrphid that can often be found on hilltops. Mount Rigaud, Quebec, 24 May 2015.

the same species again on subsequent visits once we know their tricks.

The spatter marks on this leaf are from artificial honeydew that was sprayed to attract this *Didea fuscipes*.

*Honeydew*

Another feature highly attractive to syrphids is honeydew. Aphids, scales, and relatives produce sticky, sugar-rich honeydew as a waste product, and syrphids and other insects commonly home in on these secretions to feed. If you find a natural site with honeydew speckling the leaves, watch patiently and you should see a good variety of flower flies in attendance. Of course, you can also simulate natural honeydew by making your own. Mix honey, water, and cola together to make a sweet concoction, put it in a spray bottle, and spray it on broad, exposed leaves in sunny patches in forest habitats. Slowly walk your spray route back and forth and you should encounter a variety of syrphids. Honeydew spraying works best a day or two after heavy rains, before natural honeydew builds up again. It is ineffective if there has been no recent rain and there is a lot of natural honeydew.

*Water*

Water sources are particularly attractive to syrphids in periods of drought or on very hot days. Small, shaded creeks with lots of moist, exposed rocks can attract a wide variety of flower flies under these conditions.

*Field Tools*

Binoculars that focus close (within 2 m) are useful. Roof prism binoculars (or related new designs) are the most resistant to bumps and are also typically waterproof. Light transmission may be low in some, and lens coatings may add an undesirable color cast to subjects, so be sure to test new purchases in person rather than buying them sight unseen. Full-size binoculars offer much brighter views than mid-size or compact binoculars.

At the time of writing, Nikon (Monarch 7), and Vortex (Talon) offer some of the best low-end binoculars, while Leica (Ultravid), Swarovski (EL Swarovision), and Zeiss (Victory) dominate the high end. Eight times magnification is likely to work best for most people, as 10× binoculars are harder to hold steady and do not transmit as much light. If you prefer a small binocular, the Pentax Papilio II is popular among entomologists and focuses to 0.5 m. Many review sites are available on the internet for comparing binoculars, but browsing a local store that carries a good variety of binoculars is a good start.

Syrphids can be handled for examination and later release by gently holding their legs between your fingers.

Carrying an insect net may not be for everyone, but if you want to identify all of the flower flies you see, you will need one, even for capture and release. Generally speaking, the smaller the net ring, the faster you can sweep the net and the more control you have over it. Of course, the trade-off is that you have to be very precise and adept with a very small net ring. Some of our colleagues use tiny nets that are only about 7 cm in diameter. We tend to prefer 30–38 cm diameter nets and recommend this size as a starting point. Folding nets are handy if you want to carry one in your pocket and never be caught without it, but the trade-off is that they are more flimsy than nonfolding options. There are many suppliers for entomological supplies, but Bioquip is a good starting point, as they carry a great range of nets and other supplies. Note that pole length is also open to personal preference. Longer poles work well for jumpy flies like robber flies (Asilidae), whereas standard-length 91 cm net poles work well for syrphids. We have found 152 cm net poles work well in some situations but tend to impede you when working around trees. Remarkably, some colleagues use tiny nets with 10 cm handles and rely on stealth to get them close to their targets—insect vision is based largely on movement, so a slow, steady approach may yield results if you opt for this method. Syrphids can be safely caught, handled, examined, and released if you choose not to collect them.

Make sure you have lots of clear vials to put specimens in if you need to keep them for a longer time. A high-quality hand lens is important for examining characters of specimens that you have caught. Ten times magnification should be adequate. Ensure that you evaluate lens quality, as this is the most important feature of this tool. In a pinch, look through your binoculars backward and hold the specimen very close to the ocular lens. This functions as a magnifying glass, albeit an awkward one.

### *Photography*

Options are nearly unlimited, from remarkable lens adapters for phones through to good point-and-shoot cameras with great magnification and macro potential. The problems with these options are typically low depth of field and a time lag when pressing the shutter button. The latter is most aggravating with active subjects. A dedicated DSLR camera system is thus still the best option for serious insect photographers. Canon and Nikon lead the market in this regard, but both have pros and cons. Canon has the best high-magnification insect lens available (the MP-E-65) on the market, but Nikon has more options available in remote flashes and some impressive low-light features for shooting without flash. We used both Nikon and Canon equipment for the field guide. Canon gear included EOS-50D and EOS-1D X bodies paired with either the MP-E 65 mm or the EF 100 mm f/2.8 macro. Kenko extension tubes gave us added magnification range with the latter. With this we used the MT-24EX flash in the field and a variety of flashes and slaves in the lab. Dome lighting proved to be the best for in-lab illumination, and we followed the methods outlined by Kerr *et al.* (2008). Nikon camera bodies used included the D90, D2X, D300, and D800, paired with either a 60 or 105 mm lens.

### *Collecting and Vouchers*

Not everyone reading this will want to collect syrphids, but the only way to identify and voucher some finds will be by using a specimen. Photographs will work as vouchers in many instances, but specimens are still the currency of insect taxonomists and can be preserved for hundreds of years and reexamined by anyone questioning the original identification. This section provides a quick overview of everything required to obtain, preserve, and maintain a voucher collection. For more extensive reading on this, download a free guide written by Martin (1977) at http://esc-sec.ca/aafcmonographs/insects_and_arachnids_part_1_eng.pdf. Note that collecting permits are required for most parks and protected areas.

Syrphidae are best collected by hand and Malaise trap. The latter is a tent-like trap that passively collects insects that fly into the center panel and respond by flying

Malaise traps are an effective way to collect flies, including syrphids. Like all collecting methods, they are not a panacea. Many eristaline syrphids like *Mallota* are undersampled with this method, presumably because they are able to navigate in and out of the trap head as they would a rot hole.

upward, where they then follow the roof line to the trap head. On average, the head of the trap should face the equator to maximize sun exposure to it, as many flying insects, including syrphids, are positively phototaxic and will move into the trap head more readily if it is well lit. Because flies are more active in the morning, angling the trap head to the southeast is optimal. Malaise traps work poorly if placed in shade and best if they intersect a general flight path (road, trail, woodland edge, river edge, etc.). Other options include emergence traps (best used for species that live in wood) and pan traps (yellow pans being the best color for syrphids). Pan traps work well for some groups but are generally not very effective for most syrphids.

If specimens are hand collected, a stainless-steel insect pin (#2 or 3) should be used to pierce the right-hand side of the thorax in larger specimens. Smaller specimens should be glued (glue on right side of thorax) to a point on enamel or stainless-steel pins. Labels should be printed (3.5 to 4 point Arial font) on ~80 gauge acid-free paper and include country, province/state, location, coordinates, collection date, collector, and habitat details. Unique identifiers should be added to labels if you plan to keep a database or refer photographs to individual specimens. These can be in the form of your name plus a number that is never repeated. If specimens are collected into alcohol in traps, they must be properly dehydrated and dried to prevent shriveling. Ethyl acetate (EA) drying is the easiest process for general use. For this, take specimens in 70–80% ethanol (EtOH) and pin them wet (preferably with pins that do not have enamel heads—Asta makes a cheap stainless-steel pin that has a rounded head and works well for this). When pinned, place the still-wet specimens into a 50:50 mixture of EtOH:EA and let them soak for at least four hours. Following this, move them into pure EA with a few drops of glycerin added as a softening agent. Soak for four hours or more and then lay them on a paper towel in a fume hood or well-

Some current collectors. Left: Jeff and Alexander Skevington pinning insects in Arizona. Right: Jim O'Hara and Jeff Cumming preparing the day's catch in Australia.

ventilated area. Position the wings as best you can. Although EA damages DNA, the specimens will be useful for morphological examination for hundreds of years after drying them. Note that ethanol-stored specimens are difficult to identify and degrade rapidly, particularly at room temperature.

Pinned specimens last for centuries but must be kept free of pests. Some beetles (*Anthrenus* species in the family Dermestidae) and book lice (Psocoptera) are the major threats. Keeping the specimens in sealed drawers is the best option but difficult to do without encouraging attack by fungi that can grow in high humidity. Plastic storage containers are particularly problematic for fungi, and you can lose a collection quickly if fungi start to grow. Naphthalene (mothballs) can be used to dissuade pests when nonairtight drawers are used. Freezing specimens to kill pests is a good strategy if an outbreak develops.

The other important thing to consider is the ultimate deposition of your collection. It makes sense to keep a small collection while you are learning, but once it is no longer needed it should be deposited in a permanent insect collection for long-term care, study, and storage. The best syrphid collections in North America are the Canadian National Collection (CNC) and the Smithsonian (USNM). The only other collection to rival these is the Natural History Museum (BMNH) in London, England. Donations of specimens can be made to any of these museums and also to smaller state/provincial, university, and museum collections. If your collection becomes significant, arrangements should be made with your family and a collection representative at the museum of your choice while you are in good health. Virtually everything we know about syrphids comes from these public collections, and citizen scientist donations can be significant. In addition to donations of collections, citizen scientist observations are critically needed to build our knowledge of these fantastic flies. This book aims to facilitate a new generation of research in this way.

### *Record Keeping and Databases*

It is critical to ensure that the data associated with each specimen or photo are accurate. If you cannot be sure of the data for a specimen, it should be disposed of or marked as unknown. A single specimen with erroneous data can undermine confidence in an entire collection, no matter how significant that collection is. Adding unique identifiers to specimens and photos and storing these data in a spreadsheet or database will ensure long-term confidence in your data. A spreadsheet should minimally contain a unique identifier (typically your name or a collection name followed by a unique number—if using a collection identifier, ensure that you are assigned these numbers by the collection manager; for example, CNC12345 would be a unique identifier), collection locality (including country, state/province, specific location, latitude and longitude), collector, date or date range, trap type, habitat, rearing information, identification, and identifier. All data should be on the specimen label as well as in the database. Don't take shortcuts and only put codes on labels. Those who really get serious about this hobby are welcome to contact JHS to arrange access to the CNC database. Users can be set up with access and all data can be directly deposited into the collection database. The CNC database has functions for producing specimen labels from data entered, viewing, mapping and organizing data, and more.

Many databases are available for recording natural history observations. None are yet very robust for entering flower fly records, but two stand out. iNaturalist is growing quickly and may be the best location for entering your records. A free mobile app is available for data entry, and entering single records along with photos is very easy. Entering lists of species from a single location is problematic though, and until this is addressed by developers, entry of more than a single specimen is tedious. Observation.org (http://observation.org/familie/view/15) also has potential but has a less user-friendly interface and is available only on Android at the time of writing. At this time, it seems that iNaturalist is the best bet—at least until we have a syrphid version of eBird or eButterfly.

*Rearing Larvae*

The most important thing to remember when rearing terrestrial insect larvae is that they desiccate easily, so you will find yourself trying to strike that perfect balance between too dry and too humid. Too much humidity will lead to mold problems. Predacious species are usually easy to rear on the right hosts, while saprophagous species should be maintained in the substrate in which they were collected. The easiest larvae to rear are those collected as mature or prepupal larvae, but bear in mind that the puparia too are subject to desiccation. Some saprophagous species are easily collected as puparia or mature larvae in early spring, and these often emerge soon

It may not look like much from the outside, but the Canadian National Collection of Insects, Arachnids and Nematodes housed in the Neatby Building in Ottawa is one of the largest and most significant collections of insects in the world. Collection repositories like this are critical to our understanding of insect taxonomy and phylogenetics and serve as libraries that we can visit to explore changing patterns in our fauna.

CANADA:ON:Queen's
Biological Station,
Pangman Tract;
11-24.vi.2009;
44.52°N, 76.39°W;
MM Locke

CNC DIPTERA
# 188347

Temnostoma
excentrica
Det. M.M. Locke,2013

Labels from a typical specimen. The top label contains information on locality (always include country and coordinates), date of collection (use roman numerals for month so there is no confusion), and collector. The second label is a unique specimen identifier (the prefix and number make it unique in the world). The unique identifier is important for databasing and tracking photos and published references to the specimen. Avoid abbreviations. The lower label indicates the species name and who identified the specimen. Never remove a determination label, even if incorrect. Just add another label with your identification. Never put only coded labels on specimens.

*Eristalis brousii* is one of our most rapidly declining syrphids. It has disappeared throughout most of its range and is now found in only a few places, one of which is Churchill, Manitoba.

after they are warmed up. Placing puparia in containers with moist peat moss usually works for rearing. Ensure that you include toothpicks or sticks that emergent flies can crawl onto to pump up their wings. Without this, the wings often fail to develop properly.

## Conservation

*Rare Species*

Europeans have by far the best knowledge of which syrphid species are at risk and which are not. An annual species account distributed by Martin Speight (Syrph the Net) summarizes all that is known for each European species, including larval behaviors and conservation significance. We have a long way to go to get to this level of knowledge within North America, which means that everyone reading this can contribute a lot. JHS led an Environment Canada initiative to assess the status of all flower fly species found in Canada (Canadian Endangered Species Conservation Council, 2016). This report is based on collection data and as such provides a framework but needs to be followed by rigorous fieldwork. The syrphid that appears to be in the most trouble in North America is the Hourglass Drone Fly (*Eristalis brousii*). This once-abundant species has declined considerably and is now restricted to a few sites in the western mountains and along the northern periphery of its former range. We think that its introduced relative, the European Drone Fly (*Eristalis arbustorum*), has displaced it, possibly through competitive hybridization. Other species to watch closely are extreme specialists (such as ant flies [Microdontinae]) and old-growth forest specialists (such as pine flies [*Callicera*]).

*Syritta* is an Old World genus with its center of diversity in Africa. Despite this, at least two species have made it to North America. *Syritta pipiens* larvae live in rotting material such as compost and have made their way around the world in association with people.

### *Introduced Species*

A few syrphids from other parts of the world have become established in our area. The European Drone Fly (*Eristalis arbustorum*) mentioned above is one of the most successful and can now be found on flowers throughout most of the field guide area. Other introduced species include all of the bulb flies (*Eumerus funeralis, E. narcissi, E. strigatus*, and *Merodon equestris*), Common Drone Fly (*Eristalis tenax*), Common Lagoon Fly (*Eristalinus aeneus*), two spikelegs (*Neocnemodon latitarsis* and *N. pubescens*), and Common Compost Fly (*Syritta pipiens*). Among these, the only pests are the bulb flies, which are usually only minor pests of daffodils and other bulbs.

### *Climate Change*

Insects respond quickly to environmental changes and thus serve as excellent indicators of environmental conditions. Bees and butterflies have been used to study climate change, and specimen data now available on flower flies will be a powerful additional dataset for this research. As more and more citizen scientists become interested in syrphids and contribute data to this growing dataset, flower flies will become one of the key groups for studying local and large-scale environmental changes.

### *Surveying Syrphidae*

We suggest that anyone planning to set up long-term or large-scale surveys for syrphids contact one of the authors of this guide for advice. Point counts and transects offer the best potential way to establish such long-term comparative data, but trapping projects can also work, as long as the shortcomings of traps are recognized. Malaise trap results are very hard to quantify unless many traps are run and trap setup is carefully described. Pan traps miss most of the fauna but are useful for comparing the subsets of species attracted. Hilltopping studies may provide some of the best comparative data for surveys but would need to be carefully designed for repeatability.

## History of Flower Fly Taxonomic Research in the Nearctic Region

The first recognized Nearctic syrphids were described by Linnaeus in 1758 when he created our current nomenclatural system. Ten widespread species, including the likes of *Eristalis arbustorum*, *Eristalis tenax*, *Melanostoma mellinum*, *Syritta pipiens*, and *Syrphus ribesii*, were named by the great scientist using European specimens. Fabricius conducted the first substantive work using Nearctic specimens between 1775 and 1805 when he named 28 Nearctic species, including distinctive species such as *Ornidia obesa* and *Meromacrus acutus*. The 1800s brought the first significant wave of syrphid discovery with work by Wiedemann, Meigen, Macquart, and Say (the latter introduced us to some of our most abundant species, including *Allograpta obliqua*, *Toxomerus geminatus*, and *T. marginatus*). Only three scientists have described over 100 species in our region. Curran is the clear champion, with 223 species described (724 taxa described worldwide!). Hull is a distant second at 131 species, followed by Williston at 108. Now that most of the species have been described, we are in an era of testing previous species concept hypotheses and

*Melanostoma mellinum* was described by Linnaeus in 1758. Fortunately, type specimens for most of these early species still exist in museums, as the species definitions often need to be revisited. There are undoubtedly multiple species within what we now refer to as *M. mellinum*, and it may turn out that true *mellinum* occurs only in the Old World. More research is required to solve this puzzle.

**List of people who have described more than ten species of Nearctic Syrphidae**

| Author | Years Active | # Syrphid Taxa Described | Author | Years Active | # Syrphid Taxa Described |
|---|---|---|---|---|---|
| Linnaeus | 1758 | 10 | Snow | 1892–1895 | 19 |
| Fabricius | 1775–1805 | 28 | Coquillett | 1894–1910 | 15 |
| Wiedemann | 1818–1830 | 28 | Townsend | 1895–1901 | 22 |
| Say | 1823–1829 | 24 | Hunter | 1896–1897 | 21 |
| Meigen | 1828–1829 | 15 | Johnson | 1898–1929 | 22 |
| Macquart | 1829–1855 | 57 | Jones | 1907–1922 | 21 |
| Harris | 1835–1841 | 25 | Shannon | 1915–1940 | 76 |
| Zetterstedt | 1838–1849 | 17 | Malloch | 1918–1922 | 11 |
| Loew | 1846–1876 | 95 | Lovett | 1919–1921 | 20 |
| Walker | 1849–1860 | 82 | Curran | 1921–1953 | 223 |
| Bigot | 1867–1885 | 60 | Fluke | 1922–1954 | 96 |
| Osten Sacken | 1875–1878 | 32 | Hull | 1922–1960 | 131 |
| Williston | 1882–1893 | 108 | Vockeroth | 1958–2008 | 40 |
| Giglio-Tos | 1892–1893 | 12 | Thompson | 1976– | 9* |

* 9 species described plus more undescribed species treated in this book.

describing the occasional newly discovered species. Molecular data have been a great addition to our arsenal and provide a significant addition to and major independent test of morphological species concepts. The authors of this book continue in this vein, and undoubtedly many changes will be made to the taxonomy of these flies before hypotheses are stable. A great deal of work is still required in the western and southern parts of the Nearctic Region; this the main reason that this guide focuses on the northeast, where we have the best handle on our diversity. The table to the left lists people who have described more than ten species of Nearctic Syrphidae. Note that this list includes currently valid names and synonyms. Many of these may now be considered synonyms of other species.

## Taxonomic Changes Proposed in This Guide

Many of the taxonomic concepts used in this guide are updated and have not been proposed elsewhere in the scientific literature.

*New Combinations*

Some syrphid genera have been found to be nonnatural through our research. We formally divide the genus *Lejops* into several monophyletic genera here, including *Anasimyia*, *Arctosyrphus*, *Eurimyia*, and *Polydontomyia*. We also elevate *Epistrophella* to genus (from a subgenus of *Epistrophe*), *Hammerschmidtia* to genus (from a subgenus of *Brachyopa*), *Lapposyrphus* to genus (from a subgenus of *Eupeodes*), *Megasyrphus* to genus (from a subgenus of *Eriozona*), and elevate *Meligramma* to

We hypothesize that *Lejops* in the broad sense does not form a natural group. However, the subgenera are monophyletic (natural groups that include all relatives and a single common ancestor) and are elevated to generic status here. *Anasimyia* is shown here.

genus (from a subgenus of *Melangyna*). We have also found that the bizarre species, *Merapioidus villosus*, is simply a divergent species of *Criorhina*.

*Resurrected Species*

We have re-evaluated species concepts and found that many species formerly treated as synonyms of others are actually valid species. *Anasimyia anausis* is resurrected from synonymy with *A. lunulata* (the latter is now restricted to the Old World). *Baccha cognata* is resurrected from synonymy with *B. elongata* (the latter is now restricted to the Old World, Alaska, and the Yukon). *Cheilosia albitarsis* is resurrected from synonymy with *C. bardus*. *Chrysotoxum plumeum* is resurrected from synonymy with *C. derivatum*. *Eurimyia stipata* is resurrected from synonymy with *E. lineata* (the latter is now restricted to the Old World, Alaska, the Yukon, and the Northwest Territories). *Hammerschmidtia rufa* is resurrected from synonymy with *H. ferruginea* (the latter is now restricted to the Old World). *Leucozona americana* is resurrected from synonymy with *L. lucorum* (the latter is now restricted to the Old World). *Mallota mississippensis*, *M. diversipennis* (restricted to Colorado and Utah), and *M. illinoensis* are resurrected from synonymy with *M. albipilis*. *Myolepta pretiosa* is resurrected from synonymy with *M. varipes*. *Scaeva affinis* is resurrected from synonymy with *S. pyrastri* (now restricted to the Old World). *Temnostoma excentrica* is resurrected from synonymy with *T. vespiforme* (the latter is now restricted to the Old World). *Volucella arctica*, *V. evecta*, and *V. facialis* are resurrected from synonymy with *V. bombylans* (the latter is now restricted to the Old World).

*New Synonyms*

*Anasimyia relicta* is a new synonym of *A. chrysostoma*. *Cheilosia browni* and *C. nigroapicata* are new synonyms of *C. lasiophthalma*; *C. caltha* and *C. sensua* are new synonyms of *C. comosa*; *C. consentiens* is a new synonym of *C. orilliaensis*; *C. hiawatha* is a new synonym of *C. albitarsis*; *C. nigrofasciata* is a new synonym of *C. hunteri*. *Chrysosyrphus versipellis* is a new synonym of *C. latus*. *Criorhina mystaceae* is a new synonym of *C. nigriventris*. *Chrysotoxum perplexum* is a new synonym of *C. plumeum*. *Eurimyia conostomus* is a new synonym of *E. stipata*. *Ferdinandea dives* and *F. nigripes* are synonymized with *F. buccata*. DNA barcodes are invariant between specimens of these species, and variation within series of specimens encompasses the complete range of these taxa. *Mallota palmerae* is a new synonym of *M. illinoensis*. *Neoascia distincta* is a new synonym of *N. globosa*. *Volucella sanguinea* and *V. americana* are new synonyms of *V. evecta*; *V. lateralis* and *V. rufomaculata* are new synonyms of *V. facialis*.

*New Species*

Eighteen new species are treated in the guide. They are not given formal names but are given numbers so that future publications can be cross-referenced to them. These names will be validated in a separate scientific publication. New taxa include *Anasimyia* undescribed species 1, *Anasimyia* undescribed species 2, *Brachyopa* undescribed species 17-5, *Brachyopa* undescribed species 78-2, *Cheilosia* undescribed species 17-1, *Cheilosia* undescribed species 17-3, *Cheilosia* undescribed species 76-1, *Hammerschmidtia* undescribed species 1, *Neoascia* undescribed species 1, *Neoascia* undescribed species 17-1, *Orthonevra* undescribed species 1, *Palpada* undescribed species 1, *Psilota* undescribed species 17-1, *Xylota* undescribed species 78-1, *Xylota* undescribed species 78-3, *Microdon* undescribed species 17-1, *Mixogaster* undescribed species 1, and *Trichopsomyia* undescribed species 1.

New species are sometimes genuine new discoveries made in the field, but more often they are sitting under our noses and have previously been overlooked in collections. This new species of *Palpada* was discovered when we noticed that DNA from Nearctic *Palpada furcata* was different from DNA of tropical *P. furcata* specimens. Morphological examination supported the fact that these were two species rather than one.

New research undertaken during the production of this field guide has shown that many preexisting species concepts cannot be supported. For example, molecular data and morphology refute the idea that *Mallota albipilis* is a single widespread species; instead it is four allopatric species. The midwestern species, *Mallota illinoensis,* is shown here. True *M. albipilis* is restricted to Arizona and New Mexico.

Examination of the holotype of *Criorhina mystaceae* showed that it fell within the range of typical variation of *C. nigriventris*. Having a good insect collection with series of specimens of each species allows us to examine variation within a species and refine our species hypotheses.

# Recognizing the Subfamilies of Flower Flies

## MICRODONTINAE

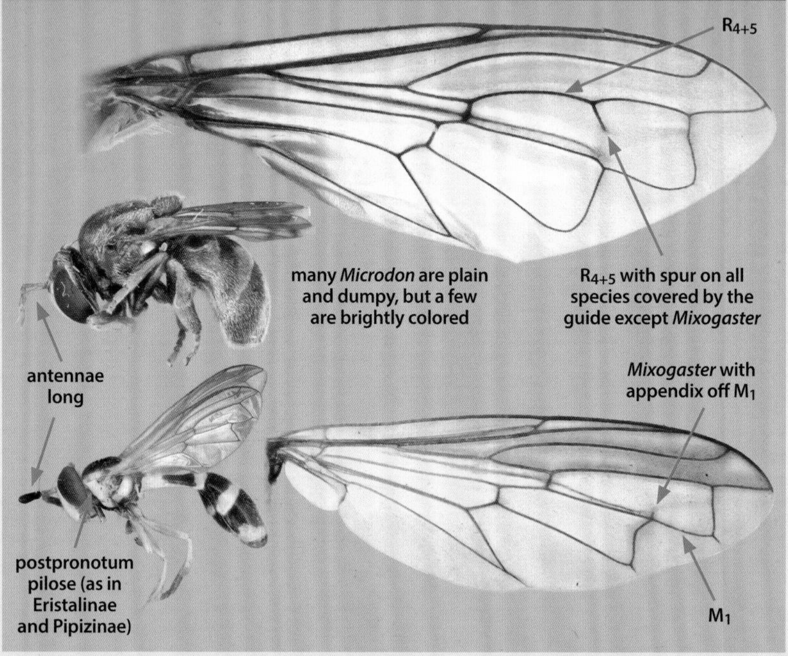

$R_{4+5}$
many *Microdon* are plain and dumpy, but a few are brightly colored
$R_{4+5}$ with spur on all species covered by the guide except *Mixogaster*
antennae long
*Mixogaster* with appendix off $M_1$
postpronotum pilose (as in Eristalinae and Pipizinae)
$M_1$

## ERISTALINAE

eristalines come in every shape, size, and color
notched oral margin (compare with Pipizinae)
facial groove (compare with Pipizinae)
antennae short or long
postpronotum pilose (as in Microdontinae and Pipizinae)

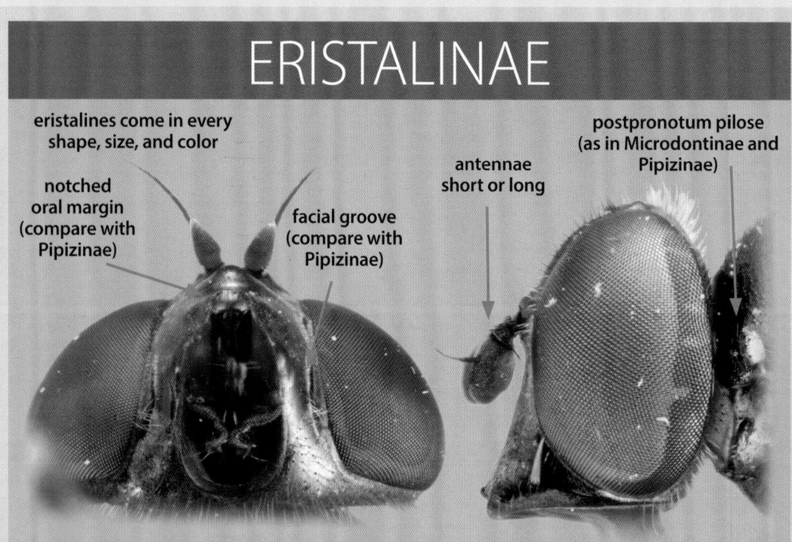

# PIPIZINAE

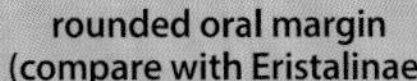

pipizines are mostly to entirely black (only some eristalines are likely to be confused with them)

# SYRPHINAE

most syrphines are slender with yellow and black patterns (but see *Chrysotoxum* below)

only syrphines have a bare postpronotum (head removed to show this); the head is usually concave posteriorly and closely appressed to the postpronotum, making it tricky to see (eristalines tend to have heads held further from the body so the postpronotum is visible)

tergite 5 is visible on male syrphines (only 4 tergites are visible in the other subfamilies)

# MICRODONTINAE

## *Omegasyrphus*

*Omegasyrphus* species are distinctive ant flies (Microdontinae) that are easily identified by the posterior appendix on wing vein $R_{4+5}$ and the long, narrow, parallel-sided abdomen. Tergite 2 has convex edges with the widest part of the tergite before the posterior margin. *Omegasyrphus* was formerly included in the genus *Microdon*. Five species of *Omegasyrphus* are known and occur in North and Central America from Montana to Guatemala. For a key to the species of North America, see Thompson (1981). This genus is in dire need of revision, and current species concepts are weak. Species diagnoses are currently based primarily on wing clouding and abdominal color. Three species have been found associating with the same species of ant (*Omegasyrphus baliopterus*, *O. coarctatus*, and *O. painteri* have all have been associated with *Monomorium minimum*).

### *Omegasyrphus coarctatus*

**Eastern Spot-winged Ant Fly**

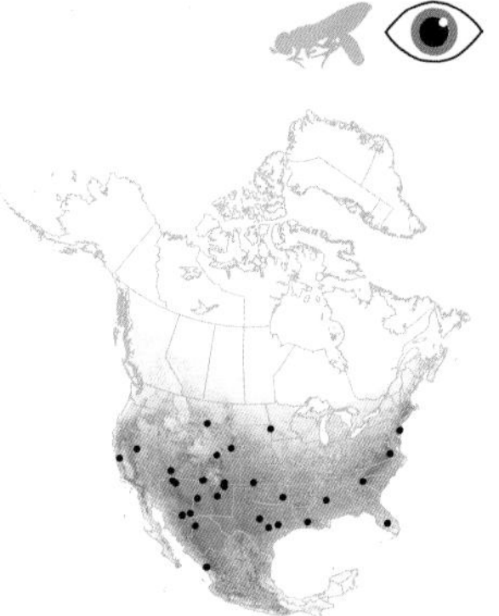

**SIZE RANGE:** 7.4–9.4 mm
**IDENTIFICATION:** The species of *Omegasyrphus* are not well understood. *Omegasyrphus coarctatus* has a bluish-tinged abdomen, while a related species, *O. baliopterus*, has a green-tinged abdomen. There are no other characters to distinguish these two species, so here they are treated as one (as Thompson suggests in his 1981 paper). **ABUNDANCE:** Rare and local. **FLIGHT TIMES:** Mid-March (Florida) to October (Oklahoma). Within the area of the field guide they have been found in late May/early June and late August. **NOTES:** Larvae of this species have been found in association with the ants *Aphaenogaster fulva* and *Monomorium minimum*. The latter was under bark and in decayed sapwood of a *Quercus montana* log.

# *Omegasyrphus coarctatus*

# *Laetodon*

*Laetodon* species are small metallic ant flies (Microdontinae) with a posterior appendix on wing vein $R_{4+5}$. The posteroapical corner of cell $R_{4+5}$ is rectangular or acute, with a small appendix at the corner. This genus used to be included within *Microdon* and was described in 2013 by Menno Reemer. For a key to Nearctic species, see Thompson (1981). The genus *Laetodon* includes five species, four of them Nearctic and one Neotropical. Only one species occurs within the area of the field guide. Larvae are presumed to be predators in ant nests but have not been described.

## *Laetodon laetus* Small Metallic Ant Fly

**SIZE RANGE:** 6.0–9.7 mm

**IDENTIFICATION:** These are small, strongly metallic flies that are green, blue, or purple. The tibiae are orange and the flagellum has a short sensory pit on the outside edge. The eye is sparsely pilose. **ABUNDANCE:** Rare and local. **FLIGHT TIMES:** Late March (Florida) to early October (Arizona), late May to late September within the area of the field guide; northern (Maryland) records are all from mid- to late July. **NOTES:** Larvae are unknown.

# *Laetodon laetus*

# *Microdon*

All *Microdon* species have a posterior appendix on wing vein $R_{4+5}$. The antennae are elongate, with the scape and flagellum longer than wide. Wing vein $M_1$ is with or without a posterior appendix. Most species are fairly robust, with an oval abdomen, and they are quite diverse in color. For a key to the species of North America, see Thompson (1981). The genus *Microdon* includes 126 species from all continents except Antarctica. Twenty-one species are Nearctic, 15 of which are included within the range of this guide. All known *Microdon* larvae are predators of ant eggs and larvae. Adults do not visit flowers and are typically found near host ant nests.

## *Microdon craigheadii* Large Metallic Ant Fly

**SIZE RANGE:** 9.1–11.2 mm

**IDENTIFICATION:** Medium-sized, strongly metallic flies that are green, blue, or purple. Size and the dark leg color are typically enough to distinguish this species from *Laetodon laetus*. Check for the long sensory pit on the antenna if in doubt.

**ABUNDANCE:** Rare and local. **FLIGHT TIMES:** Mid-June (Mississippi) to late October (Alabama); mid-July to late August in the northern part of its range.

**NOTES:** Habitat information is limited, but it has been found in oak and oak/hickory forests. Some specimens were collected at black lights. Molecular data suggest that this is a species complex.

# *Microdon*

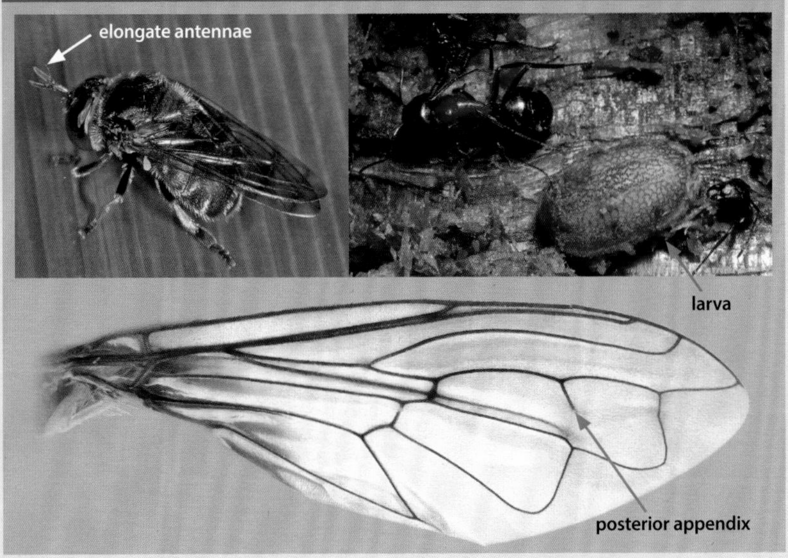

elongate antennae

larva

posterior appendix

## *Microdon craigheadii*

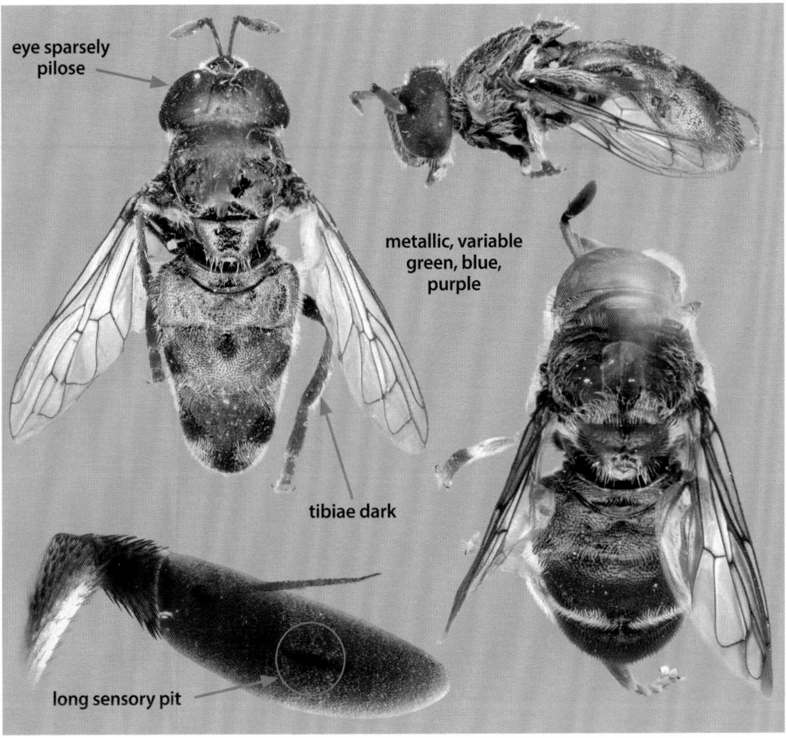

eye sparsely pilose

metallic, variable green, blue, purple

tibiae dark

long sensory pit

## *Microdon (Chymophila) fulgens*

### Rainbow Ant Fly

**SIZE RANGE:** 13.2–17.7 mm
**IDENTIFICATION:** Large, strongly metallic flies, green with some blue, purple, and orange. The abdomen is broader than the thorax, and tergite 2 is short and rectangular at the base. Wing vein $M_1$ is bent at the midpoint and has an external appendix. **ABUNDANCE:** Rare and local. **FLIGHT TIMES:** Late March (Florida) to mid-November (Florida); throughout July in northern part of range. **NOTES:** Larvae of this species have been recorded as predators of two species of formicine ants (*Camponotus atriceps* and *Polyergus lucidus*). *Camponotus floridanus* is here newly recorded as a host based on a specimen collected by R. Duffield in Pine Key, Florida (USNM_Ent 247039).

## *Microdon (Microdon) megalogaster*

### Black-bodied Ant Fly

**SIZE RANGE:** 12.5–15.7 mm
**IDENTIFICATION:** Large, pilose ant flies with a distinctive color pattern. The scutum, scutellum, and abdomen to tergite 2 are covered in yellow pile, while tergites 3–5 are covered in black pile. **ABUNDANCE:** Rare and local. **FLIGHT TIMES:** Late May to mid-July. **NOTES:** Larvae of this species are predators of *Formica subsericea* ants. Larvae and puparia were described by Greene (1923). Habitat data on labels indicate that they occur in sedge meadows and *Corema* barrens.

# *Microdon (Chymophila) fulgens*

wing vein $M_1$ is bent at the midpoint and has an external appendix

tergite 2 short and rectangular

# *Microdon (Microdon) megalogaster*

yellow pile on scutum, scutellum, and abdomen to tergite 2

tergites 3–5 black pilose

## *Microdon (Microdon) aurulentus*

Golden-haired Ant Fly

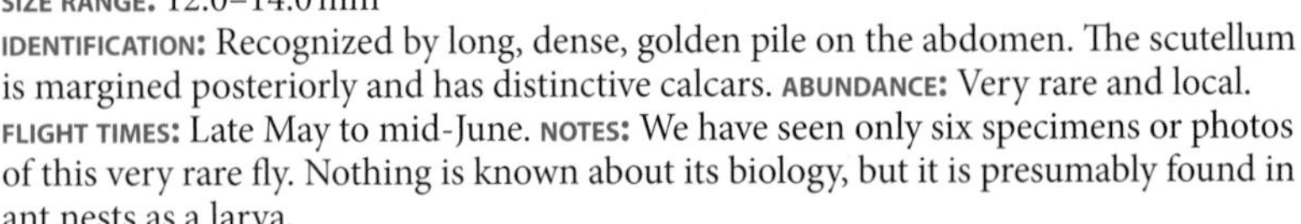

**SIZE RANGE:** 12.0–14.0 mm
**IDENTIFICATION:** Recognized by long, dense, golden pile on the abdomen. The scutellum is margined posteriorly and has distinctive calcars. **ABUNDANCE:** Very rare and local. **FLIGHT TIMES:** Late May to mid-June. **NOTES:** We have seen only six specimens or photos of this very rare fly. Nothing is known about its biology, but it is presumably found in ant nests as a larva.

## *Microdon (Microdon) manitobensis*

Greater Ant Fly

**SIZE RANGE:** 10.6–14.0 mm
**IDENTIFICATION:** Recognized by the densely yellow-pilose scutellum and yellow thorax and abdominal markings. **ABUNDANCE:** Uncommon and local. **FLIGHT TIMES:** Mid-May to mid-July plus two records from mid- and late October. **NOTES:** Larvae are known from the nests of two ant species: *Formica densiventris* and *F. neoclara*. Adults occur in many habitats including wet woods, mixed woods, wet meadows, and blueberry fields. In particular, look for them in open, sandy to gravelly sites adjacent to mixed forests.

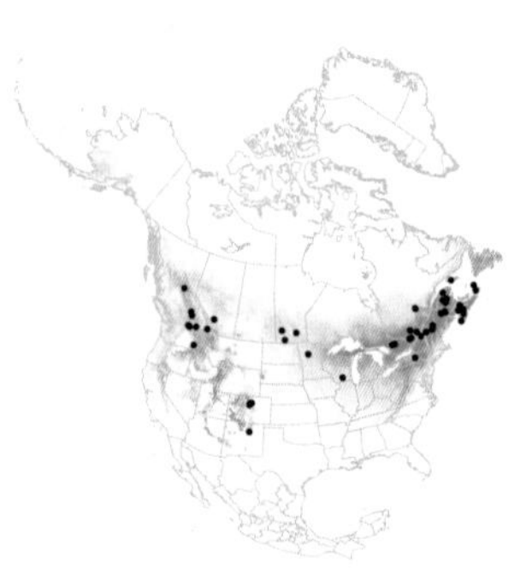

## *Microdon (Microdon) aurulentus*

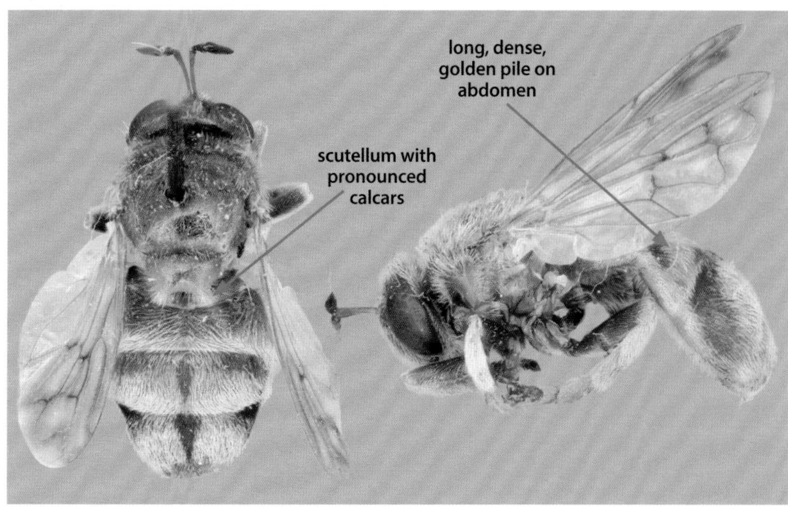

long, dense, golden pile on abdomen
scutellum with pronounced calcars

## *Microdon (Microdon) manitobensis*

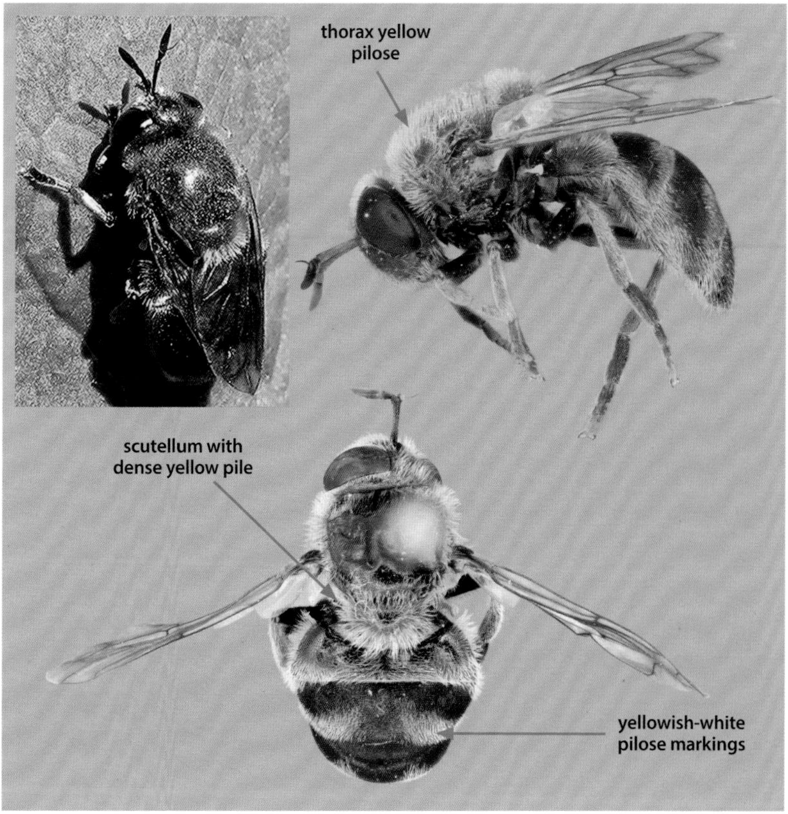
thorax yellow pilose
scutellum with dense yellow pile
yellowish-white pilose markings

## *Microdon (Microdon) tristis*

### Long-horned Ant Fly

**SIZE RANGE:** 8.5–12.5 mm

**IDENTIFICATION:** Dark flies with prominent calcars on the scutellum. The tibiae are orange and contrast with the darker femora, the flagellum is at least as long as the scape, and the wing cell bm is bare posterobasally. **ABUNDANCE:** This is arguably the most common ant fly within the range of the guide. **FLIGHT TIMES:** Mid-May to early August; one record from 6 October (Pennsylvania). **NOTES:** Larvae have been associated with the following ants: *Camponotus novaeboracensis*, *C. pennsylvanicus*, *Formica aserva*, *F. difficilis*, *F. obscuripes*, and *F. schaufussi*. Known habitats include hardwood forests, sedge meadows, open heath, and sphagnum bogs. Adults are occasionally found hilltopping. Although the western North American records are disjunct from the eastern, specimens are genetically identical. There are two undescribed species that key to *M. tristis* in existing keys; one is a Great Smoky Mountain endemic (outside the scope of the guide) and the other is from Pennsylvania (included below).

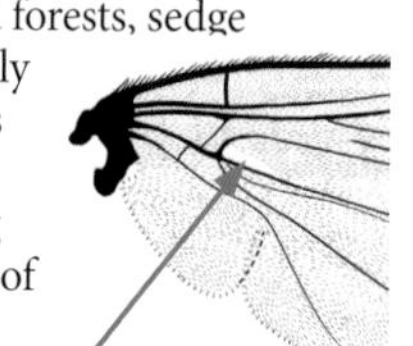

bm with small bare patch basally

## *Microdon (Microdon)* undescribed species 17-1

### Big-footed Ant Fly

**SIZE RANGE:** 7.5–11.9 mm

**IDENTIFICATION:** Similar to *M. tristis* but with swollen metabasitarsus particularly evident in the male, blocky inward-pointing calcars, short wings, longer flagellum, and completely microtrichose wings. **ABUNDANCE:** Apparently rare and local; all known specimens collected in Pennsylvania by Frank Fee. **FLIGHT TIMES:** Late August to mid-September, with one early record from 29 July. **NOTES:** Specimens have been collected in old fields and sedge meadows. Adjacent vegetation included forest dominated by white pines and marshland.

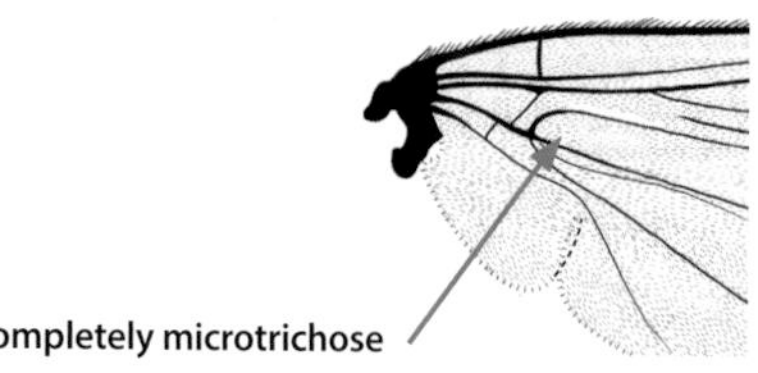

bm completely microtrichose

## *Microdon (Microdon) tristis*

scutellum with prominent outward-pointing calcars

flagellum as long as or longer than scape

♀ ♂

metabasitarsus only slightly expanded (males larger, shorter); other tarsi typically elongate

mesotibia mostly orange

## *Microdon (Microdon)* undescribed species 17-1

scutellum with prominent inward-pointing calcars

flagellum much longer than scape

♀ ♂

both sexes with metabasitarsus swollen; other tarsi very short and compact

## *Microdon (Microdon) ruficrus*

Spiny-shielded Ant Fly

**SIZE RANGE:** 9.4–11.8 mm

**IDENTIFICATION:** Dark flies with prominent calcars on the scutellum, brown-black legs, and the flagellum shorter than the scape. Wing cell bm is densely microtrichose. **ABUNDANCE:** Uncommon. **FLIGHT TIMES:** As early as mid-February in the southern part of the range; mid-May to early August in the area covered by the guide. **NOTES:** Larvae have been found in nests of *Lasius alienus*. Adults have been collected in mixed woods, burnt spruce forest, and coastal barrens and are typically associated with a sandy substrate.

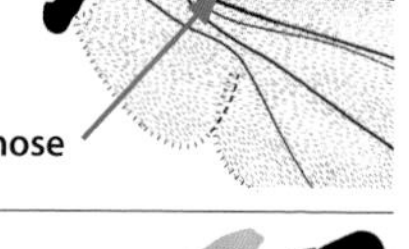

bm entirely microtrichose

## *Microdon (Microdon) ocellaris*

Hairy-legged Ant Fly

**SIZE RANGE:** 9.1–13.2 mm

**IDENTIFICATION:** Dark flies with small calcars on the scutellum and with tibiae covered in long, dense, light pile that obscures the cuticle color. Wing cell bm is bare posterobasally. **ABUNDANCE:** Rare and local. **FLIGHT TIMES:** Early May to mid-July. **NOTES:** Larvae have been found in *Formica schaufussi* ant nests. Habitat is available only from one specimen and was noted as mixed pine-oak forest.

bm bare basally

## *Microdon (Microdon) abstrusus*

Hidden Ant Fly

**SIZE RANGE:** 8.1–9.6 mm

**IDENTIFICATION:** Dark flies without scutellar calcars and with orange tibiae. Wing cell bm densely microtrichose (sometimes there are small bare areas basally). **ABUNDANCE:** Rare and local. **FLIGHT TIMES:** Late May to mid-June. **NOTES:** Larvae have been found in *Formica exsectoides* ant nests.

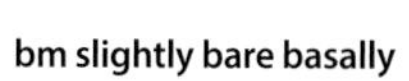

## *Microdon (Microdon) ruficrus*

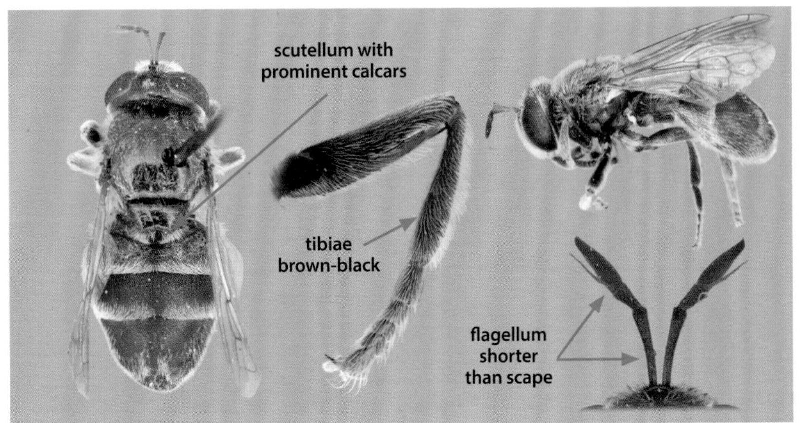


## *Microdon (Microdon) ocellaris*

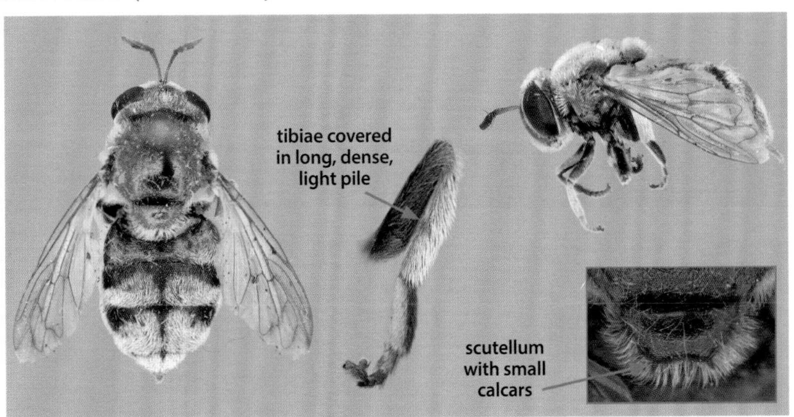


## *Microdon (Microdon) abstrusus*

# *Microdon (Microdon) albicomatus*

## White-haired Ant Fly

**SIZE RANGE:** 8.1–10.4 mm
**IDENTIFICATION:** Dark flies with small, inconspicuous scutellar calcars and brown-black tibiae. Wing cell bm is densely microtrichose. **ABUNDANCE:** Uncommon.
**FLIGHT TIMES:** Mid-May to mid-August. **NOTES:** Ant hosts include *Camponotus modoc*, *Formica accreta*, *F. aserva*, *F. fusca*, *F. neoclara*, *F. neorufibarbis*, *F. obscuripes*, and *Myrmica incompleta*. Studies have found that *Microdon albicomatus* produces and mimics the cuticular hydrocarbons of *Myrmica incompleta*. This allows ant fly larvae to move around ant nests without detection by their prey, a form of chemical camouflage. Similar findings have been made for *Microdon piperi* and their principal host *Camponotus modoc*. It is likely that many *Microdon* species have evolved this type of chemical mimicry.

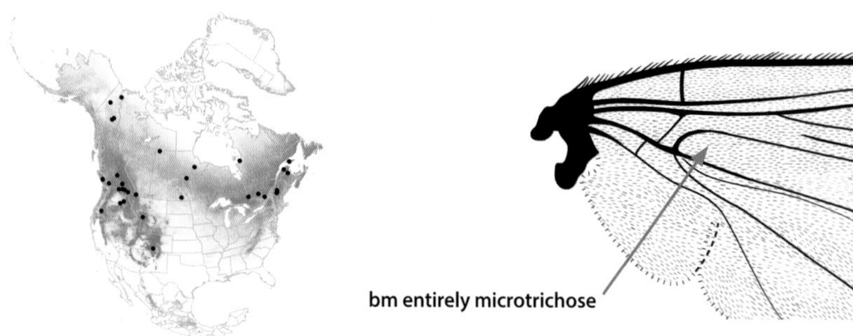


# *Microdon (Microdon) cothurnatus*

## Orange-legged Ant Fly

**SIZE RANGE:** 8.2–12.1 mm
**IDENTIFICATION:** Dark flies with small, inconspicuous scutellar calcars and orange tibiae. Wing cell bm bare posterobasally. **ABUNDANCE:** Uncommon. **FLIGHT TIMES:** Late May to late July. **NOTES:** Hosts include *Camponotus novaeboracensis*, *C. pennsylvanicus*, *C. ?vicinus*, *Formica accreta*, *F. adamsi whymperi*, *F. aserva*, *F. neoclara*, *F. obscuripes*, *F. obscuriventris*, *F. podzolica*, and *F. ravida*. Adults have been collected in mixed woods, spruce woods, and blueberry fields.

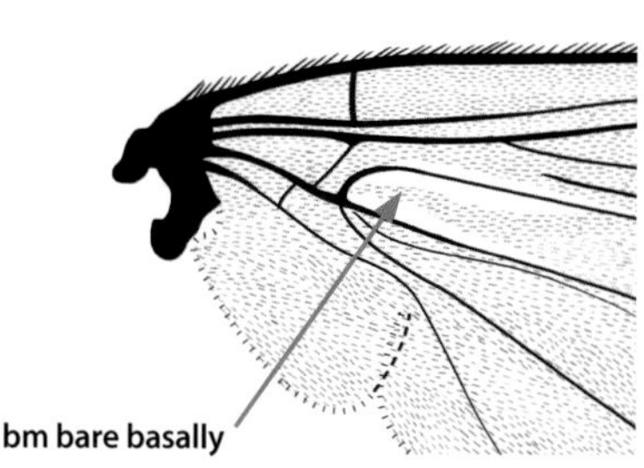

# *Microdon (Microdon) albicomatus*

tibiae brownish black

scutellum with small calcars

# *Microdon (Microdon) cothurnatus*

tibiae orange

scutellum with small calcars

## *Microdon (Dimeraspis) abditus*

### Broad-footed Ant Fly

**SIZE RANGE:** 7.4–11.8 mm
**IDENTIFICATION:** An orange-brown ant fly that is easily distinguished from other *Dimeraspis* species because the flagellum is shorter than the scape. **ABUNDANCE:** Uncommon and local. **FLIGHT TIMES:** Early June to mid-July, with a single record from 10 September (New Jersey). **NOTES:** Larvae unknown. Found in open sandy areas adjacent to mixed woods.

## *Microdon (Dimeraspis) fuscipennis*

### Short-horned Ant Fly

**SIZE RANGE:** 7.1–12.2 mm
**IDENTIFICATION:** An orange-brown *Microdon* that is easily separated from other *Dimeraspis* species because the flagellum is longer than the scape and the arista is shorter than ½ the length of the flagellum. **ABUNDANCE:** Rare and local. **FLIGHT TIMES:** Late March to mid-October; Indiana specimen from 27 September. **NOTES:** Larvae known to attack only *Forelius pruinosus*. They overwinter as both pupae and larvae in these ant nests. An average of 3.5 larvae were found per nest studied. This species may not have chemical mimicry, as over 90% of first-instar larvae are killed by host ants.

## *Microdon (Dimeraspis) globosus*

### Globular Ant Fly

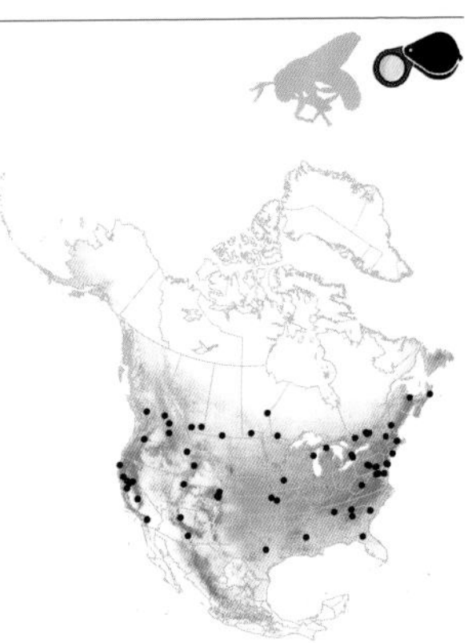

**SIZE RANGE:** 6.5–10.7 mm
**IDENTIFICATION:** An orange-brown ant fly that is easily distinguished from other *Dimeraspis* species because the flagellum is longer than the scape and the arista is longer than ½ the length of the flagellum. **ABUNDANCE:** Fairly common. **FLIGHT TIMES:** Early April to mid-October; most in August. **NOTES:** *Tapinoma sessile* is the only ant host recorded. Habitats include barrens, bogs, wet prairies, wooded swamps, and wet meadows.

## *Microdon (Dimeraspis) abditus*

## *Microdon (Dimeraspis) fuscipennis*

## *Microdon (Dimeraspis) globosus*

# *Serichlamys*

*Serichlamys* is a genus of ant flies that was recently resurrected out of synonymy with *Microdon* by Reemer and Ståhls (2013a). The best characters to differentiate this group from *Microdon* are found in the male genitalia. One wing character is diagnostic: the posteroapical corner of cell $r_{4+5}$ is rectangular (rounded in *Microdon*). *Serichlamys* is a New World genus with five known species. Three species occur in the Nearctic region, one within the range of this guide. As with other ant flies, the best key to North American species is by Thompson (1981).

## *Serichlamys rufipes* Purplish Ant Fly

**SIZE RANGE:** 8.8–10.6 mm
**IDENTIFICATION:** Recognized by the sparsely pilose eye, purple sheen, and two rounded calcars on the posterior edge of the scutellum. **ABUNDANCE:** Rare and local. **FLIGHT TIMES:** Late April to mid-September (Virginia record from 3 September). **NOTES:** The only known ant host is *Pheidole dentata*.

body with purple sheen

sparsely pilose eye

two large, rounded calcars on scutellum

# *Mixogaster*

The narrow "waist" and elongate antennae of these flies add to their wasp-like appearance. Like many other ant flies, *Mixogaster* species have an elongate scape and flagellum. However, they are the only North American microdontines to have a constricted abdomen, where tergite 1 is narrower than other tergites and tergite 2 is narrow anteriorly and widens posteriorly. They also have distinctive wing venation, with vein $M_1$ having an anterior appendix extending into cell $r_{4+5}$. For a key to the species, see Hull (1954). Four species of *Mixogaster* occur in the Nearctic, three (including one undescribed species) in the area covered by the field guide. Larval life histories for all Nearctic species are unknown, but a South American relative attacks dolichoderine ants.

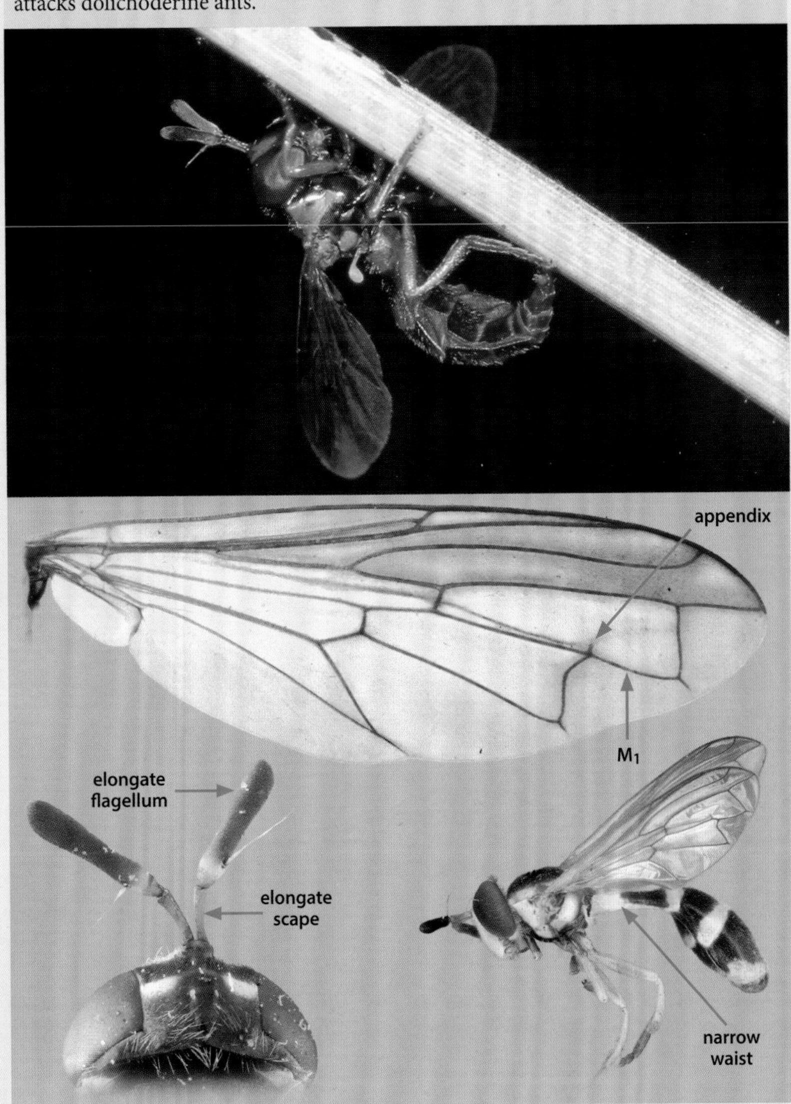

## *Mixogaster* undescribed species 1 Fattig's Ant Fly

**SIZE RANGE:** 10.3–10.8 mm
**IDENTIFICATION:** This undescribed species can be easily distinguished by its yellow vertex and black ocellar triangle. The face is also completely yellow and without a black medial stripe. **ABUNDANCE:** Rare and local. **FLIGHT TIMES:** Late June to early October. **NOTES:** The larval biology is unknown. All specimens with habitat details were found in pine or pine-oak forests.

## *Mixogaster breviventris* Slender Ant Fly

**SIZE RANGE:** 8.1–10.7 mm
**IDENTIFICATION:** This species has a vertex that is black posteriorly and yellow anteriorly and a black medial stripe on its face. The yellow lateral stripes on the scutum extend from the postpronotum to the scutellum. **ABUNDANCE:** Rare and local. **FLIGHT TIMES:** Early June to late July. **NOTES:** The larval biology is unknown.

## *Mixogaster johnsoni* Johnson's Ant Fly

**SIZE RANGE:** 9.4–10.1 mm
**IDENTIFICATION:** This species has a vertex that is mainly black, with some yellow anteriorly and a black medial stripe on its face. The yellow lateral stripes on the scutum extend from the transverse suture to the scutellum and do not contact the postpronotum. **ABUNDANCE:** Very rare and local. **FLIGHT TIMES:** The only known specimen was collected on 3 September 1935. **NOTES:** We have seen only the holotype from Dennis Port, Massachusetts, collected by Bequaert. The larval biology is unknown.

## *Mixogaster* undescribed species 1

## *Mixogaster breviventris*

## *Mixogaster johnsoni*

# ERISTALINAE

## *Helophilus*

*Helophilus* includes 40 valid species worldwide. Eleven of these are in the Nearctic and seven are within the range of the field guide. They are easy to identify to genus, but some are tricky to identify to species. Most of these flies have yellow stripes on their scutum, large yellow-orange markings on their abdomen, and an elongate pterostigma on the wing. *Helophilus* larvae are associated with wet, decaying organic material like that found in ponds, mud, manure, and silage. Larvae are still unknown for several of our species. For a key to the species of North America, see Curran and Fluke (1926).

### ***Helophilus bottnicus*** Gray-banded Marsh Fly

**SIZE RANGE:** 9.4–12.0 mm
**IDENTIFICATION:** The thoracic stripes vary from yellow to nearly indistinguishable. The abdominal banding pattern is diagnostic: bands on abdominal tergite 2 are yellow laterally and contrasting gray centrally; bands of tergites 3 and 4 are gray throughout. **ABUNDANCE:** Rare. **FLIGHT TIMES:** Mid-May to mid-July. **NOTES:** European data suggest that these flies occur in seasonally flooded alluvial grassland on sandy soil and in boreal forest. Adults have been found visiting flowers of *Ribes* and *Salix*.

# *Helophilus*

elongate pterostigma

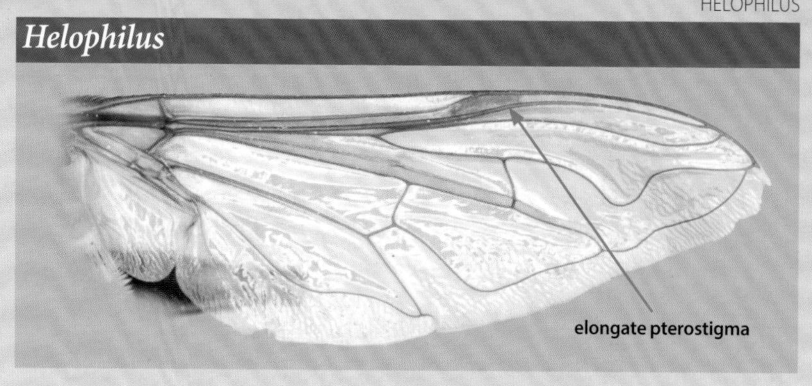

## *Helophilus bottnicus*

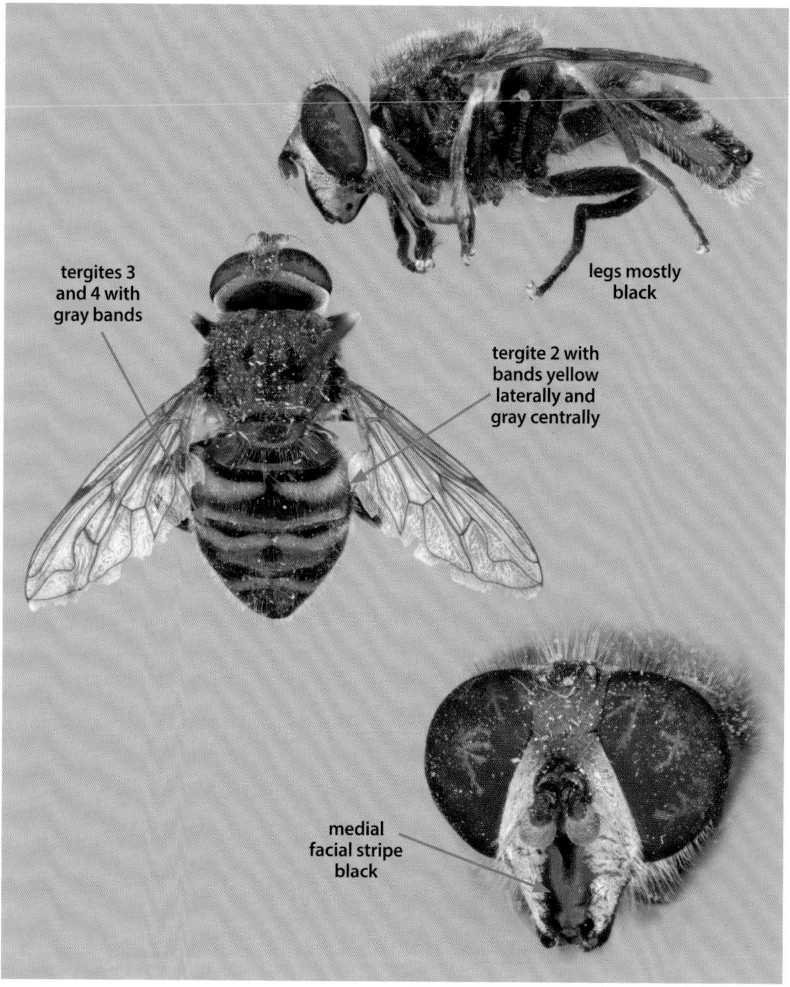

legs mostly black
tergites 3 and 4 with gray bands
tergite 2 with bands yellow laterally and gray centrally
medial facial stripe black

## *Helophilus fasciatus* Narrow-headed Marsh Fly

**SIZE RANGE:** 10.8–15.2 mm
**IDENTIFICATION:** This species is distinguished most easily by the orange antennal scape and pedicel (black in all of our other *Helophilus* species). This and *H. latifrons* have an orange medial stripe on the face. The vertex and frons are admixed black and yellow pilose and the males have a narrow vertex. **ABUNDANCE:** Abundant.
**FLIGHT TIMES:** Late March to late October.
**NOTES:** This species is one of the earliest and latest syrphids to fly every year. It is the most common *Helophilus* species south of treeline and can be confused only with *H. latifrons*. Eggs are laid in clutches on vegetation overhanging ponds. The rat-tailed larvae hatch and fall into the pond, where they develop. Flowers visited include *Barbarea*, *Cephalanthus*, *Eupatorium*, *Euthamia*, *Physocarpus*, *Prenanthes*, *Primula*, *Salix*, *Solidago*, *Symphyotrichum*, *Taraxacum*, and *Zizia*.

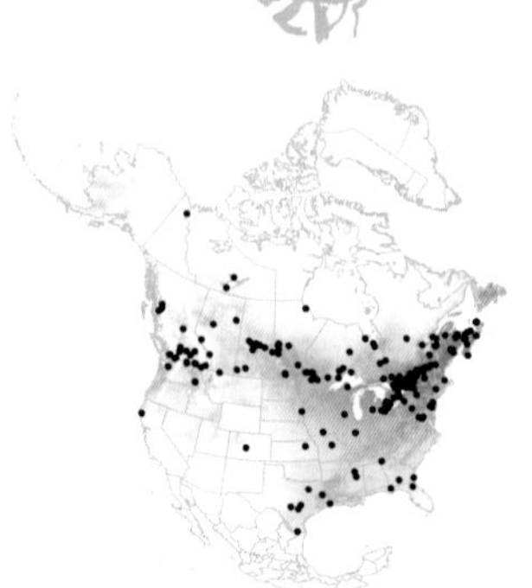

## *Helophilus latifrons* Broad-headed Marsh Fly

**SIZE RANGE:** 10.3–16.6 mm
**IDENTIFICATION:** This species is distinguished by the orange medial stripe on the face, black scape and pedicel, and black pile on the vertex of both sexes. Males have a broad vertex. **ABUNDANCE:** Formerly very common, now uncommon to rare.
**FLIGHT TIMES:** Mid-May to mid-October, with occasional records as early as mid-April.
**NOTES:** This species appears to have declined significantly. Larvae and pupae are illustrated in Jones (1922). Flowers visited include *Sonchus*, *Symphyotrichum*, and *Zizia*.

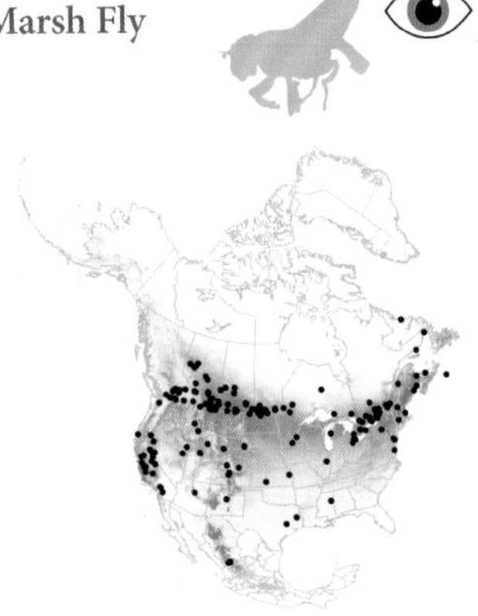

## *Helophilus fasciatus*

♂ ♀

face with shiny orange medial stripe

scape and pedicel orange (black in all other species)

♂ ♀

male vertex narrow

female vertex and frons admixed black and yellow pilose

## *Helophilus latifrons*

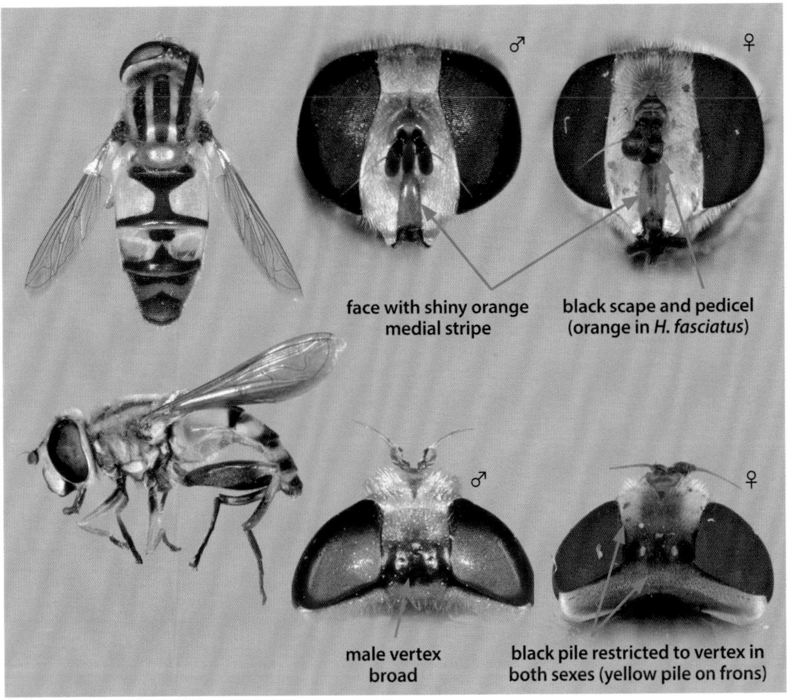

♂ ♀

face with shiny orange medial stripe

black scape and pedicel (orange in *H. fasciatus*)

♂ ♀

male vertex broad

black pile restricted to vertex in both sexes (yellow pile on frons)

## *Helophilus obscurus* Obscure Marsh Fly

**SIZE RANGE:** 10.6–16.1 mm

**IDENTIFICATION:** These flies have a shiny black medial stripe on the face and large spots on tergite 3. Males have short, typical pile on abdominal segment 8. Females have a narrow yellow margin on tergite 5, and the basal ⅙ of their femur is yellow red. **ABUNDANCE:** Common. **FLIGHT TIMES:** Early June to late September, with one central Ontario record from late April. **NOTES:** *Helophilus obscurus* is a boreo-montane species. The larvae are unknown.

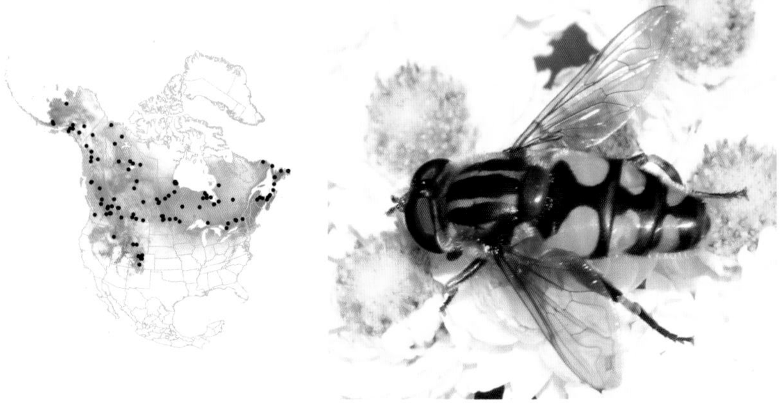

## *Helophilus hybridus* Woolly-tailed Marsh Fly

**SIZE RANGE:** 11.1–16.4 mm

**IDENTIFICATION:** These flies have a shiny black medial stripe on the face and large spots on tergite 3. Males have long, woolly pile on abdominal segment 8. Females have a broad yellow margin on tergite 5, and the base of their metafemur is black. **ABUNDANCE:** Rare in the east, common in the west. **FLIGHT TIMES:** Early July to mid-September. **NOTES:** This Holarctic species is largely western boreal in the Nearctic, with only a few records in the east. Adults feed mostly from white Apiaceae flowers and are typically found flying below 2 m height near wetlands (on *Hieracium* in photo below). Larvae are found in noneutrophic wetlands and temporary pools and were described by Hartley (1961).

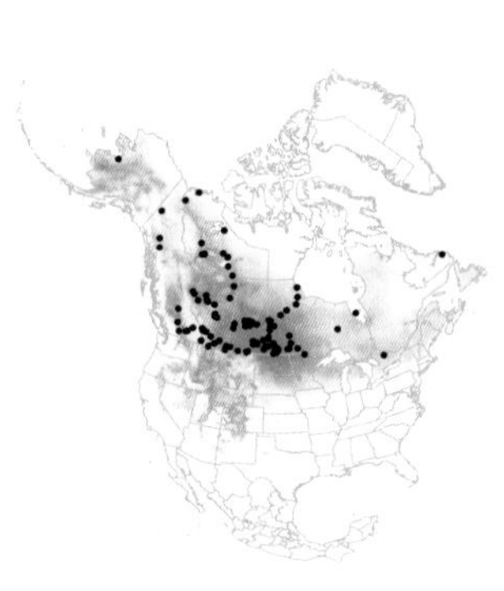

## Helophilus obscurus

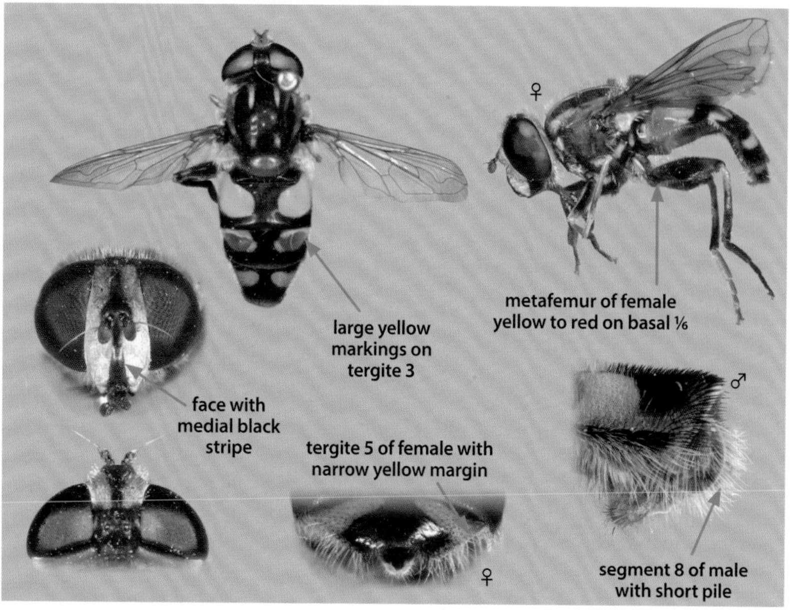

large yellow markings on tergite 3

metafemur of female yellow to red on basal ⅙

face with medial black stripe

tergite 5 of female with narrow yellow margin

segment 8 of male with short pile

## Helophilus hybridus

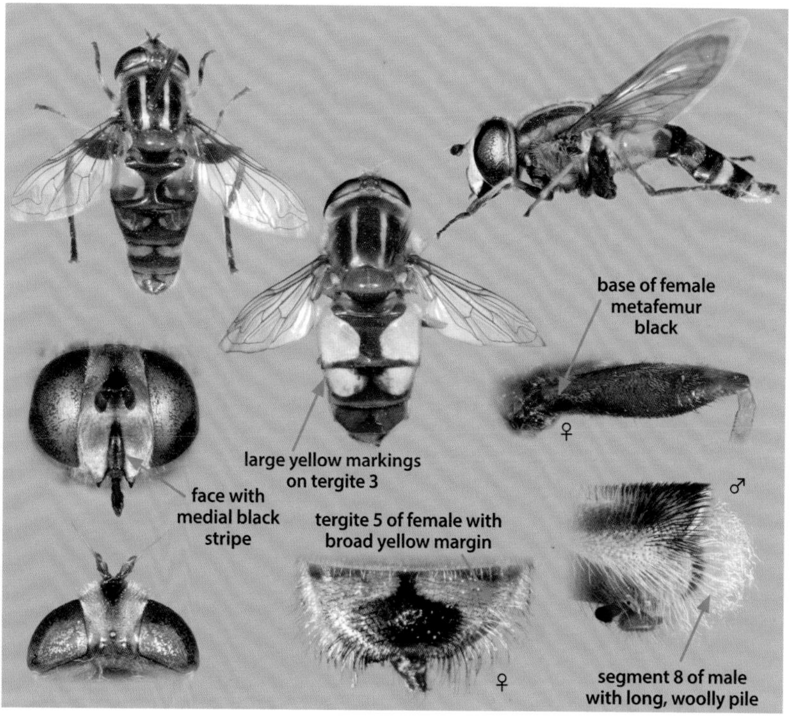

base of female metafemur black

large yellow markings on tergite 3

face with medial black stripe

tergite 5 of female with broad yellow margin

segment 8 of male with long, woolly pile

## *Helophilus lapponicus*

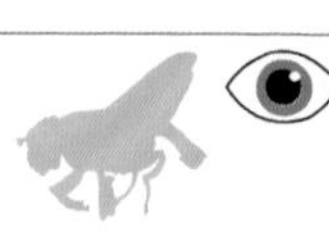

### Yellow-margined Marsh Fly

**SIZE RANGE:** 11.4–15.2 mm
**IDENTIFICATION:** These flies have a shiny black medial stripe on their face and reduced spots on the tergite 3. The lateral margins of tergite 2 are entirely yellow-orange, with no black reaching the margin. **ABUNDANCE:** Common. **FLIGHT TIMES:** Mid-May to early September. **NOTES:** *Helophilus lapponicus* is found in boreal and tundra wetlands and is Holarctic in distribution. Flowers visited include *Achillea*, *Allium*, *Caltha*, *Fragaria*, *Matricaria*, *Ranunculus*, *Rubus*, and *Sorbus*. The larvae are unknown.

## *Helophilus groenlandicus*

### Black-margined Marsh Fly

**SIZE RANGE:** 9.4–15.2 mm
**IDENTIFICATION:** These flies have a shiny black medial stripe on their face and reduced spots on the tergite 3. The lateral margins of tergite 2 are both yellow-orange and black. **ABUNDANCE:** Common. **FLIGHT TIMES:** Early June to mid-September. **NOTES:** This Holarctic species is found mostly above the treeline but occurs sporadically farther south. Adults are found in open wetlands, where they have been recorded nectaring on the following flowers: *Allium*, *Chamerion*, *Matricaria*, *Stellaria*, *Tripleurospermum*, and *Valeriana*. The larvae are unknown.

# Helophilus lapponicus

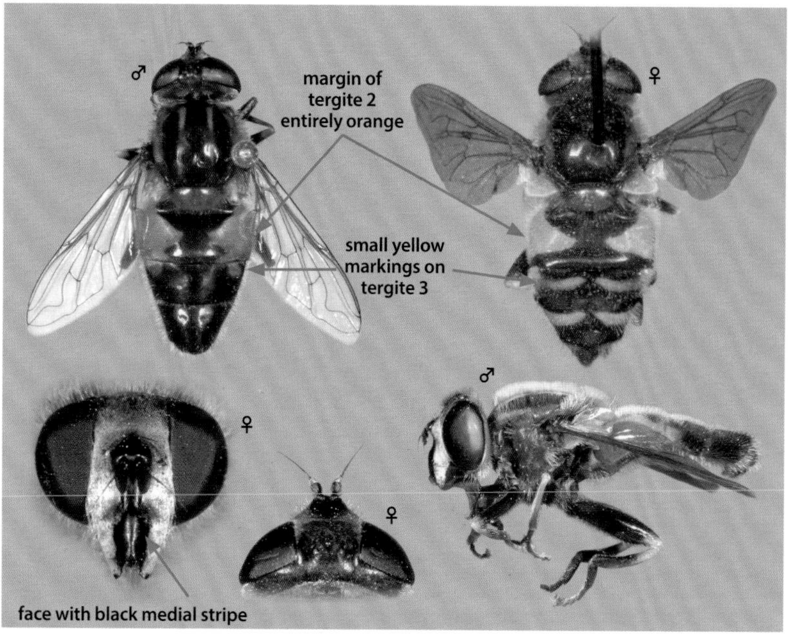

♂
margin of tergite 2 entirely orange
♀
small yellow markings on tergite 3
♀
♀
♂
face with black medial stripe

# Helophilus groenlandicus

♂
margin of tergite 2 orange and black
♀
♀
♀
♂
face with black medial stripe

# *Parhelophilus*

There are 19 species of *Parhelophilus*. Of these, nine occur in the Nearctic with eight in the area covered by the guide. Bog flies are recognized by having a short pterostigma (simulating a crossvein), stripes on the scutum (usually), oval abdomen typically with light markings, face at least partly yellow and not greatly produced into a cone (as in *Eurimyia*), and metatibiae that are truncate at the apex with a keel on the basal half (see *Anasimyia* for contrasting tibial characters). Larvae are rat tailed and are associated with decaying vegetation in ponds, wet woodlands, and slow-moving streams. No North American larvae have been described. Adults can often be found on flowers and emergent vegetation around wetlands. For a key to Nearctic species, see Curran and Fluke (1926).

## *Parhelophilus brooksi* Brooks's Bog Fly

**SIZE RANGE:** 7.7–9.2 mm
**IDENTIFICATION:** These are distinctive flies, unlike any other *Parhelophilus* species because of their mostly black abdomens. **ABUNDANCE:** Rare. **FLIGHT TIMES:** Late May to mid-July. **NOTES:** Manitoba specimens were collected around pools of saline water. Saltwater is clearly not a necessity for the species, as other specimens were collected near freshwater. For example, the Ottawa specimen is from Beechwood Cemetery, nowhere near brackish ponds.

## *Parhelophilus porcus* Black Bog Fly

**SIZE RANGE:** 8.5–10.0 mm
**IDENTIFICATION:** Among our most distinctive species of *Parhelophilus*, *P. porcus* has an entirely black abdomen with gray spots on all segments. Males have a pilose tubercle at the base of the metafemur (a character shared with only *P. divisus* and *P. flavifacies*). **ABUNDANCE:** Uncommon. **FLIGHT TIMES:** Late May to early August. **NOTES:** This predominantly boreal species often occurs in bogs but can also be found in drier taiga forests (lichen woodlands). It has been recorded nectaring on *Coreopsis* and *Potentilla*.

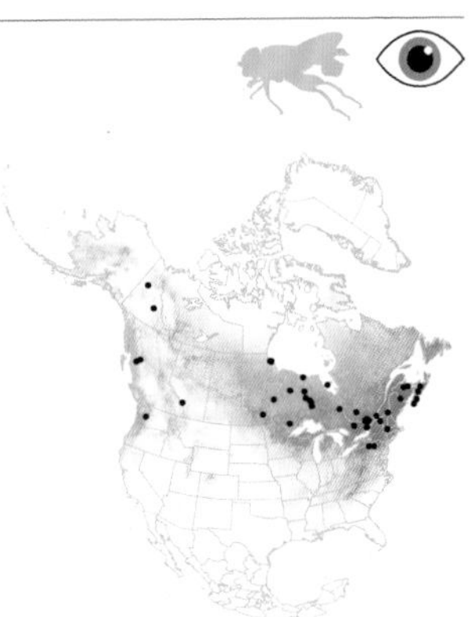

# Parhelophilus

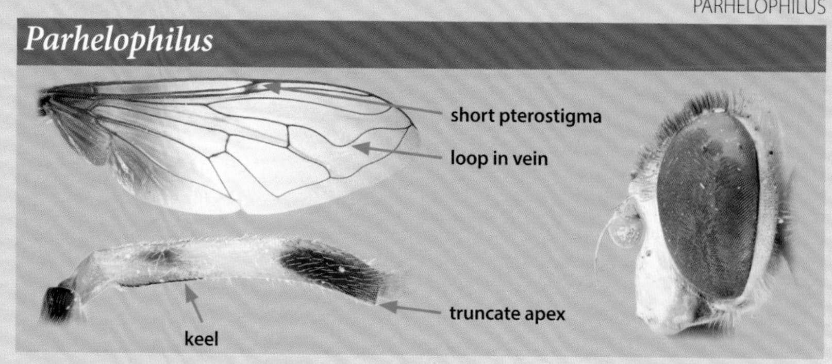


## Parhelophilus brooksi

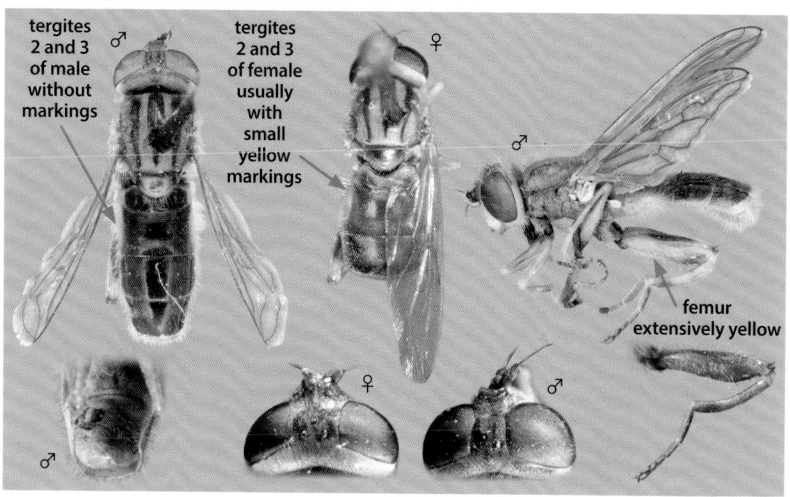


## Parhelophilus porcus

## *Parhelophilus obsoletus* Unadorned Bog Fly

**SIZE RANGE:** 7.5–9.7 mm
**IDENTIFICATION:** Although superficially similar to several other *Parhelophilus* species, *P. obsoletus* is the only one with very indistinct stripes on the thorax (sometimes even absent). The male has shiny, large pregenital segments with sternite 6 equal in length to tergite 4. Segments are variable in color. The female has a metabasitarsus with zero to three black bristles and short, appressed pile. **ABUNDANCE:** Fairly common. **FLIGHT TIMES:** Early May to early August. **NOTES:** This is predominantly a boreal species, typically found in open areas in mixed woods, with beaver ponds or other standing water sources nearby. The flower-visiting fly below is on *Cornus*.

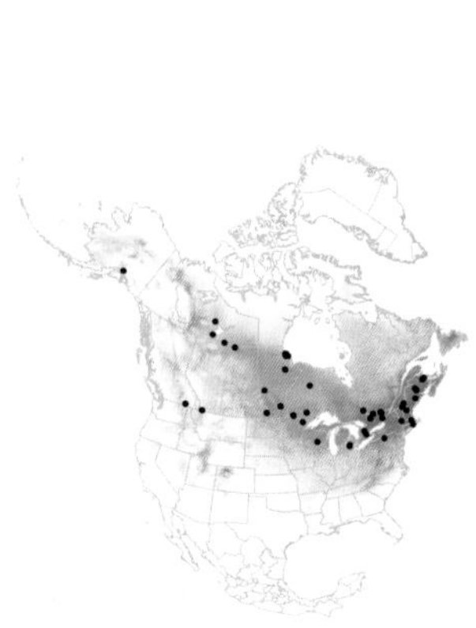

## *Parhelophilus integer* Shiny Bog Fly

**SIZE RANGE:** 8.7–10.9 mm
**IDENTIFICATION:** *Parhelophilus integer* has a distinctly striped thorax. The male of this species is distinctive because of its shiny pregenital segments, short sternite 6, and a metafemur that is broad and not arched. Females are much more difficult to identify. They are very similar to *P. divisus* and *P. laetus* and can be differentiated only by the metabasitarsus, which has zero to three black bristles and long, erect pile. **ABUNDANCE:** Rare. **FLIGHT TIMES:** Early June to mid-August, with records as early as 23 May from the southern part of its range. **NOTES:** This species is associated with large bodies of water (Great Lakes, Atlantic and Gulf coasts). In the Great Lakes region it has been found in swamps and has been recorded nectaring on *Cephalanthus* and *Salix*.

# *Parhelophilus obsoletus*

♂
stripes on thorax indistinct to absent
♀
metabasitarsus with short, appressed pile
♀
♂
metafemur of male narrow and arched
tergite 4 equal in length to sternite 6
shiny pregenital segments
♂
tergite 4
sternite 6
♂
♂

# *Parhelophilus integer*

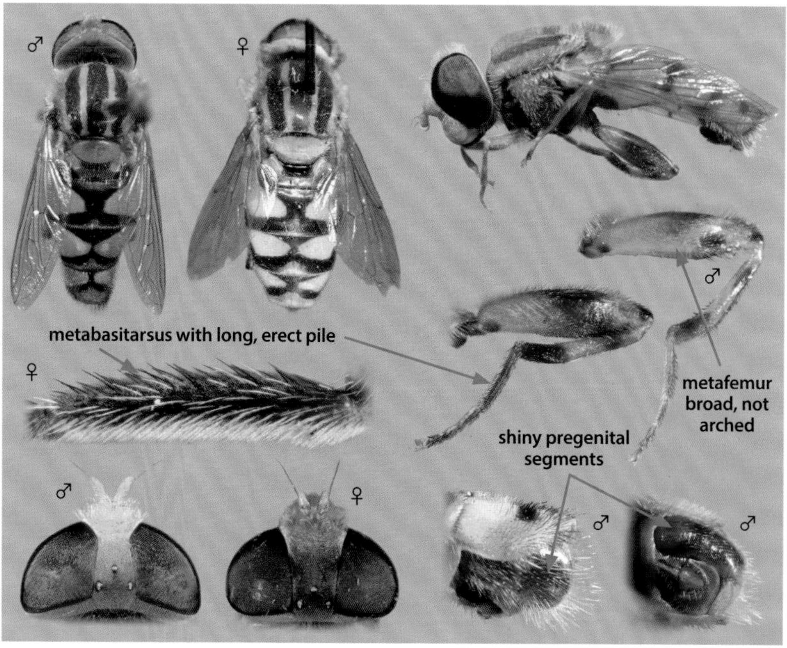

♂
♀
metabasitarsus with long, erect pile
♀
♂
metafemur broad, not arched
shiny pregenital segments
♂
♀
♂
♂

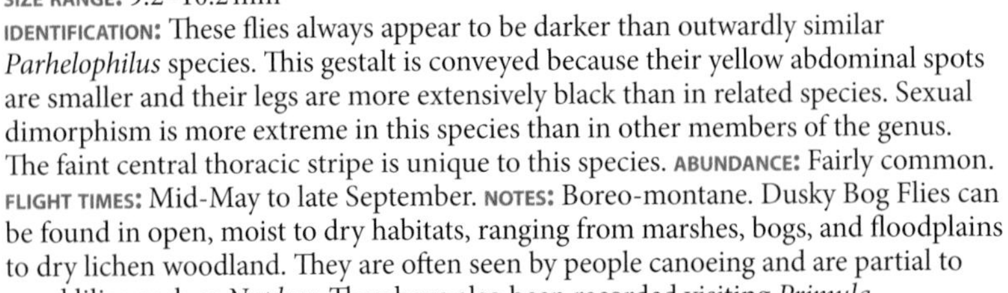

## *Parhelophilus rex* Dusky Bog Fly

**SIZE RANGE:** 9.2–10.2 mm

**IDENTIFICATION:** These flies always appear to be darker than outwardly similar *Parhelophilus* species. This gestalt is conveyed because their yellow abdominal spots are smaller and their legs are more extensively black than in related species. Sexual dimorphism is more extreme in this species than in other members of the genus. The faint central thoracic stripe is unique to this species. **ABUNDANCE:** Fairly common. **FLIGHT TIMES:** Mid-May to late September. **NOTES:** Boreo-montane. Dusky Bog Flies can be found in open, moist to dry habitats, ranging from marshes, bogs, and floodplains to dry lichen woodland. They are often seen by people canoeing and are partial to pond lilies such as *Nuphar*. They have also been recorded visiting *Primula*.

## *Parhelophilus laetus* Common Bog Fly

**SIZE RANGE:** 7.8–11.5 mm

**IDENTIFICATION:** Males of this brightly colored species can be readily identified by their elongate cerci. Barring a ventral view, a combination of leg color characters will distinguish the males. Females are much more difficult to identify and are very similar to *P. divisus* and *P. integer.* They can be differentiated only by examining the bristle pattern of the metabasitarsus with a hand lens (zero to three black bristles). **ABUNDANCE:** Common. **FLIGHT TIMES:** Mid-May to early September but have been recorded as late as early October in the south. **NOTES:** This widespread *Parhelophilus* species extends from southern boreal regions into the south-central USA and in the east seems to be largely a Great Lakes / St. Lawrence species. Specimens have been taken in cattail marshes, tamarack fens, and wet woods.

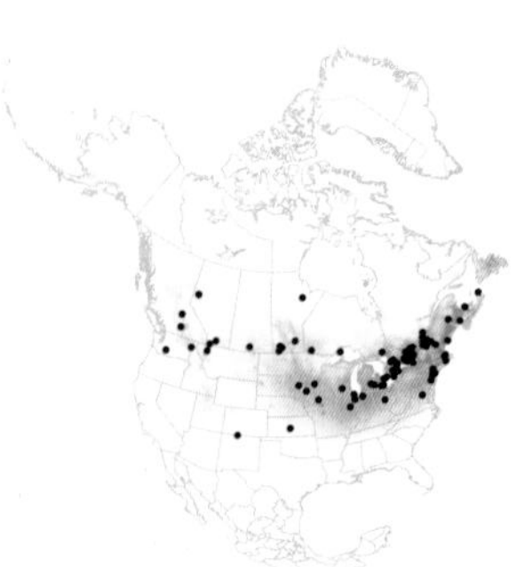

# Parhelophilus rex

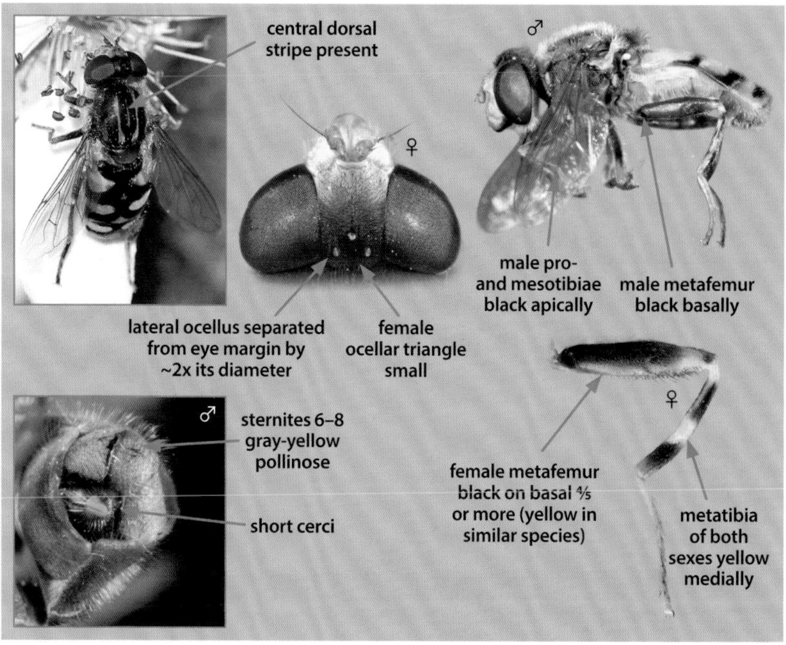

central dorsal stripe present

♂

♀

male pro- and mesotibiae black apically

male metafemur black basally

lateral ocellus separated from eye margin by ~2x its diameter

female ocellar triangle small

♂

sternites 6–8 gray-yellow pollinose

short cerci

♀

female metafemur black on basal ⅘ or more (yellow in similar species)

metatibia of both sexes yellow medially

# Parhelophilus laetus

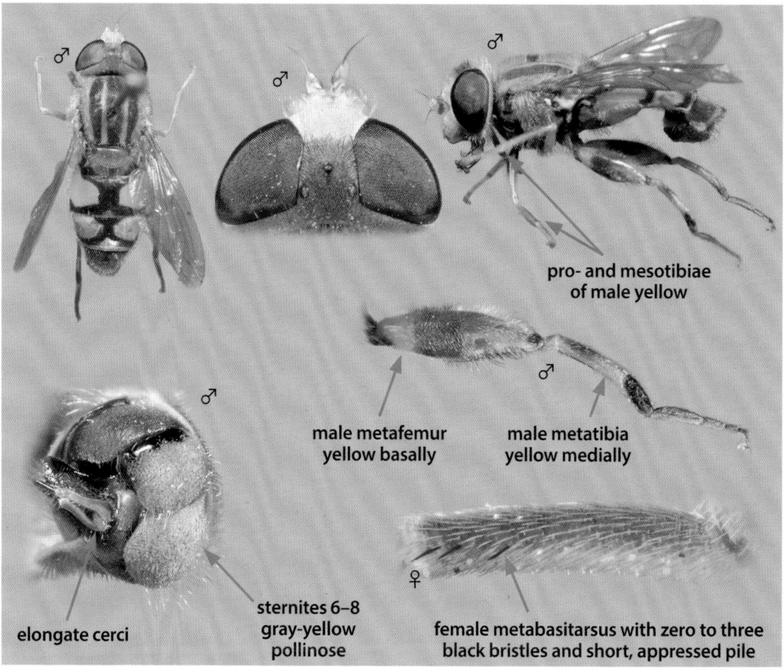

♂

♂

♂

pro- and mesotibiae of male yellow

♂

♂

male metafemur yellow basally

male metatibia yellow medially

♀

elongate cerci

sternites 6–8 gray-yellow pollinose

female metabasitarsus with zero to three black bristles and short, appressed pile

## *Parhelophilus flavifacies* Black-legged Bog Fly

**SIZE RANGE:** 9.7–11.7 mm

**IDENTIFICATION:** Males of *P. flavifacies* are most similar to *P. divisus* but can usually be distinguished by their darker thorax color. If in doubt, check the width of the vertex (narrow) and color of the metatibia (black). Females are more difficult to identify but can be readily separated from similar species by the narrow frons (wider in *P. divisus*, *P. integer*, *P. obsoletus*, and *P. laetus*). **ABUNDANCE:** Uncommon. **FLIGHT TIMES:** Late May to mid-July. **NOTES:** This species occurs south of the boreal region. DNA barcodes suggest that it is a species complex that includes two species in our region. Individuals have been recorded on *Caltha* and *Nuphar*. Specimens have been collected along shaded forest streams and in woods.

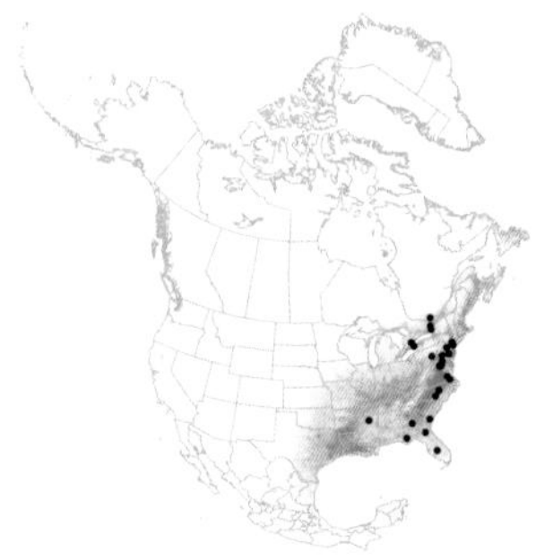

male metatrochanter without black setae

## *Parhelophilus divisus* Yellow-legged Bog Fly

**SIZE RANGE:** 8.7–10.0 mm

**IDENTIFICATION:** Males of *P. divisus* are most similar to *P. flavifacies* but can usually be distinguished by their thorax color. If in doubt, check the width of the vertex (wide) and color of the metatibia (mostly yellow). Females are very similar to *P. integer* and *P. laetus* and can be differentiated only by examining the metabasitarsus with a hand lens or microscope. **ABUNDANCE:** Rare. **FLIGHT TIMES:** Early July to late August in our area. As early as mid-May farther south. **NOTES:** Specimens have been recorded nectaring on *Nuphar* (pond-lily), apparently a preferred nectar plant for many members of this genus.

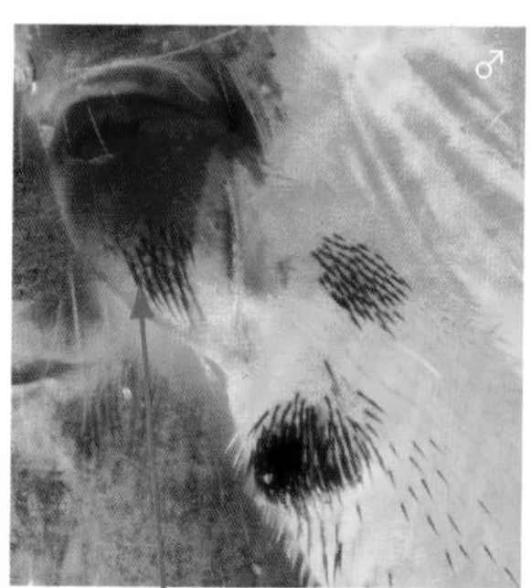


male metatrochanter with black setae

# *Parhelophilus flavifacies*

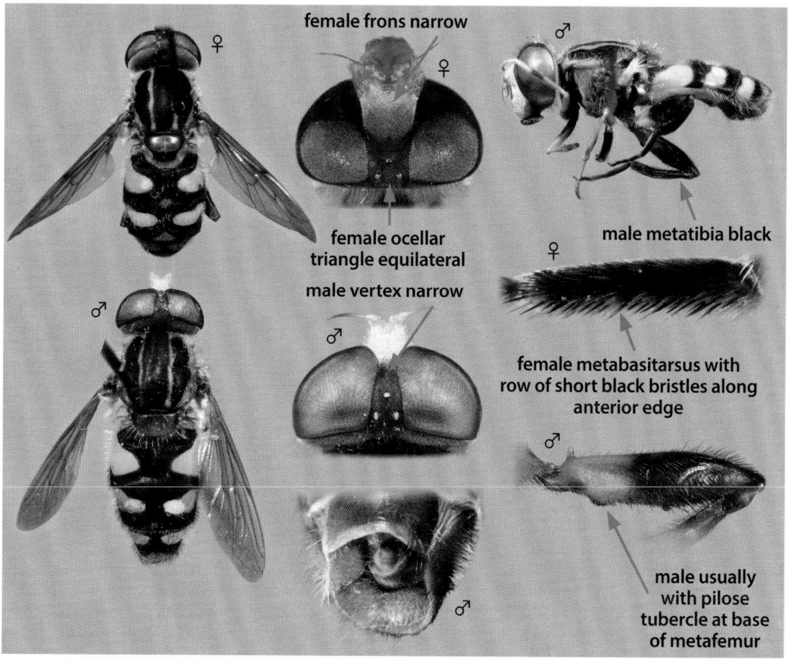

female frons narrow

♀

♂

female ocellar triangle equilateral

male metatibia black

male vertex narrow

female metabasitarsus with row of short black bristles along anterior edge

male usually with pilose tubercle at base of metafemur

# *Parhelophilus divisus* ♂

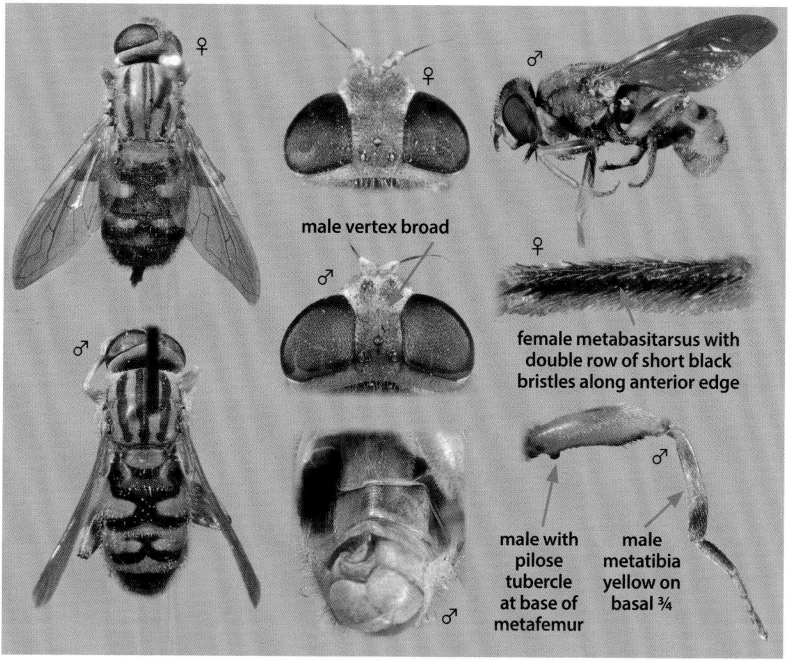

♀

♂

male vertex broad

female metabasitarsus with double row of short black bristles along anterior edge

male with pilose tubercle at base of metafemur

male metatibia yellow on basal ¾

# Eurimyia

*Lejops* has long been recognized as a confusing array of different-looking subgenera. Recent work has shown that *Lejops* is not a natural group and the subgenera should be treated separately. *Lejops* in the strict sense does not occur in our area, and we treat four of the former subgenera as genera here: *Eurimyia*, *Anasimyia*, *Arctosyrphus*, and *Polydontomyia*. The latter two genera are monotypic, while *Anasimyia* is diverse and *Eurimyia* includes only three species: one Holarctic, one Nearctic, and one found only in Japan. Despite their morphological diversity, larvae of all of these genera are associated with water.

## *Eurimyia stipata* Long-nosed Swamp Fly

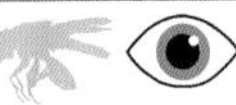

**SIZE RANGE:** 7.8–9.3 mm
**IDENTIFICATION:** This species can be easily recognized by the striped scutum and face that is produced anteroventrally into a cone. **ABUNDANCE:** Very common. **FLIGHT TIMES:** Mid-April to late September. **NOTES:** Based on DNA data and morphology, we are resurrecting *Eurimyia stipata* from synonymy with *E. lineata*. *Eurimyia lineata* is restricted to Eurasia, Alaska, the Yukon, and Northwest Territories. *Eurimyia stipata* should be used as the name for the widespread Nearctic species. Like its close relative *E. lineata*, this species presumably has aquatic larvae that are found in a variety of wetland habitats (including bogs, fens, marshes, oxbow lakes). Larvae of the closely related *E. lineata* are aquatic, found in rotting plant debris just below the water's surface. Adults are often found in moist meadows and forest gaps. Near water, they can typically be found flying low over the surface, often within stands of emergent vegetation. Floral records include *Phalaris*, *Primula*, *Sonchus*, and *Symphyotrichum*.

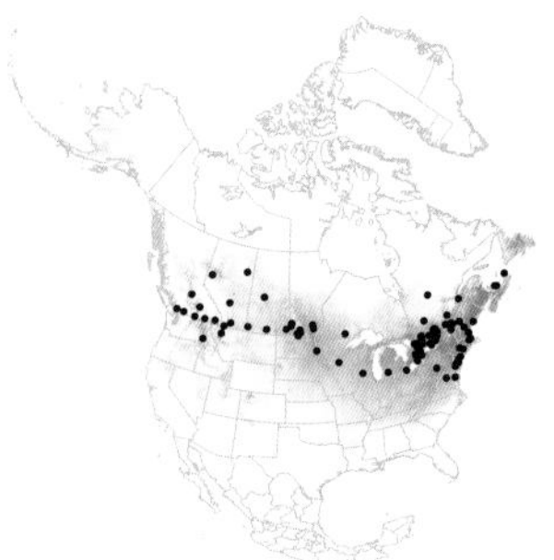

# *Eurimyia*

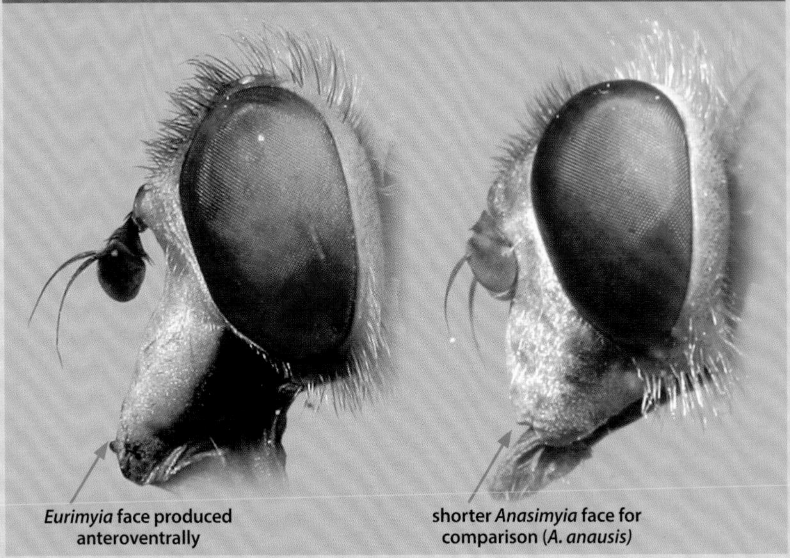

*Eurimyia* face produced anteroventrally

shorter *Anasimyia* face for comparison (*A. anausis*)

## *Eurimyia stipata*

♂

striped scutum

♀

♀

elongate face

# *Anasimyia*

In another former subgenus of *Lejops*, *Anasimyia* species are tricky to identify and the genus is in need of revision. Abdominal color varies greatly within species from mostly orange to black with yellow-orange markings. Recognizing two groups helps with identification. Males of Group 1 have no tubercle on the metatrochanter and a small to absent spur on the metatibia, while females are indistinguishable but all have a broad frons, little black on the scutellum, and a broad separation of the lateral ocelli from the eye. Males of Group 2 have metatrochanter tubercles and large metatibial spurs, while females are indistinguishable but have a narrower frons, more black on the scutellum, and a narrower separation of lateral ocelli from the eye. There are 21 species of *Anasimyia* globally, with seven in the Nearctic. We know of at least five undescribed Nearctic species in collections that were worked out by Dick Vockeroth but never described. Five of the seven described species and two of the five undescribed species occur within the range of this field guide. Male genitalia should be carefully assessed as part of a complete revision of the group in our region. The last key to *Lejops* (including *Anasimyia*) was published by Curran and Fluke (1926).

***Anasimyia*** **species are distinctive but tricky to identify to species**

# *Anasimyia* Group 1

## *Anasimyia bilinearis* Two-lined Swamp Fly

**SIZE RANGE:** 7.7–10.6 mm

**IDENTIFICATION:** Males of this species have relatively unmarked abdomens with a pair of small pollinose markings on tergite 4. The metatrochanter and tibiae have no tubercles or spurs. Group 1 females have a scutellum that is black on the anterior ⅓ or less, a broad frons, and broader separation of the lateral ocelli from the eye. **ABUNDANCE:** Uncommon. **FLIGHT TIMES:** Early April to early July. **NOTES:** One specimen was collected in sand dunes. Unlike most *Anasimyia*, this species does not vary greatly in color and patterning.

# Anasimyia

**females cannot be identified to species and can only be placed in one of two groups**

**GROUP 1**

♀

scutellum ⅓ or less black

lateral ocelli farther from eye

♀

frons broad

**GROUP 2**

♀

scutellum ⅓ or more black

lateral ocelli close to eye

♀

frons narrow

## Anasimyia bilinearis

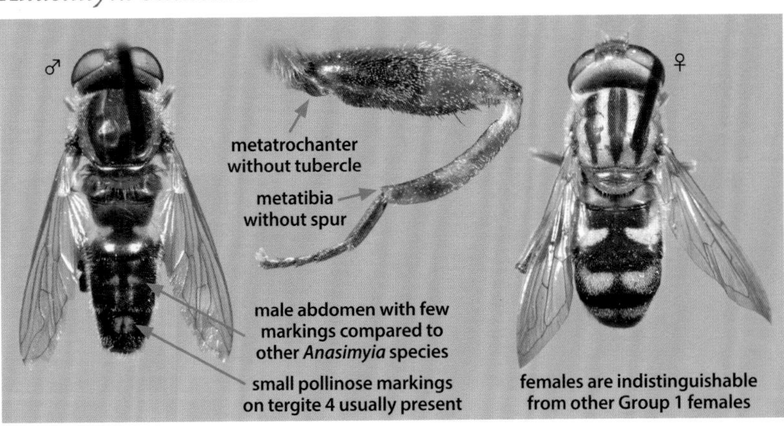

# *Anasimyia* Group 1

## *Anasimyia* undescribed species 1

### Smooth-legged Swamp Fly

**SIZE RANGE:** 7.9 mm
**IDENTIFICATION:** In males, the metatrochanter is without a tubercle and has fine, pale pile on it; the metatibia is without a spur, and tergite 4 has large pollinose markings. Group 1 females have a scutellum that is black on the anterior ⅓ or less, a broad frons, and broader separation of the lateral ocelli from the eye. **ABUNDANCE:** Very rare. **FLIGHT TIMES:** Mid-May to late August. **NOTES:** This species is known from only three specimens (from Aweme, Manitoba, and Oliver Bog, Ontario). Additional sampling in bogs may show that it is more widespread.

## *Anasimyia anausis* Moon-shaped Swamp Fly

**SIZE RANGE:** 8.0–9.7 mm
**IDENTIFICATION:** In males, the metatrochanter is without a tubercle and the metatibia has a short spur. **ABUNDANCE:** Fairly common. **FLIGHT TIMES:** Early April to mid-August. **NOTES:** We are resurrecting the name *A. anausis* from synonymy with *A. lunulata*. The latter is widespread in the Palearctic and was thought to be the same as our species. Genetic work shows clearly that they are not even sister species. *Anasimyia anausis* has been found in a wide variety of wetlands including bogs, fens, alkaline sloughs, maple-elm floodplains, tidal marshes, wet meadows, and hardwood swamps. Adults fly low among vegetation near standing water or out over the water's surface. They are usually found within a few meters of the water's edge and often land on emergent vegetation. They can be abundant on regenerating bogs that were cut over. Flowers visited include *Houstonia* and *Rosa*. The larva was described by Hartley (1961) but may have been confounded with that of another *Anasimyia* species (*A. interpuncta*), so more work is required.

## ***Anasimyia*** undescribed species 1

♂
♂
♂
♂
male with large pollinose markings on tergite 4
metatrochanter without tubercle, only fine, pale pile
metatibia without spur

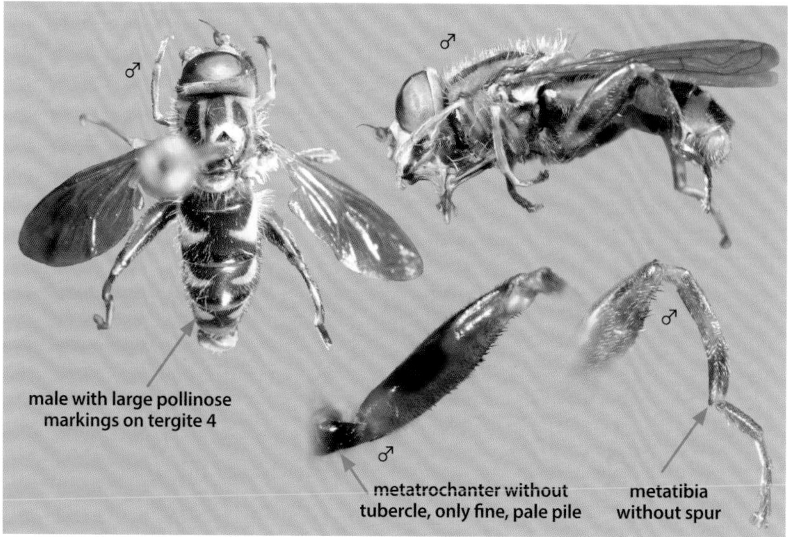

## ***Anasimyia anausis***

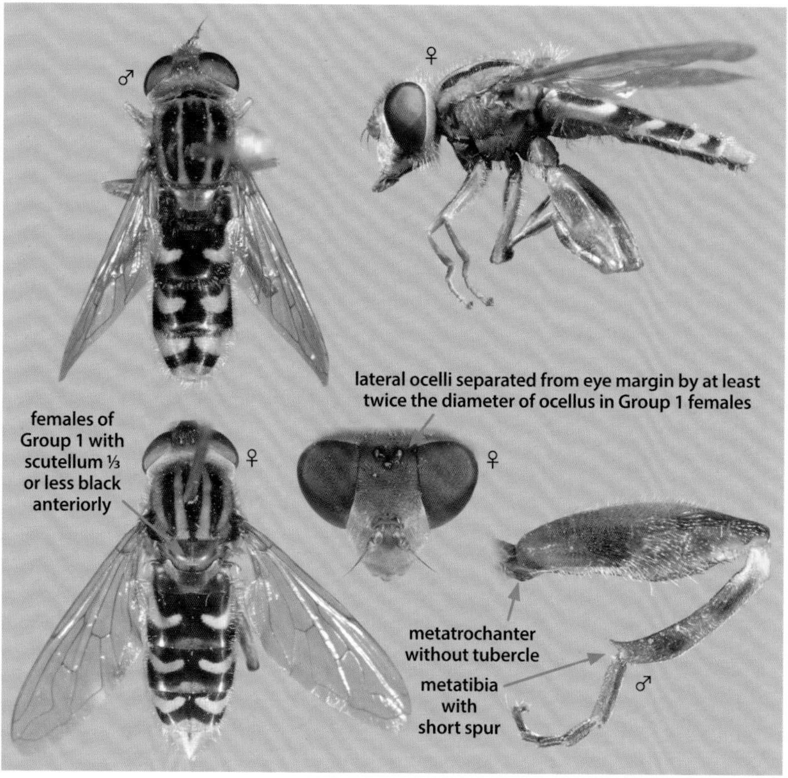

♂
♀
lateral ocelli separated from eye margin by at least twice the diameter of ocellus in Group 1 females
females of Group 1 with scutellum ⅓ or less black anteriorly
♀
♀
metatrochanter without tubercle
metatibia with short spur
♂

# *Anasimyia* Group 2

## *Anasimyia chrysostoma* Lump-legged Swamp Fly

**SIZE RANGE:** 8.2–10.6 mm
**IDENTIFICATION:** Males have a large tubercle on the metatrochanter and a large, subacute, apical metatibial spur. Group 2 females have a narrower frons, scutellum with dark on anterior ½ or more and a narrower separation of lateral ocelli from the eye. **ABUNDANCE:** Fairly common. **FLIGHT TIMES:** Mid-May to early September throughout most of its range and as late as mid-October in the south. **NOTES:** *Anasimyia relicta* is an unpublished synonym of *A. chrysostoma*. Look for these flies in swamps.

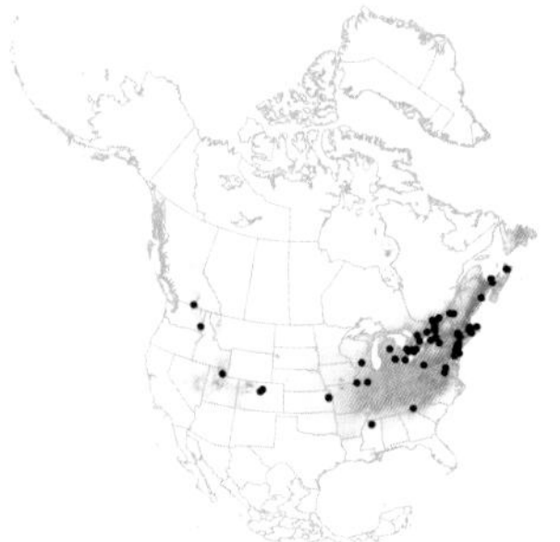

## *Anasimyia* undescribed species 2
Small-spotted Swamp Fly

**SIZE RANGE:** 9.8–10.2 mm
**IDENTIFICATION:** Males have a metatrochanter with a modest tubercle and a long, acute, apical metatibial spur. **ABUNDANCE:** Very rare. **FLIGHT TIMES:** Mid-May to mid-June. **NOTES:** Only four specimens are known, from Virginia (Chain Bridge, Great Falls, and Petersburg) and Washington, DC.

# *Anasimyia chrysostoma*

♂ ♀ ♂

scutellum with anterior ½ or
more dark in all Group 2 females

lateral ocelli separated
from eye by < width of
ocellus in Group 2 females

♂ ♀

large
tubercle

large
spur

# *Anasimyia* undescribed species 2

♂ ♂

female frons
narrow

♂ ♀

short
tubercle

long, acute
apical spur

# *Anasimyia* Group 2

## *Anasimyia distincta* Short-spurred Swamp Fly

**SIZE RANGE:** 8.4–10.5 mm
**IDENTIFICATION:** Males have a metatrochanter with a very small tubercle and a short, blunt apical spur on the metatibia. **ABUNDANCE:** Rare. **FLIGHT TIMES:** Late May to mid-September. **NOTES:** The single floral record is from *Nuphar*.

## *Anasimyia grisescens* Long-spurred Swamp Fly

**SIZE RANGE:** 7.2–10.1 mm
**IDENTIFICATION:** Males have a large, parallel-sided tubercle on the metatrochanter and a moderately long, subarcuate apical spur on the metatibia. **ABUNDANCE:** Rare. **FLIGHT TIMES:** Mid-June to early-August. **NOTES:** This appears to be a Great Lakes / Atlantic coast species and is never found far from major waterways.

## *Anasimyia distincta*

♂
♂
♂
♂
color of male abdomen variable in most species of this genus
very small tubercle
short, blunt apical spur

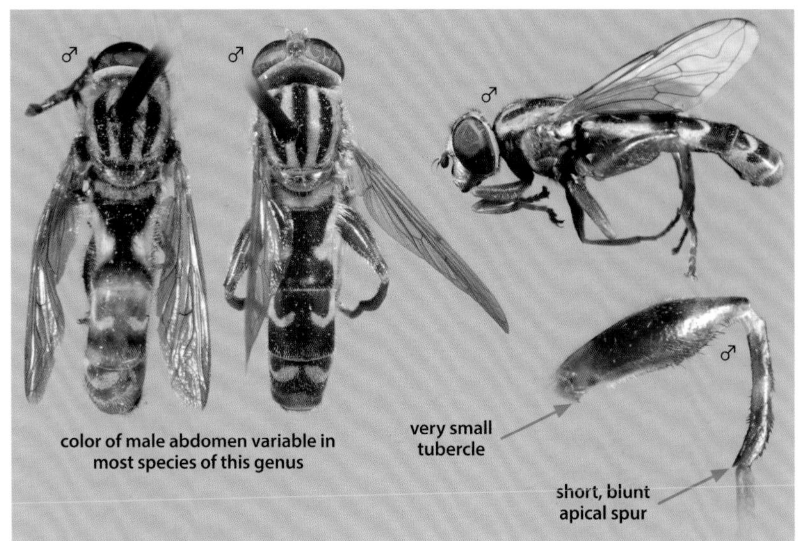

## *Anasimyia grisescens*

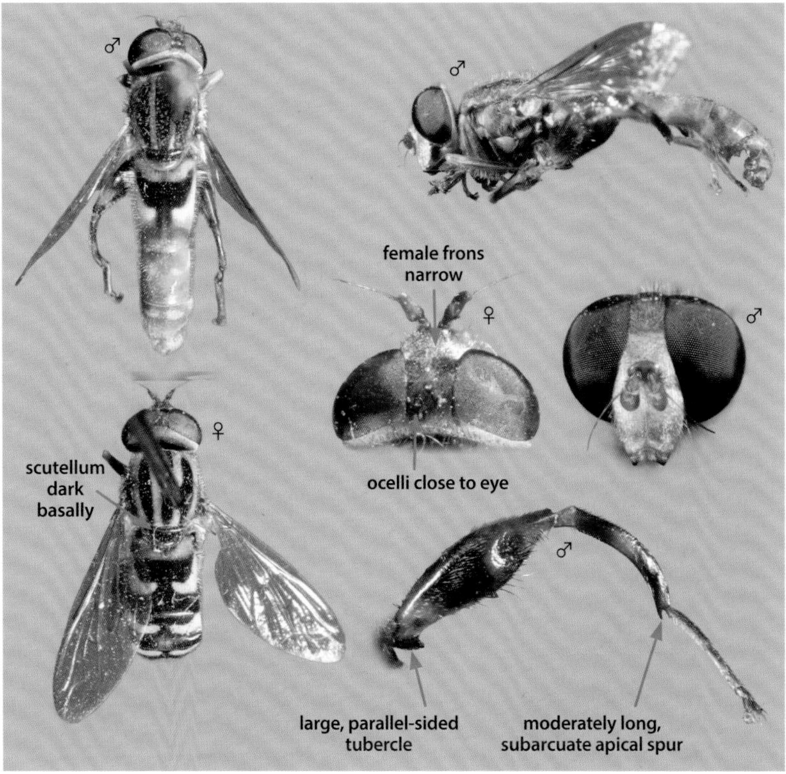

♂
♂
female frons narrow
♀
♂
♀
scutellum dark basally
ocelli close to eye
♂
large, parallel-sided tubercle
moderately long, subarcuate apical spur

# *Arctosyrphus*

*Arctosyrphus* was also recently considered a part of *Lejops*. This monotypic genus is quickly recognized by the long, nontuberculate face and white-pilose thorax.

## *Arctosyrphus willingii* Northern Longbeak

**SIZE RANGE:** 9.6–12.2 mm
**IDENTIFICATION:** This species is dark bodied with pale pile. Its face is produced anteroventrally with no tubercle. **ABUNDANCE:** Very rare in the east. **FLIGHT TIMES:** Late May to mid-August. **NOTES:** This Holarctic species has been collected near tundra wetlands, prairie sloughs, and sedge meadows. One specimen was noted nectaring on *Taraxacum*. On the tundra, adults have been recorded beside shallow freshwater pools enriched with organic material and around swampy ground with hummocks. In these habitats, females oviposit into the ground among grass roots. The larva was described by Bagatshanova (1990).

# *Polydontomyia*

*Polydontomyia* is another monotypic former member of the genus *Lejops*. Males are particularly distinctive with their enlarged metalegs and orange abdomens. Females are more subtle but also quickly field recognizable once learned. If in doubt, a quick look at the bulging sternites confirms them.

## *Polydontomyia curvipes*
Dimorphic Sickleleg

**SIZE RANGE:** 12.1–17.4 mm
**IDENTIFICATION:** This species is highly sexually dimorphic. Males have a bright orange abdomen and an enlarged, arcuate metafemur and large apical tibial spur. Females have a uniformly yellow thorax with sternites 1 and 3 swollen, sternite 1 overlapping sternite 2, and sternite 2 reduced, sunken and membranous medially. Both sexes have a black face (ground color). **ABUNDANCE:** Locally abundant in coastal salt marshes. **FLIGHT TIMES:** Late April to mid-October. **NOTES:** This attractive fly is rarely encountered away from brackish wetlands.

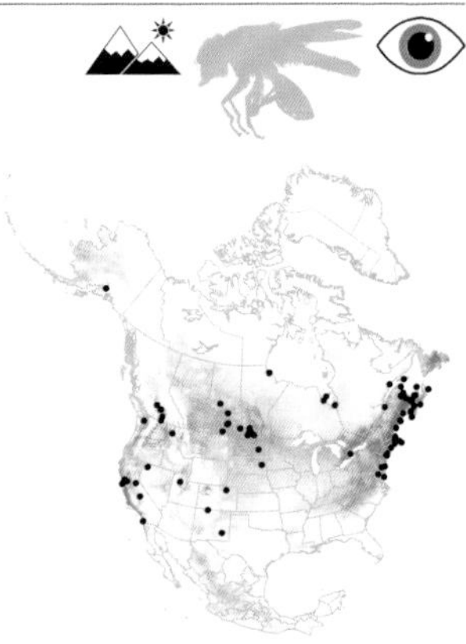

## *Arctosyrphus willingii*

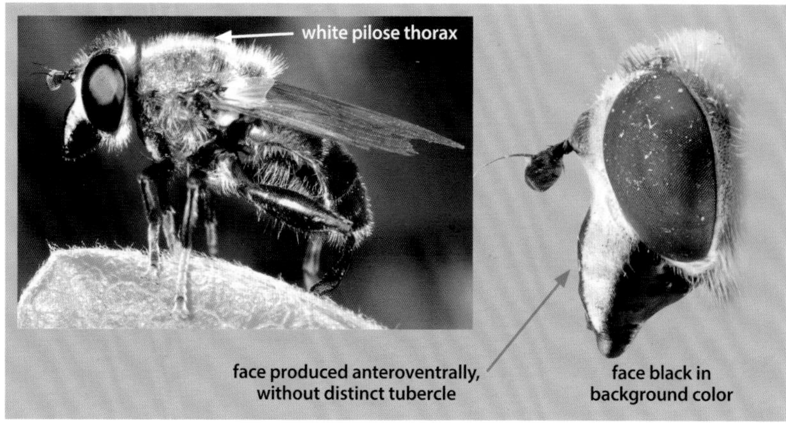


## *Polydontomyia curvipes*

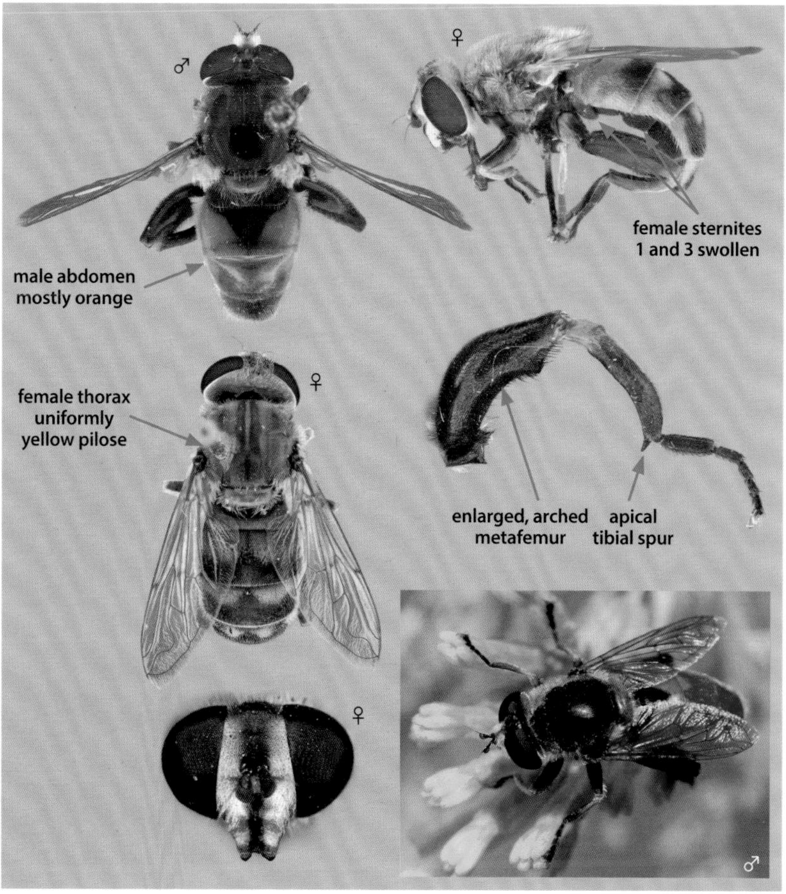

♂

♀

female sternites 1 and 3 swollen

male abdomen mostly orange

♀

female thorax uniformly yellow pilose

enlarged, arched metafemur

apical tibial spur

♀

♂

# Ferdinandea

*Ferdinandea* species can easily be distinguished by the distinct parallel stripes on the scutum, strong black bristles on the scutum and scutellum, and lack of yellow markings on the abdomen. There are 15 species globally and three in the Nearctic. The two northeastern species are distinguished by their arista color and the number of notopleural bristles. The last key to the world species of *Ferdinandea* was by Hull (1942b). Unfortunately, the taxonomy is complex and requires review. Larvae of the Nearctic species are unknown, but known *Ferdinandea* larvae filter feed on bacteria in sap runs of trees. Adults often tree sit and can be easily overlooked. There is considerable external variation between species, and molecular data should be used to support species concepts.

## *Ferdinandea buccata*

Common Copperback

**SIZE RANGE:** 7.0–12.0 mm
**IDENTIFICATION:** *Ferdinandea buccata* can be distinguished from related species by the presence of two notopleural bristles and an arista that is yellow on the basal ⅔. **ABUNDANCE:** Fairly common. **FLIGHT TIMES:** Early May to mid-September (mid-March to late October in the south). **NOTES:** Based on unpublished morphological and molecular evidence, *F. buccata* includes *F. dives* and *F. nigripes*. These new synonyms are treated here for the first time. This is by far the most common Nearctic *Ferdinandea* species. Adults have been found in many habitats including hardwood forests, spruce-birch woods, natural meadows, sphagnum-dominated fens, and creek margins. Pupae have been found in leaf mold beneath a willow.

## *Ferdinandea croesus* Golden Copperback

**SIZE RANGE:** 8.5–12.4 mm
**IDENTIFICATION:** *Ferdinandea croesus* has a black arista and four notopleural bristles. The entire abdomen is shiny. **ABUNDANCE:** Fairly common in the west, rare in the east. **FLIGHT TIMES:** April to late August in the east. Mid-April to late November in the west. **NOTES:** Eastern specimens should be treated as suspect until DNA evidence supporting their identity is available. Specimens from the east conform morphologically to *F. croesus*, but the disjunct eastern distribution is atypical among syrphids.

# Ferdinandea

## Ferdinandea buccata

## Ferdinandea croesus

# *Pterallastes*

*Pterallastes* is known only from four species: two from China, one from Japan, and our eastern Nearctic species. Thompson (1974) was the last to publish on this genus in North America. Larvae of *Pterallastes* are unknown.

## *Pterallastes thoracicus* Goldenback

**SIZE RANGE:** 10.7–13.2 mm
**IDENTIFICATION:** Our only species of *Pterallastes* is similar to *Brachypalpus* but may be recognized by differences in wing venation, the blacker abdomen, and more densely pollinose thorax. **ABUNDANCE:** Uncommon. **FLIGHT TIMES:** Mid-May to late October. **NOTES:** Most specimens have been collected in hardwood forests, with many observed hilltopping on a hardwood ridgetop in Pennsylvania. A single specimen came from a cypress swamp. Flowers visited include *Ceanothus*, *Solidago*, and *Viburnum*.

# *Merodon*

Only one introduced species of *Merodon* occurs in North America (of 159 species known). The group has its center of diversity in the Mediterranean, where the bulbs its larvae feed on are most diverse. Wing venation (sinuous $R_{4+5}$ and recurved $M_1$) combined with all black legs is diagnostic for this species.

## *Merodon equestris* Narcissus Bulb Fly

**SIZE RANGE:** 12.3–17.2 mm
**IDENTIFICATION:** Several color morphs (formerly treated as subspecies) exist. Flies can be entirely yellow pilose, to more predominantly orange, to extensively black pilose on the posterior part of thorax and anterior part of abdomen. **ABUNDANCE:** Fairly common. **FLIGHT TIMES:** Mid-March to late July (one record from 10 September). **NOTES:** The very early spring records (for example, 10 March in central Ontario) suggest that they overwinter as either pupae or adults. Despite often being hard to find, they may be locally common, and their bulb-feeding larvae may be serious pests of *Narcissus* and other Liliaceae. Adults zigzag low along the ground, often among vegetation. They frequently land on open ground and visit many species of flowers.

## *Pterallastes thoracicus*

vein $R_{4+5}$ curves into cell $r_{4+5}$
thorax covered in yellow pile and yellow pollen

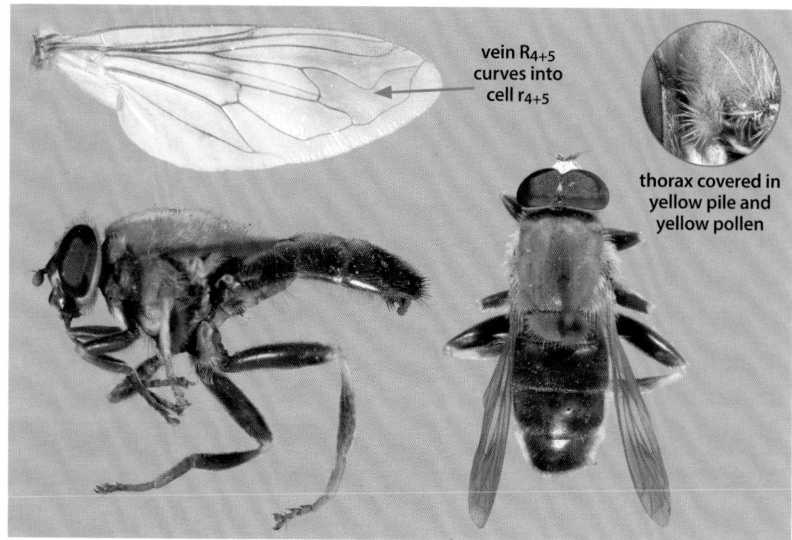

## *Merodon equestris*

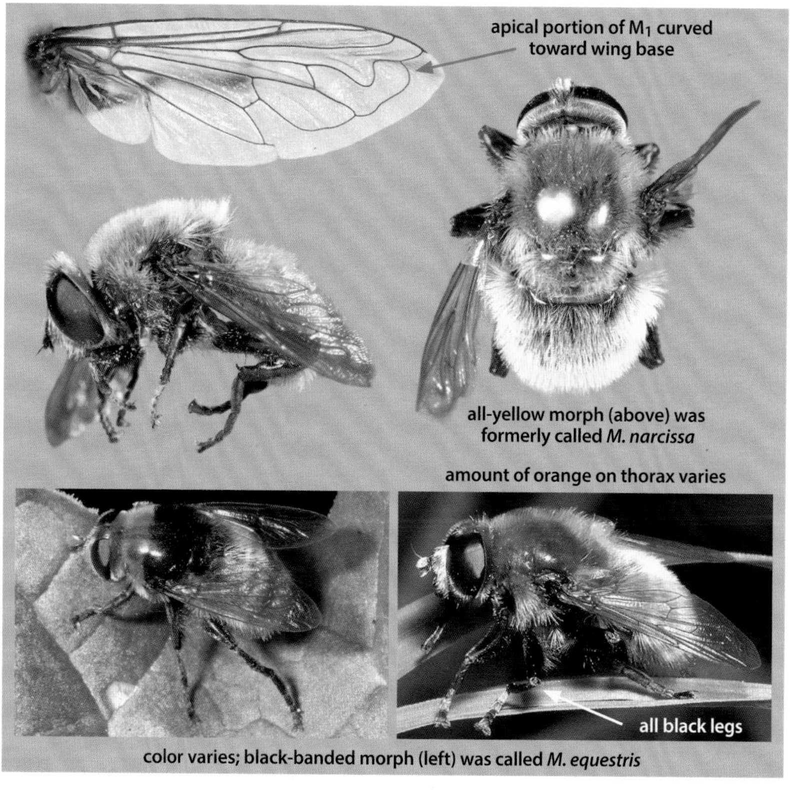
apical portion of $M_1$ curved toward wing base
all-yellow morph (above) was formerly called *M. narcissa*
amount of orange on thorax varies
all black legs
color varies; black-banded morph (left) was called *M. equestris*

# *Brachypalpus*

*Brachypalpus* species have very distinctive, triangular heads and are large, dark flies with light pile covering their body. For more information on the genus, see Hippa (1978). The 12 species in this genus occur in the Holarctic region, with half of the described species in North America and half in Europe and Asia. Only two species occur within the range of the field guide.

## *Brachypalpus (Brachypalpus) oarus*

Eastern Catkin Fly

**SIZE RANGE:** 7.5–14.2 mm

**IDENTIFICATION:** Distinguished from *B. femoratus* by the pale yellow-pilose thorax and pollinose stripes on the face. Male holoptic or with extremely thin separation. Face does not extend past antennal base. **ABUNDANCE:** Fairly common. **FLIGHT TIMES:** Late March to mid-July. **NOTES:** Larvae have been found in decaying wood. Adults often sit on logs and are commonly observed at flowers of *Amelanchier*, *Rubus*, and *Salix*.

## *Brachypalpus (Crioprora) femoratus*

Orange Catkin Fly

**SIZE RANGE:** 12.0–15.2 mm

**IDENTIFICATION:** This species appears very orange in life and can easily be distinguished from *B. oarus* by this as well as by the completely shiny face with its two yellow arcs. Males distinctly dichoptic. Face extends past antennal base. **ABUNDANCE:** Fairly common in the west but very rare in the east. **FLIGHT TIMES:** Late April to mid-August. **NOTES:** *Brachypalpus femoratus* appears to be a complex of species that is currently under revision by Kevin Moran. The putative boreal species is illustrated here.

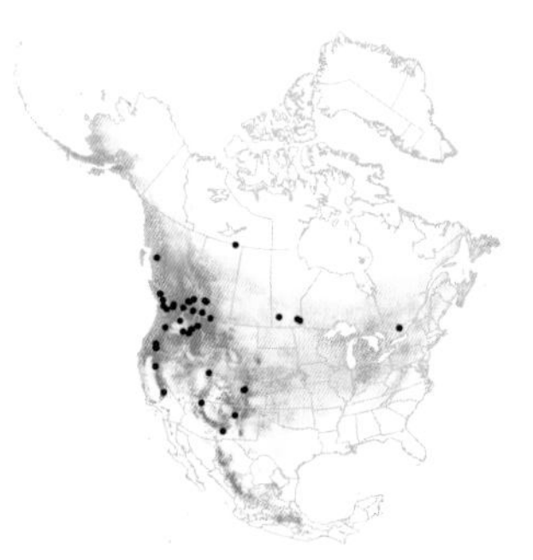

**orange pilose on thorax (fades to yellow after death)**

## *Brachypalpus (Brachypalpus) oarus*

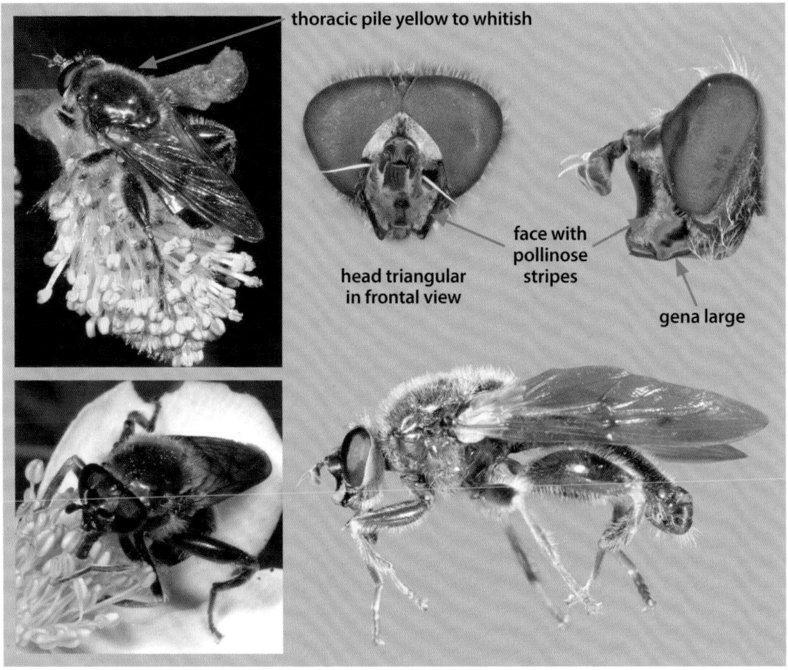

thoracic pile yellow to whitish

head triangular in frontal view

face with pollinose stripes

gena large

## *Brachypalpus (Crioprora) femoratus*

yellow arcs on face (only boreal specimens display this character)

face shiny, not pollinose

# Eristalinus

*Eristalinus* species are easily distinguished by their patterned eyes and sinuous $R_{4+5}$ vein (the latter is found in all members of the tribe Eristalini). There are 93 species globally, with two introduced species in the Nearctic, each in a different subgenus: *Eristaloides* and *Lathyropthalmus*. They are distinguished by the pattern on the eye (spotted in the latter, striped in the former). Only *Lathyropthalmus* is found in our region. Larvae occur in freshwater seepages and brackish pools as well as in sewage ponds (where they are important in the turnover of waste).

## Eristalinus (Lathyropthalmus) aeneus

Common Lagoon Fly

**SIZE RANGE:** 8.4–11.7 mm
**IDENTIFICATION:** This species has a distinctive spotted pattern on its eyes and an overall greenish sheen.
**ABUNDANCE:** Fairly common. **FLIGHT TIMES:** Late March to late November (they overwinter as adults). **NOTES:** Introduced from Europe. Adults often land on bare ground, rock, or low vegetation and may be found nectaring on yellow composites, white Apiaceae, or *Salix*.

# Mallota

*Mallota* is a paraphyletic assemblage of 72 species from at least five unrelated groups of eristaline bumblebee mimics. Current molecular phylogenetic work aims to resolve this mess that has been created by the parallel evolution of bumblebee mimicry in different regions of the world. Nearctic *Mallota* will likely retain their name since they are related to the type species upon which the generic name is founded (the European *M. fuciformis*). *Mallota* can be identified by the bare arista, sinuous $R_{4+5}$ vein, enlarged metafemur, and a face that is enlarged ventrally. Larvae are filter feeders in water-filled tree holes. Like other Eristalini species, they have a distinctive tube-like breathing siphon at their tail tip ("rat tail"). The most recent key to species of the New World was by Curran (1940).

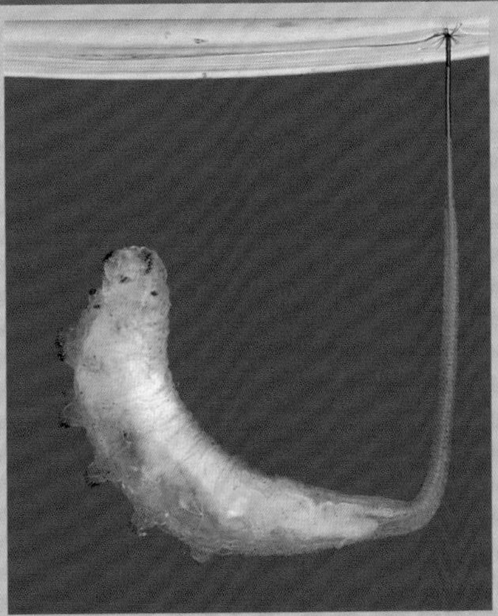

**This probable *Mallota* larva was taken from a tree hole in Backus Woods, Ontario, 10 August 2006.**

## *Eristalinus (Lathyropthalmus) aeneus*

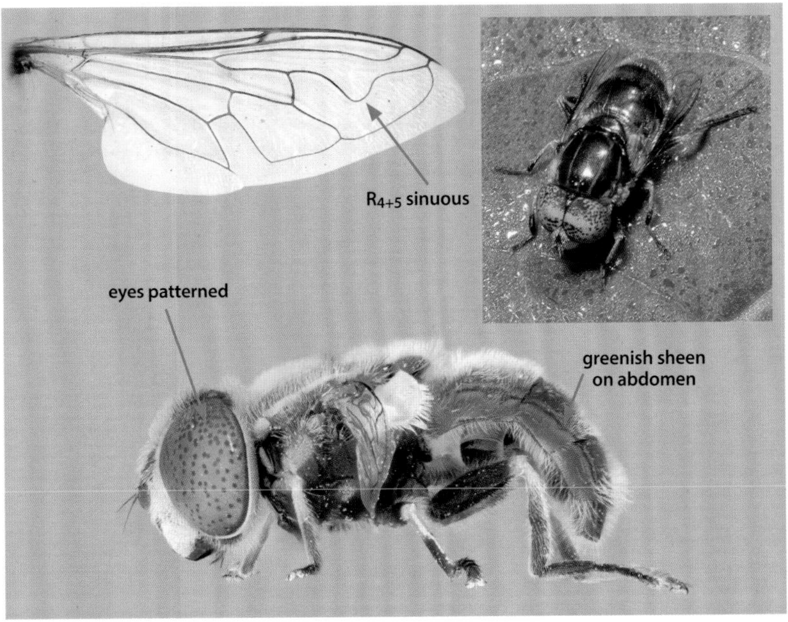

$R_{4+5}$ sinuous

eyes patterned

greenish sheen on abdomen

## *Mallota*

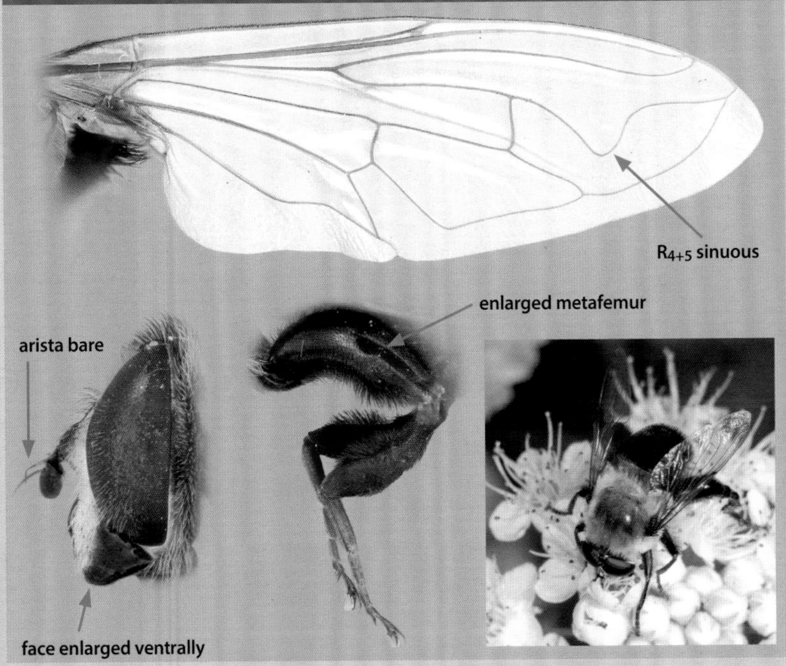

$R_{4+5}$ sinuous

enlarged metafemur

arista bare

face enlarged ventrally

## *Mallota posticata* Hairy-eyed Mimic

**SIZE RANGE:** 12.0–18.5 mm
**IDENTIFICATION:** This is the only *Mallota* species in the northeast with pilose eyes. Its abdomen is usually extensively yellow pilose. **ABUNDANCE:** Common. **FLIGHT TIMES:** Mid-May to early September (as early as late March in the southern part of its range). **NOTES:** *Mallota posticata* is relatively commonly found on flowers and has been noted visiting *Crataegus*, *Physocarpus*, and *Rubus*. *Physocarpus* has particularly great flowers for syrphids and often has many species in attendance. *Mallota posticata* adults can occasionally be found on hilltops but presumably do not have to typically visit landmarks, as they are common enough to encounter each other at flowers and suitable oviposition sites.

## *Mallota bautias* Bare-eyed Mimic

**SIZE RANGE:** 10.5–19.5 mm
**IDENTIFICATION:** This species has bare eyes and a black-pilose abdomen beyond tergite 1 and the anterolateral edges of tergite 2. Note that western specimens are yellow pilose on tergite 4. **ABUNDANCE:** Common. **FLIGHT TIMES:** Early May to mid-August (from mid-March in the south). **NOTES:** This is our most common species of *Mallota*. As with all *Mallota*, they are typically associated with mature hardwood forest and are regularly seen leaf sitting or flower visiting (*Acer*, Apiaceae, *Crataegus*, *Physocarpus*, *Rubus*, and *Viburnum*). Like *M. posticata*, they can occasionally be found hilltopping but are not as prevalent at hilltops. One specimen was collected freshly emerged from a wound on a bleeding elm, but like most *Mallota* species, they are presumed to occur in flooded rot holes.

# *Mallota posticata*

eye pilose

abdomen variable but usually yellow pilose beyond tergite 1

# *Mallota bautias*

eye bare

abdomen black pilose beyond tergite 1 in eastern specimens

## *Mallota mississipensis* Eastern Mimic

**SIZE RANGE:** 12.2–16.3 mm
**IDENTIFICATION:** Bare eyes. The abdomen is black pilose medially. Wing obfuscation does not extend much beyond crossvein r-m. Cell c is partly microtrichose. **ABUNDANCE:** Rare. **FLIGHT TIMES:** Mid-February to mid-May. **NOTES:** The taxonomy of this species has been confused. Until now, it was included under *M. albipilis*, but molecular data and genitalia differences support splitting this taxon into four. *Mallota mississipensis* is eastern, *M. illinoensis* is midwestern, *M. diversipennis* is restricted to Colorado and Utah, and *M. albipilis* is restricted to Arizona and New Mexico. Like other *Mallota* species, this species seems to have a preference for white flowers. Specimens have been found on *Cornus*, *Crataegus*, and *Prunus*.

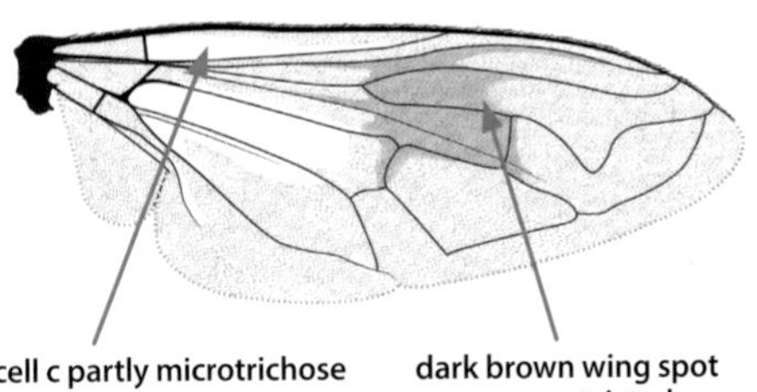

## *Mallota illinoensis* Midwestern Mimic

**SIZE RANGE:** 11.1–14.2 mm
**IDENTIFICATION:** Bare eyes. It is easily separated from the above species by the presence of a large dark brown wing spot and a completely pale-pilose abdomen. Cell c is almost completely bare. **ABUNDANCE:** Rare. **FLIGHT TIMES:** Late March to early August. **NOTES:** *Mallota palmerae* Jones, 1917 is a junior synonym. Specimens in the photos are visiting *Crataegus*.

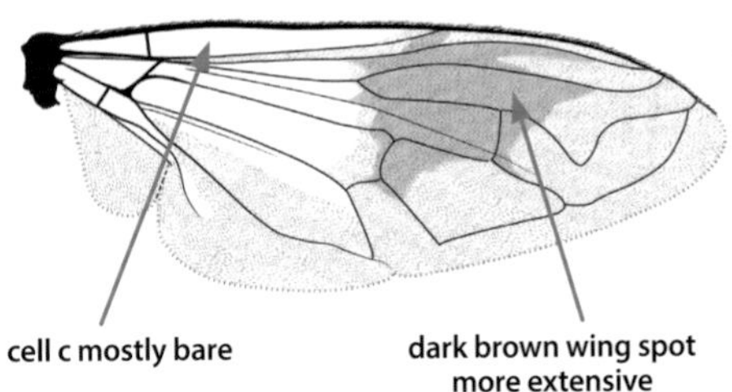

# *Mallota mississipensis*

abdomen black pilose medially

dark wing patch restricted to area around veins

eye bare

# *Mallota illinoensis*

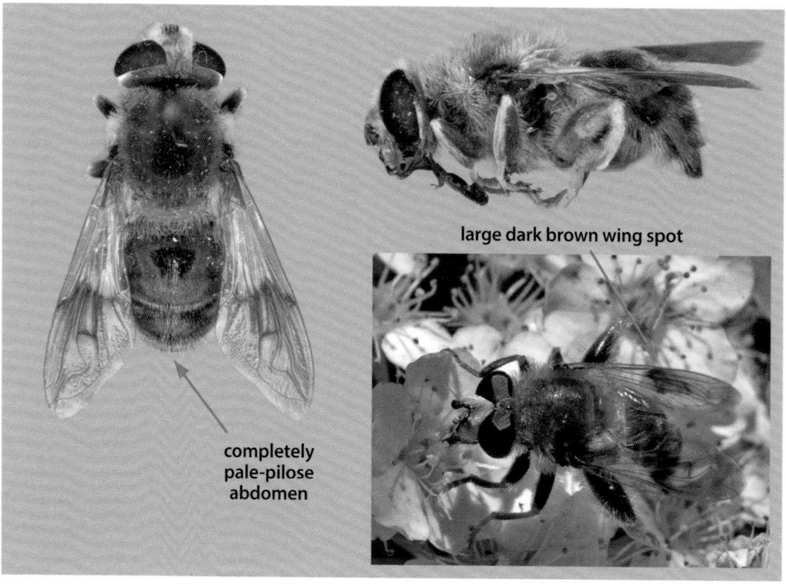

large dark brown wing spot

completely pale-pilose abdomen

# Criorhina

*Criorhina* species are mostly large bumblebee mimics. As with *Mallota*, interpretation of characters related to perfect mimicry has led to very unstable generic concepts. New data show that the 78 world *Criorhina* species should be distributed among five genera. *Criorhina* aristae are bare, their face is anteroventrally produced, and their metasternum is pilose. Pile coloring of *Criorhina* species is variable; beware of using color alone for identification. Variation shown here for *C. nigriventris* may occur in most *Criorhina* species. Three species occur in our region. The chance of finding these uncommon flies increases if you visit hilltops where males stake out territories and wait for potential mates. Larvae occur in decaying heartwood of trees, including old stumps, roots, and rot holes. Curran (1925a) published the last revision of the group, but Kevin Moran is working on a new revision.

## Criorhina nigriventris

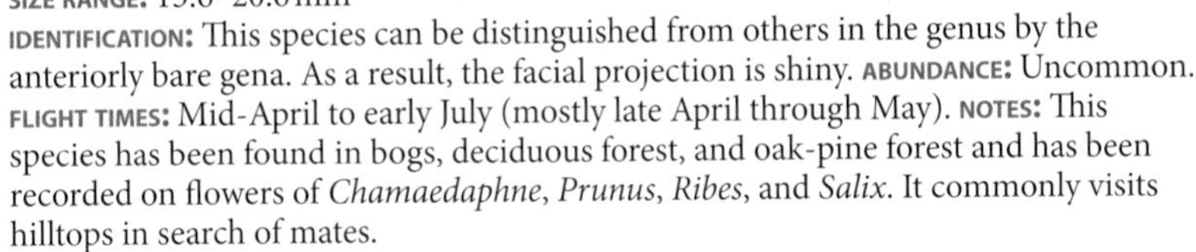

Bare-cheeked Bumblefly

**SIZE RANGE:** 13.6–20.6 mm
**IDENTIFICATION:** This species can be distinguished from others in the genus by the anteriorly bare gena. As a result, the facial projection is shiny. **ABUNDANCE:** Uncommon. **FLIGHT TIMES:** Mid-April to early July (mostly late April through May). **NOTES:** This species has been found in bogs, deciduous forest, and oak-pine forest and has been recorded on flowers of *Chamaedaphne*, *Prunus*, *Ribes*, and *Salix*. It commonly visits hilltops in search of mates.

## Criorhina verbosa

Hairy-cheeked Bumblefly

**SIZE RANGE:** 11.8–19.3 mm
**IDENTIFICATION:** The gena of this species is pilose anteriorly and the facial projection is dull. **ABUNDANCE:** Uncommon. **FLIGHT TIMES:** Late February to mid-August (mostly April and May). **NOTES:** This species may be found in association with *C. nigriventris* and has been collected in hardwood forests, pine forests, and bogs. It uses the same flowers and hilltops as *C. nigriventris*; however, it tends to be out a week or two earlier in the spring.

# *Criorhina nigriventris*

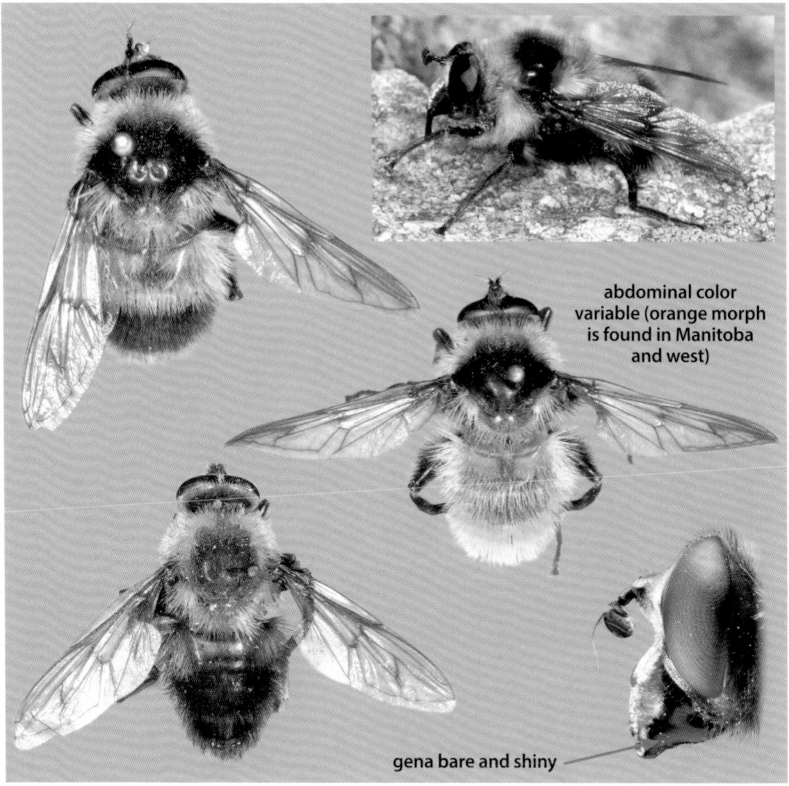
abdominal color
variable (orange morph
is found in Manitoba
and west)
gena bare and shiny

# *Criorhina verbosa*

gena pilose

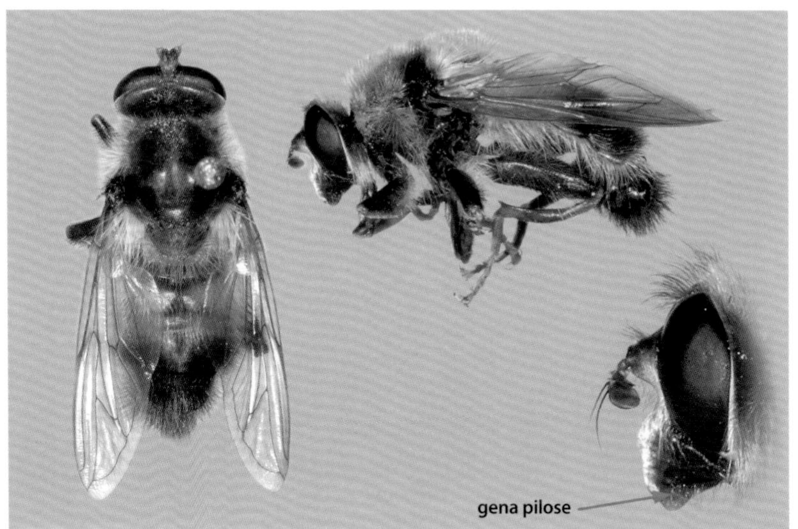

## *Criorhina villosa* Winter Bumblefly

**SIZE RANGE:** 9.8–14.2 mm
**IDENTIFICATION:** This species can be easily distinguished by the enlarged flagellum, bare eyes, and large size. **ABUNDANCE:** Very rare. **FLIGHT TIMES:** Late January to late March. **NOTES:** *Criorhina villosa* is a winter flier in old-growth habitat along waterways. Adults apparently feed on (and perhaps oviposit in) sap runs in damaged trees. They have been found in oak, maple, and beech-dominated ravines. This species was formerly treated in the monotypic genus *Merapioidus* and is easily recognizable by shape of the antennal flagellum. Molecular data have shown that *Merapioidus villosus* is simply a bizarre species of *Criorhina*. The species is apparently very rare despite being widespread. Very few specimens have been documented. The only recent records are a specimen collected in 2016 by Terry Schiefer in Mississippi (in an old-growth oak/beech ravine) and a photo from Virginia (on the right, taken in 2017 by Sheryl Pollock). Adults apparently do not visit flowers.

# *Hadromyia*

All *Hadromyia* species except the distinctive western species *H. grandis* are easily recognizable by the metallic patches on the abdomen. The rest of the abdomen is dull black with yellow pile. Only one *Hadromyia* species occurs in eastern North America. Four more occur in the west (unpublished), and one occurs in Russia.

## *Hadromyia aepalius* Sterling Quicksilver

**SIZE RANGE:** 9.5–12.2 mm
**IDENTIFICATION:** In addition to the metallic patches on the abdomen, this species is easily recognized by its yellow legs, with the femora and apical tarsi sometimes darkened. **ABUNDANCE:** Rare. **FLIGHT TIMES:** Early May to late June. **NOTES:** Adults have been collected from *Rosa multiflora* flowers and patrolling around pine stumps, where they may oviposit. The larvae are unknown.

# *Criorhina villosa*

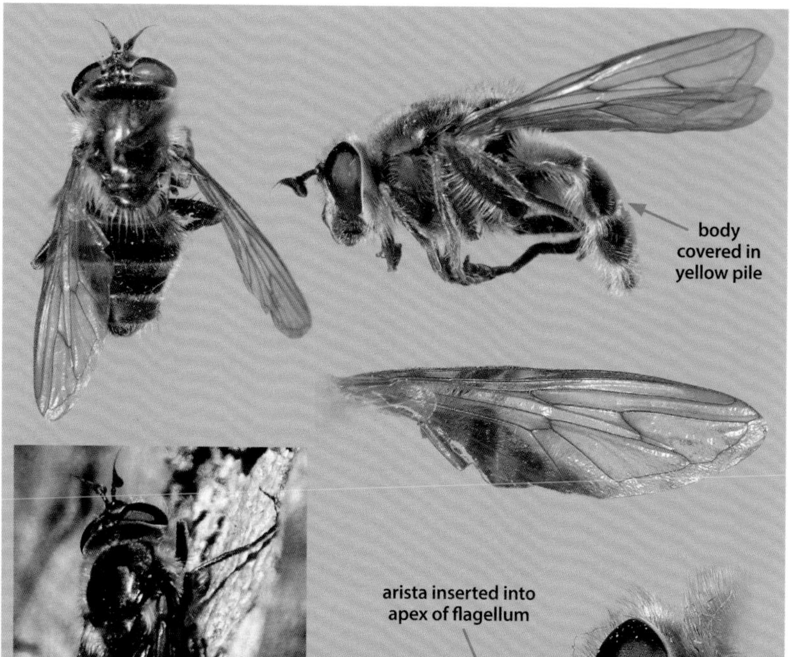
body
covered in
yellow pile
arista inserted into
apex of flagellum
distinctively shaped
antennae

# *Hadromyia aepalius*

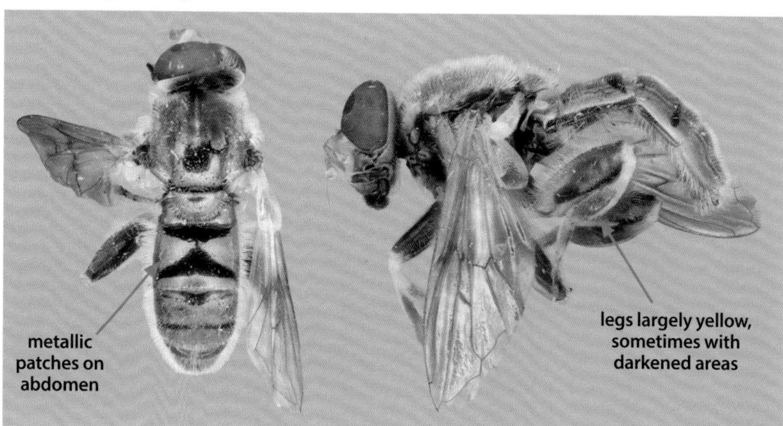

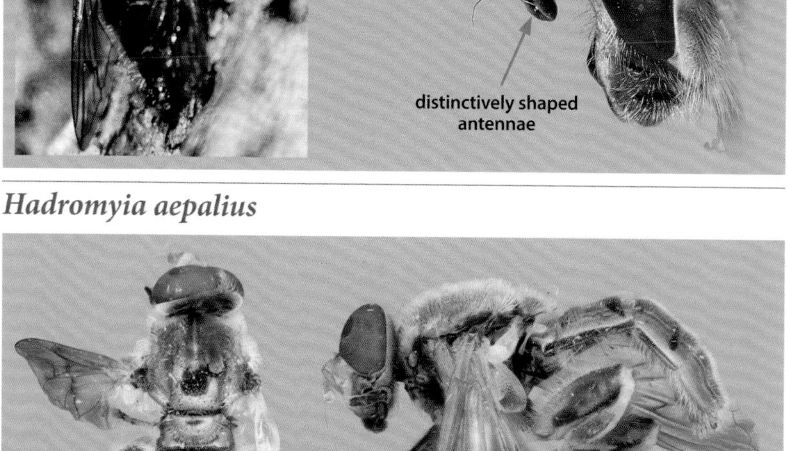
metallic
patches on
abdomen
legs largely yellow,
sometimes with
darkened areas

# *Volucella*

These large bumblebee mimics are immediately recognizable by their plumose antennae and elongate faces. The $M_1$ vein that curves back toward the wing base and the elongate flagellum also help distinguish this genus. For decades, *V. bombylans* was considered to be the only Nearctic species in this genus. Recent unpublished work by Skevington and Cheng have shown that *V. bombylans* is restricted to the Old World and there are four species in North America, three in our region. There are 50 species of *Volucella* worldwide. Larvae are found in nests of wasps and bees, where they feed on pollen, comb material, larvae, and pupae. Adults visit flowers and are also often seen perched on vegetation. For a key to Nearctic species see Curran (1930, includes *Copestylum*).

## *Volucella arctica* Arctic Swiftwing

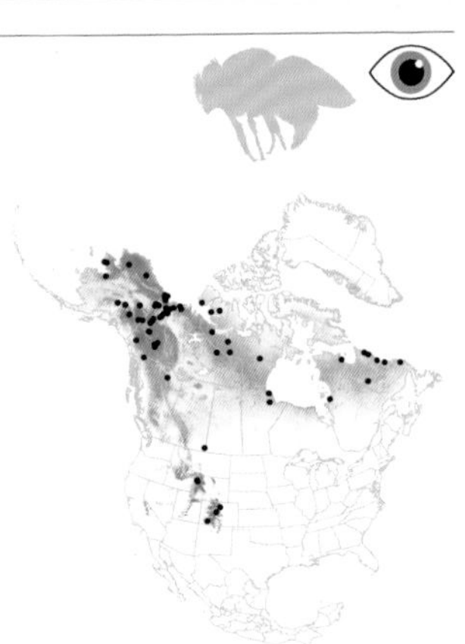

**SIZE RANGE:** 11.5–14.7 mm
**IDENTIFICATION:** Easily distinguished from other *Volucella* species by the entirely black-pilose scutum and pleuron and black face. The abdomen is commonly sparsely orange pilose but sometimes can be more yellow pilose. It is the only member of its genus that regularly occurs on the tundra. **ABUNDANCE:** Uncommon but likely underrepresented due to collecting bias. **FLIGHT TIMES:** Early June to mid-August. **NOTES:** This is an arctic alpine species, occurring only north of (or above) treeline. It has been observed visiting *Chamerion angustifolium* and *Salix*.

# *Volucella*

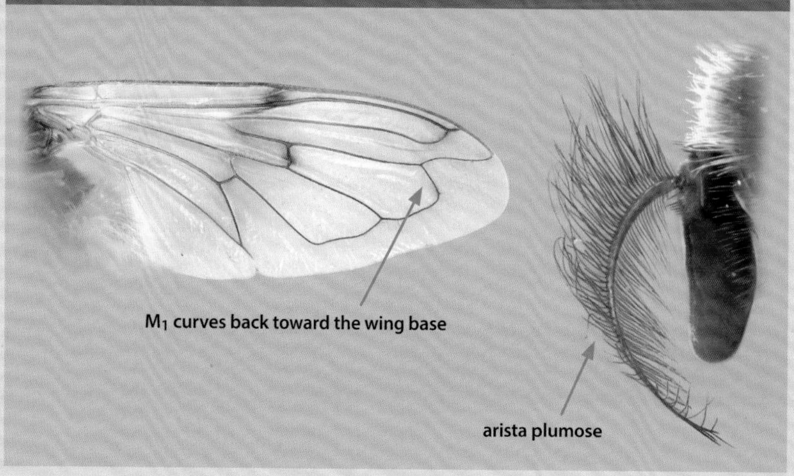

$M_1$ curves back toward the wing base

arista plumose

## *Volucella arctica*

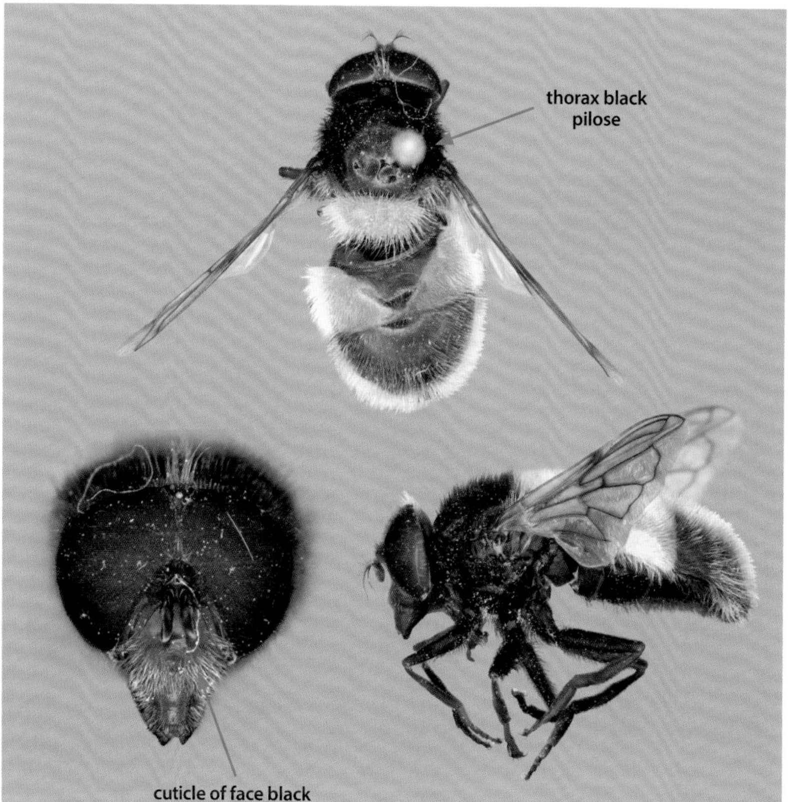

thorax black pilose

cuticle of face black

## *Volucella evecta* Eastern Swiftwing

**SIZE RANGE:** 12.6–17.9 mm
**IDENTIFICATION:** Easily distinguished from other *Volucella* species by the combination of black face and extensive yellow thoracic pile.
**ABUNDANCE:** Uncommon. **FLIGHT TIMES:** Mid-May to mid-August, up to two weeks earlier in the south.
**NOTES:** These are typically forest-dwelling flies but adults can be found anywhere from sand dunes to wetlands. Adults have been found nectaring at *Ceanothus*, *Geum*, *Rubus*, and *Viburnum*.

## *Volucella facialis* Yellow-faced Swiftwing

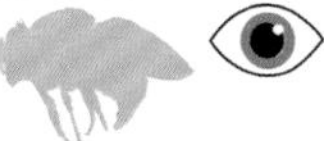

**SIZE RANGE:** 9.8–14.2 mm
**IDENTIFICATION:** Easily distinguished by the combination of yellow face and extensive yellow thoracic pile. Two color morphs, as well as intermediates, are found throughout the range. Individuals with a dark abdomen are more common in eastern North America, and individuals with an orange abdomen are more common in the west.
**ABUNDANCE:** Fairly common. **FLIGHT TIMES:** Late May to late August, up to three weeks earlier in the west. **NOTES:** As with *V. evecta*, these are typically forest-dwelling flies but adults may also be found in wetlands (particularly bogs), prairie habitats, and on the tundra. They have been recorded nectaring on *Dryas*, *Kalmia*, *Rubus*, *Taraxacum*, and *Valeriana*.

## *Volucella evecta*

thorax yellow pilose

cuticle of face black

## *Volucella facialis*

thorax yellow pilose

abdomen color variable

cuticle of face yellow

# *Eristalis*

*Eristalis* species are large flies that are easily identified at a glance with a bit of experience. If in doubt, check the antennae (not densely plumose as in *Volucella*) and the wing venation (sinuous $R_{4+5}$ and closed cell $r_1$). Species-level identification varies from easy (the first 12 species treated here, including four bumblebee mimics, are pretty straightforward) to difficult (the last five species treated require careful examination and characters are not always easy to interpret without comparative material). The bumblebee mimics are all easily distinguished from other *Eristalis* species by the dense yellow pile on the periphery of the thorax and scutellum (the pile completely obscures the ground color of the thorax). There are 103 world *Eristalis* species, of which 19 are Nearctic and 17 occur within the region covered by the field guide. For a key to Nearctic species, see Telford (1970).

## *Eristalis oestracea* Orange-tailed Drone Fly

**SIZE RANGE:** 13.9–15.2 mm
**IDENTIFICATION:** Posterior thorax, scutellum, and tergite 1 white pilose, tergites 4 and 5 orange pilose. **ABUNDANCE:** Very rare. **FLIGHT TIMES:** Early May to early September in Europe. **NOTES:** This spectacular fly is very poorly known in North America. In the area covered by the guide, it is known only from far northern Ontario with records from the Albany River area. It is on the Red List in parts of its European range. In Europe it is found in bogs, moors, and coastal dune systems where it nectars on white Apiaceae (umbellifers) and yellow composites.

## *Eristalis flavipes* Orange-legged Drone Fly

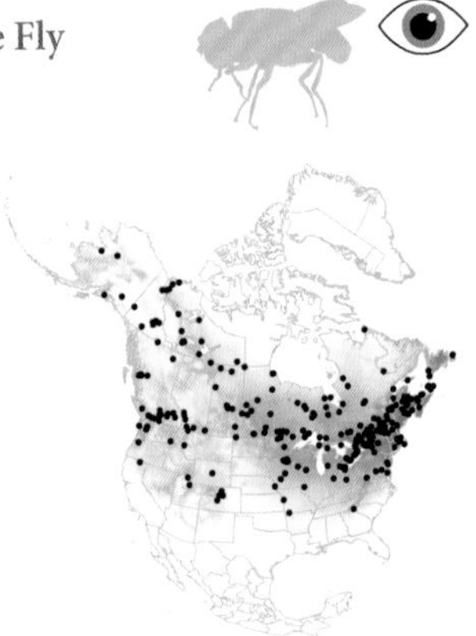

**SIZE RANGE:** 11.0–16.1 mm
**IDENTIFICATION:** Their bright orange metatarsi and pile color are distinctive. **ABUNDANCE:** Common. **FLIGHT TIMES:** Mid-May to mid-October (from early April along west coast). **NOTES:** This species has historically been referred to as *E. barda*. Adults occur in a wide variety of habitats but are most common in wetlands (in particular in swamps and bogs). Flowers visited include *Apocynum*, *Barbarea*, *Cephalanthus*, *Cirsium*, *Eupatorium*, *Heracleum*, *Malus*, *Melilotus*, *Packera*, *Physocarpus*, *Primula*, *Sedum*, *Solidago*, *Spiraea*, *Symphyotrichum*, and *Viburnum*.

# *Eristalis*

closed cell r1
sinuous $R_{4+5}$

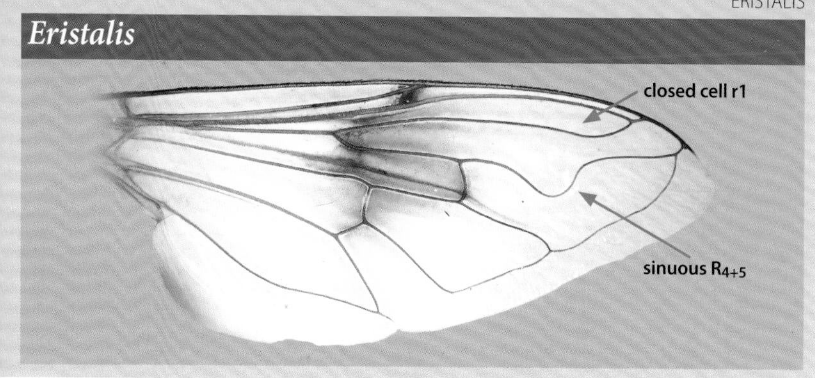

## *Eristalis oestracea*

white pilose
orange pilose

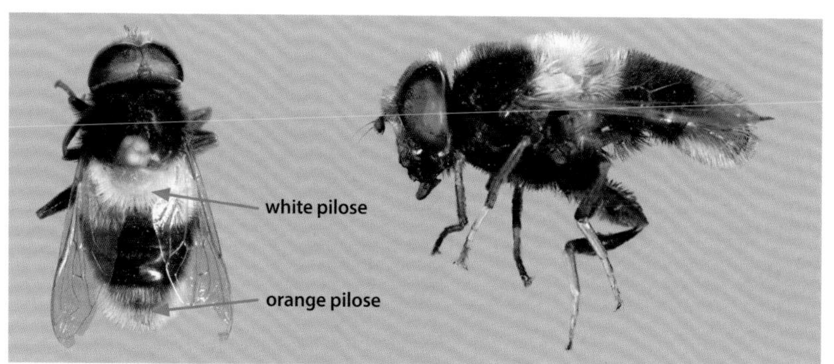

## *Eristalis flavipes*

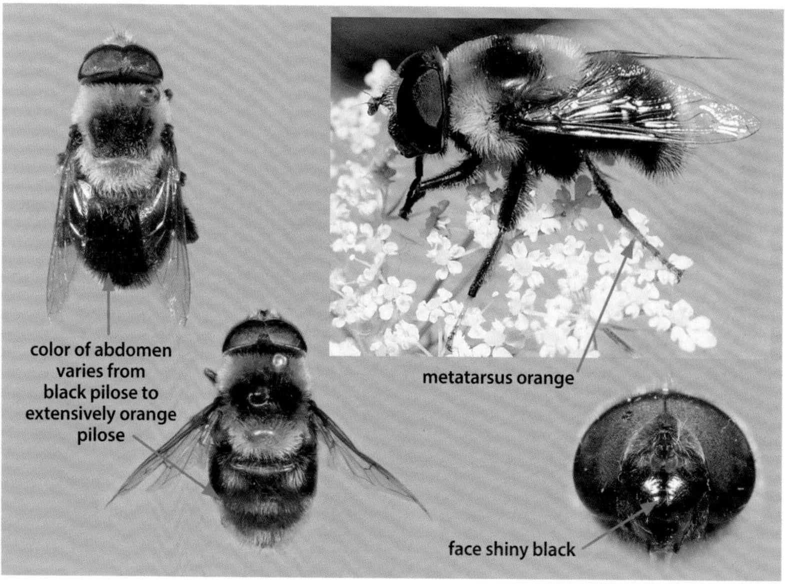

color of abdomen varies from black pilose to extensively orange pilose
metatarsus orange
face shiny black

## *Eristalis anthophorina*

### Orange-spotted Drone Fly

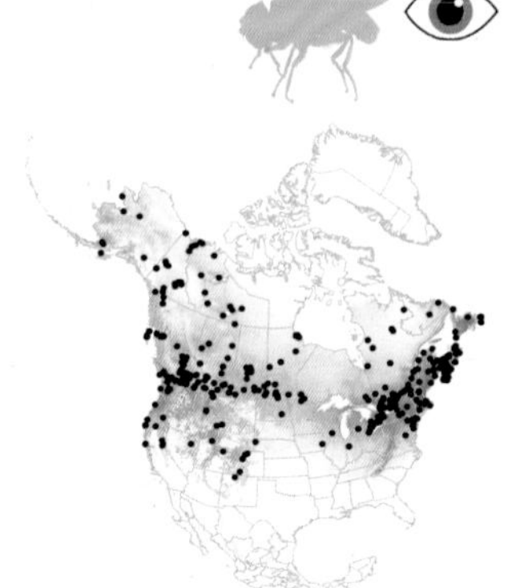

**SIZE RANGE:** 9.3–15.1 mm
**IDENTIFICATION:** Face yellow pilose, orange spots on tergite 2, tarsi black. **ABUNDANCE:** Common. **FLIGHT TIMES:** Early April to mid-November. **NOTES:** This Holarctic species has been known by several names (including *E. bastardii*, *E. mellissoides*, and *E. occidentalis*). It is common in a wide variety of wetland habitats (fens, bogs, woodland pools) and visits many flowers including *Arabis*, *Barbarea*, *Brassica*, *Caltha*, *Cardamine*, *Cirsium*, *Crataegus*, *Heracleum*, *Lycopus*, *Menyanthes*, *Nuphar*, *Physocarpus*, *Prunus*, *Ranunculus*, *Rubus*, *Salix*, *Solidago*, *Symphyotrichum*, *Taraxacum*, and *Valeriana*.

## *Eristalis fraterculus* Black-spotted Drone Fly

**SIZE RANGE:** 10.0–14.4 mm
**IDENTIFICATION:** Tarsi black, abdomen with broad, black-pilose spots laterally. **ABUNDANCE:** Rare. **FLIGHT TIMES:** Early June to late August. **NOTES:** This species has a northern Holarctic distribution. A synonym for *E. fraterculus* is *E. pilosa*. This species is found in seasonally flooded grassland with standing water in tundra and beside rivers in taiga. Adults can be found sitting on mud or visiting flowers (*Caltha*, *Matricaria*, *Packera*, *Ranunculus*, *Salix*, and *Senecio*). Larvae have been reared in a mixture of soil, water, and cow manure and are illustrated in Nielsen and Svendsen (2014).

## *Eristalis transversa* Transverse-banded Drone Fly

**SIZE RANGE:** 10.1–12.2 mm
**IDENTIFICATION:** Anterior ½ of thorax gray, black posteriorly; scutellum bright opaque yellow. **ABUNDANCE:** Abundant. **FLIGHT TIMES:** Mid-May to late October (from early April in southern part of range). **NOTES:** This is one of our most conspicuous eristaline flower flies. It is found in a wide variety of habitats and has been recorded from *Barbarea*, *Cephalanthus*, *Heracleum*, *Physocarpus*, *Primula*, *Rudbeckia*, *Sedum*, *Solidago*, *Symphyotrichum*, *Tanacetum*, and *Viburnum*.

## *Eristalis anthophorina*

## *Eristalis fraterculus*

## *Eristalis transversa*

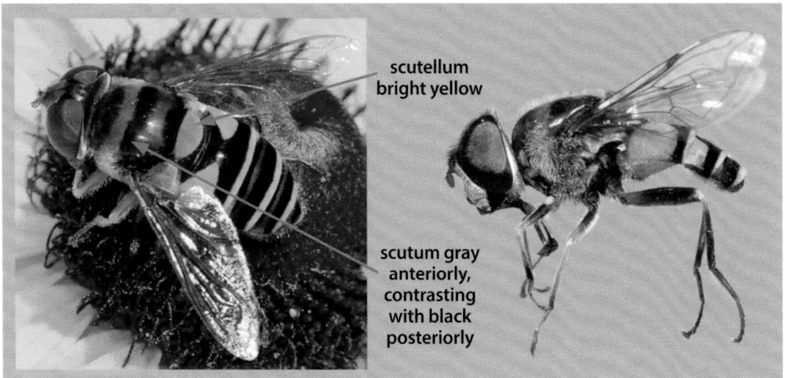

## *Eristalis tenax* Common Drone Fly

**SIZE RANGE:** 11.7–15.8 mm
**IDENTIFICATION:** Katepimeron pilose; eye with two bands of pile; orange abdominal pattern variable, but distinctive honey bee mimic. **ABUNDANCE:** Common. **FLIGHT TIMES:** Mid-March to mid-November. **NOTES:** *Eristalis tenax* is native to Europe but now is cosmopolitan. It can often be found sunning on leaves or resting on flowers and visits a very wide variety of flowers. Larvae live in a wide range of aqueous to semiaqueous habitats including dung. Common Drone Flies are known to be migratory and overwinter as adults.

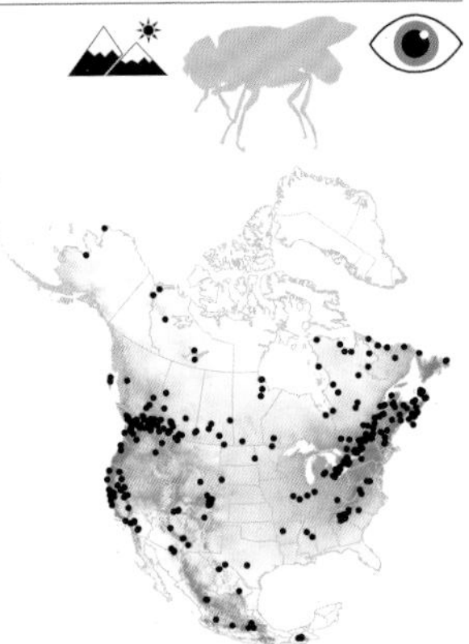

## *Eristalis saxorum* Blue-polished Drone Fly

**SIZE RANGE:** 11.3–13.7 mm
**IDENTIFICATION:** Body metallic blue, amount variable but always present. Pterostigma light brown, diffuse. **ABUNDANCE:** Uncommon. **FLIGHT TIMES:** Late May to late August (mid-March to late October in southern part of range). **NOTES:** Little is known about this species. It has been recorded visiting flowers of *Angelica*, *Baccharis*, *Daucus*, *Pastinaca*, *Solidago*, and *Spiraea*.

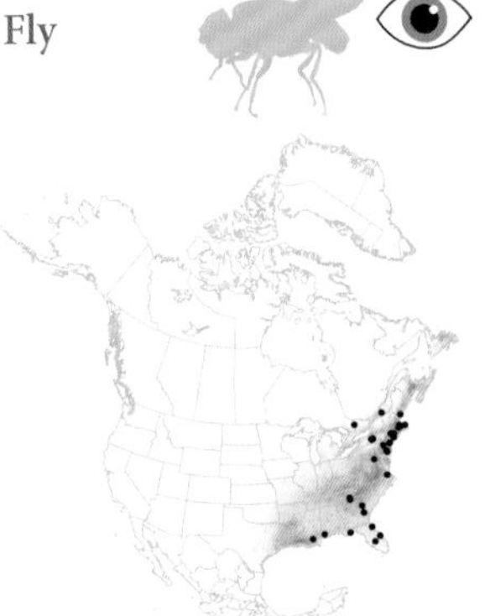

## *Eristalis cryptarum* Bog-dwelling Drone Fly

**SIZE RANGE:** 10.0–13.3 mm
**IDENTIFICATION:** Antennae orange to yellow; scutellum orange; arista with many short bristles. **ABUNDANCE:** Uncommon. **FLIGHT TIMES:** Early May to mid-September. **NOTES:** Bog-dwelling Drone Flies are bog specialists but have been found in a variety of wetlands. They have also been known as *E. compactus* and are found throughout the Holarctic region. Flowers visited include *Caltha*, *Cardamine*, *Crataegus*, *Menyanthes*, *Physocarpus*, *Potentilla*, *Ranunculus*, *Sorbus*, *Succisa*, and *Vaccinium*.

## *Eristalis tenax*

## *Eristalis saxorum*

## *Eristalis cryptarum*

## *Eristalis dimidiata* Black-shouldered Drone Fly

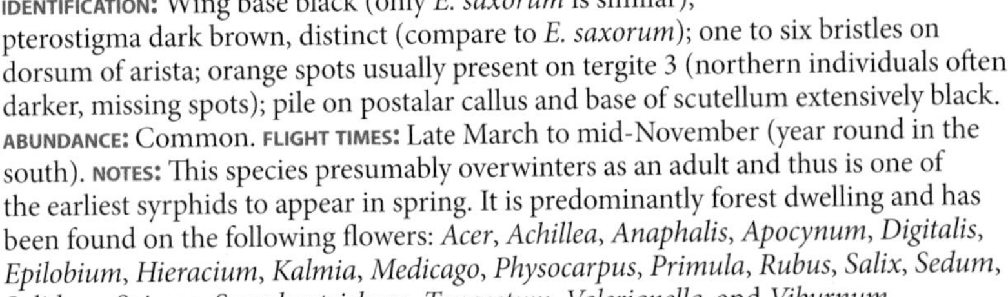

**SIZE RANGE:** 10.1–14.0 mm
**IDENTIFICATION:** Wing base black (only *E. saxorum* is similar); pterostigma dark brown, distinct (compare to *E. saxorum*); one to six bristles on dorsum of arista; orange spots usually present on tergite 3 (northern individuals often darker, missing spots); pile on postalar callus and base of scutellum extensively black. **ABUNDANCE:** Common. **FLIGHT TIMES:** Late March to mid-November (year round in the south). **NOTES:** This species presumably overwinters as an adult and thus is one of the earliest syrphids to appear in spring. It is predominantly forest dwelling and has been found on the following flowers: *Acer*, *Achillea*, *Anaphalis*, *Apocynum*, *Digitalis*, *Epilobium*, *Hieracium*, *Kalmia*, *Medicago*, *Physocarpus*, *Primula*, *Rubus*, *Salix*, *Sedum*, *Solidago*, *Spiraea*, *Symphyotrichum*, *Tanacetum*, *Valerianella*, and *Viburnum*.

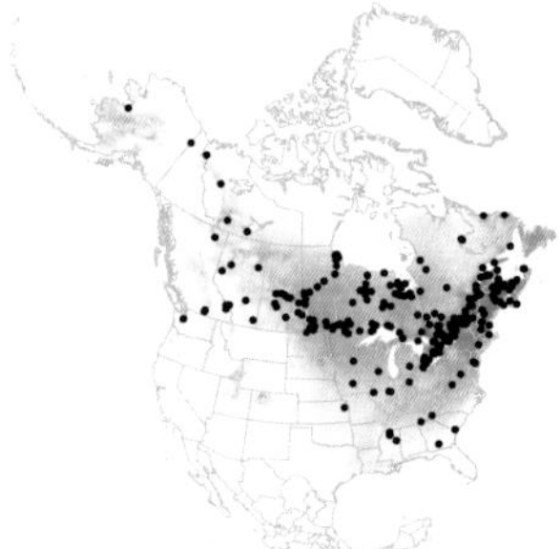

at least some black pile on postalar callus

## *Eristalis stipator*
### Yellow-shouldered Drone Fly

**SIZE RANGE:** 9.8–14.8 mm
**IDENTIFICATION:** Long white pile on tergite 4 diagnostic; tergite 3 shiny black; pile on postalar callus and scutellum completely yellow; zero to six short bristles on dorsum of arista. **ABUNDANCE:** Abundant in the west, rare in the east. **FLIGHT TIMES:** Mid-May to early November (as early as late February in the southwest). **NOTES:** This species has also been known as *E. latifrons*. Common in the west, this species has few records in the east. It is unclear if they are migrants or established within the range of the field guide. A DNA barcode for a specimen from Pukaskwa National Park in northwestern Ontario corroborates the occurrence of the species in the east. It has been documented nectaring on *Bidens*, *Cichorium*, *Grindelia*, *Helianthus*, *Medicago*, *Melilotus*, *Prunus*, *Solidago*, *Taraxacum*, and *Viburnum*.

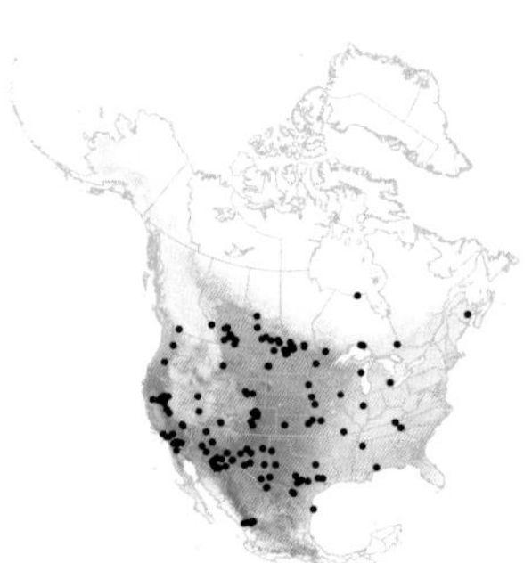

pile on postalar callus yellow

# *Eristalis dimidiata*

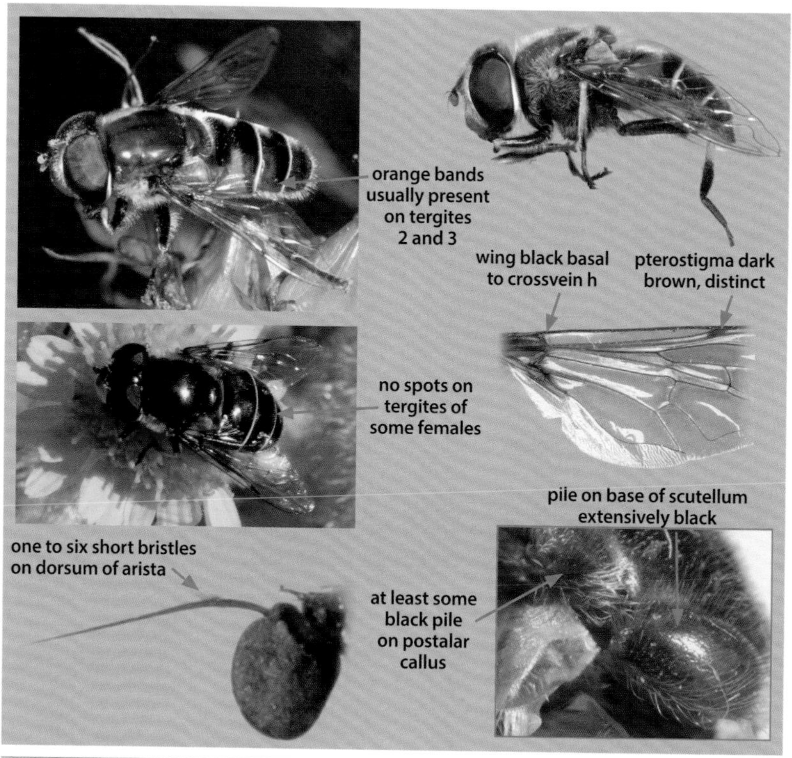
orange bands usually present on tergites 2 and 3
wing black basal to crossvein h
pterostigma dark brown, distinct
no spots on tergites of some females
pile on base of scutellum extensively black
one to six short bristles on dorsum of arista
at least some black pile on postalar callus

# *Eristalis stipator*

tergite 3 shiny black
long white pile on tergite 4
pile on postalar callus yellow
zero to six short bristles on dorsum of arista
pile on scutellum yellow

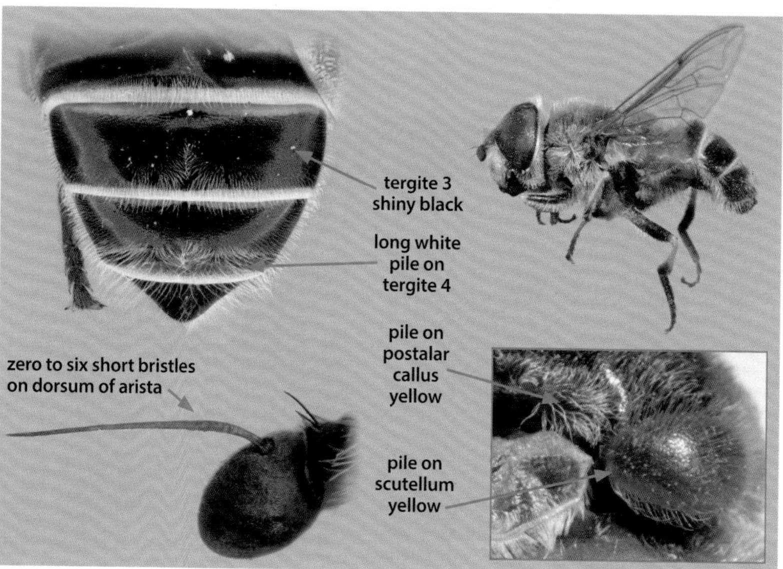

## *Eristalis arbustorum* European Drone Fly

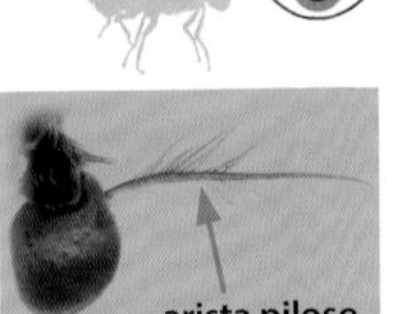

**SIZE RANGE:** 8.3–12.0 mm

**IDENTIFICATION:** To distinguish from *E. brousii*: male eyes touch for as long as or longer than length of ocellar triangle; mesotibia yellow on basal ⅔; basal tarsomere of mesoleg usually bright yellow. To distinguish from other *Eristalis*: face pilose, with at most a narrow glossy median stripe (others have a wide glossy stripe); arista pilose; male tergite 2 with black hourglass shape, female abdomen variable.


**ABUNDANCE:** Abundant. **FLIGHT TIMES:** Mid-March to late October. **NOTES:** *Eristalis arbustorum* is a Palearctic species that was introduced to North America around 1885. It has spread from its introduction point in Toronto throughout the continent. *Eristalis nemorum* is a synonym. Ubiquitous in urban and rural settings on many flowers. Migratory in Europe.

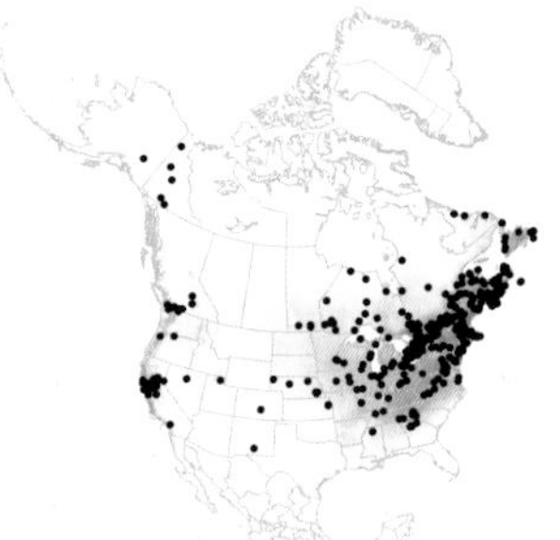

## *Eristalis brousii* Hourglass Drone Fly

**SIZE RANGE:** 9.2–12.8 mm

**IDENTIFICATION:** Like *E. arbustorum* except mesotibia yellow on basal ½ or less; basal tarsomere of mesoleg reddish brown to black. **ABUNDANCE:** Extirpated throughout most of range. Observations since 2000 are shown in red on the map. Of these, only the Manitoba, Nunavut, and Quebec populations likely remain. **FLIGHT TIMES:** Mid-April to early November. **NOTES:** This species collapsed as *E. arbustorum* spread across the continent. It persists only where *E. arbustorum* has not yet colonized and possibly in the west where *E. arbustorum* is a recent immigrant. Look for it in coastal areas along Hudson Bay and James Bay. It has been documented nectaring on *Apium*, *Chrysothamnus*, *Medicago*, and *Senecio*. Several putative hybrids with *E. arbustorum* have been observed.

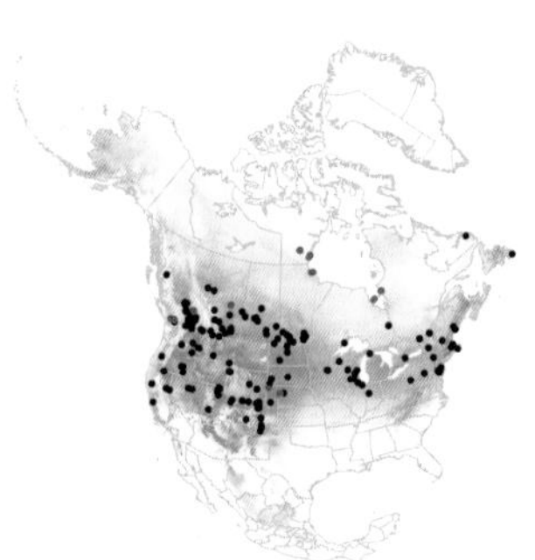

# *Eristalis arbustorum*

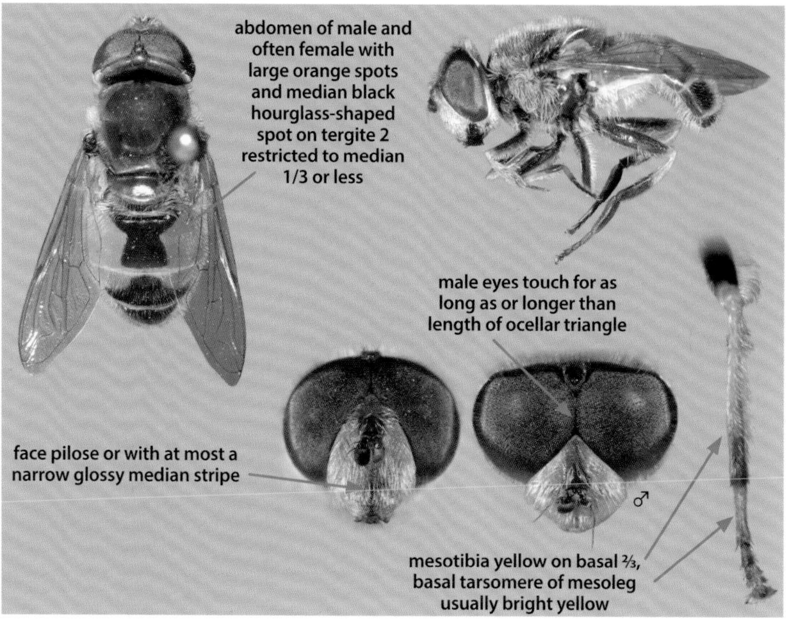

abdomen of male and often female with large orange spots and median black hourglass-shaped spot on tergite 2 restricted to median 1/3 or less

male eyes touch for as long as or longer than length of ocellar triangle

face pilose or with at most a narrow glossy median stripe

♂

mesotibia yellow on basal ⅔, basal tarsomere of mesoleg usually bright yellow

# *Eristalis brousii*

♀

female of both *E. arbustorum* and *E. brousii* variable, often without orange spots on abdomen and broader black stripe on tergite 2

♂

face mostly pilose as in *E. arbustorum*

male eyes touch for less than length of ocellar triangle

♂

mesotibia yellow on basal ½ or less, basal tarsomere of mesoleg reddish brown to black

## *Eristalis obscura* Dusky Drone Fly

**SIZE RANGE:** 10.5–14.6 mm

**IDENTIFICATION:** This is the only species of this final group of five similar *Eristalis* species with an orange face. In the rare event that the face is black, the orange antennal color (similar to *E. interrupta*) and black spines on anteroventral surface of metatarsi (also in *E. hirta* and *E. rupium*) separate it from the others below. Additionally, the basal two metatarsomeres are yellowish (similar to *E. rupium*) and the pterostigma is less than three times as long as broad (*E. interrupta* and *E. hirta* similar). **ABUNDANCE:** Common throughout the central and northern part of its range and uncommon in the south. **FLIGHT TIMES:** Early May to early October. **NOTES:** Found in marshes, fens, lake margins, and surrounding forests. Flowers visited include *Achillea*, *Barbarea*, *Caltha*, *Cicuta*, *Crataegus*, *Heracleum*, *Melilotus*, *Physocarpus*, *Prunus*, *Ranunculus*, *Rosa*, *Rubus*, *Saxifraga*, *Solidago*, *Taraxacum*, *Valeriana*, and *Viburnum*.

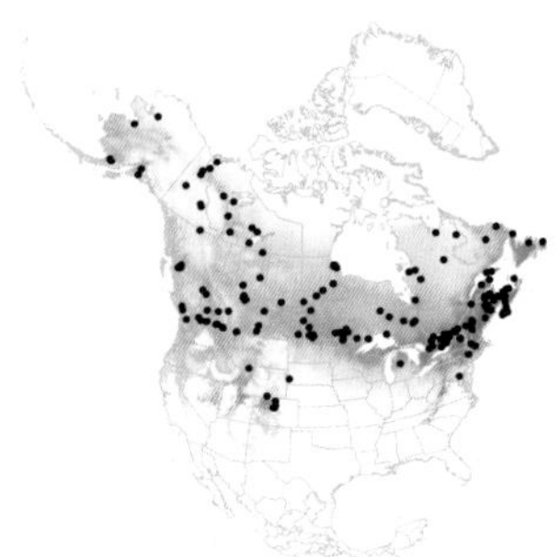

## *Eristalis interrupta* Orange-spined Drone Fly

**SIZE RANGE:** 10.4–13.7 mm

**IDENTIFICATION:** The orange antennae and dark tarsi usually clinch the identification. Check for orange spines on the metatarsi for final confirmation (black in similar species). **ABUNDANCE:** Fairly common. **FLIGHT TIMES:** Late April to late September. **NOTES:** This Holarctic species is also known as *E. nemorum*. It is found along river margins, in marshes, fens, and the edges of raised bogs. Flowers visited include *Cakile*, *Caltha*, *Cardamine*, *Cirsium*, *Crataegus*, *Eupatorium*, *Euphorbia*, *Filipendula*, *Heracleum*, *Hieracium*, *Leucanthemum*, *Malus*, *Mentha*, *Menyanthes*, *Parnassia*, *Prunus*, *Ranunculus*, *Rubus*, *Salix*, *Sorbus*, and *Succisa*. Larvae occur in streams and pools and in cow feces on waterlogged ground (Speight 2016).

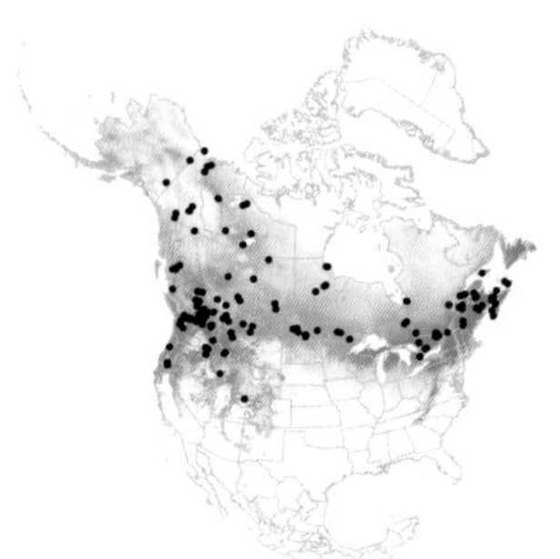

# *Eristalis obscura*

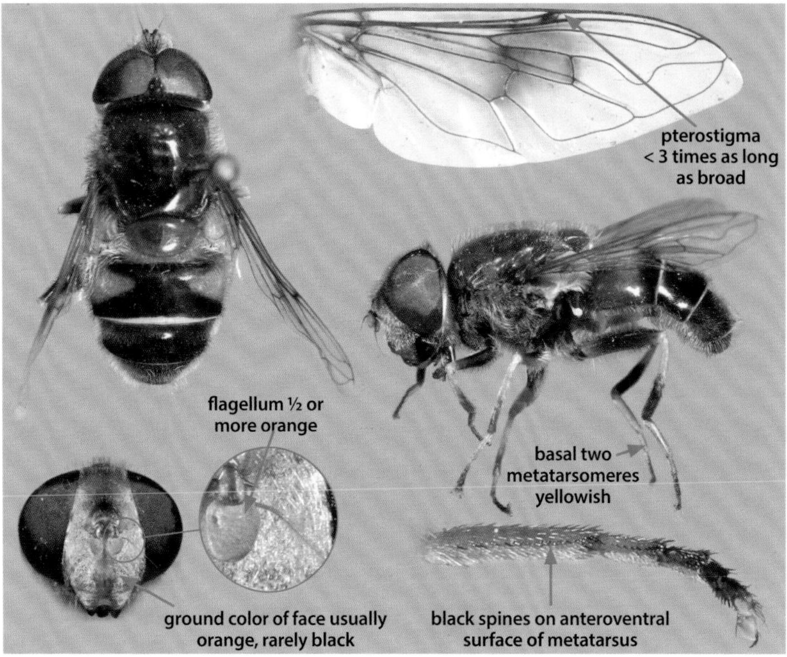

pterostigma < 3 times as long as broad

flagellum ½ or more orange

basal two metatarsomeres yellowish

ground color of face usually orange, rarely black

black spines on anteroventral surface of metatarsus

# *Eristalis interrupta*

flagellum ½ or more orange

metatarsus uniformly brown to black

ground color of face usually black, rarely orange

orange spines on anteroventral surface of metatarsus

## *Eristalis rupium* Spot-winged Drone Fly

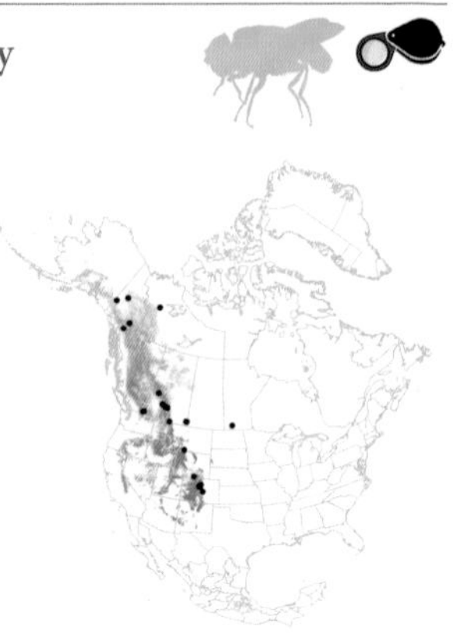

**SIZE RANGE:** 9.5–13.8 mm
**IDENTIFICATION:** The long pterostigma is diagnostic. Leg color (as in *E. obscura*), dark antennae (also in *E. hirta*), and the narrow gap between the ocelli and eyes support the identification. **ABUNDANCE:** Uncommon. **FLIGHT TIMES:** Early June to late August. **NOTES:** This Holarctic species is included in the guide because of regular confusion with eastern taxa. Dependent upon streams with clean water in humid forests and montane grassland. Flowers visited include *Anemone*, *Cardamine*, *Galearis*, *Helianthemum*, *Heracleum*, *Leucanthemum*, *Menyanthes*, *Parnassia*, *Polygonum*, *Rubus*, *Sorbus*, *Taraxacum*, and *Valeriana*.

## *Eristalis hirta* Black-footed Drone Fly

**SIZE RANGE:** 8.5–13.9 mm
**IDENTIFICATION:** Brown antennae (as in *E. rupium*), combined with uniformly colored tarsi (as in *E. interrupta*) support identification. Females can be separated from *E. rupium* by the wider gap between the ocelli and eyes. **ABUNDANCE:** Common in the west, rare in the east. **FLIGHT TIMES:** Mid-April to early November. **NOTES:** This widespread western species also occurs sporadically in the boreal forest and tundra across northern North America, Europe, and Asia. It is found in raised bogs, ditches, and temporary pools. Flowers visited include *Achillea*, *Caltha*, *Galium*, *Juncus*, *Ligusticum*, *Matricaria*, *Packera*, *Perideridia*, *Pimpinella*, *Potentilla*, *Pyrrocoma*, *Ranunculus*, *Rhododendron*, *Salix*, *Taraxacum*, *Valeriana*, and *Wyethia*.

## *Eristalis gomojunovae* Arctic Drone Fly

**SIZE RANGE:** 9.6–13.5 mm
**IDENTIFICATION:** Posterior margin of tergites black (orange or yellow in similar species) and tarsi dark. **ABUNDANCE:** Rare. **FLIGHT TIMES:** Mid-June to early August. **NOTES:** Found in boggy margins of lakes, open boreal forest, and tundra. Rests on mud at the water's edge and flies rapidly close to the water's surface. Flowers visited include *Caltha*, *Ranunculus*, *Rubus*, and *Salix*.

## *Eristalis rupium*

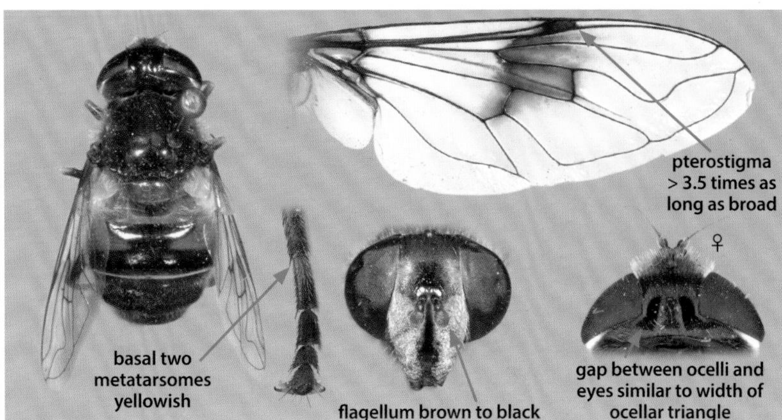


## *Eristalis hirta*

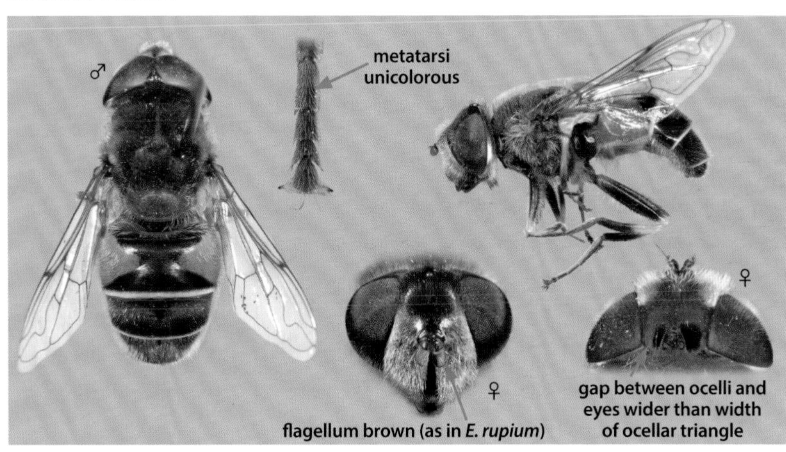


## *Eristalis gomojunovae*

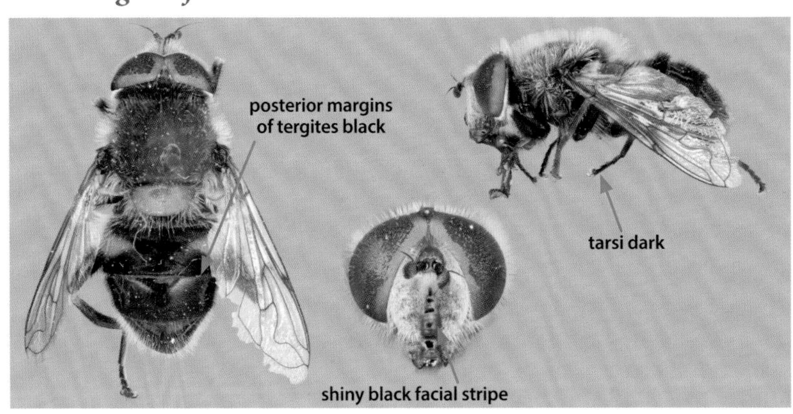

# *Palpada*

*Palpada* species can be recognized by the sinuous $R_{4+5}$ vein, closed cell $r_1$, bare arista, and pilose katepimeron. This is a distinctive New World genus of flies, generally resembling *Eristalis*, but with a characteristic color pattern consistent throughout most of the species in the genus. The larvae are filter feeders in aquatic environments. For a key to Nearctic species, see Telford (1970). There are 83 valid species, only four of which make it into our area.

## ***Palpada*** undescribed species 1

### Northern Striped Plushback

**SIZE RANGE:** 8.4–10.4 mm
**IDENTIFICATION:** This dark species has a black scutum with gray pollen forming stripes (other species have pollen forming transverse bands), and the scutellum is brownish black (in other species it's yellow). There is pronounced sexual dimorphism with females having completely dark abdomens and males having lateral yellow spots on the tergites. **ABUNDANCE:** Rare. **FLIGHT TIMES:** Mid-April to early November. **NOTES:** This new species is closely related to the tropical *P. furcata*. Flowers visited include *Baccharis*, *Patrinia*, *Symphyotrichum*, and *Verbesina*.

## ***Palpada vinetorum*** Northern Plushback

**SIZE RANGE:** 10.0–13.5 mm
**IDENTIFICATION:** This species has a pollinose face with a yellow medial stripe, and the wing is partly microtrichose apically. **ABUNDANCE:** Fairly common. **FLIGHT TIMES:** Early June to mid-October (year round in southern parts of the range). **NOTES:** Like other *Palpada* species in our area, this species may be migratory. Flowers visited include *Baccharis*, *Gymnosperma*, *Lobularia*, *Miconia*, *Serjania*, and *Solidago*.

# Palpada

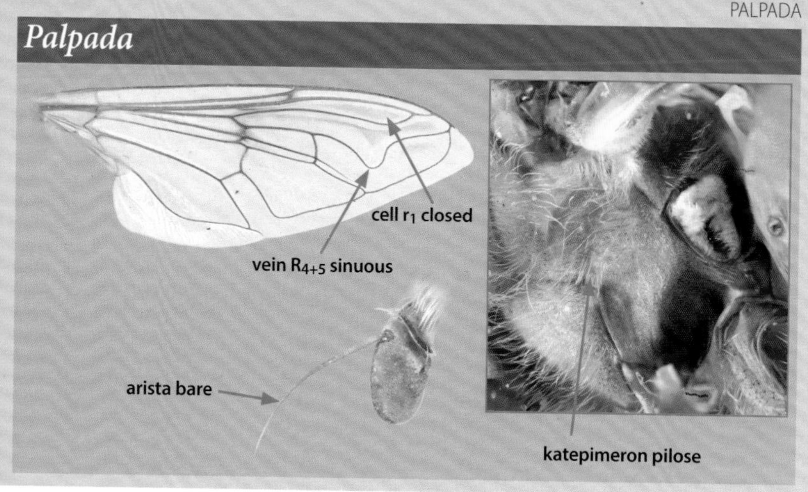


## *Palpada* undescribed species 1

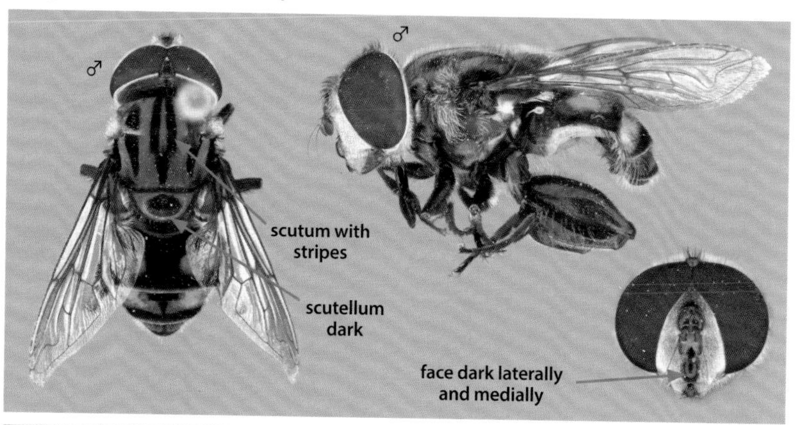


## *Palpada vinetorum*

## *Palpada pusilla* Bicolored Plushback

**SIZE RANGE:** 9.1–12.2 mm
**IDENTIFICATION:** This species has a distinctive, bicolored thorax (gray anteriorly, black posteriorly). The scutellum is yellow and tergite 2 has a central black stripe that widens posteriorly. **ABUNDANCE:** Fairly common. **FLIGHT TIMES:** Mid-February to late November. **NOTES:** Flowers visited include *Chrysanthemum*, *Rudbeckia*, and *Verbesina*. Although not in the area covered by the field guide, this species is found close to the area covered and is expected to occur based on vagrancy patterns in *Palpada*.

## *Palpada albifrons* White-faced Plushback

**SIZE RANGE:** 8.0–10.0 mm
**IDENTIFICATION:** This species has a pollinose face with a brown-black medial stripe (sometimes yellowish brown). The wing is bare and the pile on the lateral edges of tergite 2 is yellow anteriorly and black posteriorly. **ABUNDANCE:** *Palpada albifrons* is a tropical species and a very rare vagrant to the area covered by the guide (not likely established within the area covered by the guide). **FLIGHT TIMES:** Mid-April to mid-December. **NOTES:** Flowers visited include *Baccharis* and *Polygonum*.

## *Palpada pusilla*

central black stripe on tergite 2, widening posteriorly

thorax banded, gray anteriorly and black posteriorly

## *Palpada albifrons*

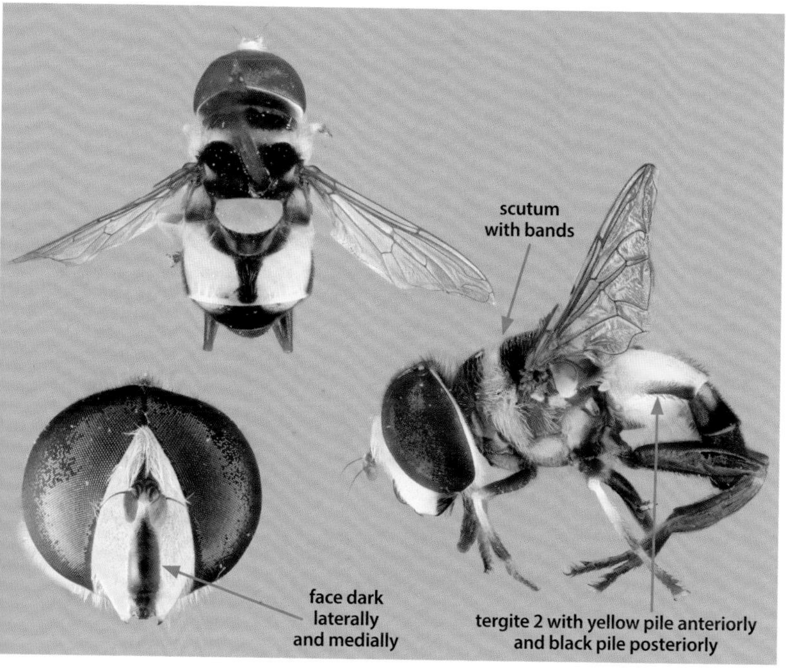

scutum with bands

face dark laterally and medially

tergite 2 with yellow pile anteriorly and black pile posteriorly

# *Copestylum*

*Copestylum* species are easily recognized by their anteroventrally produced face, plumose arista, and vein $M_1$ curving strongly back toward the wing base. This may be the most diverse genus of syrphids in the world. A huge undiscovered diversity still exists in the Neotropics and 332 valid species are already described. The genus is New World in distribution, and only four of the 41 species that inhabit the USA and Canada occur within the area covered by the field guide. Larvae of most species live in bromeliads. The most recent key is by Curran (1930).

## *Copestylum vittatum* Striped Bromeliad Fly

**SIZE RANGE:** 8.1–8.7 mm
**IDENTIFICATION:** This is the only species of *Copestylum* in our region that has a yellow face with a black medial stripe. It is also recognizable by the two round, yellow spots on the scutum anterior to the scutellum. **ABUNDANCE:** Uncommon. **FLIGHT TIMES:** Late April to mid-August (late February to mid-November in the south). **NOTES:** Flowers visited include *Helianthus*, *Licania*, *Prunus*, *Rudbeckia*, *Solidago*, *Symphyotrichum*, and *Viburnum*.

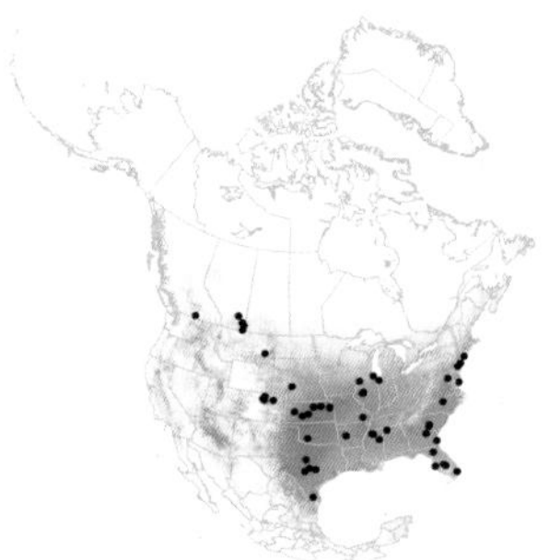

## *Copestylum sexmaculatum*
Six-spotted Bromeliad Fly

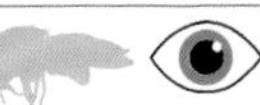

**SIZE RANGE:** 8.4–9.5 mm
**IDENTIFICATION:** This species is mostly yellow, with black lateral spots on tergites 2–4. **ABUNDANCE:** Very rare, possibly only a vagrant within the area covered by the guide. **FLIGHT TIMES:** Mid-April to mid-December. **NOTES:** Flowers visited include *Anemone*, *Eutrochium*, *Farfugium*, *Rhus*, and *Rudbeckia*. One record is from a black light.

## *Copestylum*

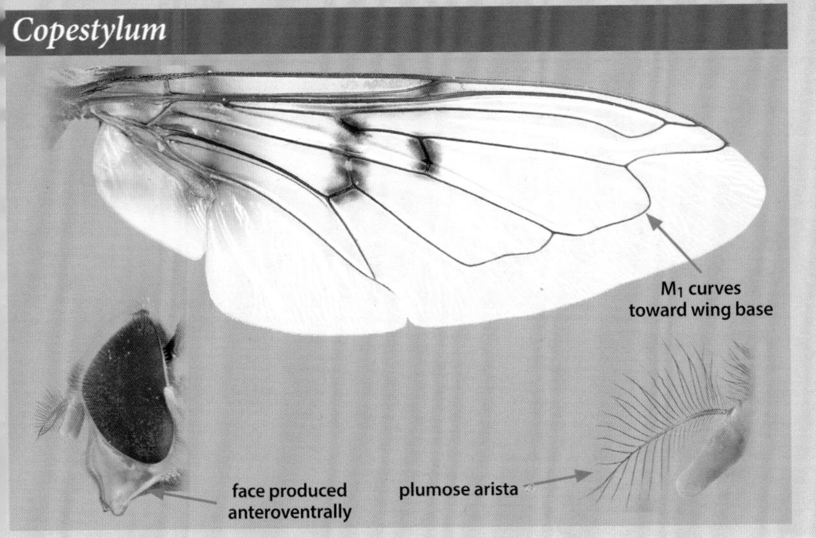


## *Copestylum vittatum*

## *Copestylum sexmaculatum*

## *Copestylum vesicularium*

### Iridescent Bromeliad Fly

**SIZE RANGE:** 7.4–10.4 mm
**IDENTIFICATION:** A dark, iridescent fly with an orange-yellow face; similar to *C. barei*. *Copestylum vesicularium* typically has a pale abdominal base and is more extensively pale pilose on the thorax (especially the anepimeron, scutellum, and katepisternum) and abdomen. **ABUNDANCE:** Uncommon. **FLIGHT TIMES:** Late May to late July (early May to early August in the south). **NOTES:** Flowers visited include *Aruncus*, *Cephalanthus*, *Cornus*, *Lycopus*, *Rubus*, *Rudbeckia*, and *Solidago*.

## *Copestylum barei* Lustrous Bromeliad Fly

**SIZE RANGE:** 9.3–9.9 mm
**IDENTIFICATION:** A dark, iridescent fly with an orange-yellow face; similar to *C. vesicularium*. *Copestylum barei* has a darker abdominal base and is more extensively black pilose on the thorax (especially the anepimeron, scutellum, and katepisternum) and abdomen. **ABUNDANCE:** Rare. **FLIGHT TIMES:** Mid-April to mid-July. **NOTES:** Very little is known about this rare fly.

# Copestylum vesicularium

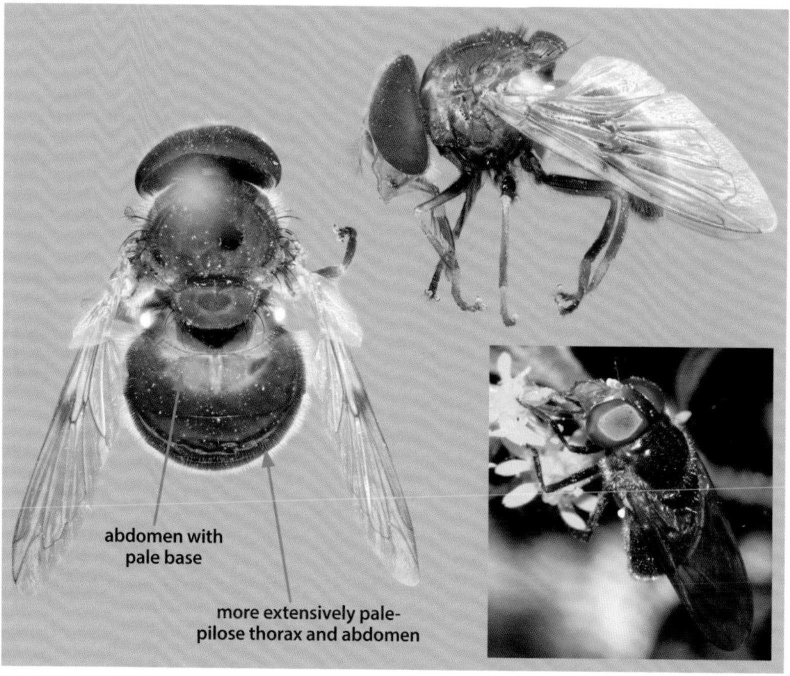

abdomen with
pale base

more extensively pale-
pilose thorax and abdomen

# Copestylum barei

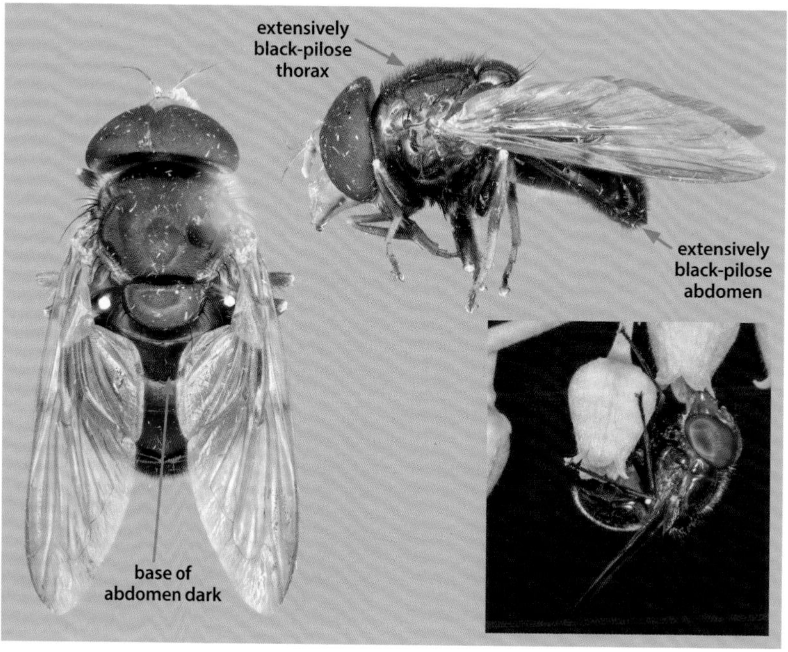

extensively
black-pilose
thorax

extensively
black-pilose
abdomen

base of
abdomen dark

# Callicera

One of three Nearctic species of *Callicera* occurs in our region. There are 18 world species of this spectacular group of flies. Most species (unlike ours) are orange or golden pilose with brightly shining or metallic bodies. The last key to Nearctic species was by Thompson (1980). Larvae are saprophagous in wet, decaying heartwood in rot holes and tree roots. In Europe, it has been shown that larvae are easier to find than adults, and MacGowan (1994) has demonstrated that it is possible to create successful artificial breeding sites.

## *Callicera erratica* Golden Pine Fly

**SIZE RANGE:** 9.5–12.1 mm
**IDENTIFICATION:** These flies are bee-like in appearance, with yellow pile covering the body. They have a pilose face and eyes and the antennae are distinctive, with the flagellum long and slightly enlarged basally. **ABUNDANCE:** Rare. **FLIGHT TIMES:** Mid-March to late June. **NOTES:** They can sometimes be found on hilltops, may visit moist soil to drink and obtain minerals, and have been found visiting *Prunus* and *Salix* flowers.

# Meromacrus

Thirty-nine species of this New World genus have been documented. Our species is one of five known from the USA and Canada. Larvae are filter feeders in rot holes in trees and, as with many rot hole species, are indicators of rich habitats. Hull (1942a) provides the last key to species. *Meromacrus* is closely related to *Mallota*.

## *Meromacrus acutus* Carolinian Elegant

**SIZE RANGE:** 13.3–18.6 mm
**IDENTIFICATION:** These wasp mimics are dark with small yellow markings on the thorax and abdomen, and the wings are darkened anteriorly, with a sinuous $R_{4+5}$ vein and cell $r_1$ closed before the wing margin. The yellow markings are made up of distinctive short, dense, yellow pile. These flies also have a black facial stripe and dorsal orange pile on the metatarsus. **ABUNDANCE:** Uncommon. **FLIGHT TIMES:** Mid-March to early October (as early as early February in the south). **NOTES:** *Meromacrus acutus* has been recorded from wet forest edges, meadows, and prairies.

# *Callicera erratica*

entire body yellow pilose

flagellum long and broad basally

# *Meromacrus acutus*

wing pattern diagnostic

closed $r_1$

sinuous $R_{4+5}$

yellow markings made of short, dense, velvet-like yellow pile

# *Milesia*

Only three of the world's 80 *Milesia* species make it into the Nearctic region and of these, only one occurs in the area covered by the guide. Larvae live in the decaying heartwood of deciduous trees, including rot holes. Adults are flower visitors and most species visit hilltops to mate.

## *Milesia virginiensis* Virginia Giant

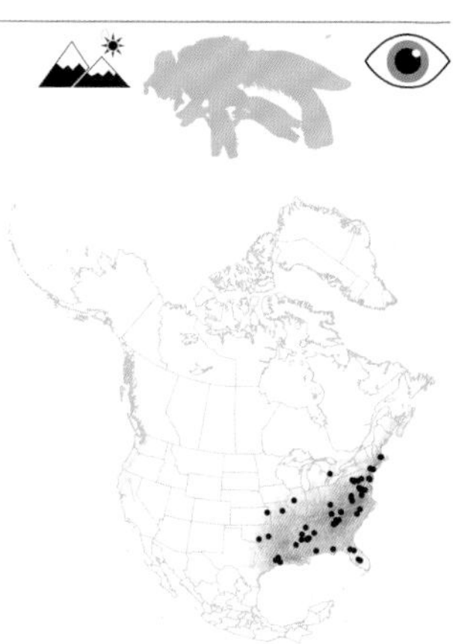

**SIZE RANGE:** 16.5–23.6 mm
**IDENTIFICATION:** *Milesia virginiensis* is a huge, unmistakable fly. **ABUNDANCE:** Uncommon.
**FLIGHT TIMES:** Early April to early October.
**NOTES:** Look for adults in sunny forest openings or along deciduous forest edges. They are often heard buzzing before being seen and sometimes are very inquisitive. *Milesia virginiensis* appeared as part of a series of insect postage stamps in the USA in 1999. This fly is a mimic of queens of the Southern Yellowjacket, *Vespula squamosa*. Flowers visited include *Cornus*, *Erigeron*, *Hydrangea*, *Ilex*, *Rosa*, *Sambucus*, *Solidago*, *Symphyotrichum*, and *Verbascum*.

# *Sphecomyia*

Eight of nine *Sphecomyia* species occur in the Nearctic region, but only one in the area of the guide. The genus concept is currently being revised by Kevin Moran, and many new species have been discovered in the west. Larvae are associated with sap runs and trunk lesions, particularly in floodplain forests.

## *Sphecomyia vittata*
Long-horned Yellowjacket Fly

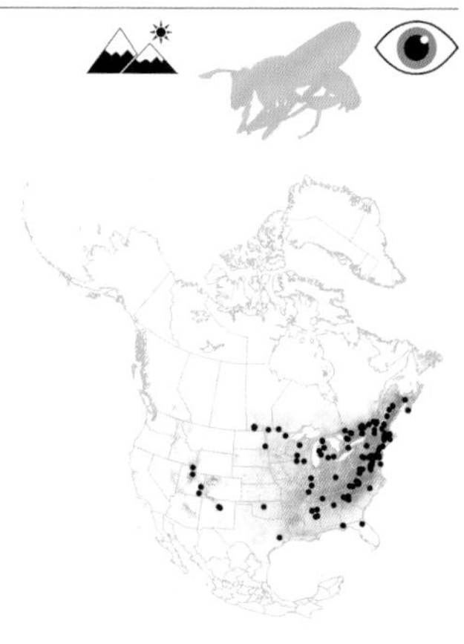

**SIZE RANGE:** 10.8–15.9 mm
**IDENTIFICATION:** *Sphecomyia vittata* is wasp-like and superficially similar to *Temnostoma* species but is easily recognized by its long antennae.
**ABUNDANCE:** Uncommon. **FLIGHT TIMES:** Early April to late June (early March to late July in the south).
**NOTES:** They are often found hilltopping and are easily overlooked since they are very similar to the yellowjackets (*Vespula* sp.). Their flight is slow, in the side-to-side habit typical of yellowjacket queens searching for nest sites. They have been recorded from damp second-growth deciduous woods, sphagnum bogs, and pastures. Flowers visited include *Acer*, *Anemone*, *Cornus*, *Crataegus*, *Prunus*, *Symplocos*, and *Viburnum*.

## *Milesia virginiensis*

## *Sphecomyia vittata*

long antennae

# Spilomyia

*Spilomyia* species are distinctive flies that mimic wasps. They are easily distinguished from other wasp mimics by their patterned eyes and preapical spur on the metafemur. There are 38 described species, 11 in North America, and four in our region. For a key to the New World *Spilomyia* species, see Thompson (1997). Larvae are found in rot holes in deciduous trees.

## *Spilomyia fusca* Bald-faced Hornet Fly

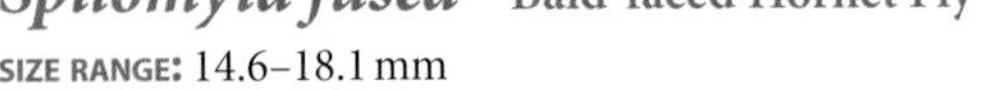

**SIZE RANGE:** 14.6–18.1 mm

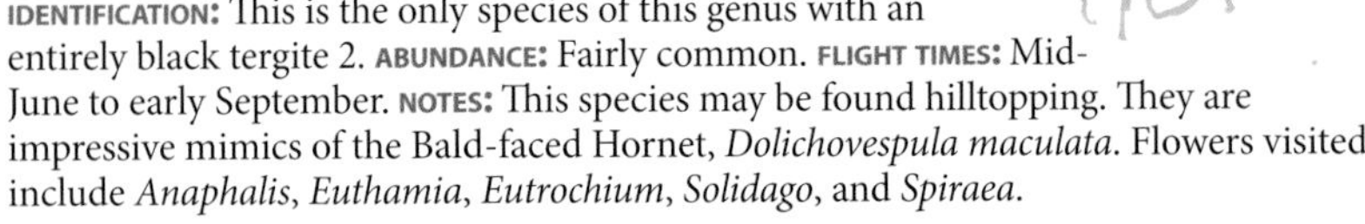

**IDENTIFICATION:** This is the only species of this genus with an entirely black tergite 2. **ABUNDANCE:** Fairly common. **FLIGHT TIMES:** Mid-June to early September. **NOTES:** This species may be found hilltopping. They are impressive mimics of the Bald-faced Hornet, *Dolichovespula maculata*. Flowers visited include *Anaphalis*, *Euthamia*, *Eutrochium*, *Solidago*, and *Spiraea*.

## *Spilomyia sayi* Four-lined Hornet Fly

**SIZE RANGE:** 10.4–16.6 mm

**IDENTIFICATION:** A single yellow band on each of tergites 2 and 3 and elongate antennae are diagnostic. **ABUNDANCE:** Fairly common. **FLIGHT TIMES:** Early July to mid-October. **NOTES:** They are sometimes encountered hilltopping and are most commonly found around wetlands. *Spilomyia sayi* is often referred to as *Spilomyia quadrifasciata* (a synonym). Flowers visited include *Anaphalis*, *Doellingeria*, *Euthamia*, and *Solidago*.

## Spilomyia

## Spilomyia fusca

## Spilomyia sayi

## *Spilomyia alcimus*

### Broad-banded Hornet Fly

**SIZE RANGE:** 13.0–20.1 mm

**IDENTIFICATION:** Two yellow bands on tergites 2 and 3 are broader than the black separating them. The scutellum is mostly yellow with a narrowly black base, and the katepimeron is black. **ABUNDANCE:** Uncommon. **FLIGHT TIMES:** Early May (late March in the south) to late July. **NOTES:** They may be found hilltopping. *Spilomyia hamifera* is a synonym. They are found in hardwood forests and have been found at the following flowers: *Achillea*, *Cornus*, *Ilex*, *Leucanthemum*, and *Sambucus*.

## *Spilomyia longicornis*

### Eastern Hornet Fly

**SIZE RANGE:** 12.4–16.2 mm

**IDENTIFICATION:** These flies have two yellow bands on tergites 2 and 3, as with *S. alcimus*. However, these bands are narrow with a broad band of black separating them. The scutellum is black on the basal ½ and yellow apically, and the katepimeron is yellow. They also have a straight face and yellow gena (a black line separates the face from the gena). **ABUNDANCE:** Uncommon. **FLIGHT TIMES:** Late May to late October. **NOTES:** *Spilomyia longicornis* is found in hardwood forests and often can be found at hilltops. Flowers visited include *Solidago* and *Symphyotrichum*.

# Spilomyia alcimus

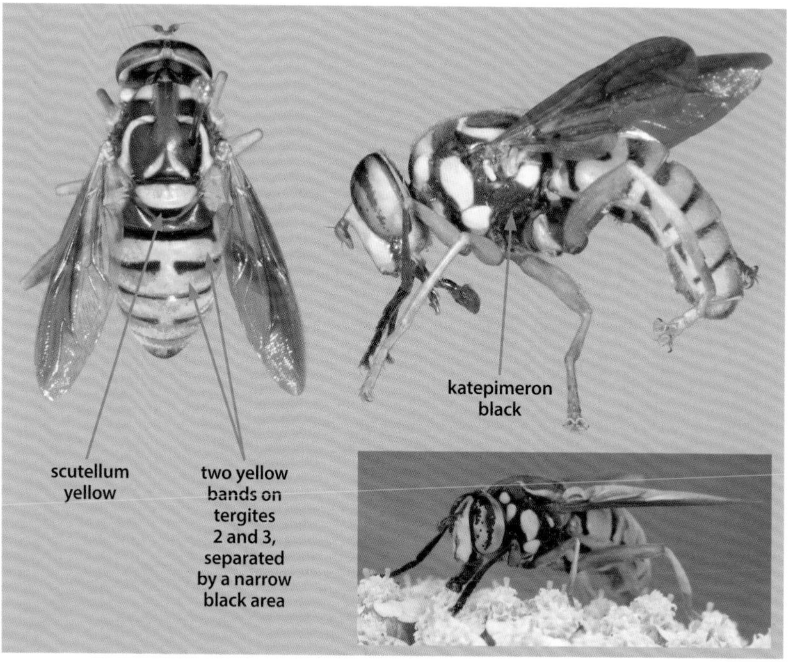

scutellum yellow

two yellow bands on tergites 2 and 3, separated by a narrow black area

katepimeron black

# Spilomyia longicornis

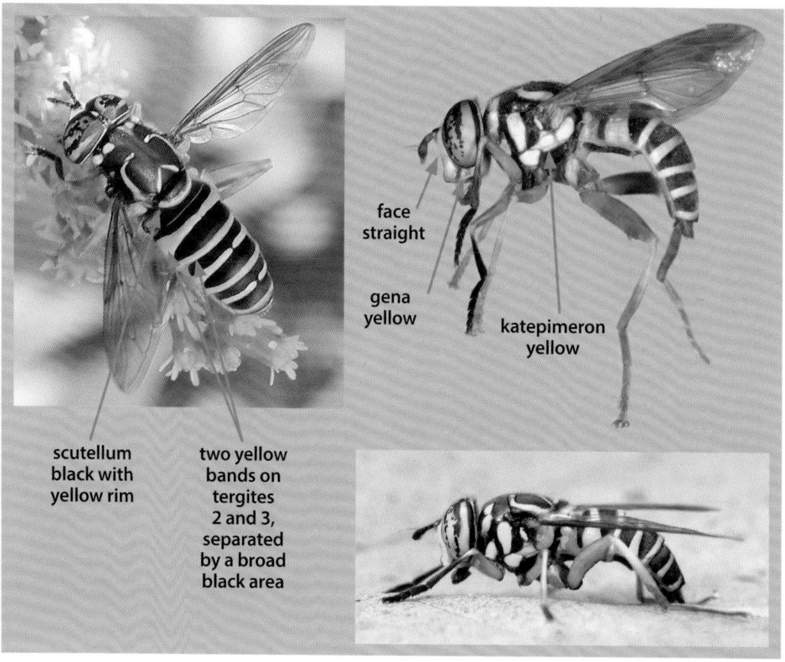

scutellum black with yellow rim

two yellow bands on tergites 2 and 3, separated by a broad black area

face straight

gena yellow

katepimeron yellow

# *Temnostoma*

All eight Nearctic species of the world's 23 *Temnostoma* species occur in the area covered by this field guide. In our region, this distinctive genus of wasp mimics contains four large species with two yellow bands on each abdominal segment and four small species with one band on each abdominal segment. The double-banded species are treated first below. *Temnostoma* antennae are never elongate and the face is never strongly produced ventrally. The katepisternum is continuously pilose on the posterior margin. Other large wasp mimics like *Spilomyia*, *Sphecomyia*, and *Doros* have separate dorsal and ventral patches of pile on the katepisternum. *Temnostoma* larvae are wood boring in solid wood within rotting stumps and logs.

## *Temnostoma alternans*

Wasp-like Falsehorn

**SIZE RANGE:** 10.5–16.5 mm
**IDENTIFICATION:** Most quickly recognized by the triangular marking above the scutellum and three small black spots on tergite 4. They also have spots on the transverse suture (as in *T. venustum*) and two yellow bands on tergites 2–4. **ABUNDANCE:** Common. **FLIGHT TIMES:** Early May to early September. **NOTES:** This species is often found hilltopping. Most are found in hardwood forests but there are also records from fens and spruce forest. Flowers visited include *Cornus*, *Prunus*, *Rubus*, *Spiraea*, *Thalictrum*, and *Viburnum*.

## *Temnostoma daochus* Yellow-spotted Falsehorn

**SIZE RANGE:** 9.5–14.5 mm
**IDENTIFICATION:** This species also has two yellow bands on tergites 2–4. The thorax has a pair of yellow prescutellar spots and the scutellum is yellow pollinose apically. **ABUNDANCE:** Rare.
**FLIGHT TIMES:** Mid-March to late June.
**NOTES:** This species has sexual dimorphism in size, with females typically being larger than males. *Temnostoma pictulum* is an old name (synonym).

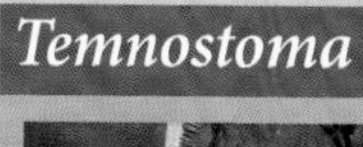

# Temnostoma

## Temnostoma alternans

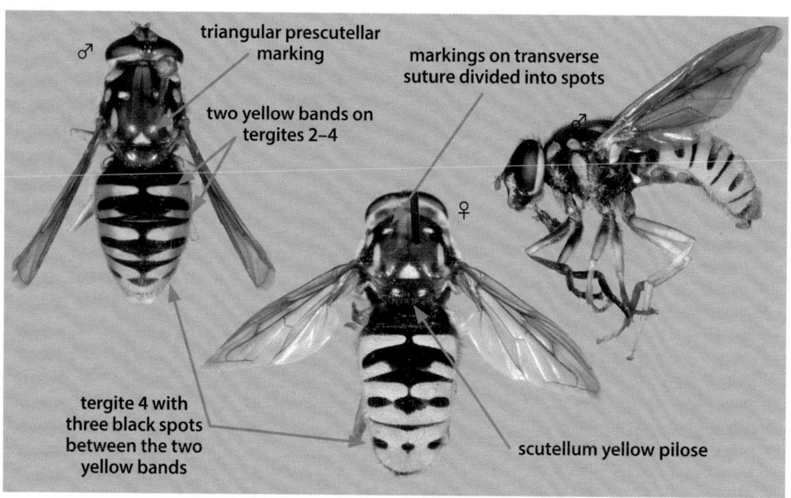


## Temnostoma daochus

# *Temnostoma excentrica*

## Black-spotted Falsehorn

**SIZE RANGE:** 12.6–17.1 mm

**IDENTIFICATION:** This species has two yellow bands on tergites 2–4. Tergite 4 has a pair of medial black bands. Yellow bands on the transverse suture are not broken into separate spots. **ABUNDANCE:** Common. **FLIGHT TIMES:** Late May to early August. **NOTES:** *Temnostoma excentrica* was treated as the Holarctic species *T. vespiforme* until recently. It is now apparent based on morphological studies and DNA data that *T. vespiforme* is a species complex and the true *vespiforme* of Linnaeus is restricted to the Palearctic. *Temnostoma aequale* is a junior synonym of *T. excentrica*. They occur in hardwood and spruce-birch forests and have been recorded visiting *Cornus*, *Kalmia*, and *Rubus*.

# *Temnostoma venustum*

## Black-banded Falsehorn

**SIZE RANGE:** 13.5–14.5 mm

**IDENTIFICATION:** This species also has two yellow bands on tergites 2–4. Tergite 4 has a single medial black band separating the two yellow bands. The scutellum is black pilose (it is yellow pilose in the preceding three species). **ABUNDANCE:** Rare. **FLIGHT TIMES:** Mid-June to early August. **NOTES:** This species is found in hardwood forests. Floral records include *Cornus* and *Polygonum*.

# *Temnostoma excentrica*

♂

♀

transverse suture with pollinose band

tergite 4 with pair of medial black bands

# *Temnostoma venustum*

markings on transverse suture divided into spots

scutellum black pilose

tergite 4 with medial black band separating two yellow bands

## *Temnostoma balyras*

### Yellow-haired Falsehorn

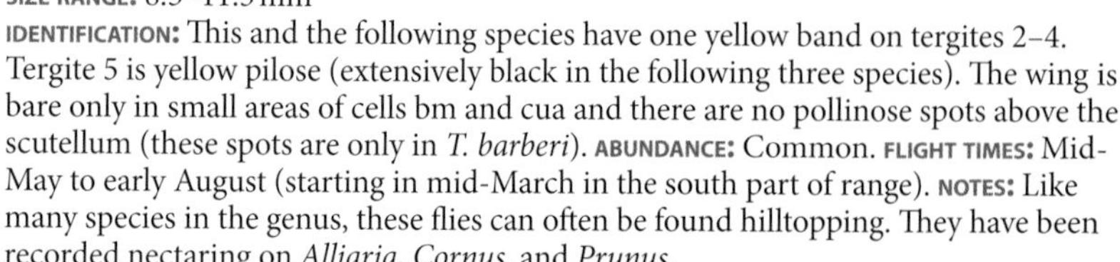

**SIZE RANGE:** 8.5–11.5 mm
**IDENTIFICATION:** This and the following species have one yellow band on tergites 2–4. Tergite 5 is yellow pilose (extensively black in the following three species). The wing is bare only in small areas of cells bm and cua and there are no pollinose spots above the scutellum (these spots are only in *T. barberi*). **ABUNDANCE:** Common. **FLIGHT TIMES:** Mid-May to early August (starting in mid-March in the south part of range). **NOTES:** Like many species in the genus, these flies can often be found hilltopping. They have been recorded nectaring on *Alliaria*, *Cornus*, and *Prunus*.

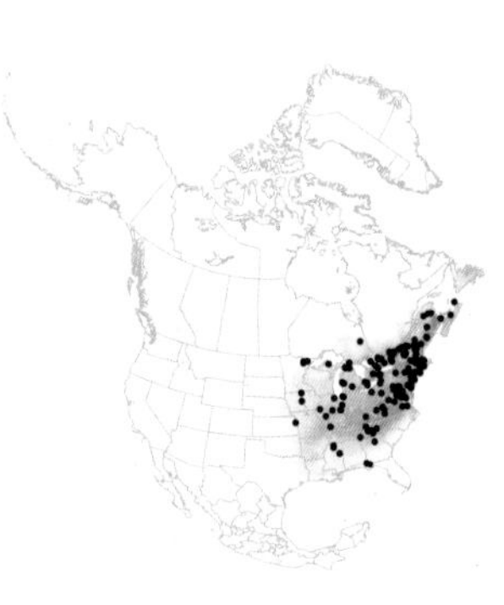

## *Temnostoma trifasciatum*

### Three-lined Falsehorn

**SIZE RANGE:** 9.4–15.8 mm
**IDENTIFICATION:** This species has one yellow band on tergites 2–4, and tergite 5 is black pilose (as in following species). Wing cells bm and cua are largely bare. **ABUNDANCE:** Rare. **FLIGHT TIMES:** Early May to early July. **NOTES:** They have been recorded nectaring on *Cornus*.

# *Temnostoma balyras*

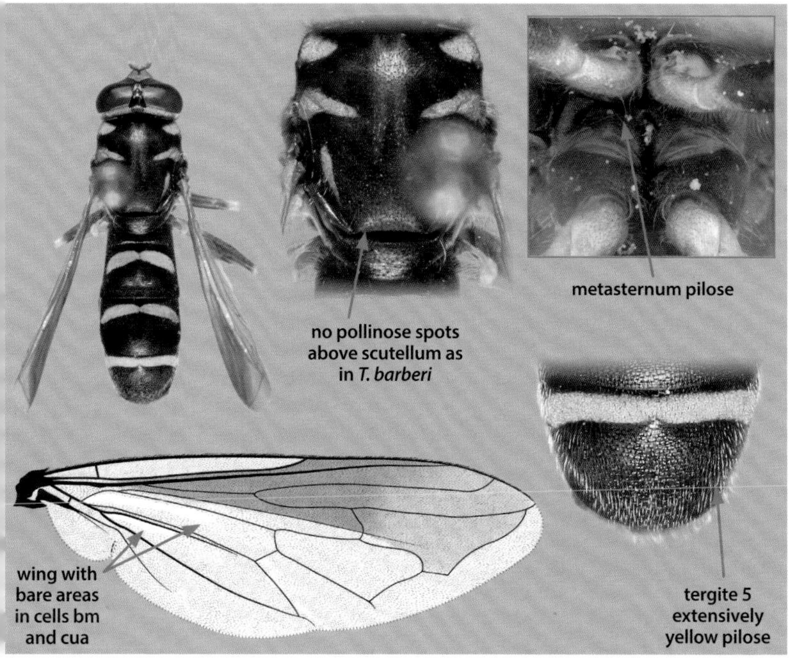

no pollinose spots above scutellum as in *T. barberi*

metasternum pilose

wing with bare areas in cells bm and cua

tergite 5 extensively yellow pilose

# *Temnostoma trifasciatum*

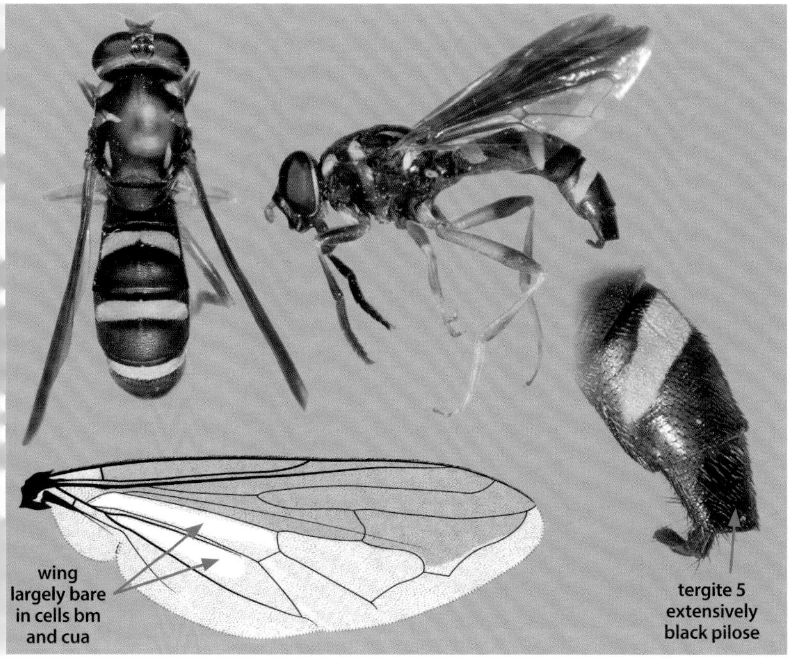

wing largely bare in cells bm and cua

tergite 5 extensively black pilose

## *Temnostoma barberi*

### Bare-bellied Falsehorn

**SIZE RANGE:** 9.0–14.5 mm
**IDENTIFICATION:** This species also has one yellow band on tergites 2–4 and black pile on tergite 5. Small pollinose spots on the dorsum of the thorax near the scutellum are diagnostic. In addition, wing cells bm, cua, and cup are extensively bare and the metasternum is bare (pilose in all other small *Temnostoma*). **ABUNDANCE:** Fairly common. **FLIGHT TIMES:** Late May to early August. **NOTES:** They are hilltoppers and may be found in a variety of habitats from hardwood forests to bogs. *Temnostoma acra* is a junior synonym of *Temnostoma barberi*. The specimen below is feeding on *Anemone*.

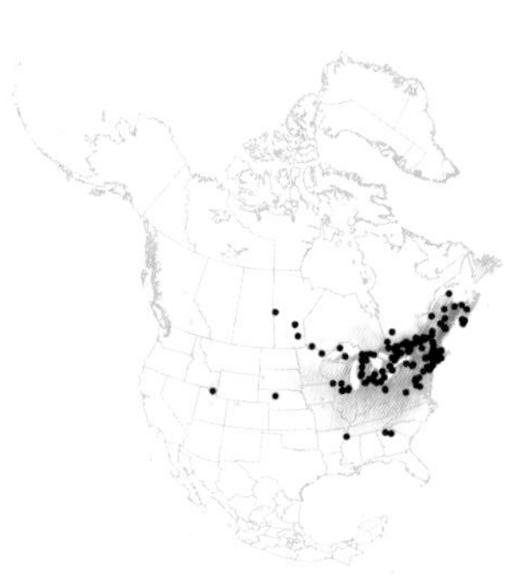

## *Temnostoma obscurum* Hairy-winged Falsehorn

**SIZE RANGE:** 10.0–13.9 mm
**IDENTIFICATION:** This species also has one yellow band on tergites 2–4 and black pile on tergite 5. Unlike in preceding species, the wing is completely microtrichose. **ABUNDANCE:** Uncommon. **FLIGHT TIMES:** Late May to mid-August. **NOTES:** They are widespread in the boreal forest where they have been recorded in spruce and spruce-birch forest and have been noted feeding on *Aruncus* and *Heracleum*.

# Temnostoma barberi

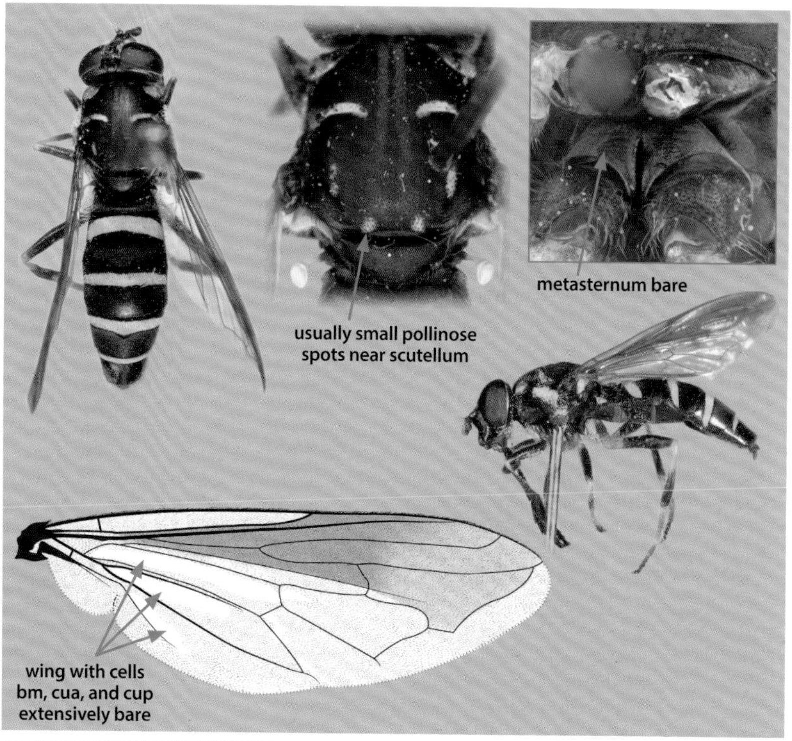

metasternum bare

usually small pollinose spots near scutellum

wing with cells bm, cua, and cup extensively bare

# Temnostoma obscurum

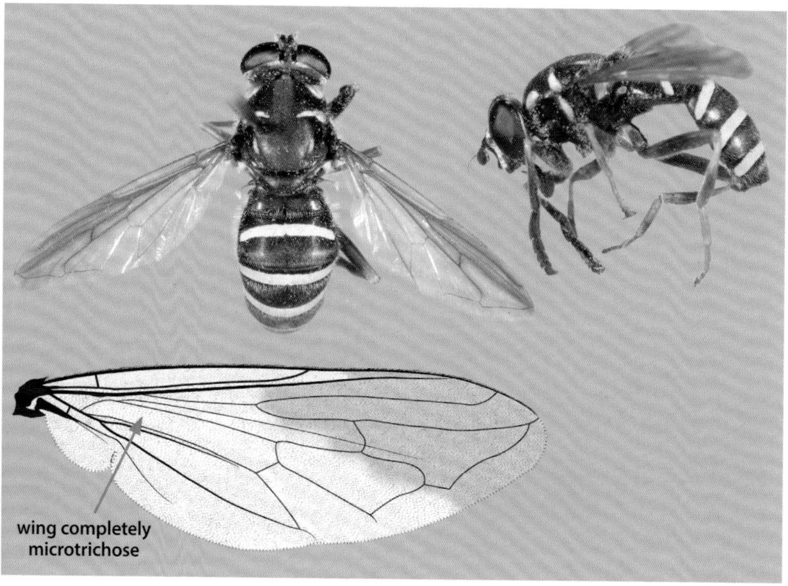

wing completely microtrichose

# Ceriana

*Ceriana* species have antennae with a terminal style (as opposed to the arista found on most syrphids). The first apparent segment of their antennae is elongate (three to four times as long as wide). In *Sphiximorpha*, this basal segment is the scape (as in most flies), whereas in the nominate subgenus, the scape inserts into a long frontal prominence. Globally, 202 species are placed in four or more subgenera within *Ceriana*. Sixteen species occur in the USA and Canada. Larvae filter feed in flowing sap in tree wounds.

## *Ceriana (Ceriana) abbreviata*

### Northern Wasp Fly

**SIZE RANGE:** 8.2–12.0 mm
**IDENTIFICATION:** This wasp mimic has elongate antennae with the frontal prominence much longer than broad. There is a terminal style on the flagellum. **ABUNDANCE:** Rare. **FLIGHT TIMES:** Late April to late July. **NOTES:** A larva was reared from *Populus balsamifera*. Adults visit flowers of *Crataegus*, *Melilotus*, *Physocarpus*, and *Spiraea*.

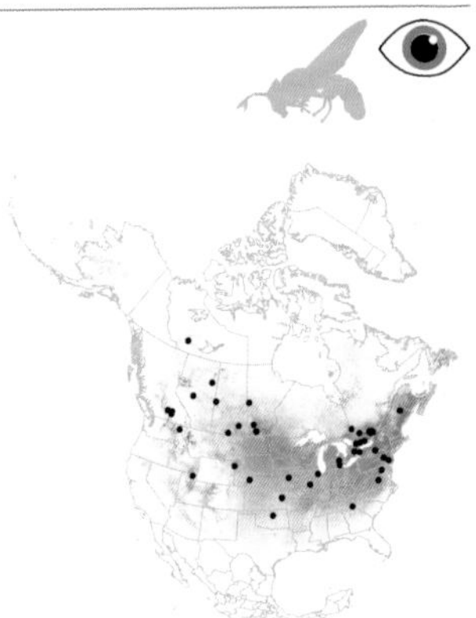

## *Ceriana (Sphiximorpha) willistoni*

### Williston's Wasp Fly

**SIZE RANGE:** 10.5–12.4 mm
**IDENTIFICATION:** This wasp mimic has elongate antennae with the frontal prominence short and the scape elongate. There is a terminal style on the flagellum. **ABUNDANCE:** Rare. **FLIGHT TIMES:** Early May (early April in the south) to early July. **NOTES:** These flies are most commonly found sitting on tree trunks on hilltops. Flowers visited include *Cornus*, *Heracleum*, and *Prosopis*. Pupae have been found under bark of *Juglans nigra* and adults have been observed around fresh sap wounds on recently fallen *Populus*.

## *Ceriana (Ceriana) abbreviata*

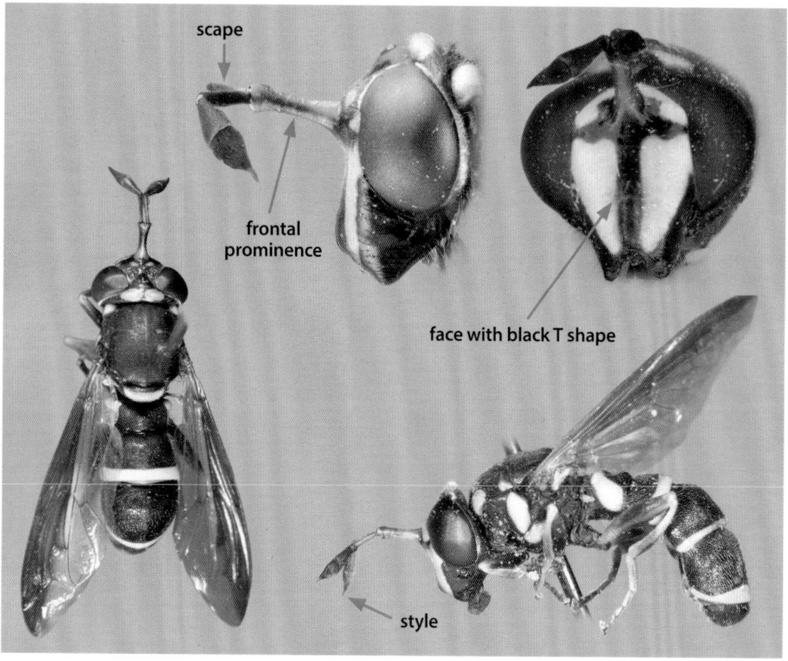

## *Ceriana (Sphiximorpha) willistoni*

# *Sericomyia*

There are 30 species of *Sericomyia*, 16 in the USA and Canada and 14 in our region. This genus is recognized by having a plumose arista, the wing vein $M_1$ directed apically, and cell $r_1$ open to the wing margin. Vein $R_{4+5}$ can be either sinuous or straight. There are two species that mimic bumblebees and 12 that are imperfect mimics of other Hymenoptera. The bumblebee mimics have been treated as a separate genus (*Arctophila*) but this and *Conosyrphus* have recently been synonymized with *Sericomyia* based on molecular data. Larvae of this genus have a long tail-like breathing tube (rat tail) and live in rotting vegetation in wetlands. For a key to species of the New World, see Skevington and Thompson (2012).

## *Sericomyia vockerothi* Vockeroth's Pond Fly

**SIZE RANGE:** 12.6–14.0 mm
**IDENTIFICATION:** This is the only eastern bumblebee mimic in the genus. It has a black face and yellow-pilose postalar callus and scutum. **ABUNDANCE:** Rare. **FLIGHT TIMES:** Mid-May to late July. **NOTES:** A few specimens have been collected in blueberry fields. Several *Sericomyia* species are associated with blueberry pollination.

## *Sericomyia transversa* Yellow-spotted Pond Fly

**SIZE RANGE:** 10.4–14.4 mm
**IDENTIFICATION:** This species is recognized by the large markings on tergite 2 and straight markings half the size on tergites 3 and 4. The legs are almost entirely black (a character shared only with *S. lata*). **ABUNDANCE:** Uncommon. **FLIGHT TIMES:** Mid-May to early August. **NOTES:** They are most often found in bogs but have also been found in fens, pine forest, and floodplain forest. Flowers visited: *Prunus* and *Spiraea*.

# *Sericomyia*

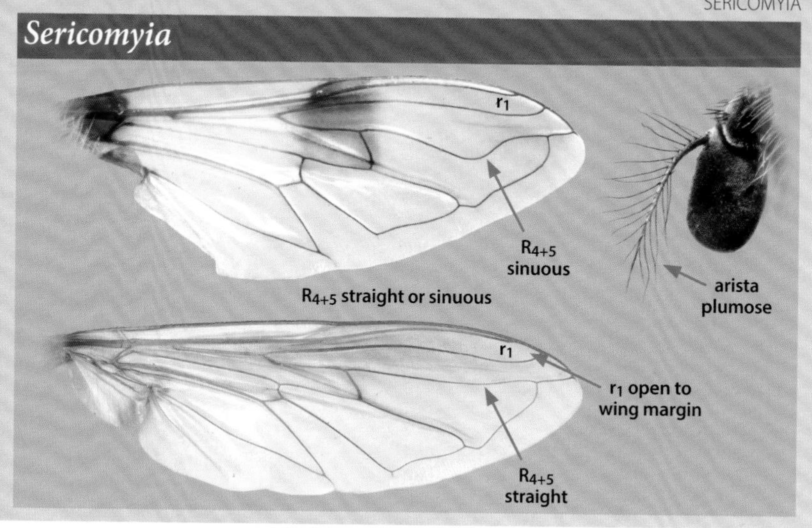


## *Sericomyia vockerothi*

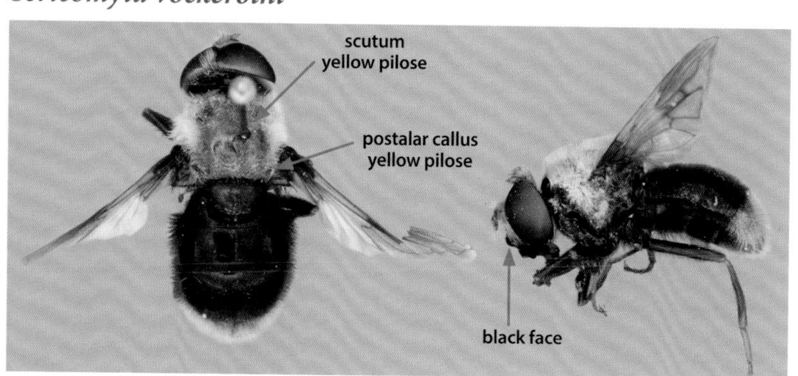


## *Sericomyia transversa*

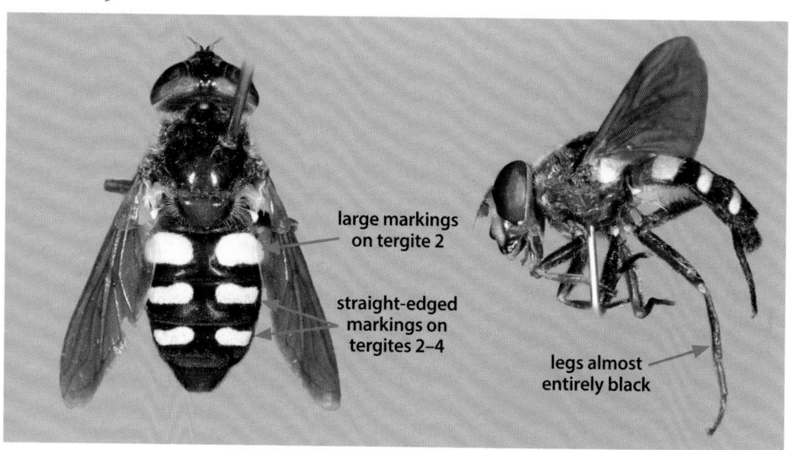

## *Sericomyia lata* White-spotted Pond Fly

**SIZE RANGE:** 11.6–15.2 mm
**IDENTIFICATION:** This species has a distinctive abdominal pattern with large spots on tergite 2 and arcuate spots (sometimes separated) on tergites 3 and 4. It is one of two *Sericomyia* species with an all-yellow face. **ABUNDANCE:** Common. **FLIGHT TIMES:** Mid-May to early September. **NOTES:** This species is occasionally found hilltopping but is more often found visiting flowers in hardwood or pine forest. It has been recorded visiting flowers of *Acer*, *Cephalanthus*, *Daucus*, *Rubus*, and *Viburnum*.

## *Sericomyia carolinensis* Two-spotted Pond Fly

**SIZE RANGE:** 11.0–12.4 mm
**IDENTIFICATION:** This and *S. lata* are the only two *Sericomyia* species with completely yellow faces. The abdomen has only one pair of narrow yellow markings on tergite 2. This species has a yellow-pilose scutellum and orange-brown basitarsi. **ABUNDANCE:** Very rare. **FLIGHT TIMES:** Early March to early July. **NOTES:** The type specimen was collected in mid-April in North Carolina from a blossoming *Pyrus* tree.

# *Sericomyia lata*

large spots
arcuate spots
yellow face

# *Sericomyia carolinensis*

one pair of yellow markings on tergite 2 only
yellow face

## *Sericomyia bifasciata* Long-nosed Pond Fly

**SIZE RANGE:** 8.5–12.7 mm
**IDENTIFICATION:** This species has a face that projects ventrally by more than ¾ the height of the eye. It also has yellow pile on the scutellum and anepimeron. Males of this species have yellow markings on tergites 2 and 3 while females have markings on tergites 2–4. The abdominal markings are widely separated.
**ABUNDANCE:** Uncommon. **FLIGHT TIMES:** Early May to late July. **NOTES:** Despite occurring in a very wide range of habitats (hardwood forest, pine forest, mixed forest, bogs, fens, coastal heath), *S. bifasciata* is most commonly associated with blueberries. They have been recorded nectaring on *Cornus*, *Prunus*, and *Vaccinium*.

## *Sericomyia chrysotoxoides*
### Oblique-banded Pond Fly

**SIZE RANGE:** 9.6–15.3 mm
**IDENTIFICATION:** This species has yellow pile on its scutellum and anepimeron as well, but its face projects ventrally by less than 1/2 the height of the eye. Both sexes have pairs of widely separated yellow markings on tergites 2–4. **ABUNDANCE:** Common. **FLIGHT TIMES:** Early May to late October (from late March in the south). **NOTES:** This species commonly visits hilltops and is found in a variety of habitats (most commonly in hardwood forest, but also pine forest, spruce forest, bogs, and fens). Flowers visited include *Acer*, *Apocynum*, *Centaurea*, *Cornus*, *Eupatorium*, *Euthamia*, *Eutrochium*, *Fragaria*, *Rubus*, *Solidago*, *Spiraea*, *Tanacetum*, *Thalictrum*, and *Viburnum*.

## *Sericomyia bifasciata*

## *Sericomyia chrysotoxoides*

## *Sericomyia militaris*

Narrow-banded Pond Fly

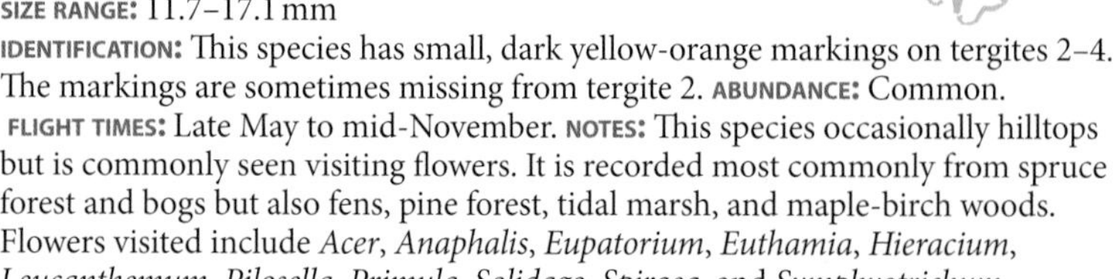

**SIZE RANGE:** 11.7–17.1 mm
**IDENTIFICATION:** This species has small, dark yellow-orange markings on tergites 2–4. The markings are sometimes missing from tergite 2. **ABUNDANCE:** Common. **FLIGHT TIMES:** Late May to mid-November. **NOTES:** This species occasionally hilltops but is commonly seen visiting flowers. It is recorded most commonly from spruce forest and bogs but also fens, pine forest, tidal marsh, and maple-birch woods. Flowers visited include *Acer*, *Anaphalis*, *Eupatorium*, *Euthamia*, *Hieracium*, *Leucanthemum*, *Pilosella*, *Primula*, *Solidago*, *Spiraea*, and *Symphyotrichum*.

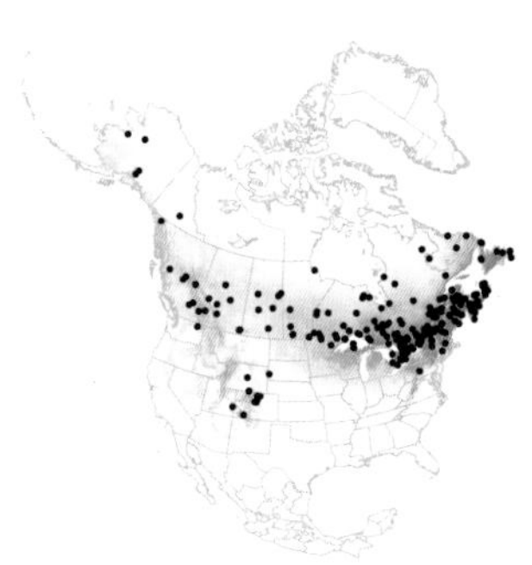

## *Sericomyia nigra* Polar Pond Fly

**SIZE RANGE:** 12.5–15.6 mm
**IDENTIFICATION:** There are narrowly separated, straight yellow markings on tergites 2–4. The femora of this species are extensively orange and the proepimeral and procoxal pile is yellow. **ABUNDANCE:** Rare. **FLIGHT TIMES:** Early June to mid-August. **NOTES:** Found in Europe, Asia, and North America in tundra as well as open areas within boreal coniferous and deciduous forests. They have been found visiting *Achillea*, *Crataegus*, *Malus*, *Prunus*, *Pyrus*, *Rosa*, *Rubus*, *Spiraea*, *Taraxacum*, *Tilia*, *Vaccinium*, and *Valeriana* flowers.

# Sericomyia militaris

abdominal markings occasionally absent on tergite 2

markings usually small, very narrow

# Sericomyia nigra

femora orange

proepimeron yellow pilose

procoxa yellow pilose

## *Sericomyia sexfasciata* Six-banded Pond Fly

**SIZE RANGE:** 12.6–13.6 mm

**IDENTIFICATION:** This species has pairs of widely separated yellow markings on tergites 2–4. These markings have distinctive concave margins. The pile on the scutellum is yellow and black (centrally) and the anepimeron is yellow pilose. **ABUNDANCE:** Uncommon. **FLIGHT TIMES:** Late May to late August. **NOTES:** Records are from spruce-aspen forest and willows in coastal tundra. They have been recorded visiting *Melilotus* (sweet clover) flowers.

## *Sericomyia slossonae* Slosson's Pond Fly

**SIZE RANGE:** 11.4–13.7 mm

**IDENTIFICATION:** This is the only species of *Sericomyia* with pile of both the scutellum and anepimeron extensively black. The femora are black basally and there are pairs of straight yellow markings on tergites 2–4 that are narrowly separated (similar to *S. nigra*). **ABUNDANCE:** Rare. **FLIGHT TIMES:** Late May to late June. **NOTES:** These flies have been collected in bogs and found visiting flowers of several typical bog plants including *Chamaedaphne*, *Rhododendron*, and *Vaccinium*.

## *Sericomyia sexfasciata*

concave margin

scutellum pile yellow and black

anepimeron pile yellow

## *Sericomyia slossonae*

femora black basally

narrowly separated

scutellum black pilose

anepimeron black pilose

**Sericomyia arctica and S. jakutica are indistinguishable except for differences in male genitalia. The hypandrium is most distinctly different.**

## *Sericomyia arctica* Arctic Pond Fly

**SIZE RANGE:** 13.4–14.8 mm
**IDENTIFICATION:** This species and *S. jakutica* are very similar. Females are indistinguishable and males can be differentiated only by dissection and examination of the male hypandrium (part of the genitalia). The two species can be differentiated from the other similar northern species (*S. sexfasciata*) by the presence of black pile on tergite 3 (all black in males and mixed black and yellow in females—*S. sexfasciata* has yellow pile on tergite 3). The scutellum is reddish on *S. arctica* and *S. jakutica* and black on *S. sexfasciata*. The shape of the band on tergite 4 also differs from *S. sexfasciata*. **ABUNDANCE:** Rare. **FLIGHT TIMES:** Mid-June to mid-August. **NOTES:** These flies are found near pools in humid spruce forest, forested bogs, and tundra where they have been recorded visiting *Achillea* and *Rubus* flowers. Larvae are unknown but are likely associated with small pools. This species is found throughout northern Europe, Asia, and North America.

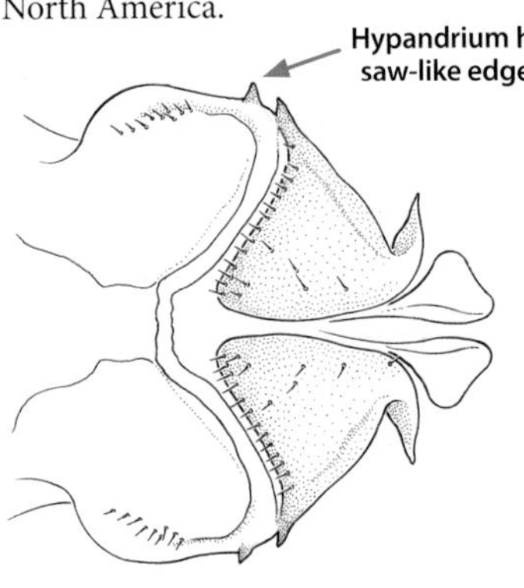


## *Sericomyia jakutica* Northern Pond Fly

**SIZE RANGE:** 12.6–15.9 mm
**IDENTIFICATION:** See above for identification tips. Also see Skevington and Thompson (2012) for more details on genitalic structures.
**ABUNDANCE:** Rare. **FLIGHT TIMES:** Early June (most from late June) to early August.
**NOTES:** Very little is known about this tundra species. Larvae and nectar plants have not been documented. They are found across Arctic Eurasia and North America where they often overlap with *S. arctica*.

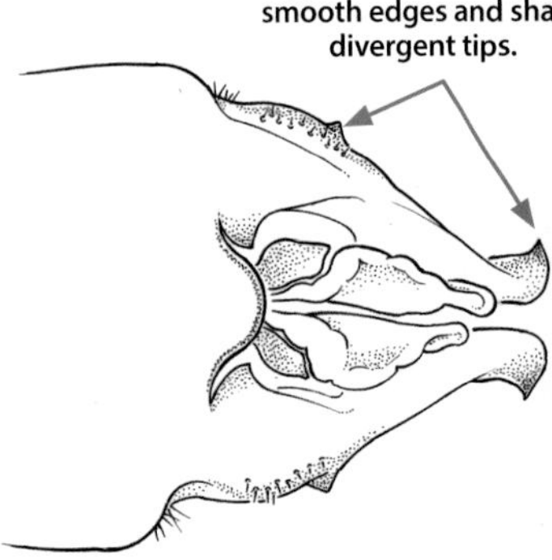

## *Sericomyia arctica*

scutellum reddish (black in *S. sexfasciata*)

In this and *S. jakutica*, the band on tergite 4 is narrower than those on tergites 2 and 3 (all similar in *S. sexfasciata*).

## *Sericomyia jakutica*

scutellum reddish (black in *S. sexfasciata*)

In this and *S. arctica*, the band on tergite 4 is narrower than those on tergites 2 and 3 (all similar in *S. sexfasciata*).

# Somula

*Somula* species are very distinctive flies. They have a projecting antennal base and distinctive spots on the abdomen. There are only two species worldwide, both in (or near) our region. The last key to the species of this genus was by Curran (1925a). Larvae live in decaying wood. These flies are mimics of scoliid wasps. *Scolia nobilitata* and some *Campsomeris* species are very similar.

## ***Somula decora*** Spotted Wood Fly

**SIZE RANGE:** 10.7–18.8 mm

**IDENTIFICATION:** This species has a completely yellow face. The yellow abdominal spots are large. **ABUNDANCE:** Uncommon.

**FLIGHT TIMES:** Late April to late July (as early as late March in the south).

**NOTES:** This species is regularly encountered on hilltops. It is found most commonly in bogs but may regularly be found along roadsides and even in city gardens. Frank McAlpine reared an adult from a pupa he found in a large, rotten poplar (*Populus*) stump in Carp, Ontario. This is the only definitive evidence that the larvae live in decaying wood, although Curran hypothesized this in 1925. They have been found on *Cornus* and *Crataegus* flowers.

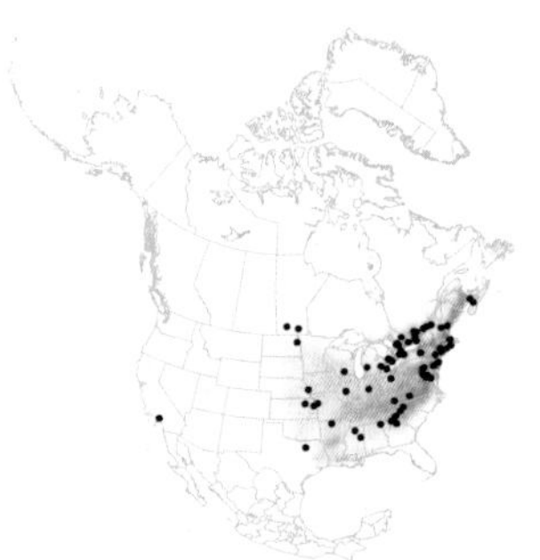

## ***Somula mississippiensis*** Banded Wood Fly

**SIZE RANGE:** 14.5–16.7 mm

**IDENTIFICATION:** This species can be easily separated from *S. decora* by the black medial stripe on the face. The yellow abdominal spots are noticeably smaller than in *S. decora*. **ABUNDANCE:** Very rare.

**FLIGHT TIMES:** Late March to early July. **NOTES:** They have been caught visiting *Crataegus* and *Prunus* flowers. Oddly, two specimens were caught inside a laundromat and two were caught in a spiderweb. No specific habitat data are available.

## *Somula decora*

yellow face

large spots

## *Somula mississippiensis*

antennal base long in this species and *S. decora*

smaller markings

yellow face with black medial stripe

# Blera

*Lejota*, *Brachypalpus*, and *Chalcosyrphus* include species that may be mistaken for *Blera* but the latter always have some pale markings on either the abdomen or face. *Blera* larvae feed on decaying plant matter and typically live in dead wood. They are thus forest dwellers, often living in older growth forests where there is more fallen dead timber. There are 30 valid species of *Blera*, 16 of them in the Nearctic region and eight in our area. For a key to Nearctic species, see Curran (1953, treated as *Cynorhina*). Barkalov and Mutin (1991) also provide a useful key that includes many Nearctic species.

## *Blera armillata* Orange-faced Wood Fly

**SIZE RANGE:** 10.5–13.6 mm

**IDENTIFICATION:** The abdomen is entirely black. The frons, face (at least lower ½), and legs (basal and apical ⅓ of tibiae and basal 3 tarsomeres) are yellow orange. **ABUNDANCE:** Uncommon. **FLIGHT TIMES:** Mid-May to early August. **NOTES:** This species has a boreo-montane distribution and can be expected in many areas where it has not yet been collected. Habitats recorded include hardwood forest, fir-poplar forest, and rocky slopes. Flowers visited include *Amelanchier*, *Fragaria*, and *Rhododendron*.

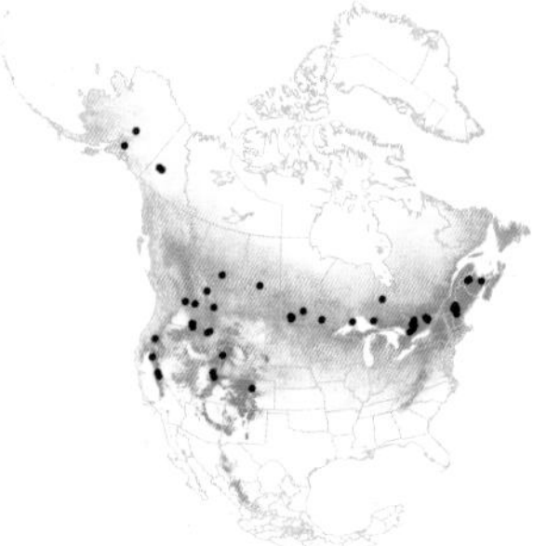

## *Blera nigra* Golden-haired Wood Fly

**SIZE RANGE:** 7.2–10.1 mm

**IDENTIFICATION:** The abdomen, face (at least the lower ½), and legs are extensively black. The frons and leg color separate this from *Blera armillata*. **ABUNDANCE:** Fairly common. **FLIGHT TIMES:** Early May to early August. **NOTES:** Habitats recorded include birch-fir forests, hardwood forests, and sphagnum fens. Has been recorded visiting *Fragaria* flowers.

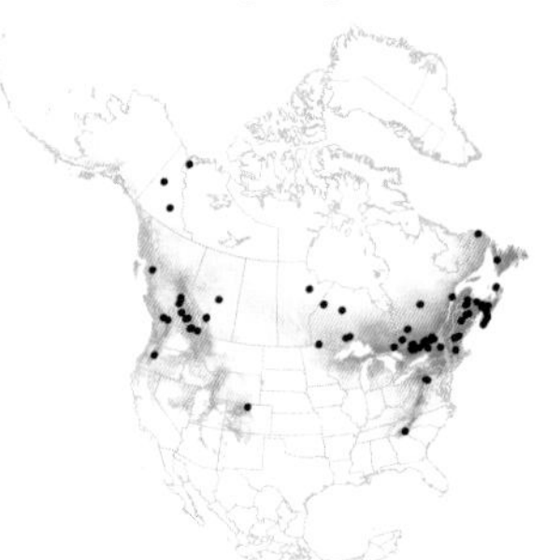

# Blera

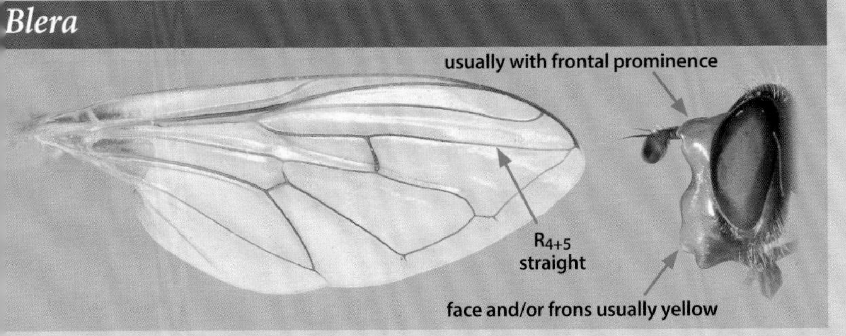


The two species shown on this page are our only *Blera* that have an entirely black abdomen.

## *Blera armillata*

## *Blera nigra*

## *Blera badia* Common Wood Fly

**SIZE RANGE:** 8.4–11.6 mm
**IDENTIFICATION:** This species has a yellow face with a black medial stripe and orange abdominal markings. The color of the legs separates this species from *B. confusa*. The basal ⅓ or more of the metafemur is yellow and the apical three tarsomeres of the protarsus are black. If in doubt, check the color of the pile above the wings, which is yellow. **ABUNDANCE:** Common. **FLIGHT TIMES:** Early May to late July. **NOTES:** Found in hardwood and mixed forests and bogs. Flowers visited include *Rubus* and *Viburnum*.

## *Blera confusa* Confusing Wood Fly

**SIZE RANGE:** 10.0–12.0 mm
**IDENTIFICATION:** Like *B. badia*, this species has a yellow face with a black medial stripe and orange abdominal markings. The extensively black metafemur, two black apical protarsomeres, and black pile above the wing distinguish it. **ABUNDANCE:** Common. **FLIGHT TIMES:** Mid-May to late July. **NOTES:** Most records are from bogs but they may also be found in forests. Flowers visited include *Houstonia* and *Rubus*.

## *Blera umbratilis* Hairy Wood Fly

**SIZE RANGE:** 8.4–11.0 mm
**IDENTIFICATION:** Although similar in outward appearance to the above species, they are more robust, more brightly colored and have denser pile. If in doubt, the black color of the face will quickly distinguish this species. The metafemur is yellow on the basal ⅔. **ABUNDANCE:** Uncommon. **FLIGHT TIMES:** From late March in the south, early May to late June in the north. **NOTES:** Flowers visited include *Crataegus*, *Physocarpus*, and *Prunus*.

## *Blera badia*

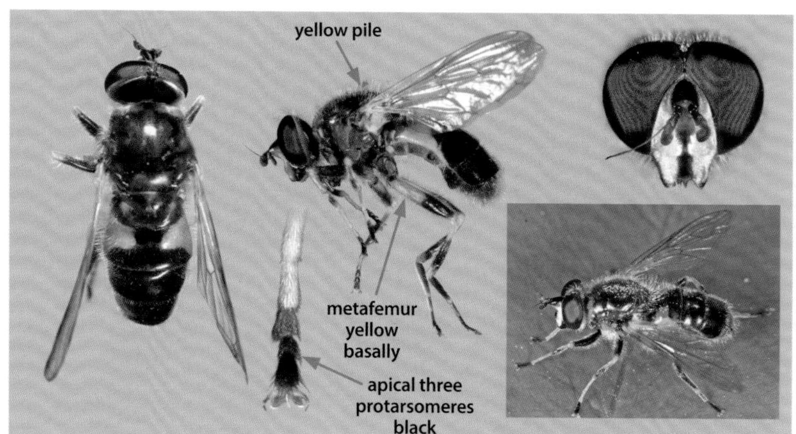


## *Blera confusa*

## *Blera umbratilis*

## *Blera analis* Orange-tailed Wood Fly

**SIZE RANGE:** 10.4–12.1 mm
**IDENTIFICATION:** This species has an entirely yellow-orange face and frons and is easily identified by the bright orange-tipped abdomen. **ABUNDANCE:** Uncommon. **FLIGHT TIMES:** Mid-May to mid-July. **NOTES:** This unmistakable species is found in hardwood forests and bogs and regularly visits hilltop mating sites. Flowers visited include *Crataegus*, *Prunus*, and *Viburnum*.

## *Blera pictipes* Painted Wood Fly

**SIZE RANGE:** 7.6–10.3 mm
**IDENTIFICATION:** This species also has an entirely yellow-orange face and frons. It is a distinctive fly, with yellow markings on at least tergite 2, but also sometimes 3 and 4. The scutum is yellow above the wings and the scutellum is yellow apically. **ABUNDANCE:** Rare. **FLIGHT TIMES:** From early March in the south; mid-May to early July in the north. **NOTES:** They are found in hardwood forests and have been documented visiting *Physocarpus* and *Prunus* flowers.

## *Blera notata* Ornate Wood Fly

**SIZE RANGE:** 13.4 mm
**IDENTIFICATION:** This species also has an entirely yellow-orange face and frons. It shares some characteristics with *B. pictipes*, like the apically yellow scutellum, but the abdominal markings are differently shaped and the scutum is entirely black. **ABUNDANCE:** Very rare. **FLIGHT TIMES:** Early March to mid-April. **NOTES:** We have seen specimens from only a few sites in the southeastern USA but there are historical records from New Jersey and Pennsylvania. Flowers visited include *Cornus*, *Crataegus*, and *Prunus*.

## *Blera analis*

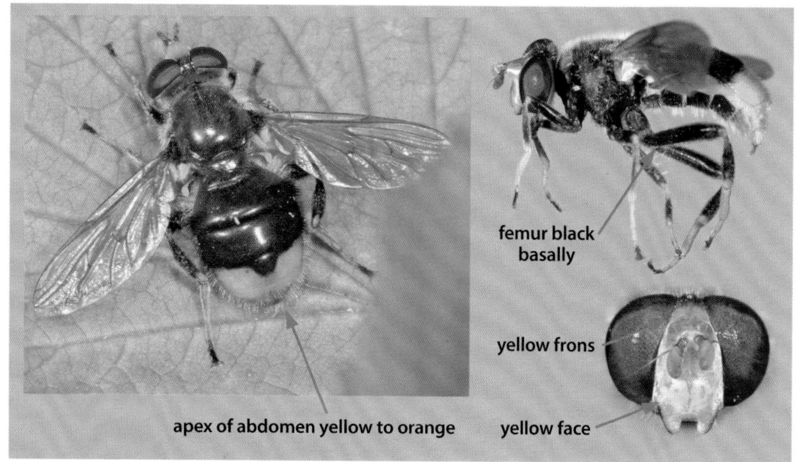


## *Blera pictipes*

## *Blera notata*

# *Teuchocnemis*

Both known species of *Teuchocnemis* occur in the northeast. They are distinctive and unlikely to be mistaken for other flies. When in doubt, use wing characters to identify this genus. It has a sinuous $R_{4+5}$ vein, cell $r_1$ is open to the wing margin, and CuA + CuP is elongate and straight, not reaching the wing margin. *Teuchocnemis* males have a strong ventral spur on their metatibia. The larvae are unknown.

## *Teuchocnemis bacuntius* Orange Spur Fly

**SIZE RANGE:** 13.5–18.1 mm

**IDENTIFICATION:** This species is easily identifiable by the large apical orange markings on the tergites, and the orange proleg. **ABUNDANCE:** Rare. **FLIGHT TIMES:** Late March in the south; early May to early July in the north. **NOTES:** Adults have been found near seeps along sand-bottomed streams, patrolling rotting stumps and logs, and visiting *Crataegus* flowers.

## *Teuchocnemis lituratus* Black Spur Fly

**SIZE RANGE:** 10.5–15.4 mm

**IDENTIFICATION:** This species has a black abdomen and black-pollinose stripes on the scutum. The proleg is mostly black. **ABUNDANCE:** Uncommon. **FLIGHT TIMES:** Late April to late June. **NOTES:** They have been documented from rich deciduous woodlands and fens.

# *Teuchocnemis*

## *Teuchocnemis bacuntius*

## *Teuchocnemis lituratus*

# *Chalcosyrphus*

*Chalcosyrphus* species are long, slender black flies, sometimes with orange legs and abdomens or with yellow abdominal markings. They have an enlarged metafemur and a fairly straight $R_{4+5}$ vein. These forest-dwelling flies are typically encountered sitting on leaves or logs or hilltopping. Larvae occur in sap runs and under bark in accumulations of decaying sap. There are 107 species of *Chalcosyrphus* known, 22 of these in the Nearctic and 16 in our region. *Chalcosyrphus* is divided into 10 subgenera, two of which occur in our area. The nominate subgenus is recognized by a flattened area on the scutum anterior to the scutellum, and crossvein r-m situated before the middle of cell dm. *Xylotomima* has a uniformly convex scutum, and crossvein r-m is situated either at or beyond the middle of cell dm. For a key to species see Curran (1941). Many *Xylota* are very similar. See the *Xylota* section for notes on how to differentiate them.

## *Chalcosyrphus (Chalcosyrphus) aristatus*

Black-dented Leafwalker

**SIZE RANGE:** 7.9–8.3 mm
**IDENTIFICATION:** Male narrowly dichoptic, with less than the width of an ocellus separating the eyes; sternite 4 produced apicomedially; female is not distinguishable from female *C. depressus*. **ABUNDANCE:** Very rare. **FLIGHT TIMES:** Late April to mid-May. **NOTES:** One record notes that the fly was collected from a tree trunk. Another specimen was collected on *Carex* flowers.

## *Chalcosyrphus (Chalcosyrphus) depressus*

Wide-eyed Leafwalker

**SIZE RANGE:** 8.4–9.1 mm
**IDENTIFICATION:** Male broadly dichoptic, with more than the width of an ocellus separating the eyes; sternite 4 not produced apicomedially; female is not distinguishable from female *C. aristatus*. **ABUNDANCE:** Very rare. **FLIGHT TIMES:** Early May to mid-June. **NOTES:** Nothing is known about this species but it appears to be boreo-montane in distribution based on the few records, so it should be widespread.

## *Chalcosyrphus (Chalcosyrphus)* | *Chalcosyrphus (Xylotomima)*

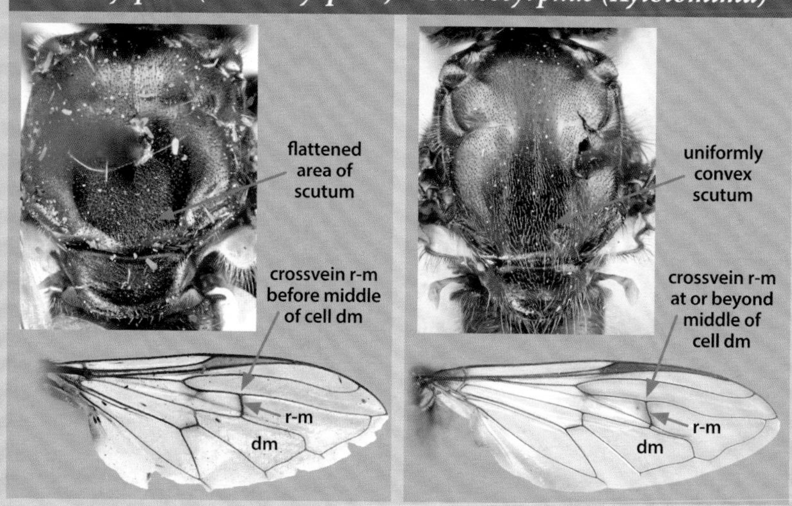


## *Chalcosyrphus (Chalcosyrphus) aristatus*

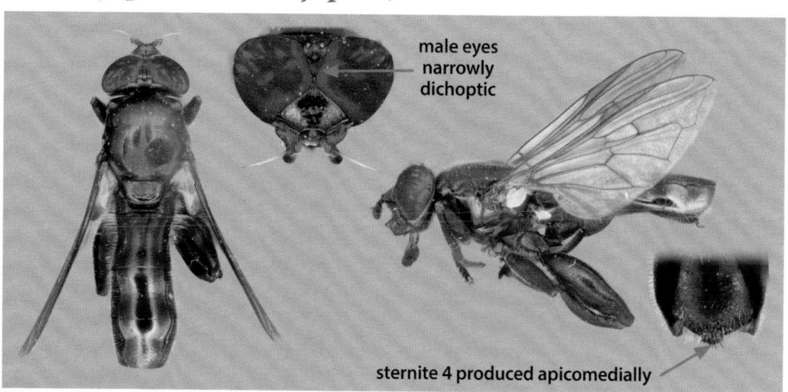


## *Chalcosyrphus (Chalcosyrphus) depressus*

## *Chalcosyrphus (Xylotomima) plesius*

### Black-hipped Leafwalker

**SIZE RANGE:** 9.5–11.5 mm
**IDENTIFICATION:** Orange legged with brown halter and partly black metacoxa; typically smaller than the other orange-legged species.
**ABUNDANCE:** Uncommon. **FLIGHT TIMES:** Early May to mid-July. **NOTES:** These attractive flies are most commonly encountered at hilltopping sites but differ in their hilltopping behavior from the other orange-legged species. On hilltops, the males repeatedly hover over a landmark for short periods of time. *Chalcosyrphus vecors* and *C. curvarius* are both perchers and rarely hover like this. Away from hilltops, *C. plesius* is most commonly encountered in hardwood forests. Single records are from a sphagnum bog and a stand of *Juniperus virginiana*. Our orange-legged *Chalcosyrphus* species are mimics of *Sphex nudus*.

## *Chalcosyrphus (Xylotomima) vecors*

### Orange-hipped Leafwalker

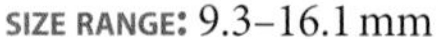

**SIZE RANGE:** 9.3–16.1 mm
**IDENTIFICATION:** Orange legged with brown halter like *C. plesius* but with entirely orange metacoxa. They are typically much larger than the preceding species.
**ABUNDANCE:** Uncommon. **FLIGHT TIMES:** Early May to late July. **NOTES:** They are also regularly encountered on hilltops, sometimes with both their Black-hipped and Yellow-haltered relatives. On hilltops, the males are typically leaf and twig perchers. They are typically encountered in hardwood and mixed forests, but may occur in spruce forests, including bogs. Flowers visited include *Acer*, *Anemone*, *Heracleum*, *Rhus*, *Sambucus*, and *Viburnum*.

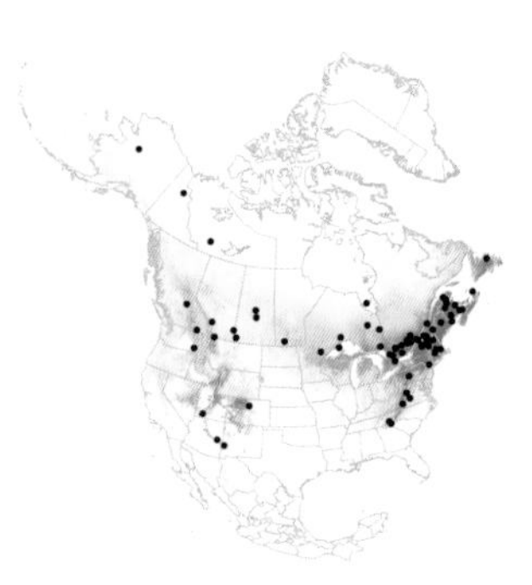

## *Chalcosyrphus (Xylotomima) plesius*

halter brown

pro- and mesolegs orange

metacoxa partially black

## *Chalcosyrphus (Xylotomima) vecors*

halter brown

pro- and mesolegs orange

metacoxa entirely orange

## *Chalcosyrphus (Xylotomima) curvarius*

### Yellow-haltered Leafwalker

**SIZE RANGE:** 8.5–16.5 mm
**IDENTIFICATION:** With its bright yellow halteres, this is the most distinctive of the orange-legged *Chalcosyrphus* species. It also has an entirely black metacoxa. **ABUNDANCE:** Common. **FLIGHT TIMES:** Mid-May to late August. **NOTES:** Like the two preceding species, they can often be found on hilltops. On hilltops, the males more often land on the ground rather than on leaves or twigs. They are mostly found in hardwood forests but there are a few records from the tundra. There is no genetic variation between Arctic and eastern specimens. One specimen was collected on a large fallen *Populus* (aspen) log that had been on the ground for about one year.

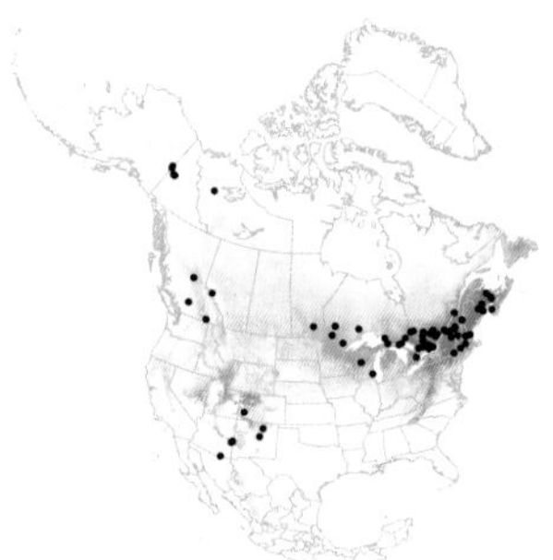

## *Chalcosyrphus (Xylotomima) chalybeus*

### Violet Leafwalker

**SIZE RANGE:** 12.4–16.1 mm
**IDENTIFICATION:** This all black *Chalcosyrphus* is very distinctive because of the metallic purple sheen to its body. Its legs are entirely black, and unlike the wings of other black *Chalcosyrphus*, the wings are largely dark brown. **ABUNDANCE:** Fairly common. **FLIGHT TIMES:** Mid-May to mid-August. **NOTES:** These hardwood forest flies are often seen around fallen dead tree trunks. They are spectacular and glisten with purplish iridescence on a sunny day. They only occasionally visit hilltops. Flowers visited include *Rubus* and *Spiraea*. These flies mimic solitary wasps such as *Sphex pensylvanicus* and *Chalybion californicum*.

## *Chalcosyrphus (Xylotomima) curvarius*

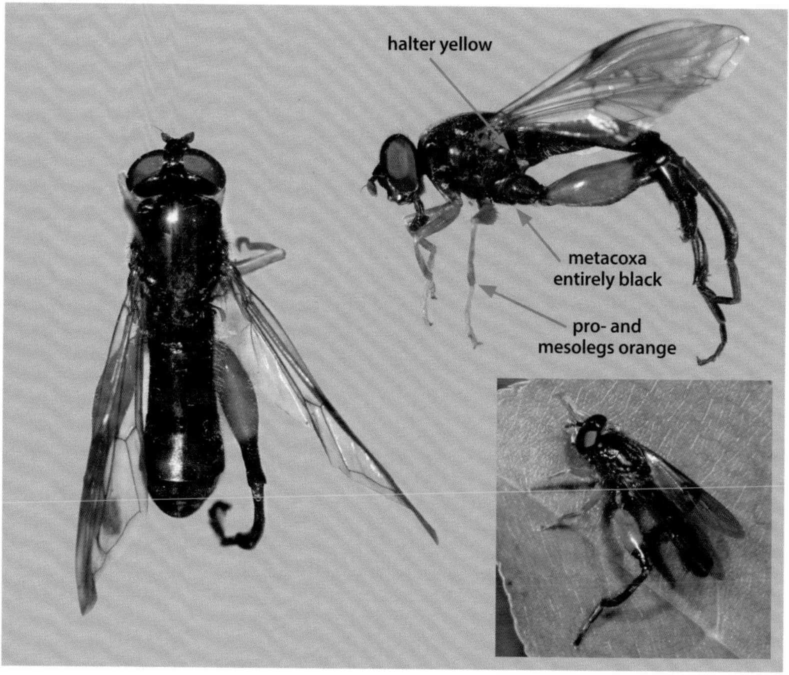

halter yellow

metacoxa entirely black

pro- and mesolegs orange

## *Chalcosyrphus (Xylotomima) chalybeus*

body black, iridescent purple in good light

legs entirely black

wing dark brownish black, frequently iridescent

## *Chalcosyrphus (Xylotomima) libo*

### Long-haired Leafwalker

**SIZE RANGE:** 8.5–10.5 mm
**IDENTIFICATION:** This is one of two *Chalcosyrphus* species with a reddish-orange abdomen. Unlike its counterpart, *Chalcosyrphus libo* has long, erect pile on the thorax and the metatarsus is yellow on the basal three tarsomeres. **ABUNDANCE:** Uncommon. **FLIGHT TIMES:** Early May (early April in southern part of range) to late July. **NOTES:** This species may be found around rotting logs or trees in a variety of habitats (deciduous or coniferous forests, fens, or *Corema* barrens). Males are most reliably found on hilltops. A single floral record is from *Salix* (opposite).

## *Chalcosyrphus (Xylotomima) piger*

### Short-haired Leafwalker

**SIZE RANGE:** 8.5–12.5 mm
**IDENTIFICATION:** *Chalcosyrphus piger* is similar to the above species but has short, appressed pile on the thorax and an entirely black metatarsus. **ABUNDANCE:** Uncommon. **FLIGHT TIMES:** Early May to late September. **NOTES:** This species is widespread through Eurasia and North America. In western Europe it has declined dramatically and is now approaching extirpation. Most of what we know about it is from the Old World, where it is found in old-growth spruce and pine forest. The flies can be found on cut trunks and fallen trees in small patches of sunlight, on the lower leaves of large-leaved plants growing beneath shrubs in partial sunlight, at the edge of small forest openings, at forest edges, or on hilltops. When it is hot they visit streams to drink, choosing locations where the margin is sandy or muddy. Flowers visited include *Crataegus*, *Grindelia*, *Heracleum*, *Leucanthemum*, *Potentilla*, *Prunus*, *Ranunculus*, *Rubus*, *Seseli*, and *Verbascum*. Larvae and pupae have been found in sappy hollows beneath bark of *Larix* and *Pinus*.

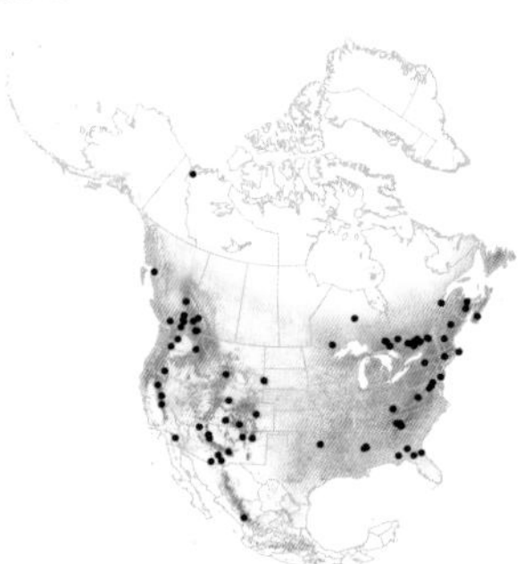

## *Chalcosyrphus (Xylotomima) libo*

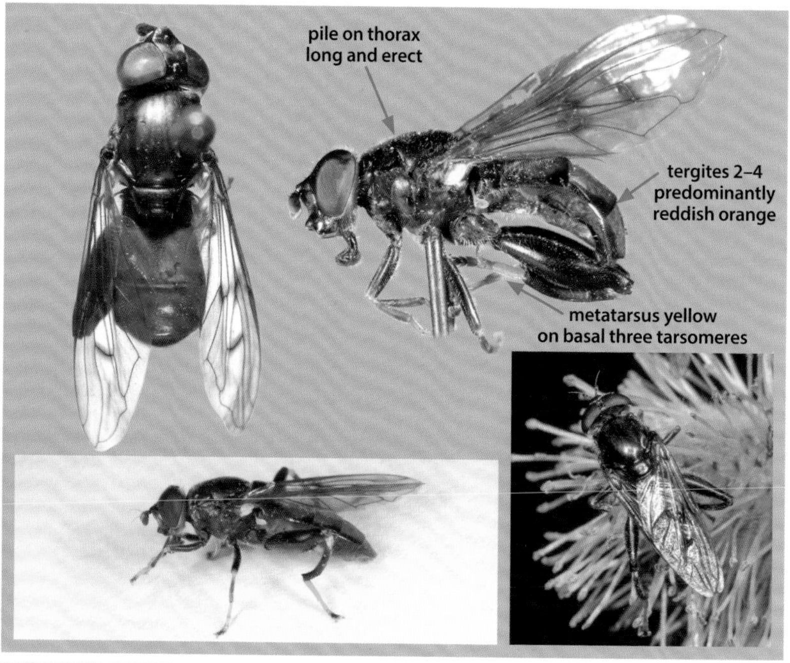

pile on thorax long and erect

tergites 2–4 predominantly reddish orange

metatarsus yellow on basal three tarsomeres

## *Chalcosyrphus (Xylotomima) piger*

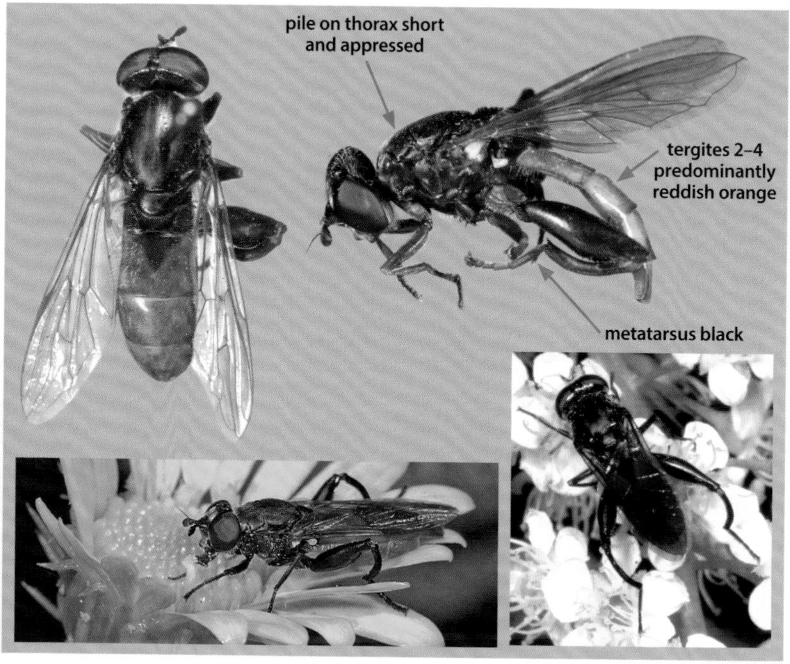

pile on thorax short and appressed

tergites 2–4 predominantly reddish orange

metatarsus black

# *Chalcosyrphus (Xylotomima) anthreas*

## Yellow-banded Leafwalker

**SIZE RANGE:** 9.5–11.5 mm
**IDENTIFICATION:** This is one of three *Chalcosyrphus* species that have yellow spots on the abdomen. This species has a yellow face (on the lower ⅓), and a metatibia that is yellow on at least the basal ¼. Beware of *Xylota* species as many have this general gestalt but none have a yellow face. **ABUNDANCE:** Rare. **FLIGHT TIMES:** Late May to mid-July (late March to early September in southern parts of range). **NOTES:** This is a hardwood forest species. We have specific records from a spring seep along a sand-bottomed stream, alluvial forest, and damp second-growth maple-birch woods. As with most *Chalcosyrphus* species, it is a hilltopper. It has been noted visiting flowers of *Cornus*, *Heracleum*, and *Polygonum*.

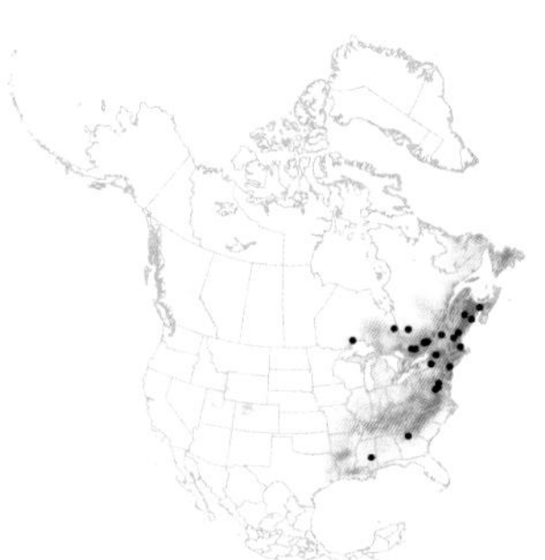

# *Chalcosyrphus (Xylotomima) nemorum*

## Dusky-banded Leafwalker

**SIZE RANGE:** 6.5–9.2 mm
**IDENTIFICATION:** This species has yellow spots on the abdomen, a black face (ground color, under white pollinosity), and a metatibia that is yellow only at the extreme base. **ABUNDANCE:** Common. **FLIGHT TIMES:** Mid-April to mid-September (from mid-March in the south). **NOTES:** This is a moist forest and wetland species often associated with alder and willow. Adults are often found on sunlit foliage overhanging water, on trunks of fallen trees beside water, and on bare, damp mud or sand at the water's edge. Flowers visited include *Anemone*, *Caltha*, *Euphorbia*, *Heracleum*, *Potentilla*, *Pycnanthemum*, *Ranunculus*, *Rubus*, *Sorbus*, and *Taraxacum*. Larvae occur under bark of waterlogged deciduous trees and stumps (often in practically submerged logs) and in damp tree rot holes in deciduous trees such as *Betula*, *Fagus*, *Populus*, *Quercus*, *Salix*, and *Ulmus*. They have also been recorded under the bark of stumps of *Larix* and *Pinus*. Larvae overwinter and have a short (~10 day) pupal stage in the spring before emerging.

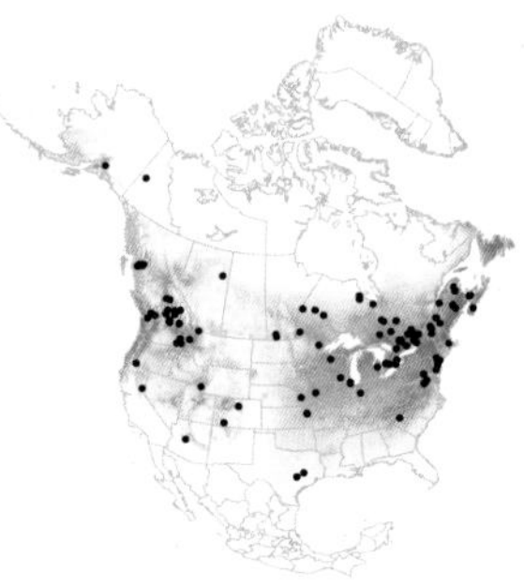

## *Chalcosyrphus (Xylotomima) anthreas*

tergites 2 and 3 black with yellow spots

ground color of face yellow on lower ⅓

metatibia yellow white on basal ¼ or more

## *Chalcosyrphus (Xylotomima) nemorum*

tergites 2 and 3 black with dull yellow spots

metatibia yellowish only at extreme base

ground color of face (under pollinose white pile) entirely black

## *Chalcosyrphus (Xylotomima) metallicus*

### Yellow-legged Leafwalker

**SIZE RANGE:** 8.1–10.0 mm
**IDENTIFICATION:** This species is most likely to be confused with the two preceding species or with species of *Xylota*. The yellow color of the protibia easily distinguishes it from the previous species (which have partially black protibiae). The bicolored metafemur distinguishes it from any *Xylota* species. **ABUNDANCE:** Uncommon. **FLIGHT TIMES:** Early March to early November. **NOTES:** This species is associated with wetlands, including stream edges and swamp forests. They have been recorded visiting *Baccharis*, *Cephalanthus*, *Ilex*, *Pontederia*, *Prunus*, *Solidago*, *Symphyotrichum*, and *Verbesina*.

## *Chalcosyrphus (Xylotomima) inarmatus*

### Yellow-haired Leafwalker

**SIZE RANGE:** 10.7–14.1 mm
**IDENTIFICATION:** This entirely dark brown to black species of *Chalcosyrphus* is recognized by the yellow-pilose thorax and basally orange tibiae and tarsi. It is most similar to *Brachypalpus* species. Check the shape of the face to be sure it is not this more common fly. *Chalcosyrphus* species have oval heads in frontal view whereas *Brachypalpus* species have triangular heads. **ABUNDANCE:** Uncommon. **FLIGHT TIMES:** Mid-May and early August (one outlier from early April on Long Island, New York). **NOTES:** This is a wetland species (mostly in sphagnum bogs) and is strongly associated with *Caltha* flowers. The specimen photographed is on *Vaccinium*.

## *Chalcosyrphus (Xylotomima) metallicus*

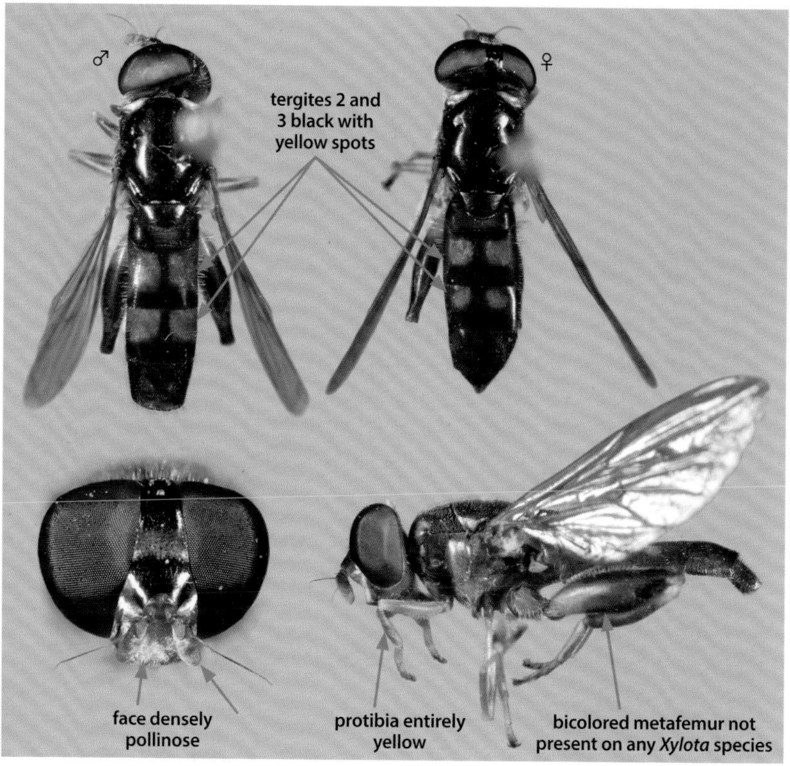

## *Chalcosyrphus (Xylotomima) inarmatus*

# *Chalcosyrphus (Xylotomima) anomalus*

## Long-tailed Leafwalker

**SIZE RANGE:** 8.3–9.0 mm
**IDENTIFICATION:** This entirely black species of *Chalcosyrphus* is recognized by the depression near the posterior edge of the scutellum, the black, sparsely pollinose face, and the apicomedially produced tergite 4 of the male, which extends over the genitalia. **ABUNDANCE:** Rare. **FLIGHT TIMES:** Early May to mid-June. **NOTES:** One specimen was collected on a *Larix* trunk in wet woods. Another was noted as being in hardwood forest while another was found in a rich fen.

# *Chalcosyrphus (Xylotomima) metallifer*

## Orange-horned Leafwalker

**SIZE RANGE:** 9.4–10.6 mm
**IDENTIFICATION:** This all black species has its wing darkened at crossvein r-m, and the arista is orange on at least the basal ⅔. The wing is entirely microtrichose. **ABUNDANCE:** Rare. **FLIGHT TIMES:** Early April to early June (from late February in the south). **NOTES:** This species is out very early in the year and may thus be overlooked to some extent. It is found in wetlands (wooded swamps, seepage areas) where it is associated with wet, rotting logs.

## *Chalcosyrphus (Xylotomima) anomalus*

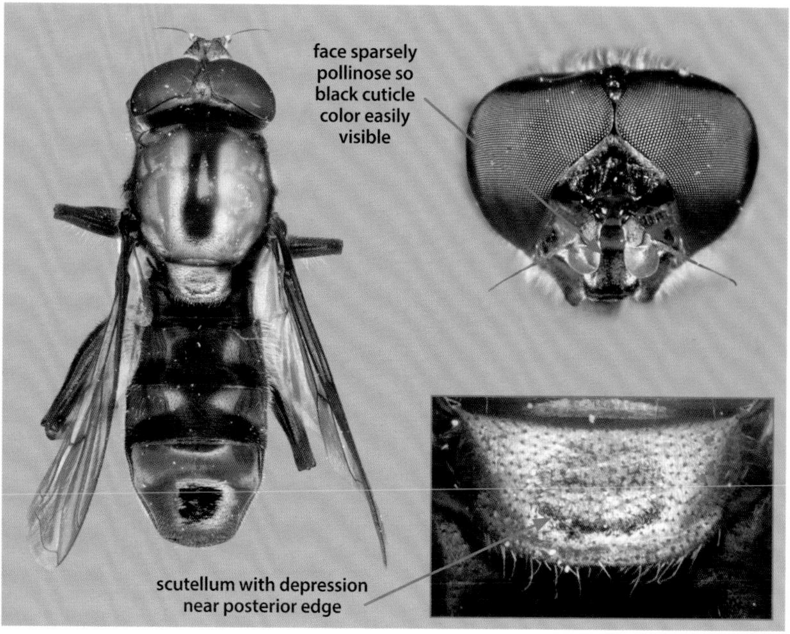

face sparsely pollinose so black cuticle color easily visible

scutellum with depression near posterior edge

## *Chalcosyrphus (Xylotomima) metallifer*

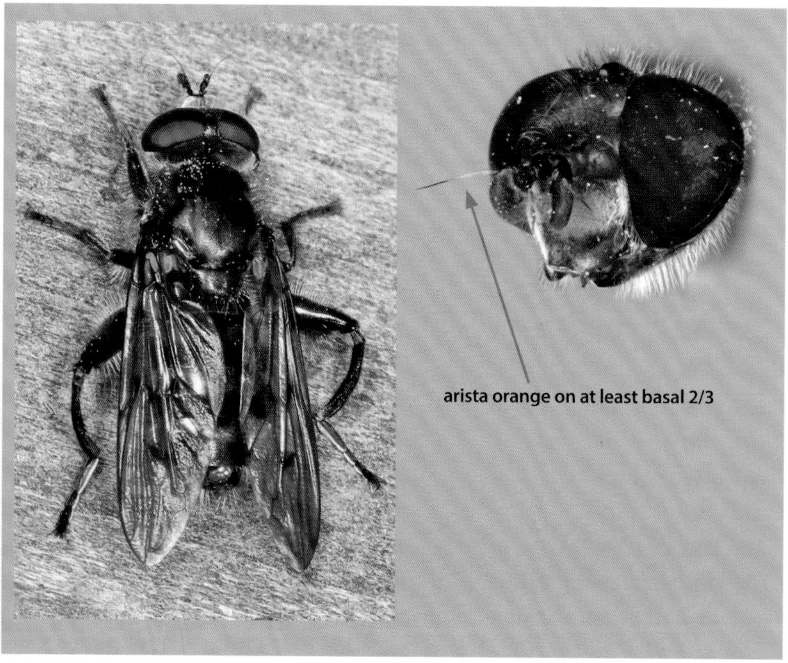

arista orange on at least basal 2/3

## *Chalcosyrphus (Xylotomima) ontario*

### Bare-winged Leafwalker

**SIZE RANGE:** 7.3–9.4 mm
**IDENTIFICATION:** This all black species has its wing darkened at crossvein r-m, and the arista is entirely brown black. The wing has bare areas basally in cells bm and cua. **ABUNDANCE:** Rare. **FLIGHT TIMES:** Late April to early June. **NOTES:** Nothing is known about the habitat of this species but it is presumably in wet hardwood forests. Two specimens were found nectaring on *Salix*.

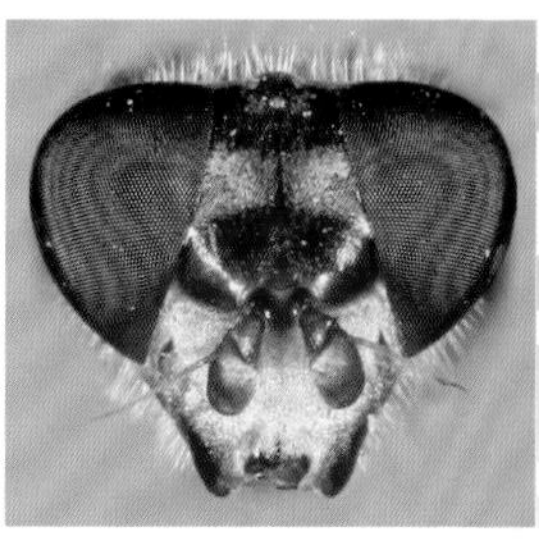

## *Chalcosyrphus (Xylotomima) sacawajeae*

### Hairy-winged Leafwalker

**SIZE RANGE:** 7.9–10.2 mm
**IDENTIFICATION:** As with *C. ontario*, this all black species has its wing darkened at crossvein r-m, and the arista is entirely brown black. The key difference from *C. ontario* is that the wing has no bare areas and is entirely microtrichose. **ABUNDANCE:** Rare. **FLIGHT TIMES:** Mid-April to mid-May. **NOTES:** Nothing is known about the life history of this species.

## *Chalcosyrphus (Xylotomima) ontario*

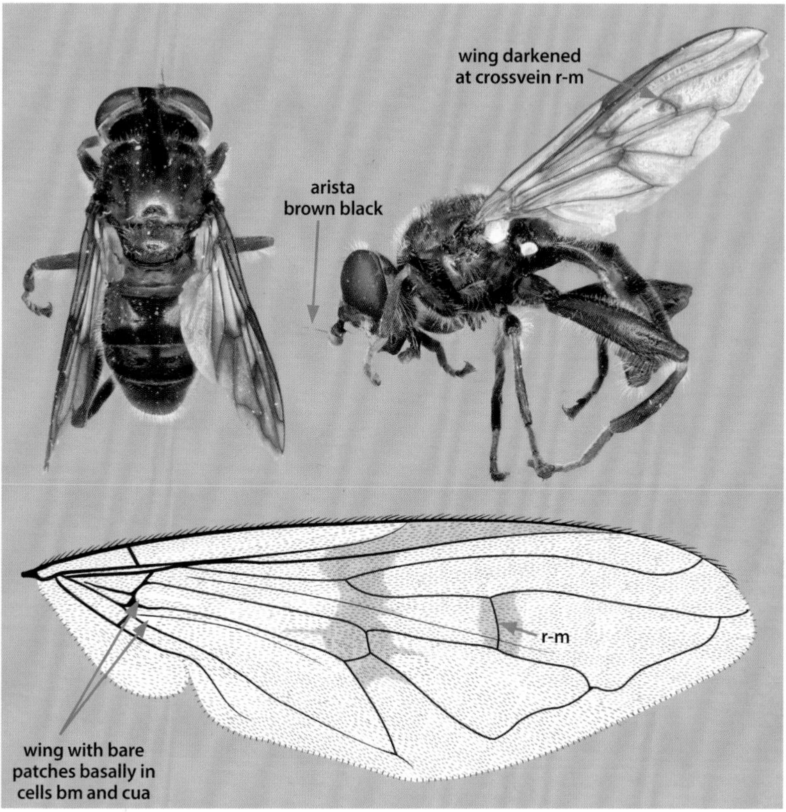
wing darkened
at crossvein r-m
arista
brown black
r-m
wing with bare
patches basally in
cells bm and cua

## *Chalcosyrphus (Xylotomima) sacawajeae*

wing darkened at
crossvein r-m
wing entirely
microtrichose
arista
brown black

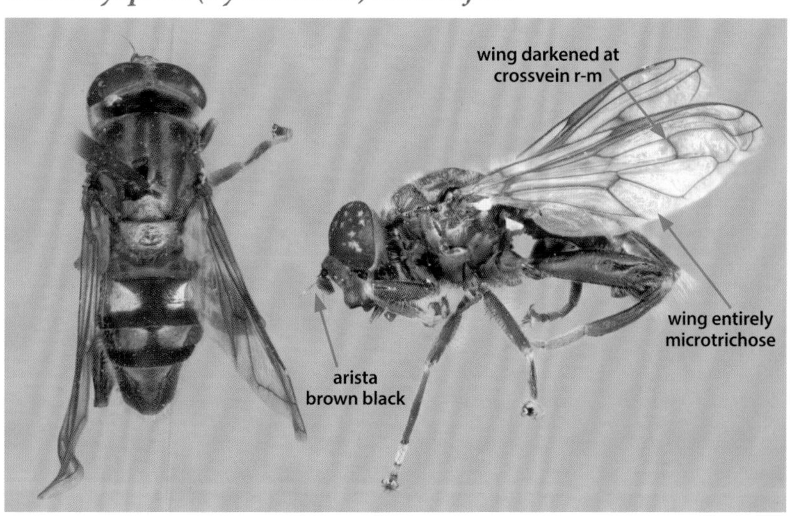

# *Xylota*

*Xylota* species are very similar to *Chalcosyrphus* species in general gestalt, habitat, and behavior. To distinguish them, look between the metalegs at the metasternum. This area is covered with very short pile in *Xylota* (considered "bare" in all of the literature and keys) and has long pile in *Chalcosyrphus*. There are 132 species of *Xylota*, 25 in the Nearctic and 17 in our area of coverage. For keys to species, see Curran (1941) and Shannon (1926b).

## *Xylota (Ameroxylota) flukei*

Fringeless Leafwalker

**SIZE RANGE:** 7.2–9.4 mm

**IDENTIFICATION:** Unlike in other *Xylota* species, the yellow spots on tergite 2 reach the anterior margin of the tergite. Also unique among our *Xylota* species, the subscutellar fringe is absent. **ABUNDANCE:** Rare. **FLIGHT TIMES:** Early June to mid-July. **NOTES:** Very little is known about this species. One specimen was collected in a dry sphagnum bog, another in *Corema* barrens.

subscutellar fringe absent

# *Xylota*

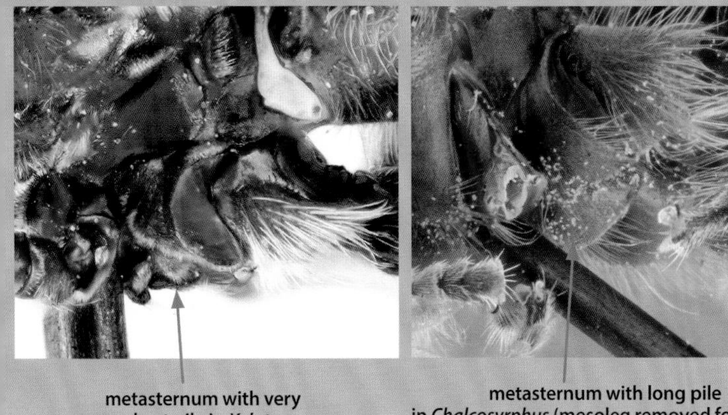

metasternum with very short pile in *Xylota*

metasternum with long pile in *Chalcosyrphus* (mesoleg removed for photo)

## *Xylota (Ameroxylota) flukei*

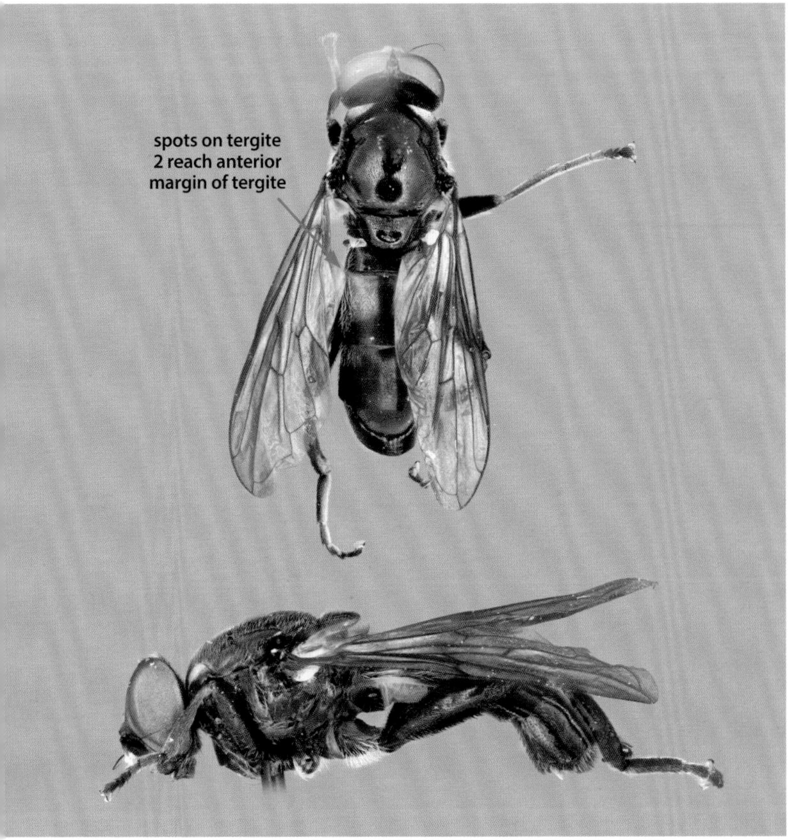

spots on tergite 2 reach anterior margin of tergite

## *Xylota (Xylota) flavitibia*

### Yellow-footed Leafwalker

**SIZE RANGE:** 11.0–14.5 mm

**IDENTIFICATION:** This and the following two *Xylota* species have almost entirely orange tergites 2 and 3 (with variation in the amount of orange on tergites 4 and 5). In *X. flavitibia* the first three tarsomeres of the metaleg are bright yellow. Beware of similar *Chalcosyrphus libo*. **ABUNDANCE:** Uncommon. **FLIGHT TIMES:** Late May to mid-August (to late September in the west). **NOTES:** They have been collected along creek and river margins and are known to visit flowers of *Fragaria* and *Heracleum*.

## *Xylota (Xylota) bicolor*

### Eastern Orange-tailed Leafwalker

**SIZE RANGE:** 13.8–14.7 mm

**IDENTIFICATION:** This orange-abdomened fly is similar to *X. flavitibia* and *X. segnis* but the first three tarsomeres of the metaleg are brown to black and the arista is yellow basally. **ABUNDANCE:** Rare. **FLIGHT TIMES:** Mid-May to early August. **NOTES:** *Xylota bicolor* has been collected visiting flowers of *Actaea*, *Rosa*, and *Viburnum*.

# *Xylota (Xylota) flavitibia*

♂
♀
tarsomeres 1–3 of metaleg bright yellow
tergites 2 and 3 orange

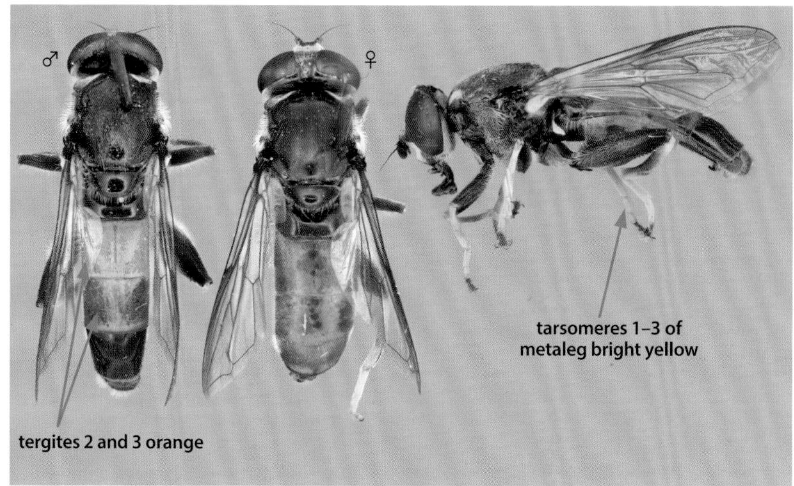

# *Xylota (Xylota) bicolor*

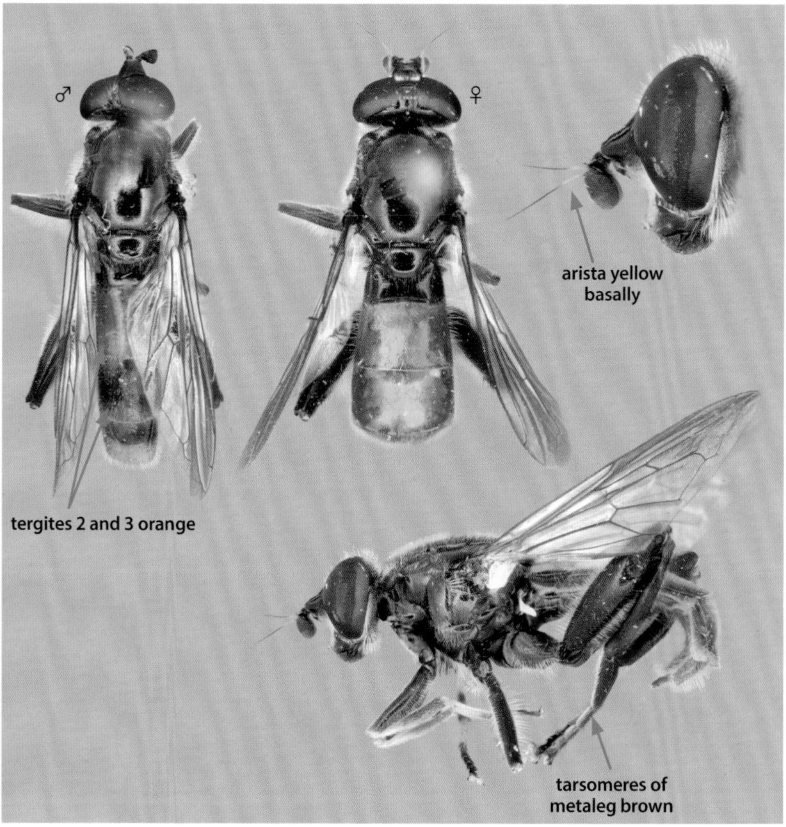
♂
♀
arista yellow basally
tergites 2 and 3 orange
tarsomeres of metaleg brown

## *Xylota (Xylota) segnis* Brown-toed Leafwalker

**SIZE RANGE:** 9.1–14.6 mm
**IDENTIFICATION:** Orange abdomen as in the preceding two species but with the first three tarsomeres of the metaleg brown dorsally and the arista brown black. **ABUNDANCE:** Uncommon. **FLIGHT TIMES:** Late May to mid-August. **NOTES:** This is our only *Xylota* species that is also found in Europe and Asia, and the distribution suggests that it is introduced in the Nearctic. Most of what we know about it is from Europe, where it is much more common. These flies are found in most types of coniferous and deciduous forests and are also common in gardens. They have been recorded visiting flowers of *Corylus*, *Crataegus*, *Heracleum*, *Solidago*, *Sorbus*, *Tilia*, and *Viburnum*. Adults may be found running on leaves of shrubs and on or near fallen trunks and stumps of both deciduous and coniferous trees. Larvae are found under bark of rotten stumps and logs of both deciduous trees and conifers, in damp rot holes; and in sap runs on living trees and in rotting plant debris (such as decomposing silage, rotting sawdust, and rotting potatoes). Larvae have even occasionally been found in rotting mammalian carcasses. The species overwinters as larvae.

## *Xylota (Xylota) angustiventris*
Two-spotted Leafwalker

**SIZE RANGE:** 12.2–14.4 mm
**IDENTIFICATION:** Male with two yellow-orange spots on tergite 2; tergite 3 typically dark, sometimes with small spots; with anepisternum entirely pale pilose. Female abdomen immaculate and dark. Alula extensively bare medially in both sexes (entirely microtrichose in all of the following species). **ABUNDANCE:** Uncommon. **FLIGHT TIMES:** Late May to mid-August. **NOTES:** *Xylota angustiventris* has been collected at flowers of *Ceanothus* and *Rhododendron*. One specimen was recorded from the edge of mixed woods and a shrubby lakeside meadow.

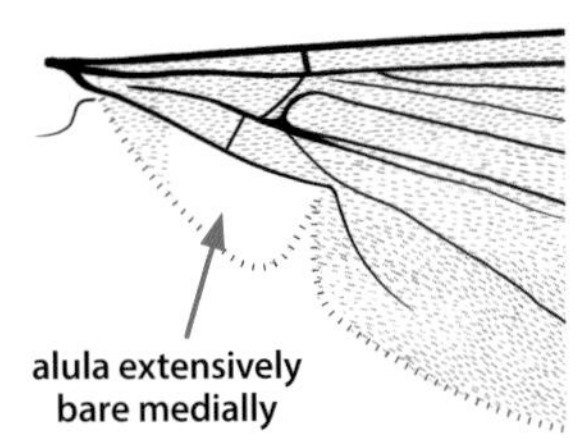

# *Xylota (Xylota) segnis*

# *Xylota (Xylota) angustiventris*

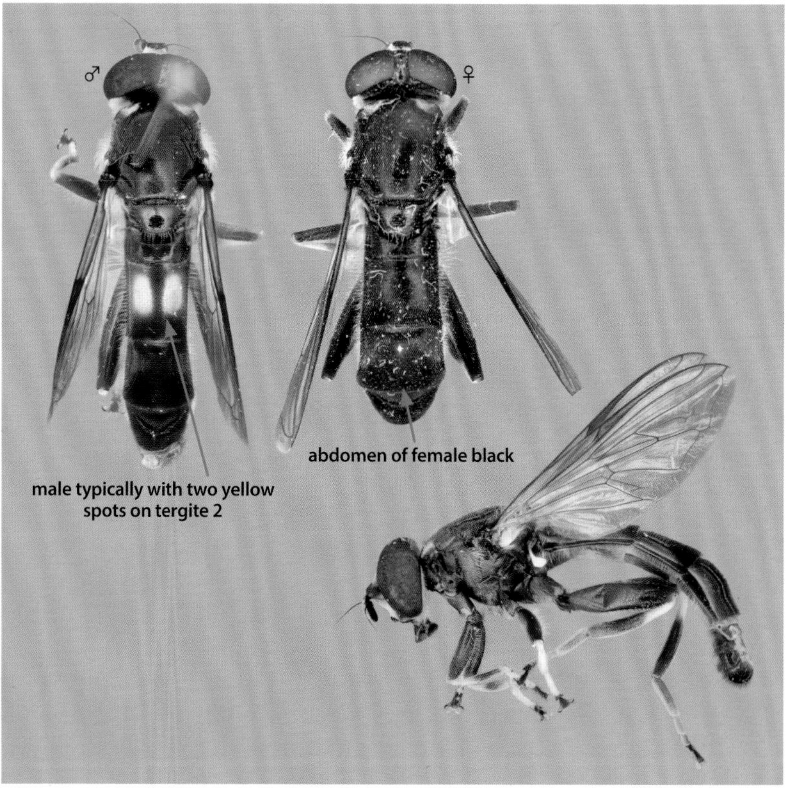

♂

♀

abdomen of female black

male typically with two yellow spots on tergite 2

# *Xylota (Xylota) subfasciata*

## Large-spotted Leafwalker

**SIZE RANGE:** 8.7–12.5 mm

**IDENTIFICATION:** Tergites 2 and 3 with yellow-orange spots. Spots are large but do not reach the anterior margin of the tergite (as in *X. flukei*) and are narrowly divided medially or meet narrowly. Alula densely microtrichose in this and all six of the yellow-spotted species that follow. Arista pubescent, with fine pile (bare in all other species—this character is difficult to see as pile is very short). Males with wing cell c entirely microtrichose. Apical two tarsomeres black. **ABUNDANCE:** Uncommon. **FLIGHT TIMES:** Early June to late August. **NOTES:** *Xylota subfasciata* has been recorded in hardwood and mixed forests and near rivers. Flowers visited include *Cicuta*, *Heracleum*, *Rubus*, *Sambucus*, *Tragopogon*, and *Viburnum*.

# *Xylota (Xylota) ejuncida* Polished Leafwalker

**SIZE RANGE:** 9.6–10.9 mm

**IDENTIFICATION:** Tergites 2 and 3 with yellow-orange spots; alula densely microtrichose; katepisternum entirely pollinose; pro- and mesotibiae yellow reddish. Male probasitarsomere without a long apicolateral seta, anepisternum entirely pale pilose; cell c bare at base. Female with pale pile on postalar callus, no more than three black bristles. **ABUNDANCE:** Rare. **FLIGHT TIMES:** Early July to mid-August in the north; early March to late October in the south. **NOTES:** *Xylota ejuncida* has been recorded in pine forests and cultivated cotton fields and has been noted visiting *Cornus* and *Prunus* flowers.

## *Xylota (Xylota) subfasciata*

♂

♀

tergites 2 and 3 with large yellow-orange spots

arista pubescent

## *Xylota (Xylota) ejuncida*

♂

♀

frontal triangle extensively shiny

tergites 2 and 3 with yellow-orange spots

mesonotal pile short and erect

pro- and mesotibiae yellow red

## *Xylota (Xylota)* undescribed species 78-3

### Black-backed Leafwalker

**SIZE RANGE:** 8.1–11.2 mm
**IDENTIFICATION:** This species is most similar to *Xylota ejuncida*. Mesonotum with mostly short, appressed pile; frontal triangle entirely pollinose basolaterally; tergite 4 black pilose on medial ⅔. Male probasitarsomere with a long apicolateral seta. **ABUNDANCE:** Uncommon. **FLIGHT TIMES:** Early May to mid-August. **NOTES:** Flowers visited include *Aronia*, *Chrysanthemum*, *Clematis*, *Cornus*, *Crataegus*, *Daucus*, *Hydrangea*, *Kalmia*, *Physocarpus*, *Rosa*, *Rubus*, *Sambucus*, *Spiraea*, *Thalictrum*, and *Viburnum*. Often found around moist soil or rotting logs.

## *Xylota (Xylota) quadrimaculata*

### Four-spotted Leafwalker

**SIZE RANGE:** 8.2–11.3 mm
**IDENTIFICATION:** Tergites 2 and 3 with yellow-orange spots; alula densely microtrichose; katepisternum shiny on ventral ⅔ (note that no other *Xylota* or *Chalcosyrphus* species show this character). Male probasitarsomere with a long apicolateral seta, anepisternum entirely pale pilose; cell c entirely microtrichose, frontal triangle usually entirely pollinose; wing with extensive bare area in front of $A_1$. Female with pale pile on postalar callus, no more than three black pili. **ABUNDANCE:** Common. **FLIGHT TIMES:** Early May to late September. **NOTES:** These flies have been found in a wide range of habitats including hardwood forests, bald cypress swamps, mixed woods, open black spruce forests, and meadows (often wet and scrubby). Flowers visited: *Begonia*, *Cornus*, *Diervilla*, *Rubus*, *Solidago*, and *Tragopogon*. Rarely recorded hilltopping.

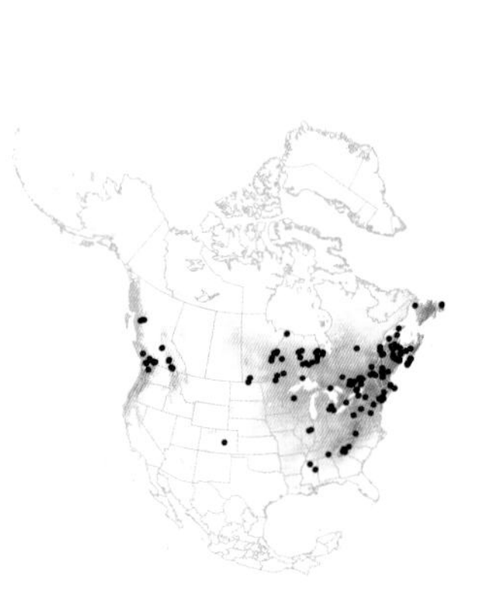

# *Xylota (Xylota)* undescribed species 78-3

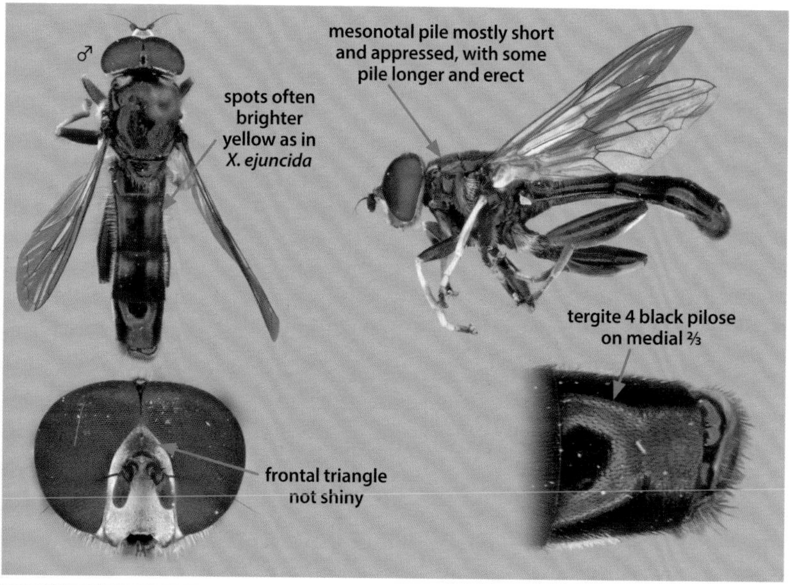

# *Xylota (Xylota) quadrimaculata*

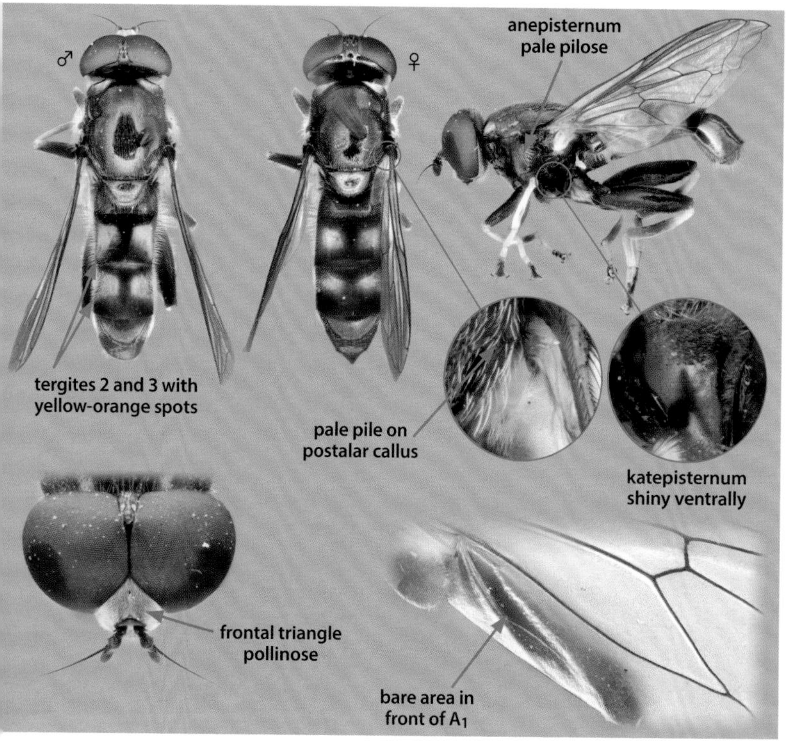

## *Xylota (Xylota) hinei* Hine's Leafwalker

**SIZE RANGE:** 7.8–12.9 mm
**IDENTIFICATION:** Tergites 2 and 3 with yellow-orange spots; alula densely microtrichose. Male probasitarsomere with a long apicolateral seta, anepisternum entirely pale pilose; cell c bare on basal ⅓ to entirely microtrichose; frontal triangle with bare area above antenna; wing with area in front of $A_1$ microtrichose. Female with black bristle-like pile on anterior ½ or more of postalar callus. **ABUNDANCE:** Uncommon. **FLIGHT TIMES:** Late May to early September. **NOTES:** Found in mixed woods, hardwood forest, and burnt spruce forest. Flowers visited include *Cicuta* and *Heracleum*.

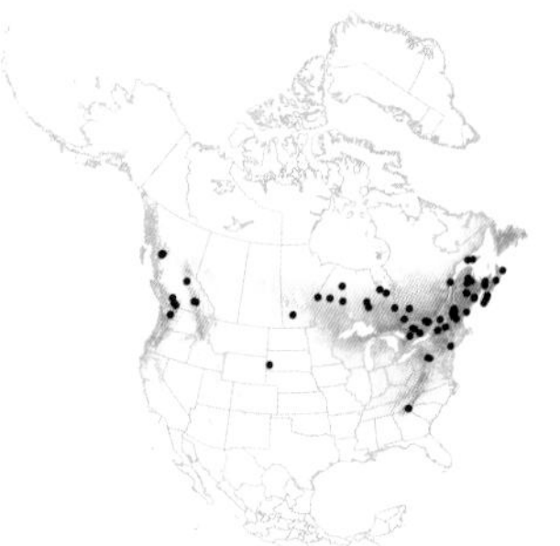

## *Xylota (Xylota) annulifera*

Longspine Leafwalker

**SIZE RANGE:** 7.5–11.8 mm
**IDENTIFICATION:** Tergites 2 and 3 with yellow-orange spots; metatrochanter with long spine (longer than its base is wide); alula densely microtrichose; katepisternum entirely pollinose. Male probasitarsomere with a long apicolateral seta, anepisternum entirely pale pilose; cell c entirely microtrichose, frontal triangle usually entirely pollinose; wing with extensive bare area in front of $A_1$; tarsi with apical three tarsomeres black. Female with pale pile on postalar callus, no more than three black pili; pro- and mesotibiae dark apically. **ABUNDANCE:** Uncommon. **FLIGHT TIMES:** Mid-May to mid-September. **NOTES:** Found in hardwood and mixed forests. They have been recorded visiting *Eutrochium*, *Ilex*, and *Sambucus* flowers. This is the only *Xylota* species to regularly visit hilltops for mating.

# Xylota (Xylota) hinei

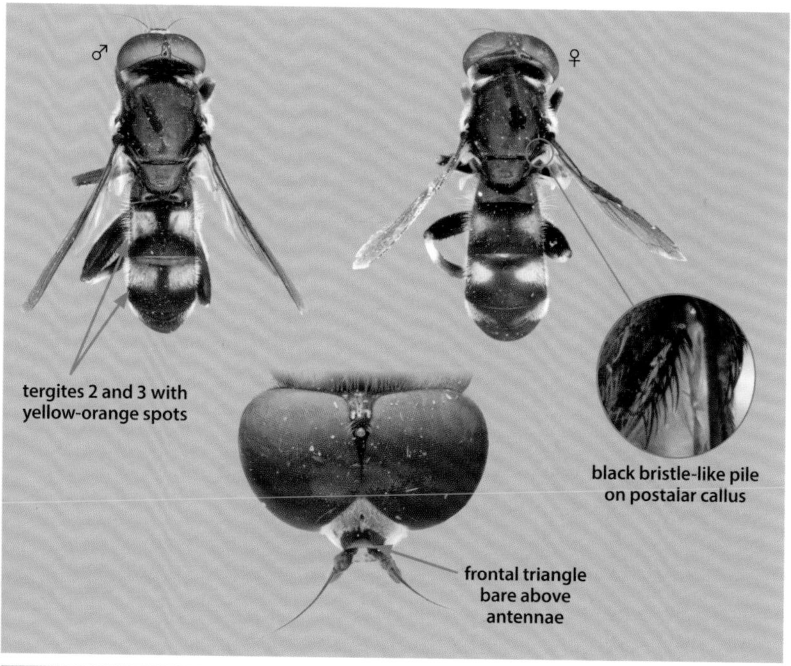

♂ ♀

tergites 2 and 3 with yellow-orange spots

black bristle-like pile on postalar callus

frontal triangle bare above antennae

# Xylota (Xylota) annulifera

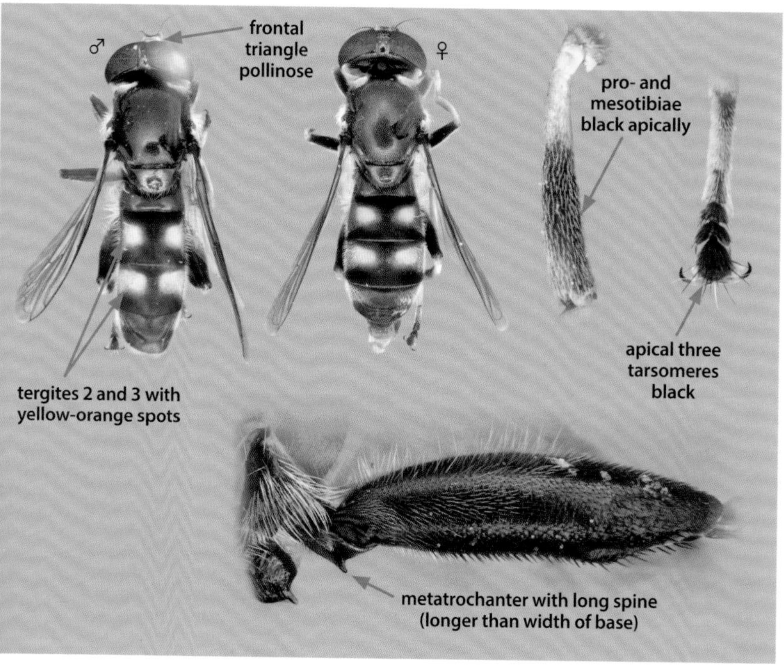

♂ ♀

frontal triangle pollinose

pro- and mesotibiae black apically

tergites 2 and 3 with yellow-orange spots

apical three tarsomeres black

metatrochanter with long spine (longer than width of base)

## *Xylota (Xylota) confusa* Confusing Leafwalker

**SIZE RANGE:** 9.3–12.6 mm
**IDENTIFICATION:** Male with tergites 2 and 3 with yellow-orange spots; female with tergites 2 and 3 dark, with dark, shiny spots; alula densely microtrichose. Male probasitarsomere with a long apicolateral seta, anepisternum entirely pale pilose; cell c bare on basal ⅓ or less; pro- and mesotibiae black apically; frontal triangle shiny above antenna; arista short, shorter than facial width; katepisternum sometimes entirely pollinose; apical two tarsomeres black. Female frons with large rectangular area above antennae shiny. **ABUNDANCE:** Uncommon. **FLIGHT TIMES:** Mid-May to mid-August. **NOTES:** This appears to be primarily a wetland species with records from oak-poplar-hemlock swamps, maple-elm floodplains, and floodplains associated with tamarack fens and open bogs. More general records are from coniferous forests (including burnt spruce forest) and mixed forests. Flowers visited include *Heracleum*, and *Viburnum*.

## *Xylota (Xylota)* undescribed species 78-1
### Appalachian Leafwalker

**SIZE RANGE:** 9.7–13.7 mm
**IDENTIFICATION:** This species is most similar to *Xylota confusa*. Cell c bare on basal ⅔ or more; arista long (longer than facial width); frontal triangle entirely pollinose; probasitarsus without long pile. Male probasitarsomere with a long apicolateral seta. **ABUNDANCE:** Uncommon. **FLIGHT TIMES:** Early May to late September. **NOTES:** Adults are often found on moist soil in wet forests or around potential oviposition sites (such as excavated stumps, dry logs, and fresh cut pine logs). Flowers visited include *Actaea*, *Clematis*, *Cornus*, *Crataegus*, *Ilex*, *Ranunculus*, *Rhus*, *Rosa*, *Rubus*, *Sambucus*, *Spiraea*, *Thalictrum*, and *Viburnum*.

# *Xylota (Xylota) confusa*

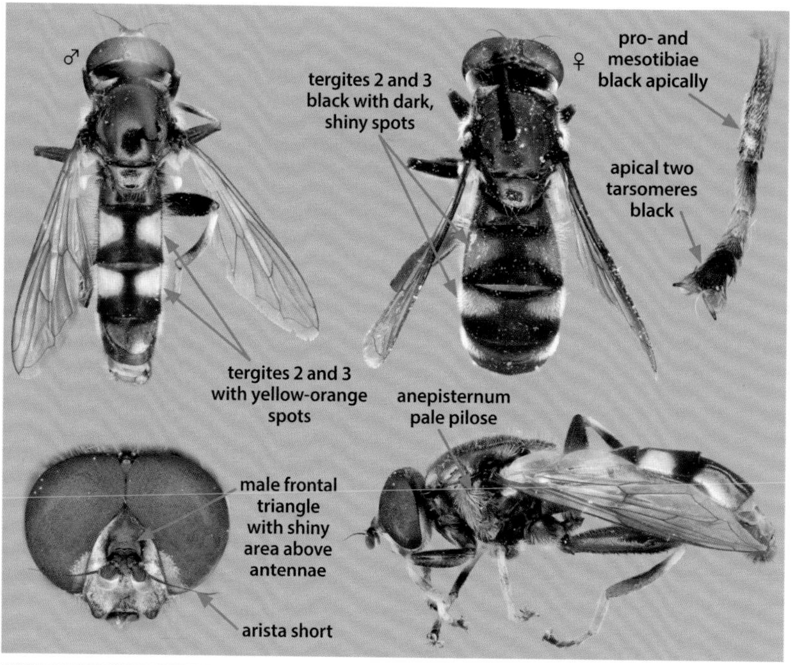

♂

tergites 2 and 3 black with dark, shiny spots

♀

pro- and mesotibiae black apically

apical two tarsomeres black

tergites 2 and 3 with yellow-orange spots

anepisternum pale pilose

male frontal triangle with shiny area above antennae

arista short

# *Xylota (Xylota)* undescribed species 78-1

♂

male frontal triangle entirely pollinose

arista long

## *Xylota (Xylota) naknek* Naknek Leafwalker

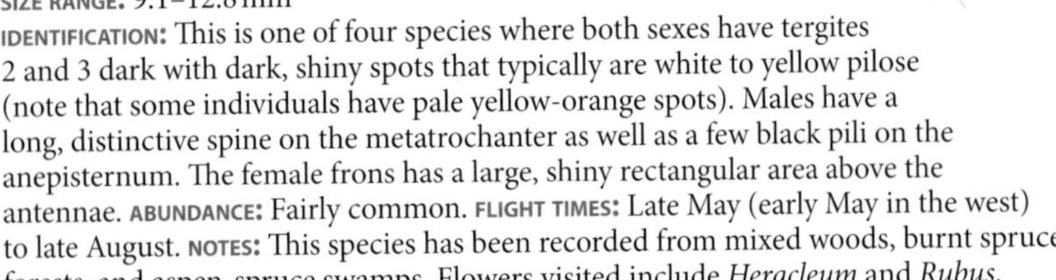

**SIZE RANGE:** 9.1–12.8 mm
**IDENTIFICATION:** This is one of four species where both sexes have tergites 2 and 3 dark with dark, shiny spots that typically are white to yellow pilose (note that some individuals have pale yellow-orange spots). Males have a long, distinctive spine on the metatrochanter as well as a few black pili on the anepisternum. The female frons has a large, shiny rectangular area above the antennae. **ABUNDANCE:** Fairly common. **FLIGHT TIMES:** Late May (early May in the west) to late August. **NOTES:** This species has been recorded from mixed woods, burnt spruce forests, and aspen-spruce swamps. Flowers visited include *Heracleum* and *Rubus*.

## *Xylota (Xylota) tuberculata*
Short-spined Leafwalker

**SIZE RANGE:** 9.7–12.8 mm
**IDENTIFICATION:** This species is very similar to *X. naknek*. Males differ by having a short, indistinct spine on the metatrochanter. Females are indistinguishable from *X. naknek*. **ABUNDANCE:** Rare. **FLIGHT TIMES:** Late May to late August (one outlier in early October). **NOTES:** *Xylota tuberculata* is a wetland species and has been recorded from maple-elm floodplains, in swampy oak-poplar-hemlock woods, and on beaver lodges in wetlands. It has been found visiting *Sambucus* and *Symphyotrichum* flowers.

# *Xylota (Xylota) naknek*

♂ ♀

tergites 2 and 3 black with dark, shiny spots or pale yellow-orange spots

long spine on male metatrochanter

anepisternum partly black pilose

# *Xylota (Xylota) tuberculata*

♂ ♀

tergites 2 and 3 black with dark, shiny markings

short, indistinct spine on metatrochanter

anepisternum partly black pilose

## *Xylota (Xylota) flavifrons* Northern Leafwalker

**SIZE RANGE:** 9.4–14.7 mm

**IDENTIFICATION:** Tergites 2 and 3 black with dark, shiny spots, which typically are white to yellow pilose. Male probasitarsomere without a long apicolateral seta, anepisternum entirely pale pilose; postalar callus mainly pale pilose. Female frons extensively pollinose, with small triangular area above antennae shiny. **ABUNDANCE:** Uncommon. **FLIGHT TIMES:** Late May to mid-August. **NOTES:** *Xylota flavifrons* has been collected in poplar parkland, clearings in spruce forest, beaver meadows, and on beaver lodges. There is a single floral record from *Fragaria*.

## *Xylota (Xylota) ouelleti* Black-haired Leafwalker

**SIZE RANGE:** 8.5–9.6 mm

**IDENTIFICATION:** This species is very similar to *X. flavifrons* and males can be differentiated only by the mainly black-pilose postalar callus. Females are indistinguishable from *X. flavifrons*. **ABUNDANCE:** Rare. **FLIGHT TIMES:** Late May to late July. **NOTES:** This species has been recorded in wooded bogs, open bogs, *Larix* bogs, and Christmas tree plantations.

## *Xylota (Xylota) flavifrons*

♂

♀

postalar callus mainly pale pilose

tergites 2 and 3 black with dark, shiny spots

♂

female frons extensively pollinose

## *Xylota (Xylota) ouelleti*

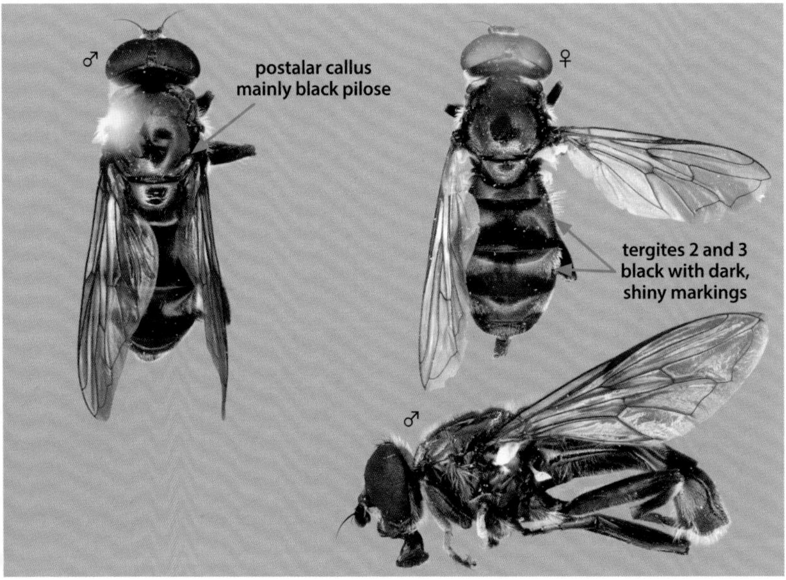

♂

postalar callus mainly black pilose

♀

tergites 2 and 3 black with dark, shiny markings

♂

# *Tropidia*

Four of the world's 22 species of *Tropidia* occur in our area. Nine species are found in the Nearctic region. *Tropidia* species have an enlarged metafemur that has an apicoventral triangular process. The face is carinate, coming to a sharp ridge medially. *Tropidia* larvae feed on rotting organic matter in wet environments. The last key to the Nearctic species of this genus was by Shannon (1926b).

## *Tropidia albistylum* Yellow-thighed Thickleg

**SIZE RANGE:** 10.9–13.4 mm
**IDENTIFICATION:** This species is recognized by the yellow-orange base of the metafemur (black in other *Tropidia*) and orange band on the metatibia (black in other species). **ABUNDANCE:** Rare. **FLIGHT TIMES:** Mid-March to mid-November (mid-May to late September in the north). **NOTES:** The only habitat data available for this species is one record from upland forest. Flowers visited include *Baccharis*, *Prunus*, and *Symphyotrichum*.

## *Tropidia quadrata* Common Thickleg

**SIZE RANGE:** 10.0–13.7 mm
**IDENTIFICATION:** This species is recognized by the combination of an entirely black metafemur and a yellow face with a black medial stripe. **ABUNDANCE:** Common. **FLIGHT TIMES:** Late May (mid-April in the south) to mid-October. **NOTES:** *Tropidia quadrata* is never found far from water (records are from fens, beaver meadows, and alluvial forests). Adults wander from wet areas to edges of fields and openings in forests. Flowers visited include *Barbarea*, *Ceanothus*, *Eupatorium*, *Nyssa*, and *Persicaria*.

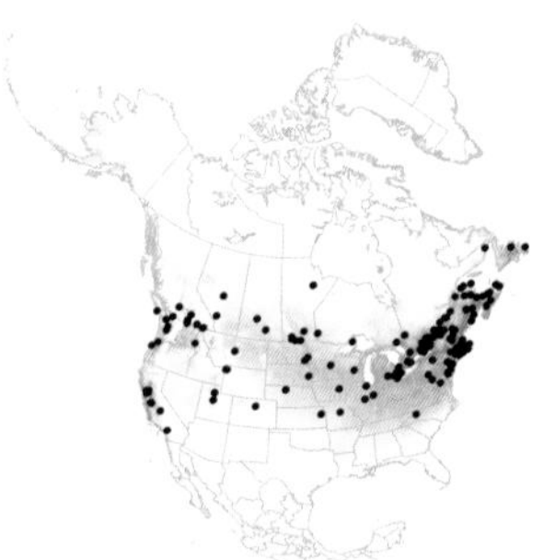

metatibia entirely black

# Tropidia

## Tropidia albistylum

## Tropidia quadrata

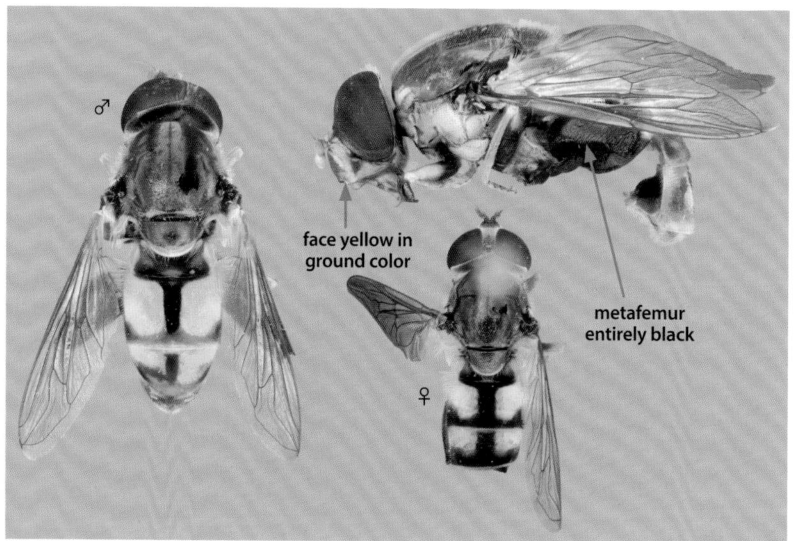

## *Tropidia calcarata* Lily-loving Thickleg

**SIZE RANGE:** 10.4–12.5 mm
**IDENTIFICATION:** This and the following species have black faces (in ground color) and black metafemora. *Tropidia calcarata* can be distinguished from *T. mamillata* by the gray pollinosity on tergite 4. **ABUNDANCE:** Rare. **FLIGHT TIMES:** Early May to mid-October. **NOTES:** This is a wetland species, apparently specializing on water lilies. Adults nectar on *Nuphar* (known as spatterdock or yellow pond-lily), and larvae have been reared from rotting rhizomes of *N. advena*.

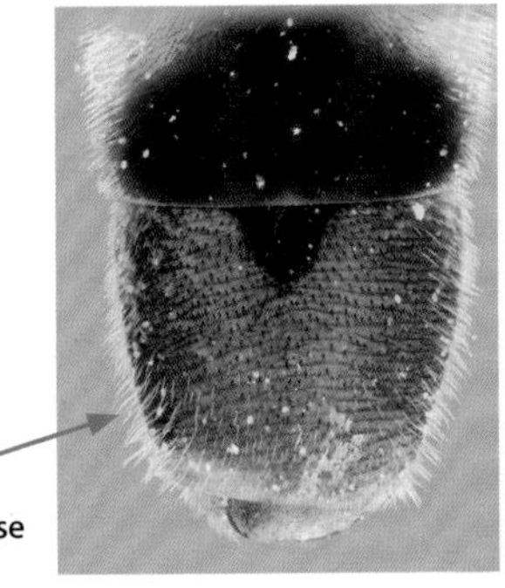


## *Tropidia mamillata* Shiny Thickleg

**SIZE RANGE:** 8.2–9.7 mm
**IDENTIFICATION:** As with *Tropidia calcarata*, this species has a black face (in ground color) and black metafemur. It can be distinguished from *T. calcarata* by the shiny brown appearance of tergite 4. **ABUNDANCE:** Rare. **FLIGHT TIMES:** Late April to mid-September. **NOTES:** This species appears to be a tallgrass prairie specialist. As such, it is likely of conservation concern.

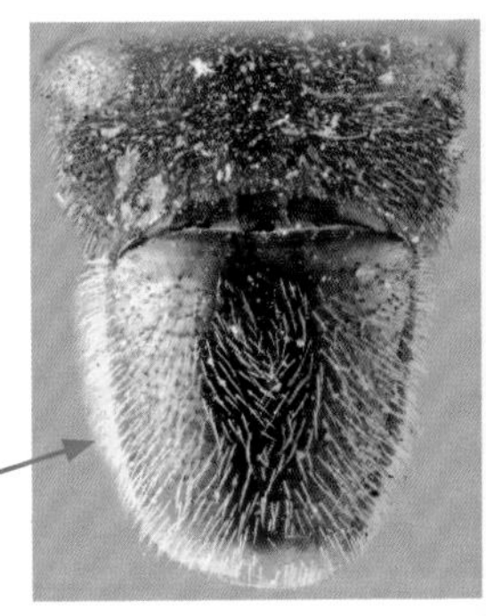

## Tropidia calcarata

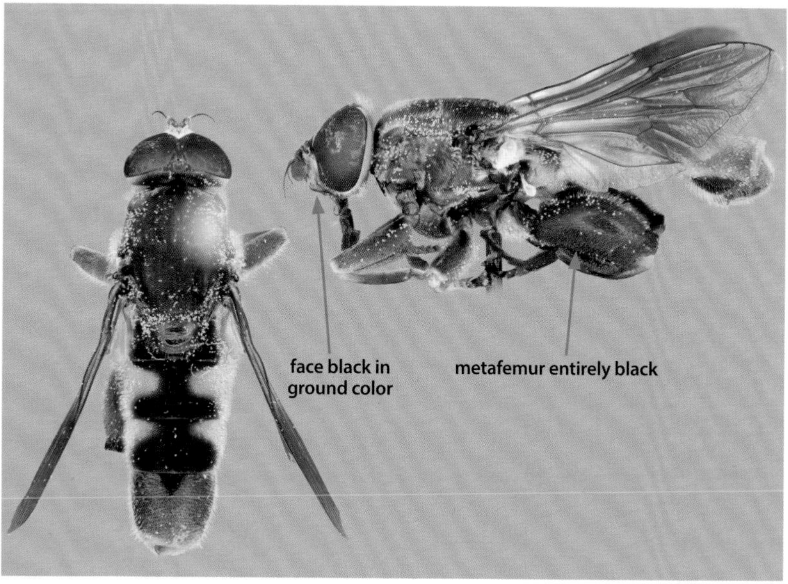

face black in ground color

metafemur entirely black

## Tropidia mamillata

metafemur entirely black

face black in ground color

# *Syritta*

*Syritta* is a predominantly Afrotropical genus of flower flies. There are 61 species currently recognized, 60 of them treated in a world revision by Lyneborg and Barkemeyer (2005). Our two Nearctic species are introduced. This genus is easily identified by its distinctive metafemora. *Syritta* larvae feed on decaying plant material (compost, manure, silage).

## *Syritta pipiens* Common Compost Fly

**SIZE RANGE:** 6.5–9.5 mm

**IDENTIFICATION:** Distinguished from *S. flaviventris* by the absence of gray-pollinose spots on the thorax and the lack of a peg on the inner metafemur. Tergites 2 and 3 have rounded markings, the spurious vein is present, and there is a brownish stripe near the notopleuron. **ABUNDANCE:** Abundant. **FLIGHT TIMES:** Mid-April to mid-November (from mid-February in the south). **NOTES:** *Syritta pipiens* has spread over much of the world and has been established in North America since about 1895 (first records from Ontario and Pennsylvania). They are abundant along water bodies as well as in farmland and urban areas and visit a wide range of flowers including *Achillea*, *Allium*, *Anaphalis*, *Begonia*, *Cardamine*, *Chrysanthemum*, *Cirsium*, *Convolvulus*, *Coreopsis*, *Cornus*, *Crataegus*, *Daucus*, *Doellingeria*, *Epilobium*, *Euphorbia*, *Euthamia*, *Galium*, *Leontodon*, *Leucanthemum*, *Persicaria*, *Potentilla*, *Prunus*, *Ranunculus*, *Rhexia*, *Rosa*, *Senecio*, *Solidago*, *Sorbus*, *Spiraea*, *Symphyotrichum*, *Thalictrum*, and *Tussilago*.

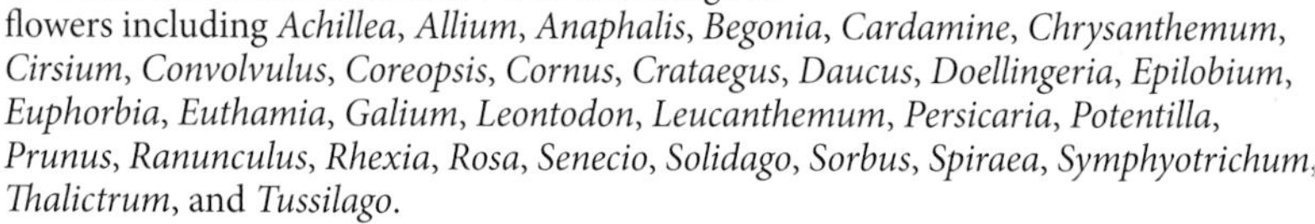

## *Syritta flaviventris* Peg-legged Compost Fly

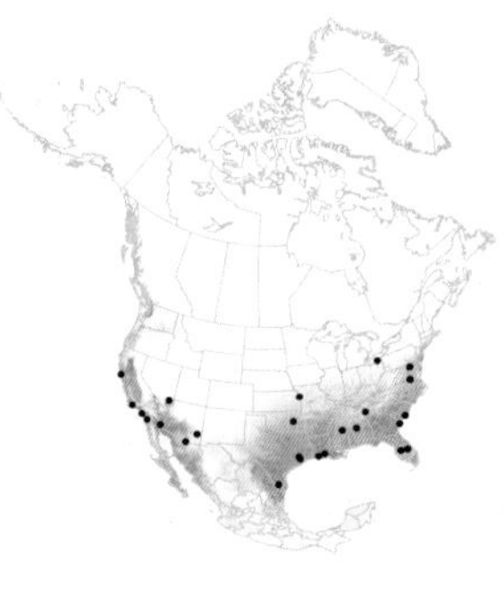

**SIZE RANGE:** 6.9–9.4 mm

**IDENTIFICATION:** This is one of the only flower flies in our region with no spurious vein. It can be distinguished from *S. pipiens* by the presence of two white-pollinose spots on the thorax at the transverse suture and by the cone-shaped peg on the inner metafemur. Additionally, the markings on tergites 2 and 3 are quadrate and the notopleural stripe is white. **ABUNDANCE:** Rapidly increasing range and abundance. **FLIGHT TIMES:** Mid-June to late October (early March to mid-December in the south). **NOTES:** *Syritta flaviventris* was discovered in the USA on 24 March 1988 at Bentsen-Rio Grande Valley State Park in Texas. It is found near water where adults can be found flying low among sparse vegetation. The specimen on the right is visiting *Gnaphalium*.

## *Syritta pipiens*

thorax without pollinose spots

metafemur enlarged in both species but missing basal peg in *S. pipiens*

spots on tergites 2 and 3 rounded

## *Syritta flaviventris*

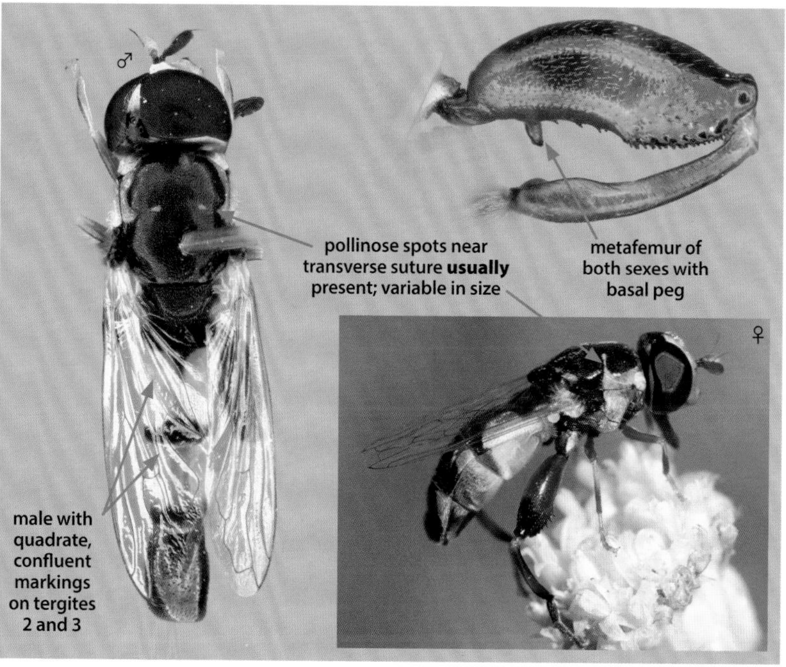

♂

pollinose spots near transverse suture **usually** present; variable in size

metafemur of both sexes with basal peg

♀

male with quadrate, confluent markings on tergites 2 and 3

# *Pelecocera*

Mengual *et al.* (2015) revised *Pelecocera* and provided a useful summary of known biology for this poorly known genus. *Pelecocera* species are small with a flat, elongate abdomen and are most easily characterized by the broad and basally expanded antennal flagellum. Their larval biology is unknown but species tend to frequent dry habitats like dunes or open ground in conifer forests. They are often found on yellow composites such as *Hieracium* and *Hypochaeris* along forest edges, but also visit other genera of plants. *Chamaesyrphus* is sometimes treated as a subgenus of *Pelecocera* but is treated as a genus by Mengual *et al.* There are eight species of *Chamaesyrphus* (two in the western Nearctic) and four species of *Pelecocera* (only one in the Nearctic).

## ***Pelecocera pergandei*** Eastern Bighorn Fly

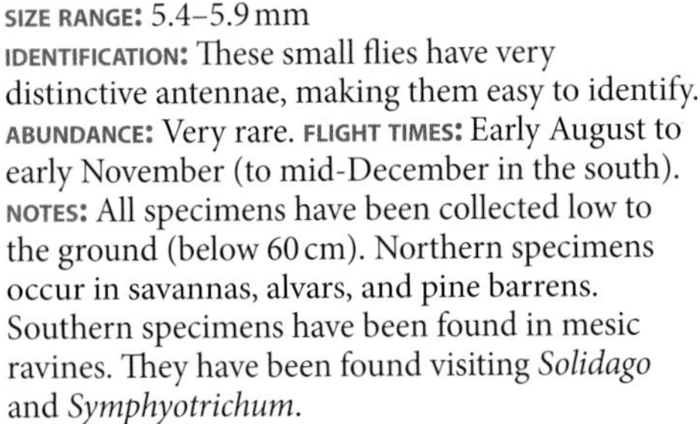

**SIZE RANGE:** 5.4–5.9 mm
**IDENTIFICATION:** These small flies have very distinctive antennae, making them easy to identify. **ABUNDANCE:** Very rare. **FLIGHT TIMES:** Early August to early November (to mid-December in the south). **NOTES:** All specimens have been collected low to the ground (below 60 cm). Northern specimens occur in savannas, alvars, and pine barrens. Southern specimens have been found in mesic ravines. They have been found visiting *Solidago* and *Symphyotrichum*.

# *Rhingia*

Only one of the 47 species of *Rhingia* occurs in the Nearctic. Larvae are likely all associated with dung of large herbivores. For example, larvae of European *R. campestris* live in cow dung, while *R. rostrata* eggs are laid on the underside of leaves overhanging dung. Larvae hatch and drop to the dung.

## ***Rhingia nasica*** American Snout Fly

**SIZE RANGE:** 6.0–9.0 mm
**IDENTIFICATION:** This species is recognized by the long snout that is directed forward. **ABUNDANCE:** Common. **FLIGHT TIMES:** Late April to mid-October. **NOTES:** Most specimens of this species have been collected in hardwood forests. A few records are also from bogs and sphagnum-dominated fens. Flowers visited include *Centaurea*, *Euthamia*, *Galax*, *Geranium*, *Impatiens*, *Prenanthes*, *Rubus*, *Solidago*, *Symphyotrichum*, and *Viola*.

# *Pelecocera pergandei*

# *Rhingia nasica*

*Rhingia* species tend to specialize on tubular flowers where their long mouthparts can reach the nectar.

face produced into snout

long mouthparts

# *Brachyopa*

*Brachyopa* species are mostly orange-brown syrphids. They look similar to *Hammerschmidtia* (sometimes considered a subgenus) but are stubbier and lack strong bristles on the thorax and scutellum. *Brachyopa* larvae occur in sap runs or in pools of decaying sap under bark. There are 42 recognized species of *Brachyopa*: 29 Palearctic, 1 Indomalayan, and 12 Nearctic. Eight species occur within the area of the guide. Finding *Brachyopa* species often requires a special effort. They are small and rather inconspicuous. Even when they visit hilltops or flowers they can be hard to find. Search for them in forests by looking for trees with bleeding wounds, particularly when the wound is in the sun. They are also often found around recently fallen trees where they sit inconspicuously on the trunks. They may also sometimes be found in dappled sunlight on foliage along small streams. Most species are likely much more common than the records indicate because of the special effort required to find them. The last key to *Brachyopa* was published by Curran (1922).

Some *Brachyopa* and *Hammerschmidtia* species can appear quite dark. Here is a typical *Brachyopa perplexa* specimen (left) and a very dark *Hammerschmidtia rufa* specimen (below).

## *Brachyopa* undescribed species 78-2

### Plain-winged Sapeater

**SIZE RANGE:** 4.8–6.3 mm

**IDENTIFICATION:** Katepisternum pilose on dorsal ½. Arista pubescent. Abdomen and legs extensively yellow. Scutellum yellow pilose.

**ABUNDANCE:** Uncommon. **FLIGHT TIMES:** Mid-May to late July (from early March in the southwest).

**NOTES:** Habitats noted on specimen labels include black spruce bog and alder-poplar-spruce forest. Flowers visited include *Heracleum* and *Rubus*. This species is part of the *B. perplexa* complex of five species.

## *Brachyopa* and *Hammerschmidtia*

## *Brachyopa* undescribed species 78-2

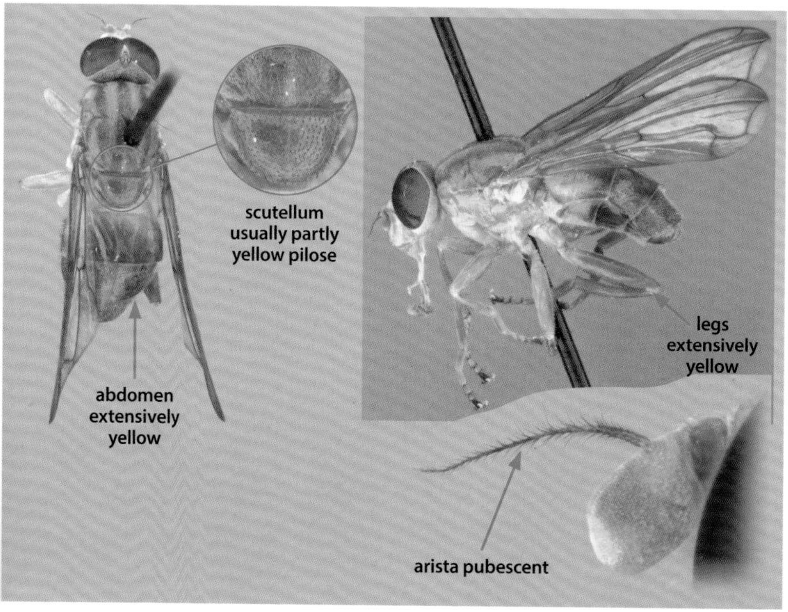

## *Brachyopa flavescens* Yellow Sapeater

**SIZE RANGE:** 4.1–7.0 mm
**IDENTIFICATION:** Katepisternum bare. Arista bare. Abdomen and legs extensively yellow. Scutellum yellow pilose. Thorax variable, entirely yellow or gray and black pollinose. **ABUNDANCE:** Fairly common. **FLIGHT TIMES:** Early May to early July. **NOTES:** This is a relatively common hardwood forest species. Some data suggest that they may prefer maple-birch forest. One specimen was collected visiting *Physocarpus* flowers.

## *Brachyopa notata* Black-banded Sapeater

**SIZE RANGE:** 5.5–7.0 mm
**IDENTIFICATION:** Katepisternum bare. Arista plumose. Abdomen mostly orange, with narrow black apical band and medial stripe on tergites. Black pile on scutellum. **ABUNDANCE:** Fairly common. **FLIGHT TIMES:** Late April to mid-August. **NOTES:** This is primarily a coniferous forest species (bogs, hemlock and spruce forests). Flowers visited include *Anemone*, *Cornus*, *Heracleum*, *Prunus*, and *Rubus*.

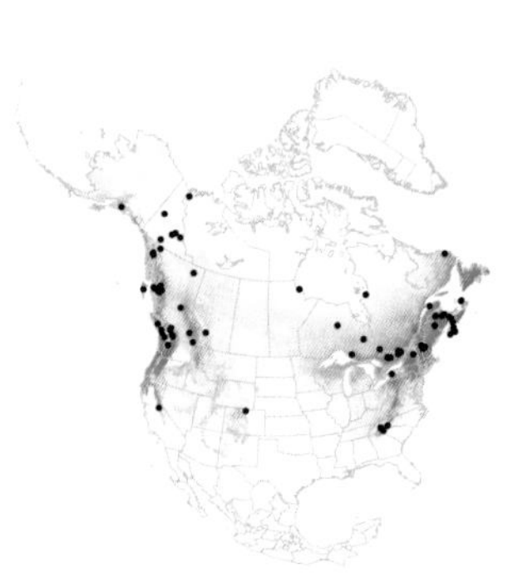

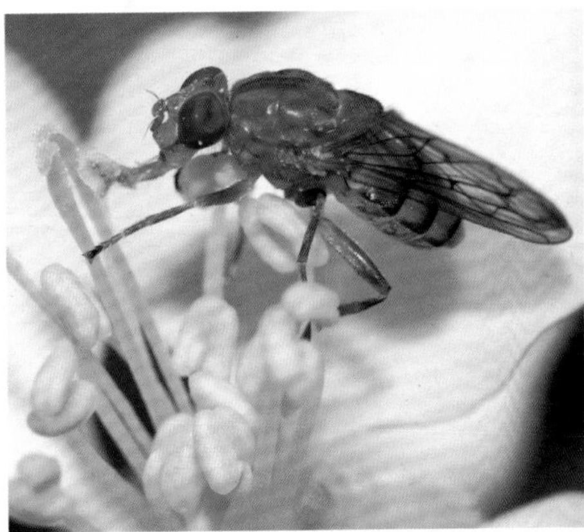

## Brachyopa flavescens

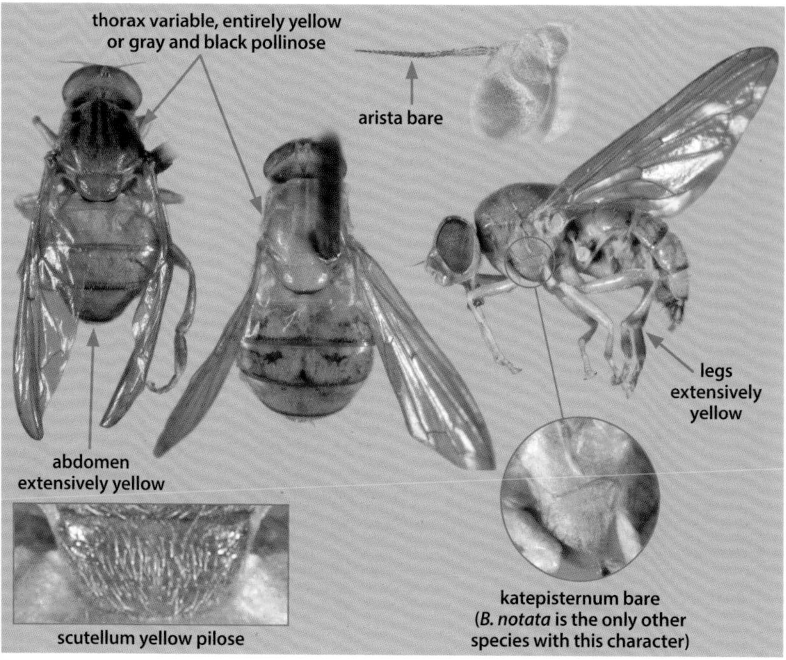

thorax variable, entirely yellow or gray and black pollinose

arista bare

legs extensively yellow

abdomen extensively yellow

katepisternum bare (*B. notata* is the only other species with this character)

scutellum yellow pilose

## Brachyopa notata

arista plumose

abdomen mostly orange, with narrow black apical band and medial stripe on tergites

scutellum black pilose

katepisternum bare (*B. flavescens* is the only other species with this character)

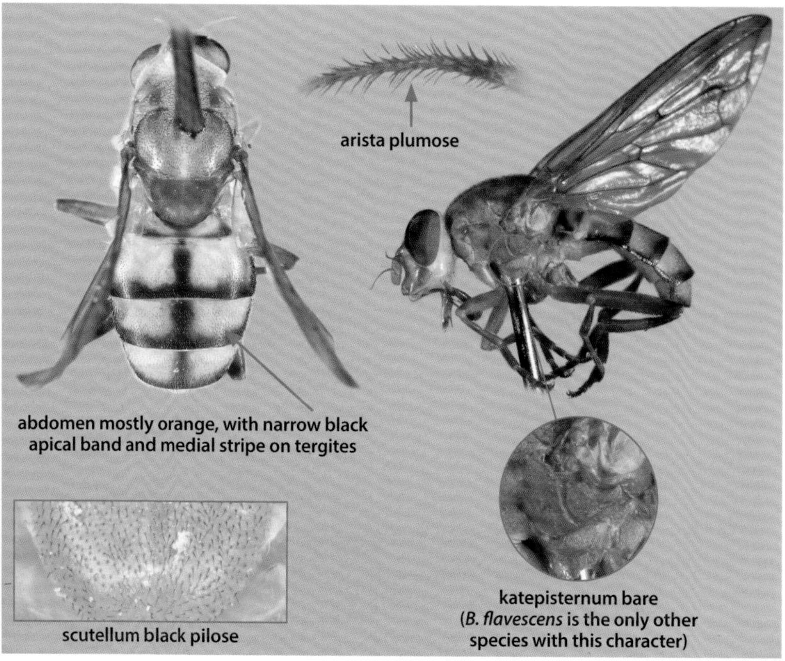

## *Brachyopa perplexa* Hairy-striped Sapeater

**SIZE RANGE:** 5.8–7.7 mm

**IDENTIFICATION:** Katepisternum pilose on dorsal ½. Arista pubescent. Abdomen mostly orange with narrow black apical band and medial black stripe on tergites (similar to *B. notata*). Black pile on scutellum. **ABUNDANCE:** Uncommon. **FLIGHT TIMES:** Early May to late June with a single record for late July. **NOTES:** This is a hardwood forest species. Several specimens have been collected at sap on maple trunks. Flowers visited include *Aruncus*, *Physocarpus*, and *Saxifraga*.

## *Brachyopa daeckei* Black-tailed Sapeater

**SIZE RANGE:** 6.1–6.9 mm

**IDENTIFICATION:** Katepisternum pilose on dorsal ½. Scutellum pale pilose. Abdomen uniformly reddish brown to black. **ABUNDANCE:** Rare. **FLIGHT TIMES:** Early April to mid-June. **NOTES:** Several specimens were captured at hilltops in mixed, predominantly deciduous forests.

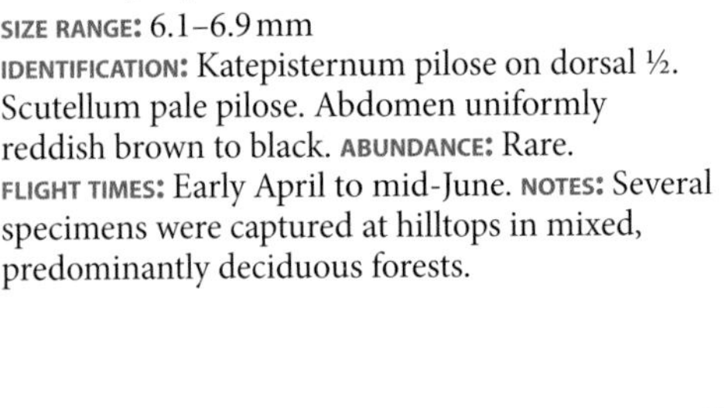

# Brachyopa perplexa

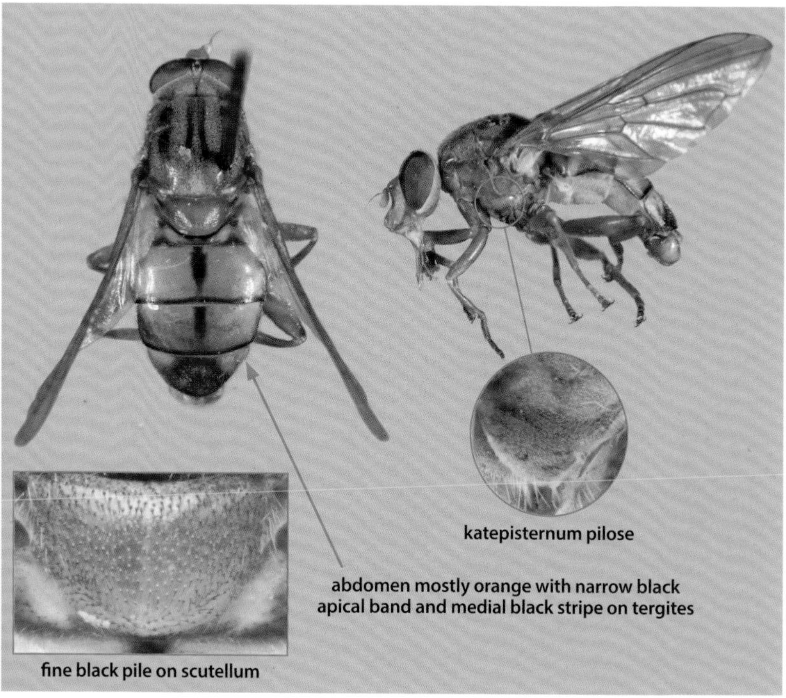

katepisternum pilose

abdomen mostly orange with narrow black apical band and medial black stripe on tergites

fine black pile on scutellum

# Brachyopa daeckei

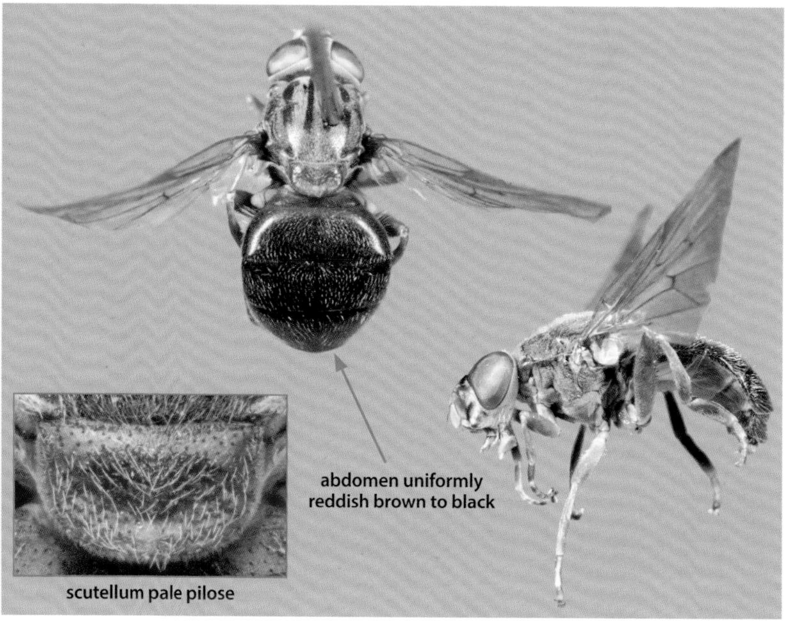

abdomen uniformly reddish brown to black

scutellum pale pilose

## *Brachyopa diversa* Pale-banded Sapeater

**SIZE RANGE:** 6.8–8.0 mm

**IDENTIFICATION:** Katepisternum pilose on dorsal ½. Abdomen with pollinose markings. Tergite 2 is pale on the posterior margin, dark along the anterior margin. Scutellum with black pile. **ABUNDANCE:** Rare. **FLIGHT TIMES:** Late April to late June. **NOTES:** Several specimens were captured at hilltops in mixed, predominantly deciduous forests. Many were found at the same locations as *B. daeckei*.

## *Brachyopa vacua* Yellow-spotted Sapeater

**SIZE RANGE:** 7.4–10.0 mm

**IDENTIFICATION:** Katepisternum pilose on dorsal ½. Scutellum pale pilose. Large yellow markings on tergite 2; tergite 3 usually with basolateral corner yellow. **ABUNDANCE:** Rare. **FLIGHT TIMES:** Early May to early July (mid-March to late July in southern parts of range). **NOTES:** One specimen was collected on sap on a maple trunk and many were collected at bleeding elms. Flowers visited include *Aruncus*, *Cornus*, and *Heracleum*.

## *Brachyopa* undescribed species 17-5

### Somber Sapeater

**SIZE RANGE:** 5.5 mm

**IDENTIFICATION:** Arista pubescent but not plumose. Male flagellum with distinct sensory pit (*B. daeckei* without). Wing hyaline. Katepisternum pilose on dorsal ½. Scutellum pale pilose; completely pollinose. Abdomen uniformly light brown; completely pollinose. **ABUNDANCE:** Very rare. **FLIGHT TIMES:** Mid-May. **NOTES:** This species is known from a single specimen collected in Ottawa on 14 May 1986 by Jeff Cumming.

## *Brachyopa diversa*

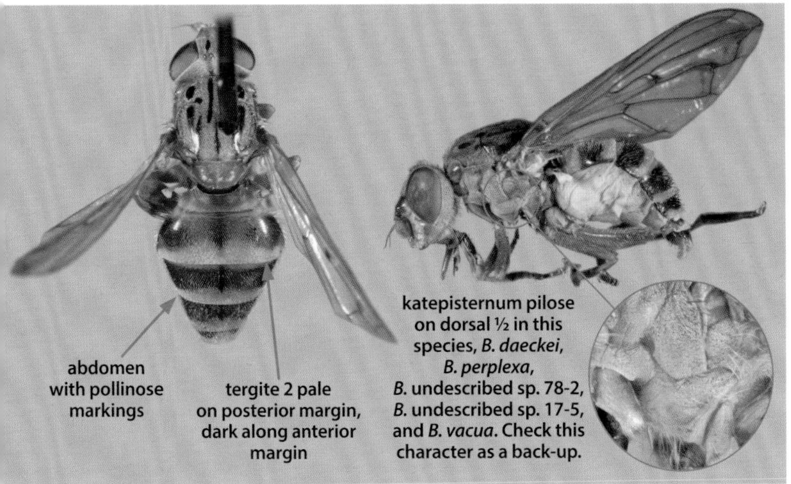


## *Brachyopa vacua*

## *Brachyopa* undescribed species 17-5

# *Hammerschmidtia*

This genus has been considered a subgenus of *Brachyopa*, but our unpublished evidence shows that they are not closely related to each other despite appearing similar. *Hammerschmidtia* can be distinguished from *Brachyopa* by the presence of strong bristles on the thorax and scutellum, short, strong black spines on the anterior surface of the metatibia, larger size, and more elongate gestalt. Three species have been recognized in this genus. *Hammerschmidtia ferruginea* was thought to be Holarctic. DNA evidence shows that it is Palearctic. Instead, the more northern Eurasian species is actually Holarctic. It is called *H. ingrica* in the Old World but should be called *H. rufa* (the North American name is older and takes precedence). We have one additional endemic species here that has not yet been described, despite having been recognized for decades. Larvae of this genus live in recently fallen logs of poplar, elm, walnut, and willow.

## ***Hammerschmidtia rufa*** Black-bristled Logsitter

**SIZE RANGE:** 7.6–11.8 mm
**IDENTIFICATION:** The abdominal color of this species can vary from orange to black. The anepisternal bristles are black and the katepisternum is densely pilose.
**ABUNDANCE:** Uncommon. **FLIGHT TIMES:** Mid-May to late July (to mid-October in the west). **NOTES:** This Holarctic species is widespread in North America but apparently rare in the Great Plains region. Flowers visited include *Anemone*, *Cicuta*, *Heracleum*, *Physocarpus*, *Prunus*, and *Rubus*. Larvae have been found under the bark of recently fallen aspen, elm, walnut, and willow.

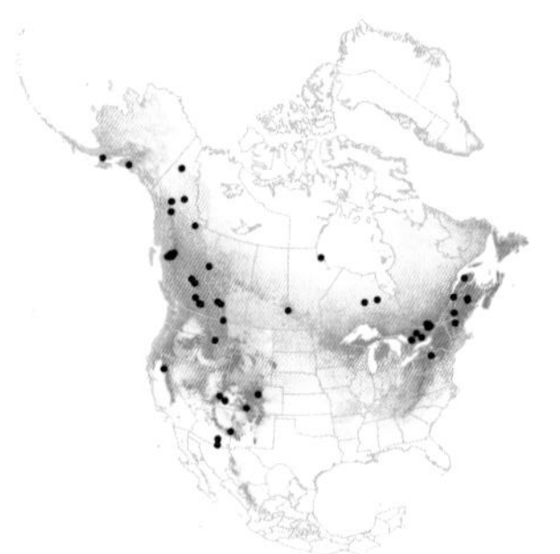

## ***Hammerschmidtia*** undescribed species 1

### Pale-bristled Logsitter

**SIZE RANGE:** 7.5–9.5 mm
**IDENTIFICATION:** Always orange with pale anepisternal bristles and a sparsely pilose katepisternum. **ABUNDANCE:** Uncommon.
**FLIGHT TIMES:** Mid-May to mid-July. **NOTES:** This species has been recognized for years but never described. It has been found in bogs and mixed woods and ovipositing in recently fallen aspen logs. Flowers visited include *Cornus*, *Heracleum*, and *Physocarpus*.

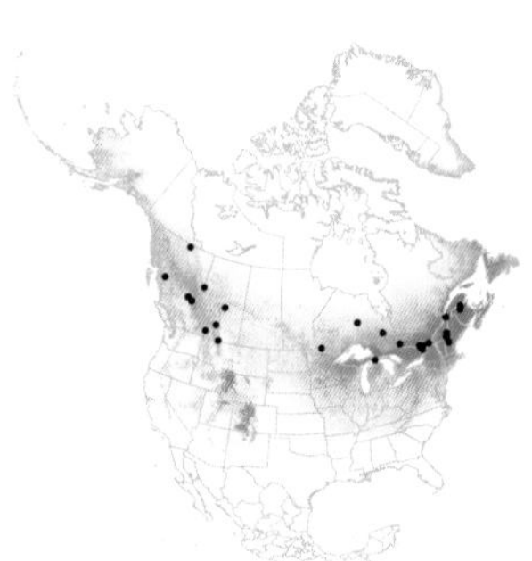

## *Hammerschmidtia rufa*

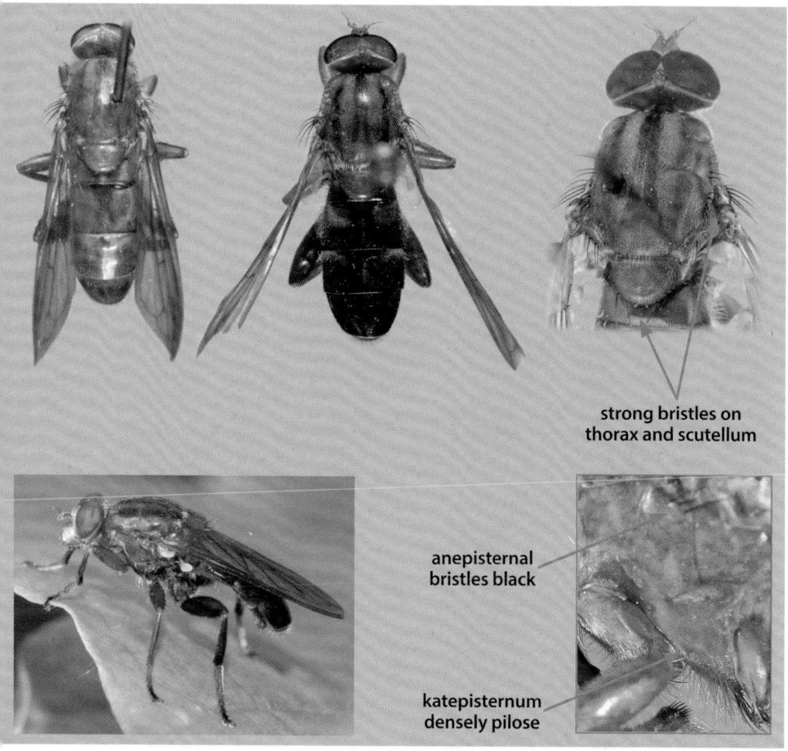

strong bristles on
thorax and scutellum
anepisternal
bristles black
katepisternum
densely pilose

## *Hammerschmidtia* undescribed species 1

short, strong black spines on anterior
surface of metatibia found on both of
these species but not *Brachyopa*
anepisternal
bristles pale
katepisternum
sparsely pilose

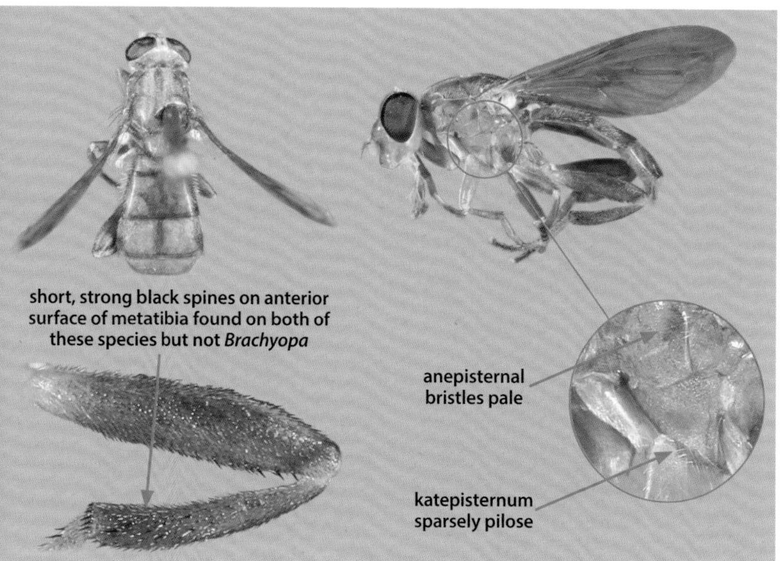

# Sphegina

There are 125 species of *Sphegina* worldwide, 20 of which are Nearctic and 11 of which are found in our region. Dozens of undescribed species are still awaiting discovery and description in the Old World, and the taxonomy of western Nearctic species is in need of revision. Fortunately, the genus was revised recently in our region (see Coovert and Thompson [1977] for a key to eastern Nearctic species). These small flies have a distinctly petiolate abdomen, enlarged metafemora, and a concave face.

The genus is divided into two subgenera, *Sphegina* and *Asiosphegina*. The subgenera can be distinguished by looking at the first sternite. It is well developed and sclerotized in the nominate subgenus and reduced or absent in *Asiosphegina*. The body color tends to vary within each species, so color patterns should not be used to distinguish species unless otherwise stated.

## *Sphegina (Asiosphegina) petiolata*

### Long-spined Pufftail

**SIZE RANGE:** 6.8–7.8 mm

**IDENTIFICATION:** This species has a shiny scutum (sometimes with orange stripes on the cuticle, not pollinose), a pair of apical bristles on the scutellum, and black apical tarsomeres on the pro- and mesolegs. **ABUNDANCE:** Uncommon. **FLIGHT TIMES:** Mid-May to late August. **NOTES:** Specimens have been collected in hardwood forest, in a floodplain near a tamarack fen, in a riparian meadow, along a weedy roadside in mixed forest, from a drained beaver pond, from a marsh, and along a small, cold, moss-edged stream. Flowers visited include *Blephilia*, *Chrysanthemum*, *Cryptotaenia*, *Leucanthemum*, *Ranunculus*, *Sanicula*, *Spiraea*, and *Veratrum*.

## *Sphegina (Asiosphegina) biannulata*

### Banded Pufftail

**SIZE RANGE:** 5.3–7.2 mm

**IDENTIFICATION:** This species has a shiny scutum, has no bristles on the scutellum, and the apical two tarsomeres on the pro- and mesolegs are yellow brown. **ABUNDANCE:** Rare. **FLIGHT TIMES:** Mid-May to early August. **NOTES:** Flowers visited include *Galax* and *Ilex*.

Sphegina (Sphegina)

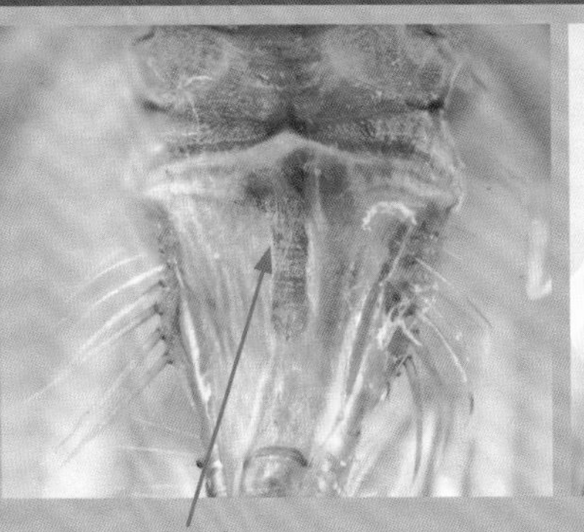


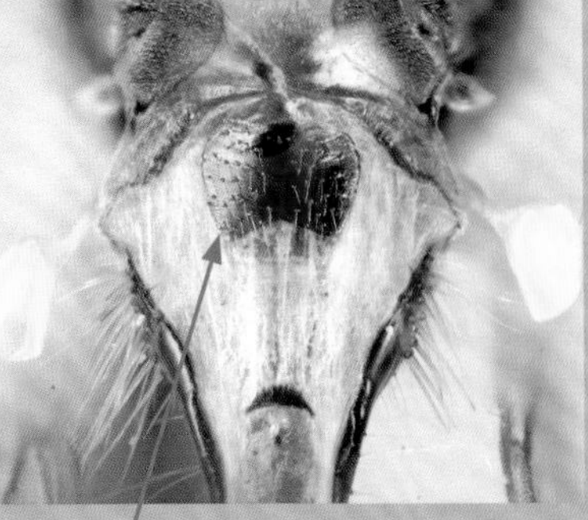


## Sphegina (Asiosphegina) petiolata

## Sphegina (Asiosphegina) biannulata

## *Sphegina (Asiosphegina) campanulata*

### Orange-horned Pufftail

**SIZE RANGE:** 6.7–7.6 mm
**IDENTIFICATION:** This species has two pollinose stripes on its scutum, a flagellum that is at least partly orange, and a completely microtrichose wing. **ABUNDANCE:** Fairly common. **FLIGHT TIMES:** Early May to early August. **NOTES:** *Sphegina campanulata* is a hardwood forest species. Flowers visited include *Actaea*, *Aruncus*, *Cryptotaenia*, *Hydrangea*, *Hydrophyllum*, *Osmorhiza*, *Rubus*, *Sambucus*, *Thalictrum*, *Viburnum*, and *Viola*.

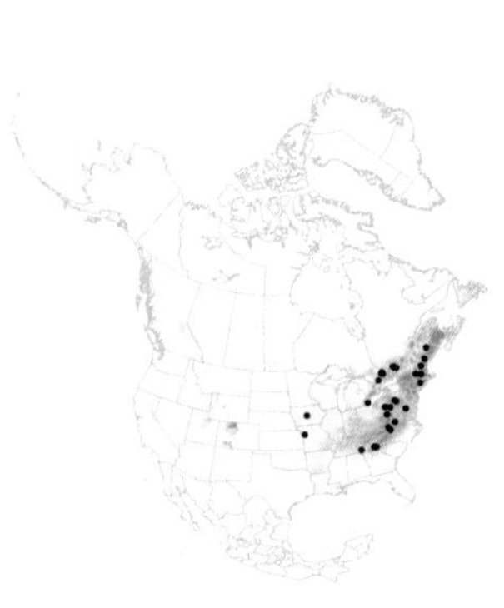

## *Sphegina (Asiosphegina) rufiventris*

### Black-horned Pufftail

**SIZE RANGE:** 6.3–8.8 mm
**IDENTIFICATION:** This species has two pollinose stripes on its scutum, a flagellum that is dark brown to black, and a wing that is partly bare basally. **ABUNDANCE:** This is by far our most common *Sphegina* species. **FLIGHT TIMES:** Late April to mid-August. **NOTES:** The number of males collected outnumbers the females by over six times, a much higher ratio than any other eastern *Sphegina*. Records are from hardwood forests, spruce forests, and bogs. Flowers visited include *Actaea*, *Anemone*, *Aralia*, *Aruncus*, *Castanea*, *Ceanothus*, *Clintonia*, *Conium*, *Cornus*, *Cryptotaenia*, *Daucus*, *Heracleum*, *Hydrangea*, *Prunus*, *Rubus*, *Saxifraga*, *Spiraea*, *Viburnum*, and *Washingtonia*.

## *Sphegina (Asiosphegina) campanulata*

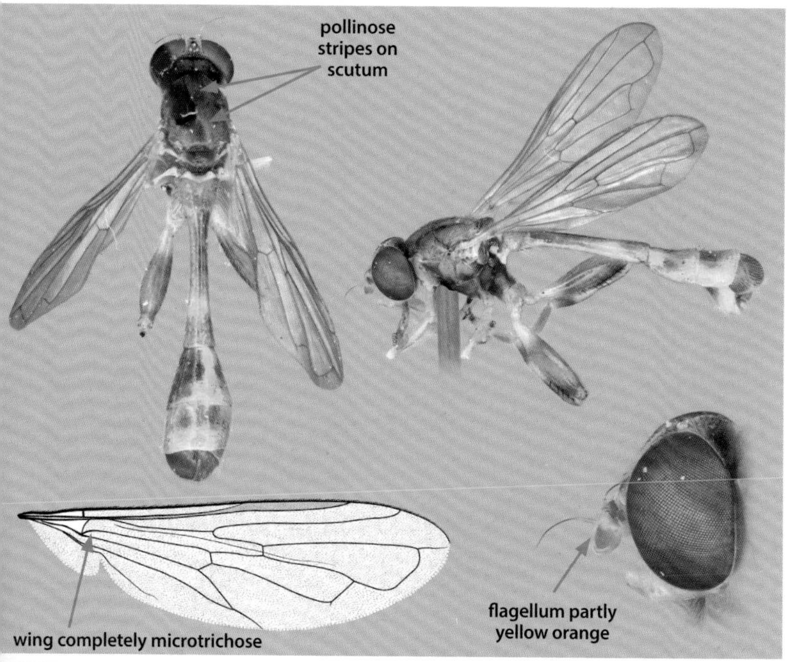

pollinose stripes on scutum

wing completely microtrichose

flagellum partly yellow orange

## *Sphegina (Asiosphegina) rufiventris*

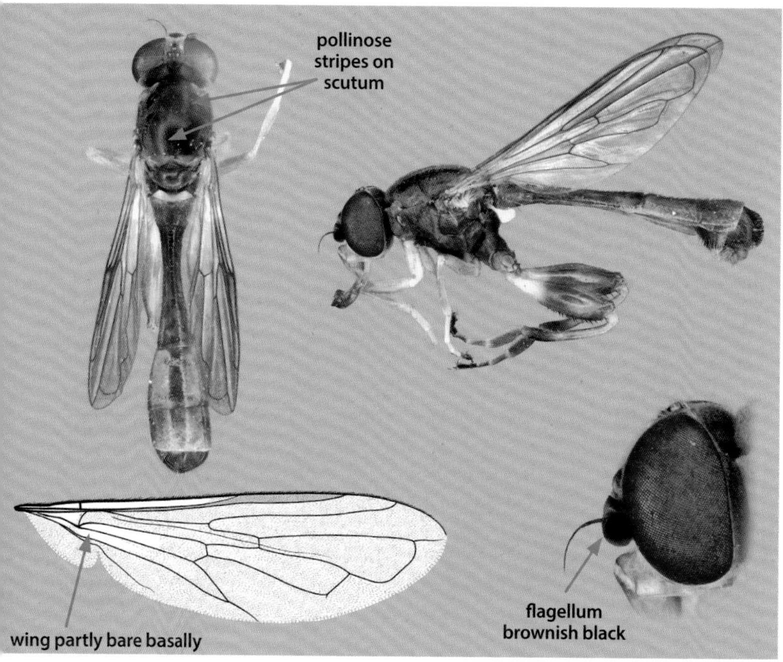

pollinose stripes on scutum

wing partly bare basally

flagellum brownish black

## *Sphegina (Sphegina) lobata*

### Yellow-lobed Pufftail

**SIZE RANGE:** 6.0–7.6 mm
**IDENTIFICATION:** Male sternite 4 asymmetrical, with black bristles apically and lobe usually yellow. Female with katepisternum shiny, tarsomeres 2 and 3 lighter than 4 and 5 and tergite 4 flared dorsoapically. **ABUNDANCE:** Uncommon.
**FLIGHT TIMES:** Mid-May to early August.
**NOTES:** This hardwood forest species has been recorded visiting flowers of *Anemone*, *Cicuta*, *Cornus*, and *Ranunculus*.

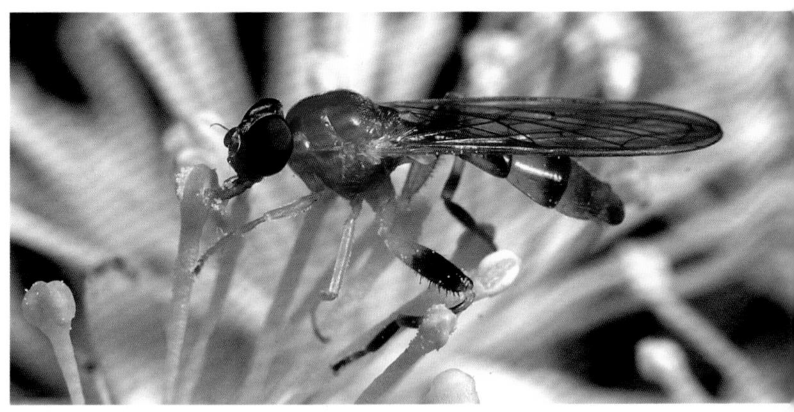

## *Sphegina (Sphegina) lobulifera*

### Black-lobed Pufftail

**SIZE RANGE:** 4.3–8.1 mm
**IDENTIFICATION:** Male sternite 4 asymmetrical, with only fine pile and lobe usually brown to black. Female with katepisternum shiny, tarsomeres 2 and 3 lighter than 4 and 5 and tergite 4 not flared apically. **ABUNDANCE:** Uncommon.
**FLIGHT TIMES:** Late May to mid-July (from early April in the south). **NOTES:** Flowers visited include *Aruncus*, *Caulophyllum*, *Heracleum*, and *Rubus*.

# Sphegina (Sphegina) lobata

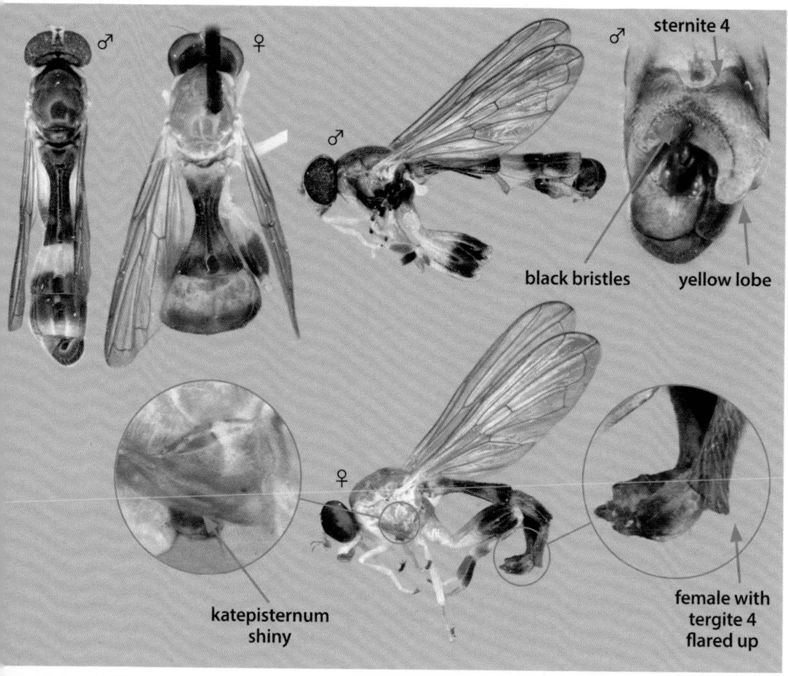

# Sphegina (Sphegina) lobulifera

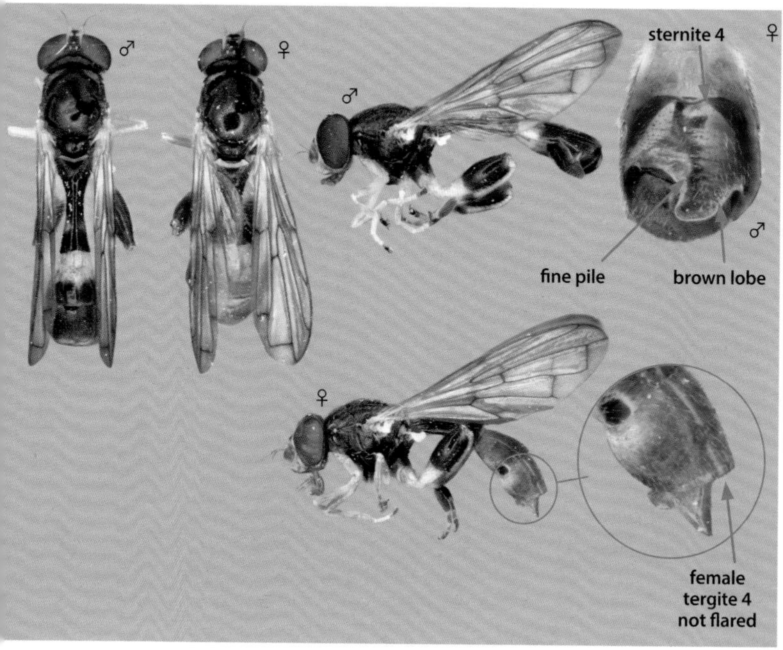

## *Sphegina (Sphegina) appalachiensis*

### Appalachian Pufftail

**SIZE RANGE:** 4.8–6.0 mm

**IDENTIFICATION:** Male sternite 4 symmetrical, with black bristles apically. Both sexes with katepisternum shiny, apical two tarsomeres of pro- and mesotarsi yellow brownish and metatibia simple apically. **ABUNDANCE:** Uncommon. **FLIGHT TIMES:** Mid-May to early July. **NOTES:** As is common for species of this genus, *S. appalachiensis* flies in association with other *Sphegina* species. This species is most likely to be confused with *S. keeniana*. Flowers visited include *Aruncus*, *Hydrangea*, and *Liriodendron*.

## *Sphegina (Sphegina) keeniana*

### Peg-legged Pufftail

**SIZE RANGE:** 4.4–6.3 mm

**IDENTIFICATION:** This species is likely to be confused only with *S. appalachiensis*, from which it can be separated by the apicoventral scoop-like tooth on the metatibia and the structure of the male genitalia. This tooth, unlike that of *S. flavomaculata*, is present in both sexes, but is more prominent in the male. Male sternite 4 symmetrical with black bristles apically. Both sexes with shiny katepisternum, black apical pro- and mesotarsi. **ABUNDANCE:** Fairly common. **FLIGHT TIMES:** Late April to mid-July (from early March in the south with outliers to late August throughout range). **NOTES:** This is another hardwood forest species. Flowers visited include *Aruncus*, *Camassia*, *Cicuta*, *Conium*, *Cryptotaenia*, *Daucus*, *Galax*, *Geranium*, *Heracleum*, *Hydrangea*, *Rubus*, and *Sanicula*.

## *Sphegina (Sphegina) appalachiensis*

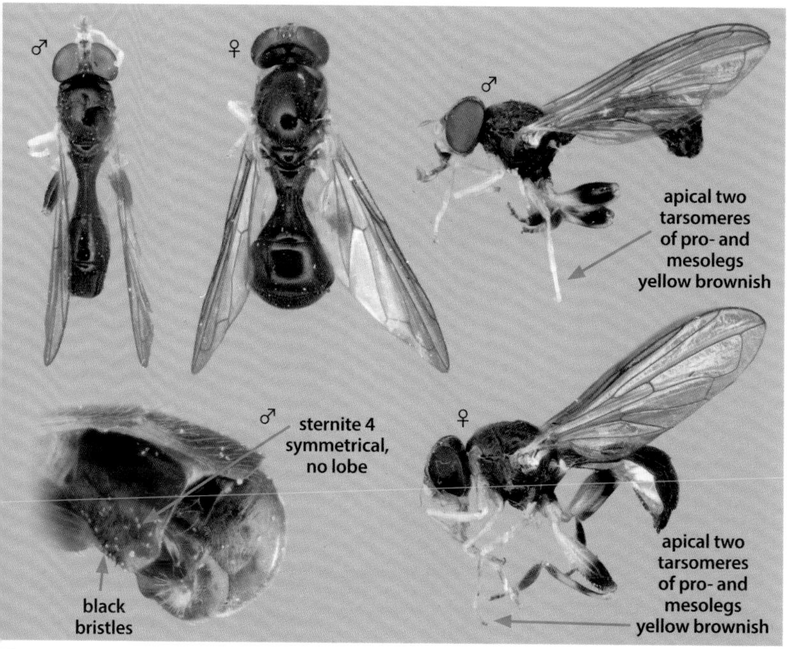

## *Sphegina (Sphegina) keeniana*

## *Sphegina (Sphegina) flavimana*

### Tuberculate Pufftail

**SIZE RANGE:** 4.7–6.6 mm
**IDENTIFICATION:** Male sternite 4 symmetrical, with fine pile. Seventh segment (epandrium) with tubercle on apicomedial margin. Metafemur bicolored in both sexes. Female with katepisternum pollinose. Tarsomeres 2–5 all dark. Tergite 5 with two apicolateral clefts and postpronotum yellow. **ABUNDANCE:** Fairly common. **FLIGHT TIMES:** Early May to early August (to mid-September in the south). **NOTES:** This is a hardwood forest species. Flowers visited include *Aruncus*, *Ceanothus*, *Chrysanthemum*, *Cicuta*, *Conium*, *Cryptotaenia*, *Hydrangea*, *Lysimachia*, *Thalictrum*, and *Viburnum*.

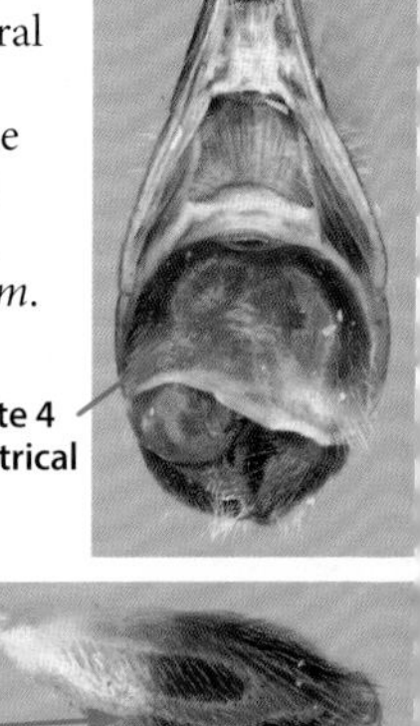


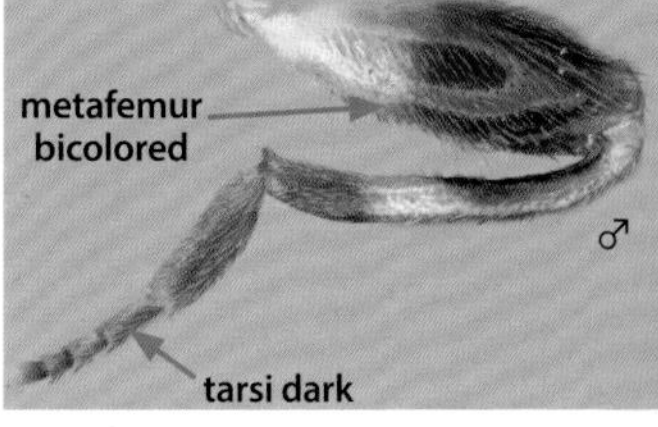

## *Sphegina (Sphegina) flavomaculata*

### Tooth-legged Pufftail

**SIZE RANGE:** 4.4–5.6 mm
**IDENTIFICATION:** Male sternite 4 symmetrical, with fine pile. Seventh segment without tubercle on apicomedial margin. Metafemur bicolored in both sexes. Female with katepisternum pollinose. Tarsomeres 2–5 all dark. Tergite 5 without apicolateral clefts, and postpronotum dark. At its narrowest, tergite 2 narrower than scutellum. The characteristic apicoventral tooth on the male metatibia is lacking in the female. **ABUNDANCE:** Uncommon. **FLIGHT TIMES:** Late April to late July (from late March in the south). **NOTES:** Flowers visited include *Anemone*, *Caltha*, *Cardamine*, *Caulophyllum*, *Prunus*, *Rubus*, and *Tiarella*.

♂

sternite 4
symmetrical

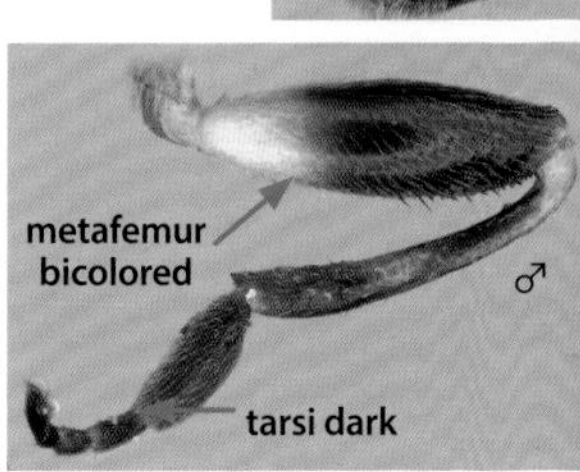

# *Sphegina (Sphegina) flavimana*

♂
postpronotum yellow
♀
♂
tarsomeres 2–5 dark
tergite 5
♀
♂
♀
fine pile
no lobe
tubercle
apicolateral clefts
metafemur bicolored

# *Sphegina (Sphegina) flavomaculata*

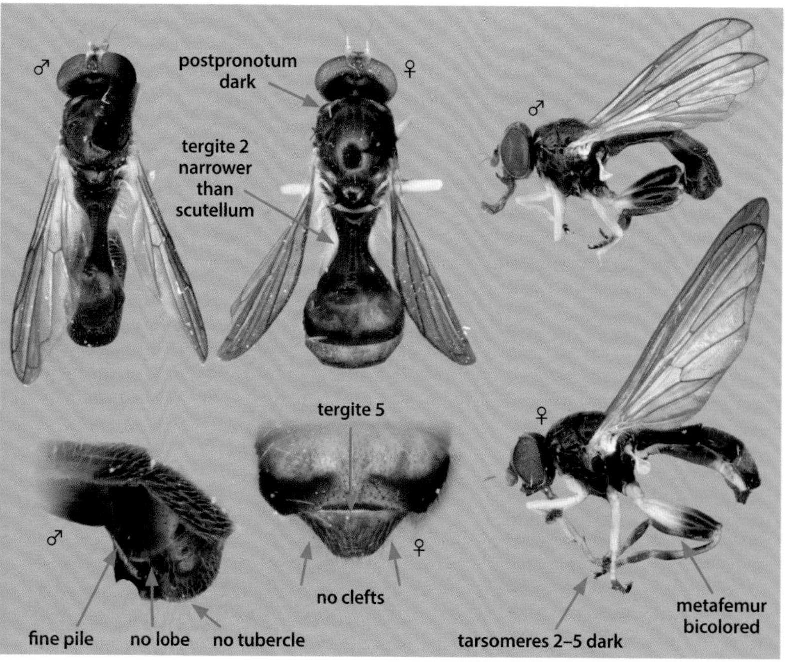

♂
postpronotum dark
♀
♂
tergite 2 narrower than scutellum
tergite 5
♀
♂
♀
fine pile
no lobe
no tubercle
no clefts
tarsomeres 2–5 dark
metafemur bicolored

## *Sphegina (Sphegina) brachygaster*

### Thick-waisted Pufftail

**SIZE RANGE:** 4.4–7.2 mm
**IDENTIFICATION:** Male sternite 4 symmetrical, with fine pile. Epandrium without tubercle on apicomedial margin. Metafemur uniformly yellow brown. Female with katepisternum pollinose. Tarsomeres 2–5 all dark. Tergite 5 without apicolateral clefts, and postpronotum dark. At its narrowest, tergite 2 is wider than the scutellum.
**ABUNDANCE:** Fairly common. **FLIGHT TIMES:** Late April to late July. **NOTES:** This species uses a range of habitats, including hardwood forest, mixed forest, spruce forest, sedge meadows, and bogs. It is often found along rivers and lakes. Flowers visited include *Anemone*, *Caltha*, *Daucus*, and *Lindera*.

# *Neoascia*

*Neoascia* species are small, narrow-waisted syrphids with distinctively flared faces and no facial tubercle. There are 29 described world species, with eight currently recognized from North America. Four of these are in our region along with two additional undescribed species. Two subgenera are recognized. There are nine species in the nominate subgenus (two in our region) and 20 in *Neoasciella*. Some (*N. globosa* and *N. willistoni*, for example) are incorrectly assigned to the nominate subgenus in online databases and the online syrphid generic key (Miranda *et al.* 2013). This is symptomatic that *Neoascia* is in need of revision. We have unofficially revised the Nearctic species and include our changes here. *Neoascia* species are tiny wetland-dwelling flies and require special effort to find as they fly among the grasses and sedges in wetlands. They are often best found by sweeping grasses and sedges rather than by direct observation. Larvae are aquatic or subaquatic and have been found in water-soaked or submerged plant debris and dung. For a key to species, see Curran (1925b). Examination of male genitalia is necessary to confirm identifications. Females cannot be reliably distinguished.

**distinctively flared face is found in all *Neoascia***

# *Sphegina (Sphegina) brachygaster*

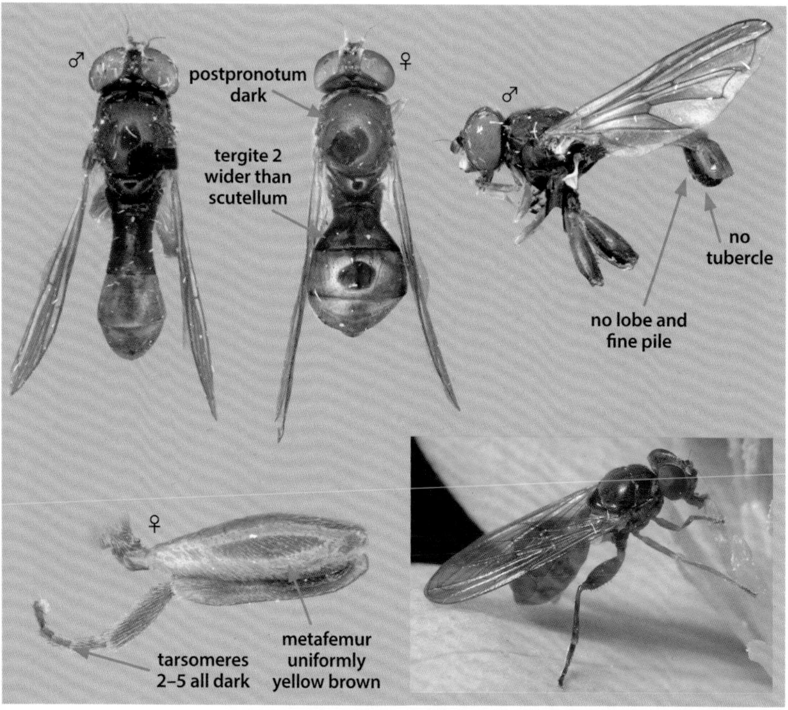


## *Neoascia (Neoascia)*

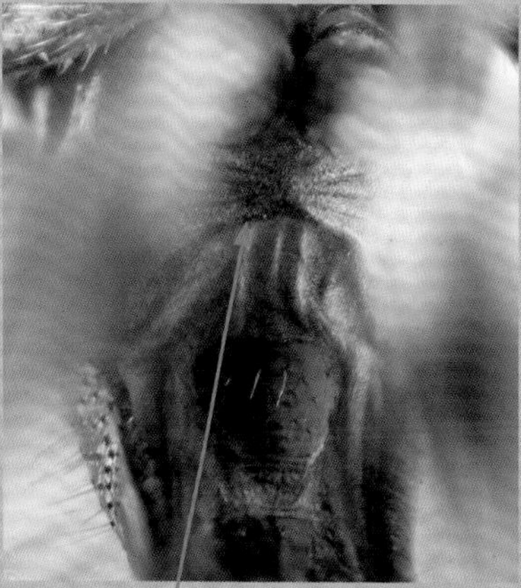

complete sclerotized bridge behind the metacoxae (view of underside of insect with head at top of page)

## *Neoascia (Neoasciella)*

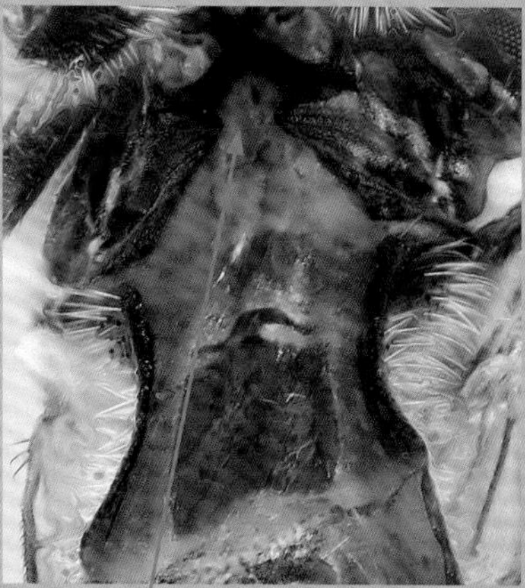

membranous gap behind the metacoxae

## *Neoascia (Neoascia) metallica*

### Double-banded Fen Fly

**SIZE RANGE:** 4.0–6.2 mm

**IDENTIFICATION:** Postmetacoxal bridge complete. Face concave, short, not extending past flagellum. Flagellum oval. Paired yellow abdominal bands on tergites 2 and 3, sometimes indistinct in females. Yellow tip of metafemur of male usually separates this from *N. tenur*. Genitalia must be checked to be certain. **ABUNDANCE:** Common. **FLIGHT TIMES:** Late April to late August (early April to late September in the south). **NOTES:** These flies have been found in a variety of wetlands including wet meadows, reedbeds, marshes, and bogs. Flowers visited include *Heracleum* and *Ranunculus*.

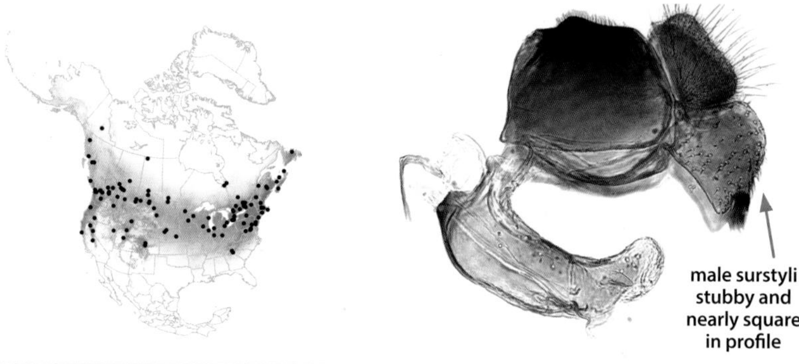


## *Neoascia (Neoascia) tenur* Black-kneed Fen Fly

**SIZE RANGE:** 3.7–6.3 mm

**IDENTIFICATION:** Postmetacoxal bridge complete. Face concave, short, not extending past flagellum. Flagellum oval. Paired yellow abdominal markings on tergites 2 and 3 (sometimes indistinct in females, and male abdomen entirely black in some western specimens). Black tip of metafemur usually separates this from *N. metallica*. Genitalia must be checked to be certain. **ABUNDANCE:** Common. **FLIGHT TIMES:** Mid-May to early September. **NOTES:** These flies are found around streams in bogs, around the periphery of raised bogs, in fens, in humid, oligotrophic grasslands, and along pond and lake margins and small streams. They fly low among dense vegetation close to water, usually landing within the vegetation. Flowers visited include *Caltha*, *Cicuta*, *Filipendula*, *Potentilla*, *Ranunculus*, *Rhododendron*, *Rubus*, *Salix*, and *Tragopogon*. Larvae and pupae have been collected from within stem sheaths of dead *Typha*, around the level of the water surface (illustrated in Maibach and Goeldlin [1993]). This species was formerly thought to occur only in Eurasia but new data show that it is also common in the Nearctic region.

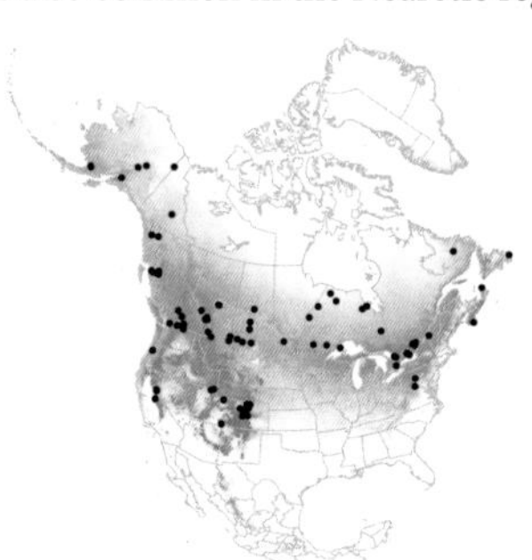

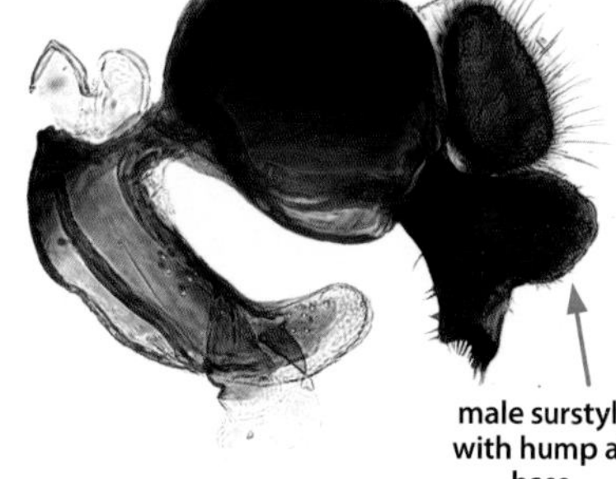

***Neoascia metallica* and *N. tenur* are our only *Neoascia* species with a completely sclerotized bridge behind the metacoxae (see photo to the right of generic account on page 231).**

## *Neoascia (Neoascia) metallica*

♂

male metafemur usually yellow at tip

♀

both sexes with yellow bands on tergites 2 and 3

most females indistinguishable from *N. tenur*

## *Neoascia (Neoascia) tenur*

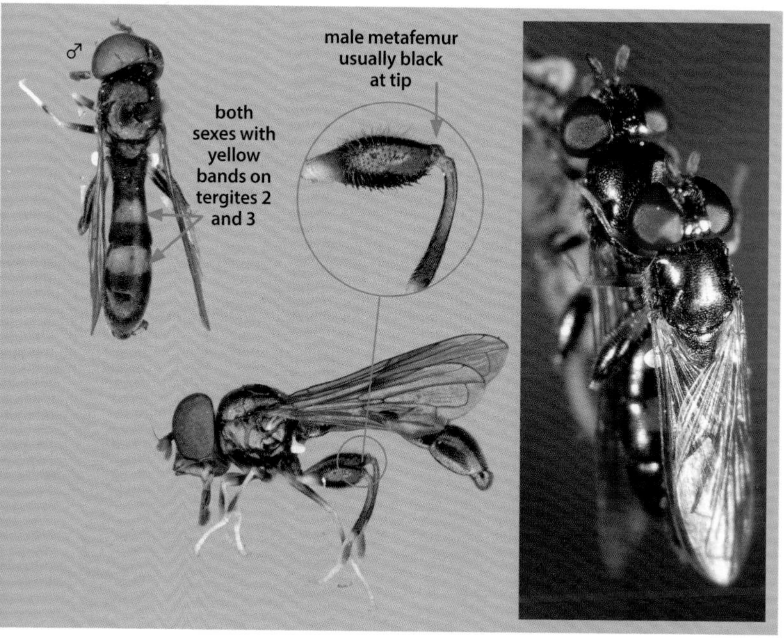

♂

male metafemur usually black at tip

both sexes with yellow bands on tergites 2 and 3

# *Neoascia (Neoasciella) globosa*

## Black-margined Fen Fly

**SIZE RANGE:** 4.0–5.9 mm
**IDENTIFICATION:** Postmetacoxal bridge absent. Face concave, short, not extending past flagellum. Flagellum elongated, oval. Single abdominal band on tergite 3 usually not meeting lateral margin. Syntergosternite 8 pale pilose. Genitalia must be checked to be absolutely certain. **ABUNDANCE:** Fairly common. **FLIGHT TIMES:** Late April to late July. **NOTES:** This species has previously been placed in the subgenus *Neoascia* but lacks the sclerotized bridge behind the metacoxae that is found in that subgenus. *Neoascia distincta* has been considered separate by previous authors but recent work by Thompson (unpublished conspectus) suggests that *distincta* is a junior synonym of *globosa*. Most have been found in bogs.

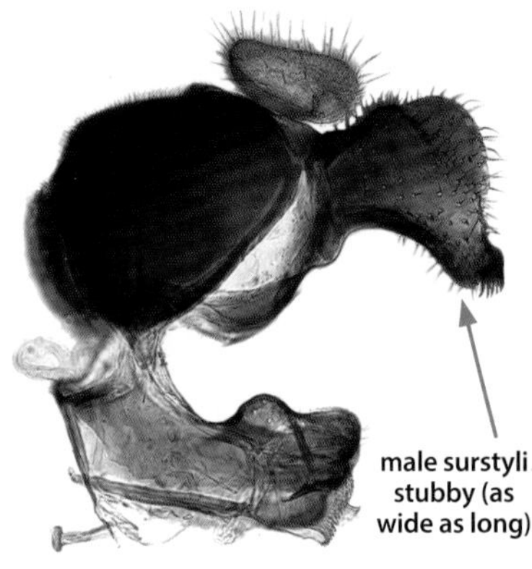


# *Neoascia (Neoasciella)* undescribed species 1

## Sands's Fen Fly

**SIZE RANGE:** 4.0–6.0 mm
**IDENTIFICATION:** Postmetacoxal bridge absent. Face concave, short, not extending past flagellum. Flagellum oval. Single abdominal band on tergite 3 usually meeting lateral margin. Syntergosternite 8 black pilose. Genitalia must be checked to be absolutely certain. **ABUNDANCE:** Fairly common. **FLIGHT TIMES:** Late April to late June. **NOTES:** This species is found in wetlands in hardwood forests and coniferous forests, including bogs. The synonymy of *N. distincta* with *N. globosa* leaves this widespread eastern species without a name (this was thought to be *distincta* until the type was examined).

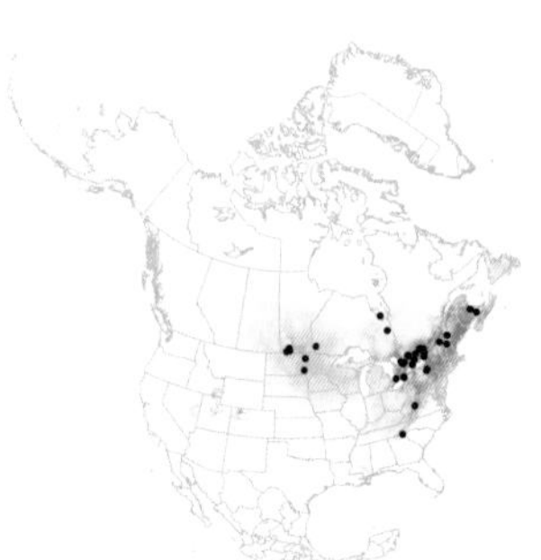

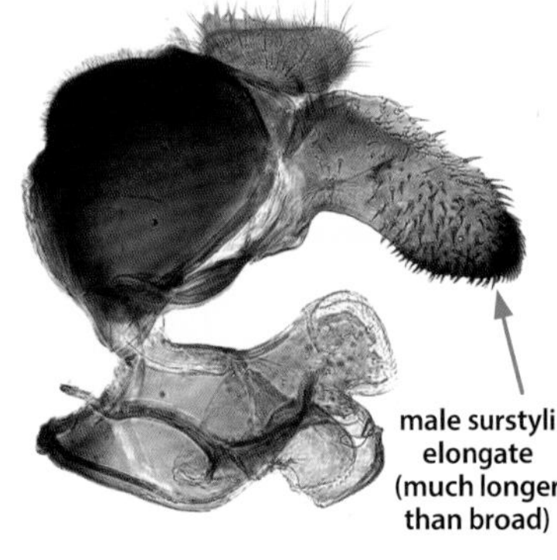

# *Neoascia (Neoasciella) globosa*

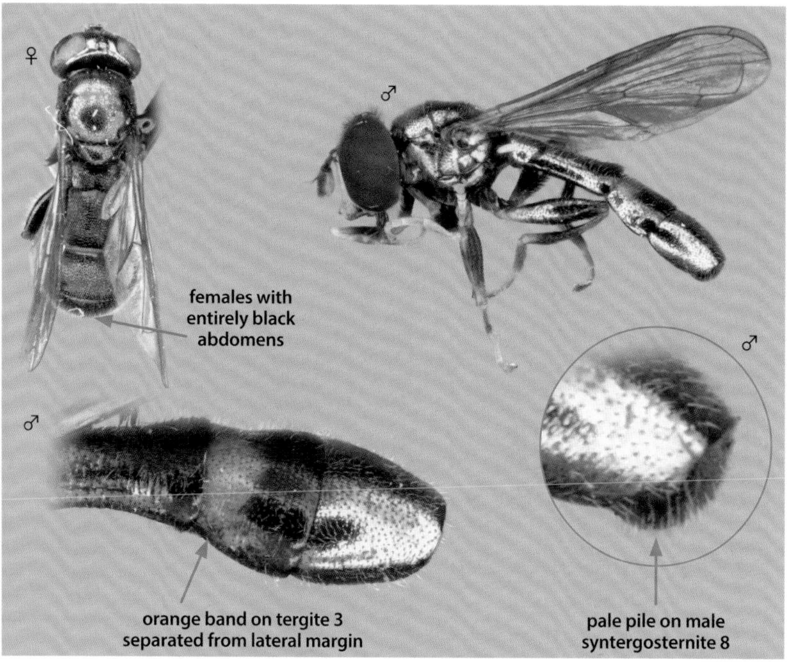

♀

♂

females with entirely black abdomens

♂

♂

orange band on tergite 3 separated from lateral margin

pale pile on male syntergosternite 8

# *Neoascia (Neoasciella)* undescribed species 1

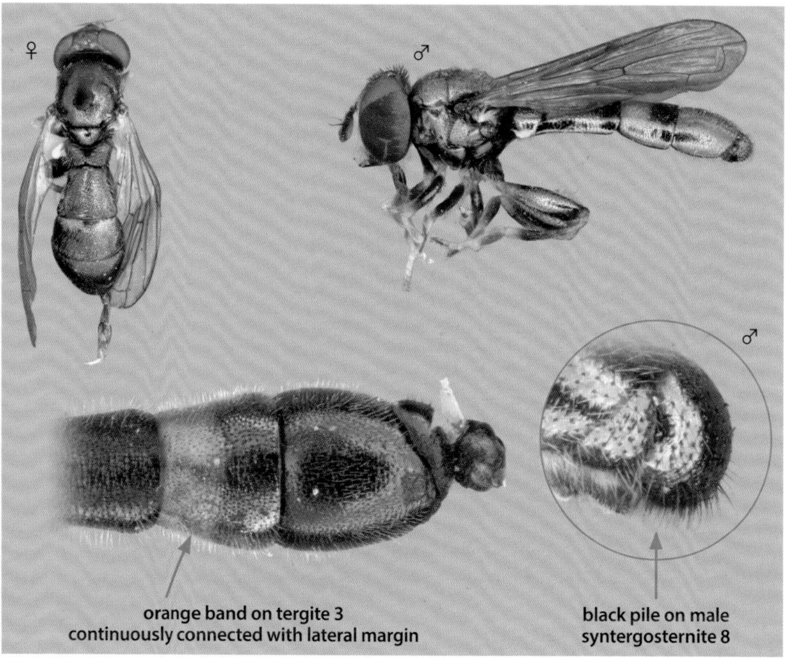

♀

♂

♂

orange band on tergite 3 continuously connected with lateral margin

black pile on male syntergosternite 8

## *Neoascia (Neoasciella)* undescribed species 17-1

### Spotted Fen Fly

**SIZE RANGE:** 4.7–4.8 mm
**IDENTIFICATION:** Postmetacoxal bridge absent. Paired spots on tergite 4 of male. Face not concave, long, extends past oval flagellum. Females are similar to female *N. subchalybea* in that the abdomen is entirely black, but can be distinguished by a nonconcave face. Dissect males to check identification. **ABUNDANCE:** Very rare. **FLIGHT TIMES:** Early June to mid-July. **NOTES:** The only known male specimen is from 19–20 July 1967 in Lawrencetown, Nova Scotia. Two females collected in Alberta in June appear to be this species (from Banff and Jumpingpound Creek).

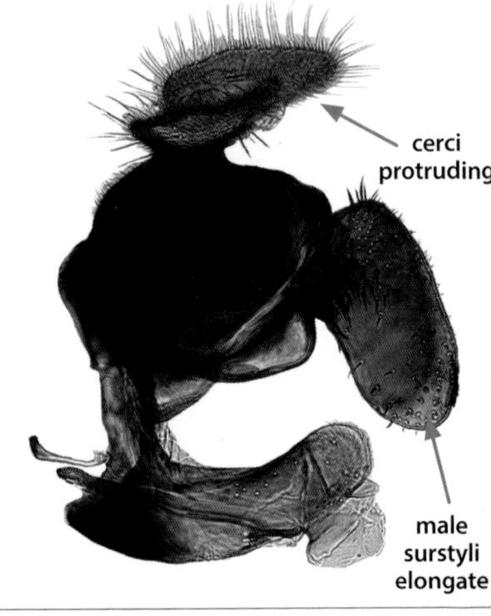


## *Neoascia (Neoasciella) subchalybea* Black Fen Fly

**SIZE RANGE:** 4.6–5.6 mm
**IDENTIFICATION:** Postmetacoxal bridge absent. Face concave, long, extends past flagellum. Flagellum oval. Abdomen, procoxae, and trochanters all black in males, usually also in females. Genitalia must be checked to be absolutely certain. **ABUNDANCE:** Rare. **FLIGHT TIMES:** Mid-April to late June. **NOTES:** This species is found across Eurasia and North America in boreal wetlands. Some of the eastern records are based on females and may be misidentified. All records from within the range of the field guide should be carefully checked. Flowers visited include *Caltha*, *Salix*, and *Stachys*.

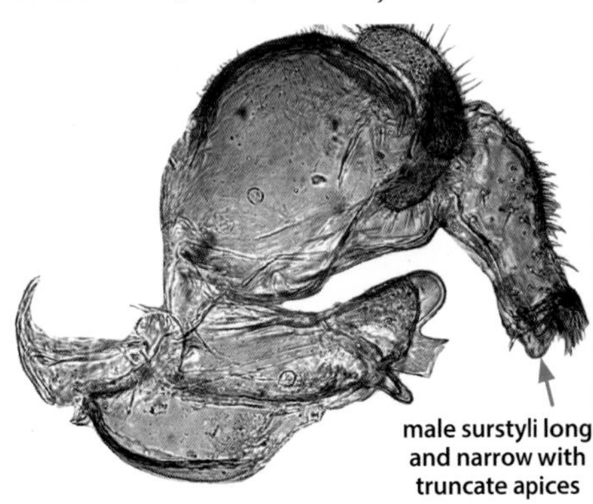

# *Neoascia (Neoasciella)* undescribed species 17-1

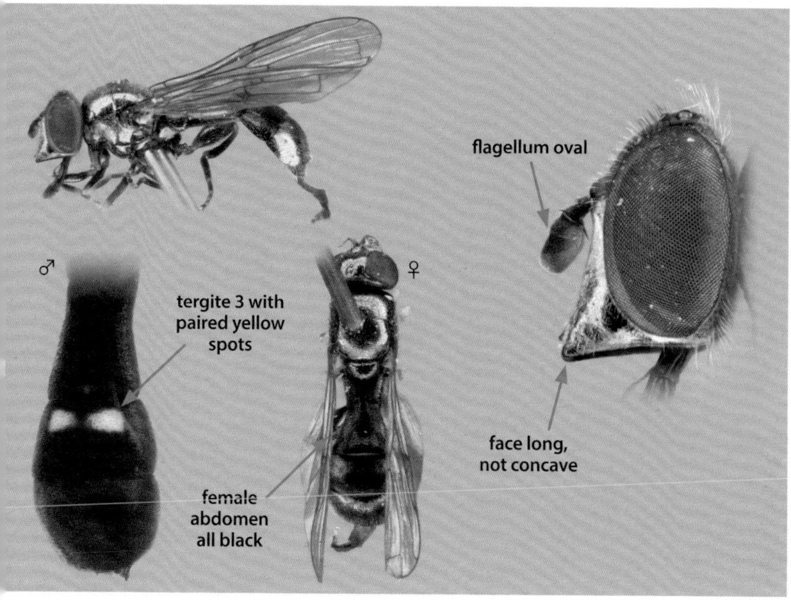

flagellum oval

♂

tergite 3 with paired yellow spots

♀

face long, not concave

female abdomen all black

# *Neoascia (Neoasciella) subchalybea*

♂

♀

face concave

♂

abdomen all black in males and usually in females

♀

procoxae and trochanters usually dark brown to black

# Eumerus

This diverse Old World genus of 266 species has at least 76 species awaiting description from Africa based on unfinished work by Lyneborg. Only three species have made it to the New World. All are introduced and found within the area of the field guide. *Eumerus* larvae feed on bulbs and roots, and some have achieved pest status. Males are best distinguished by examining the shape of sternite 4 with a hand lens or microscope. For a key to the species of North America, see Latta and Cole (1933) or Speight *et al.* (2013).

## *Eumerus funeralis* Lesser Bulb Fly

**SIZE RANGE:** 5.4–8.4 mm

**IDENTIFICATION:** Metafemur with slight but distinct basal tubercle. Facial pile yellow. Male with sternite 4 cleft and lobed. Female with longitudinal ridge along lateral margin of tergite 5. **ABUNDANCE:** Fairly common. **FLIGHT TIMES:** Mid-May to late August (outliers from early March to late November). **NOTES:** These flies are found around open ground, dry grassland, and clearings in dry woodland as well as suburban gardens and horticultural land. They fly close to the ground in sparsely vegetated grassland and woodland clearings as well as over flowerbeds in gardens. Like most *Eumerus*, they are fast flying but frequently land on bare ground or stones. Flowers visited include *Convolvulus*, *Euphorbia*, *Fragaria*, *Leucanthemum*, and *Ranunculus*. Larvae feed on damaged bulbs of *Allium*, *Amaryllis*, *Hyacinthus*, *Iris*, and *Narcissus*. They are classed as a minor pest, but larvae are unable to complete their development in the absence of fungi infesting the bulb (especially *Fusarium* basal-rot fungus and yeasts, *Saccharomyces* spp.). They overwinter as larvae. The species likely originated in the Mediterranean but is now nearly cosmopolitan. *Eumerus tuberculatus* is a common synonym for this species.

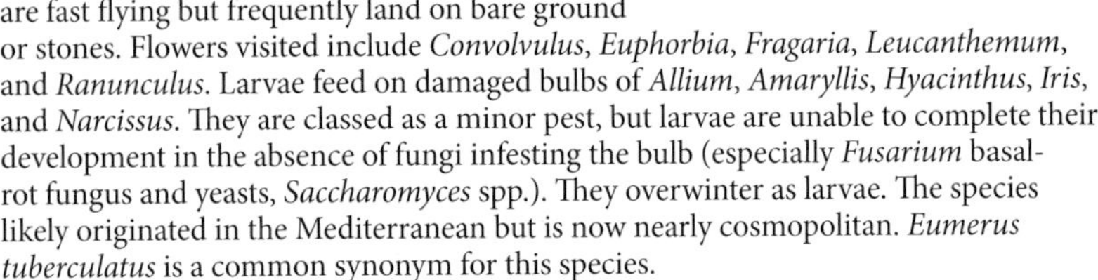

# *Eumerus*

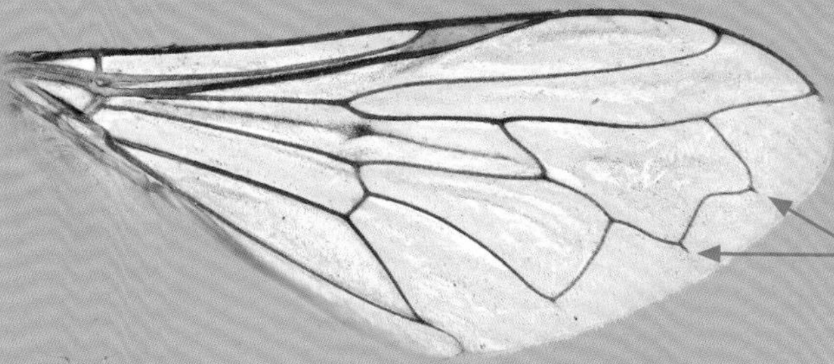

vein $M_1$ biangulate with two posterior spurs

oblique white-pollinose bands on tergites 2–4

## *Eumerus funeralis*

♀

face yellow pilose

metafemur with tubercle

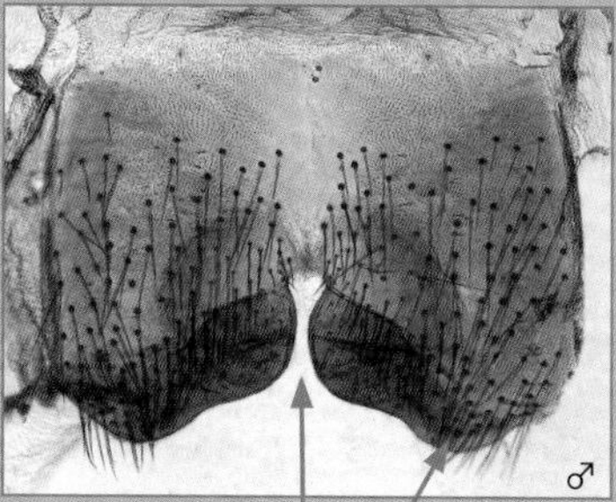

♂

sternite 4 cleft, with lobes

## *Eumerus narcissi* Daffodil Bulb Fly

**SIZE RANGE:** 6.6–8.1 mm
**IDENTIFICATION:** Metafemur simple, without tubercle. Facial pile white. Male with sternite 4 not cleft. Female tergite 5 partly black pilose. **ABUNDANCE:** Rare. **FLIGHT TIMES:** Mid-June to early October. **NOTES:** These flies occur in forested areas, small open areas in evergreen oak shrubland, and around bulb farms. Larvae are frequently found in bulbs of *Narcissus* but have also been reared from *Allium* and *Hippeastrum*. The species is native to Europe and has become naturalized in California. The single Ontario record may not represent an established population. The notation on the specimens from Ottawa suggests that they were reared from bulbs introduced from France, and it is unclear if they were intercepted in a shipment or found in the wild. It is included in the guide because of the possibility of the species spreading to other areas in North America.

## *Eumerus strigatus* Onion Bulb Fly

**SIZE RANGE:** 4.4–8.8 mm
**IDENTIFICATION:** Metafemur simple, without tubercle. Pile of face white. Male with sternite 4 cleft and not lobed. Female with tergite 5 yellow pilose. **ABUNDANCE:** Fairly common. **FLIGHT TIMES:** Late April to early October (outliers from late March to early November). **NOTES:** This species is found in a variety of habitats including wet forests; humid, seasonally flooded grassland; deciduous forests; open, dry unimproved pastures; and dune grasslands. They sometimes can be found in horticultural land and suburban gardens. Adults can be found among thick vegetation of humid grassland, along the edges of clearings and tracks. They fly low, usually keeping within vegetation, but may land on open patches of ground or on stones. Flowers visited include *Allium*, *Convolvulus*, *Eschscholzia*, *Euphorbia*, *Fragaria*, *Leontodon*, *Papaver*, *Potentilla*, *Ranunculus*, *Sonchus*, and *Taraxacum*. Larvae are regarded as a minor pest of horticulture, attacking a variety of rotting bulbs and roots (including artichoke roots [*Cynara scolymus*], asparagus, carrot, *Iris*, *Narcissus*, onion, parsnip, and potato). They overwinter as larvae.

# Eumerus narcissi

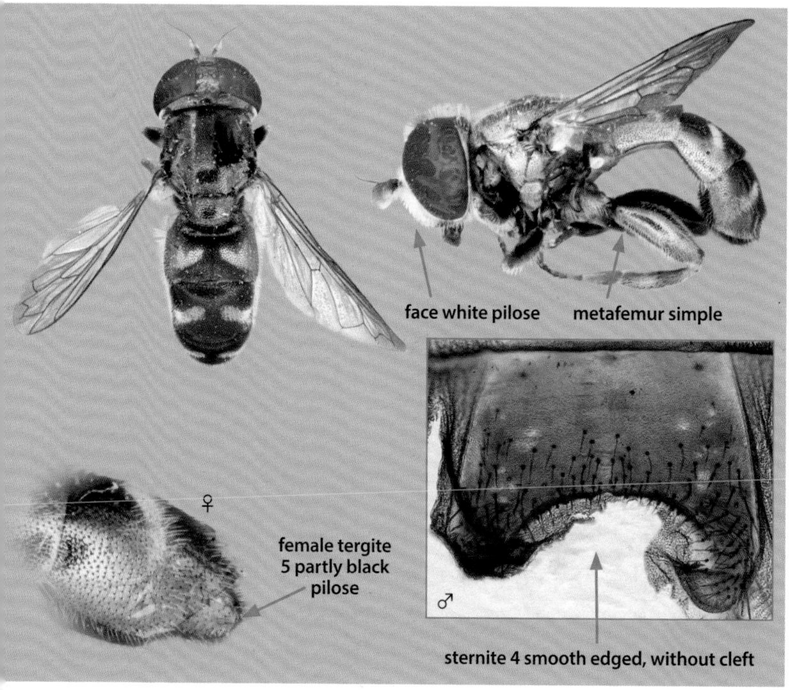

face white pilose

metafemur simple

♀

female tergite 5 partly black pilose

♂

sternite 4 smooth edged, without cleft

# Eumerus strigatus

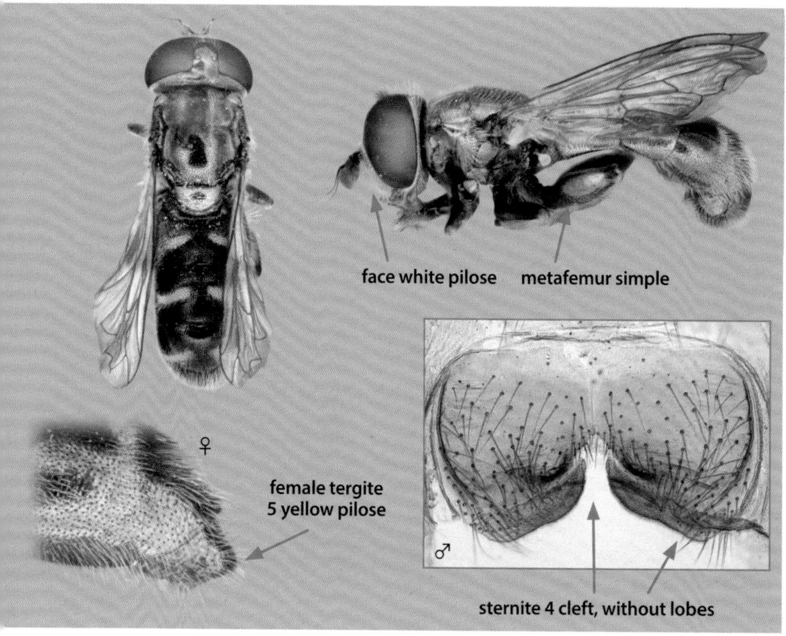

face white pilose

metafemur simple

♀

female tergite 5 yellow pilose

♂

sternite 4 cleft, without lobes

# *Myolepta*

Four of eight Nearctic *Myolepta* species occur in our region (of 42 world species). These small mostly black syrphids are easy to distinguish from each other but may be confused with other small black syrphids. *Myolepta* can be differentiated from other small, black syrphids by the anteroventral and posteroventral spines on the pro- and mesofemora. *Lepidomyia* species share this character but they have elongate antennae and are not found in our region. Larvae live in rot holes in deciduous trees.

## *Myolepta nigra* Black Spineleg

**SIZE RANGE:** 7.6–10.0 mm

**IDENTIFICATION:** This species has an abdomen that is entirely black dorsally, tibiae that are entirely black and a brown-black flagellum. **ABUNDANCE:** Uncommon. **FLIGHT TIMES:** Early May to mid-July. **NOTES:** Records are from hardwood and alluvial forests, and bog edges. Flowers visited include *Cornus*, *Crataegus*, *Prunus*, and *Rosa*.

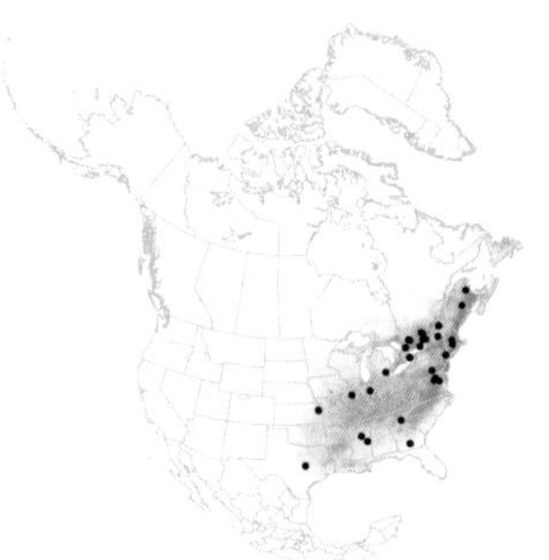

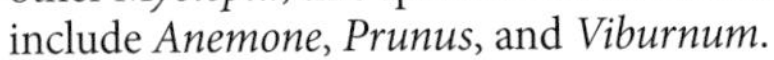

## *Myolepta strigilata* Scaled Spineleg

**SIZE RANGE:** 5.7–9.8 mm

**IDENTIFICATION:** This species has distinctive scale-like pile. **ABUNDANCE:** Uncommon. **FLIGHT TIMES:** Late April to late June (from late March in the south). **NOTES:** As with our other *Myolepta*, this species is found in rich deciduous woodlands. Flowers visited include *Anemone*, *Prunus*, and *Viburnum*.

# *Myolepta*

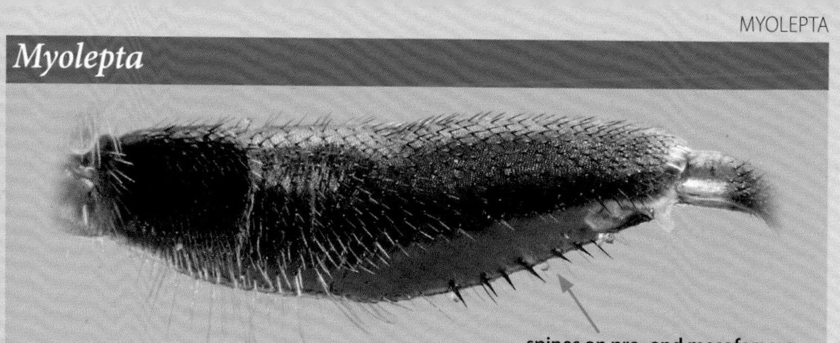


## *Myolepta nigra*

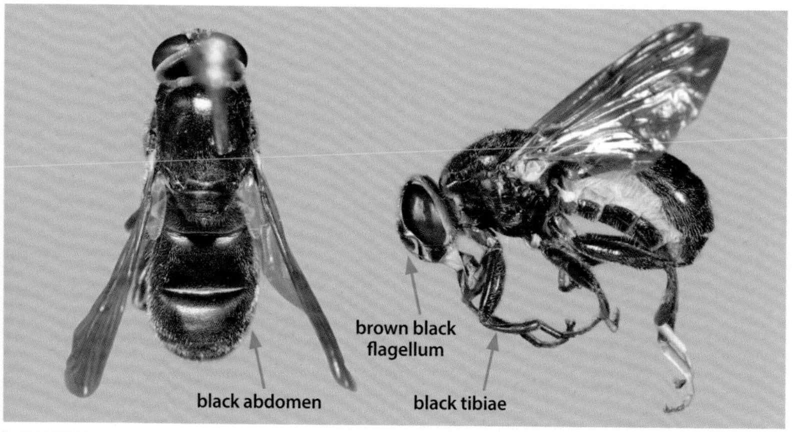


## *Myolepta strigilata*

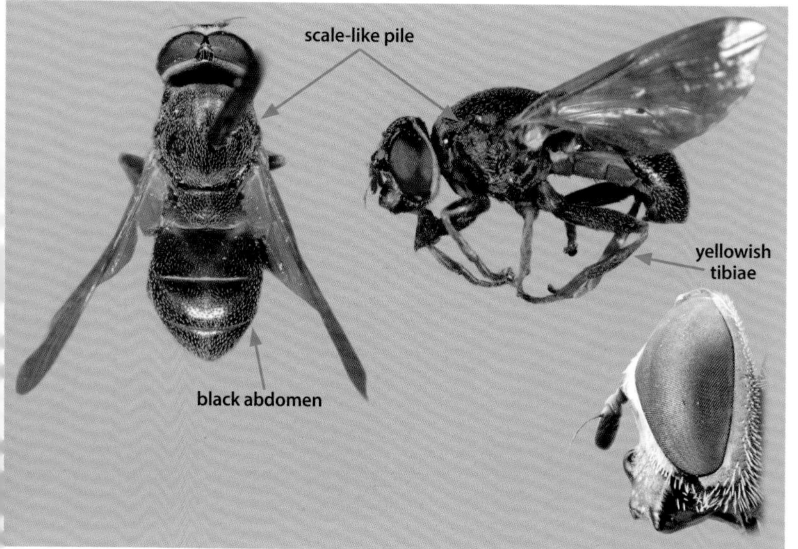

## *Myolepta varipes* Orange-banded Spineleg

**SIZE RANGE:** 7.5–9.4 mm

**IDENTIFICATION:** Orange on tergites 1–3 often extensive but varies individually and between sexes. Tibiae yellow, at least basally. Flagellum orange. Male face with broad shiny midstripe. Occiput shiny in males. Procoxa shiny medially in both sexes. **ABUNDANCE:** Uncommon. **FLIGHT TIMES:** Early May to early July (from late March in the south). **NOTES:** Found in rich deciduous woodlands. Males have been observed patrolling maple rot holes. The only floral record is from *Cornus*.

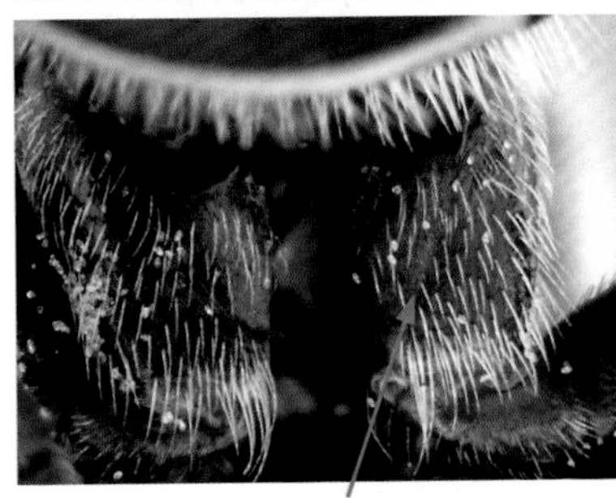

procoxa shiny medially in both sexes

## *Myolepta pretiosa* Dusted Spineleg

**SIZE RANGE:** 6.8–9.9 mm

**IDENTIFICATION:** Orange on tergites 1–3 often extensive but varies individually and between sexes. Tibiae yellow, at least basally. Flagellum orange. Male face with narrow, shiny midstripe. Occiput pollinose in males. Procoxa pollinose in both sexes. **ABUNDANCE:** Uncommon. **FLIGHT TIMES:** Early April to early July. **NOTES:** Found in rich deciduous woodlands. Males have been observed patrolling maple rot holes. Flowers visited include *Salix* and *Spiraea*.

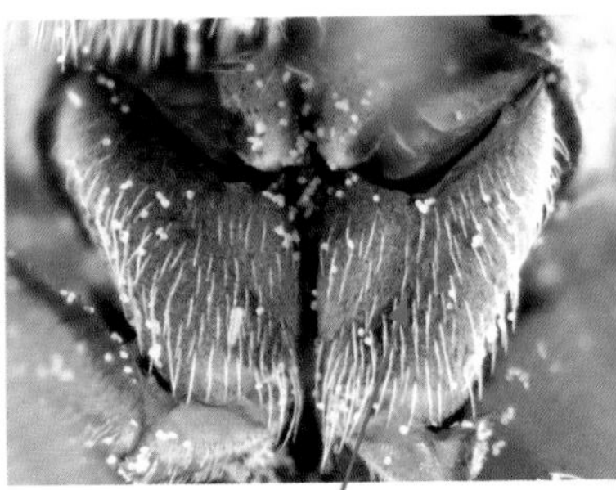

procoxa completely pollinose in both sexes

# *Myolepta varipes*

♂

occiput shiny in males

♂

orange on abdomen

♂

♀

black shiny midstripe on face broad in males

at least some orange on female abdomen as in *M. pretiosa*

# *Myolepta pretiosa*

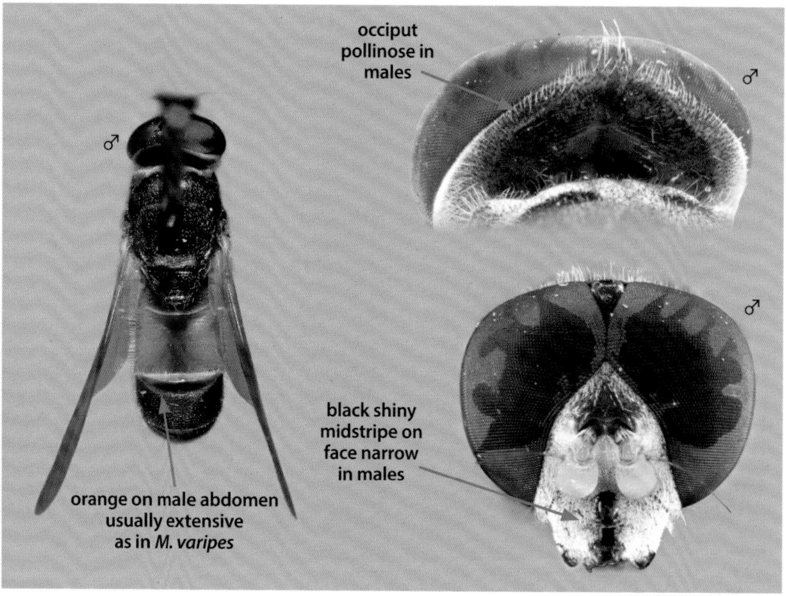

occiput pollinose in males

♂

♂

♂

black shiny midstripe on face narrow in males

orange on male abdomen usually extensive as in *M. varipes*

# Orthonevra

Seven of the 16 Nearctic *Orthonevra* species occur in our region (59 worldwide). *Orthonevra* are small dark flies that have wing vein $M_1$ straight or angled slightly toward the wing base. All of them lack the subscutellar fringe of pile found in *Chrysosyrphus* species, and most species have patterned eyes. *Orthonevra nitida* is the only easy species to identify in the field. Close examination of the other six is required for identification. *Orthonevra* are all wetland associates and most live around forested wetlands and springs. Larvae of some European species live in organically enriched mud and have been described and illustrated by Maibach and Detiefenau (1994). There is no single key to the Nearctic species of *Orthonevra*, but a combination of keys can be used to distinguish species (see Shannon [1916], Sedman [1964, 1966]). There are three groups in *Orthonevra*, one with complex patterns on the eye, one with a horizontal line through the eye, and one with no markings on the eye.

## *Orthonevra nitida* Wavy Mucksucker

**SIZE RANGE:** 4.5–5.5 mm
**IDENTIFICATION:** This species is easily identifiable by the complex pattern on the eyes; other species have a single medial band or no eye patterning. **ABUNDANCE:** Fairly common. **FLIGHT TIMES:** Early June to late October (from mid-March in the south). **NOTES:** *Orthonevra nitida* has been recorded in marshes, bogs, and wet prairies on flowers of *Anemone*, *Baccharis*, *Chrysanthemum*, *Fragaria*, *Leucanthemum*, *Pastinaca*, *Prunus*, *Salix*, and *Viburnum*.

## *Orthonevra*

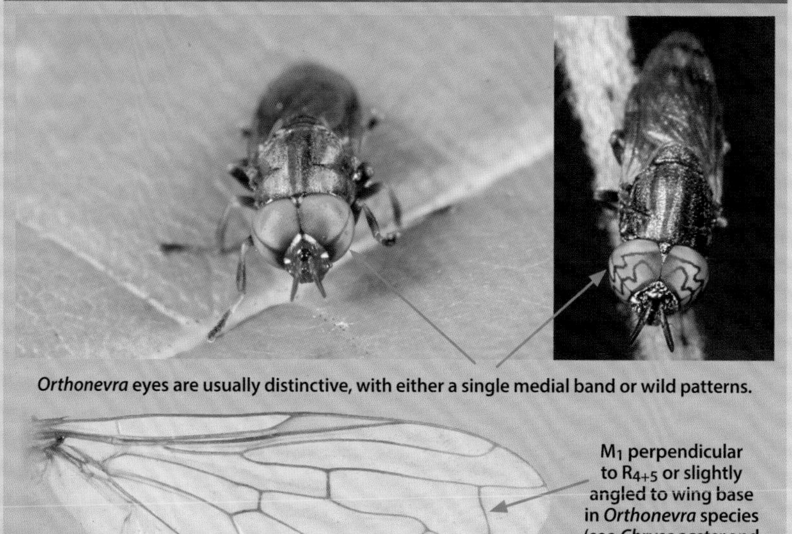

*Orthonevra* eyes are usually distinctive, with either a single medial band or wild patterns.

$M_1$ perpendicular to $R_{4+5}$ or slightly angled to wing base in *Orthonevra* species (see *Chrysogaster* and *Chrysosyrphus* for comparison)

## *Orthonevra nitida*

This species has distinctively patterned eyes (all other species in our region have single-banded or unmarked eyes).

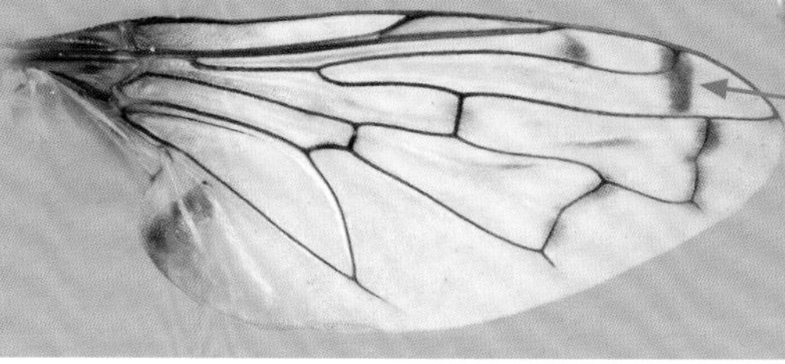

wing darkened around crossveins in *O. nitida* and two other eastern species

## *Orthonevra anniae* Shiny-sided Mucksucker

**SIZE RANGE:** 5.4–6.5 mm
**IDENTIFICATION:** Eye with single medial band in this and following three species. This species is most like *O. pulchella* and has a hyaline wing. Unlike *pulchella*, the thorax is shiny and bluish, the facial pollinose spot is large, and the pro- and mesotibiae are yellowish. **ABUNDANCE:** Rare. **FLIGHT TIMES:** Late April to late June. **NOTES:** *Orthonevra anniae* likely lives around wetlands in hardwood forests. The only floral records are from *Prunus* and *Salix*.

## *Orthonevra pulchella* Dusky Mucksucker

**SIZE RANGE:** 5.4–6.5 mm
**IDENTIFICATION:** This species has the eye with a single medial band, a hyaline wing, small pollinose markings on the face (reaching less than halfway to the base of the antennae), and the pro- and mesotibiae are black medially. **ABUNDANCE:** Fairly common. **FLIGHT TIMES:** Early May to early September (from early March in the west). **NOTES:** This species has been found in a variety of habitats, from bogs and forest seeps to wetlands in deciduous and mixed forest, pine forest, and prairie. Flowers visited include *Physocarpus* and *Viburnum*.

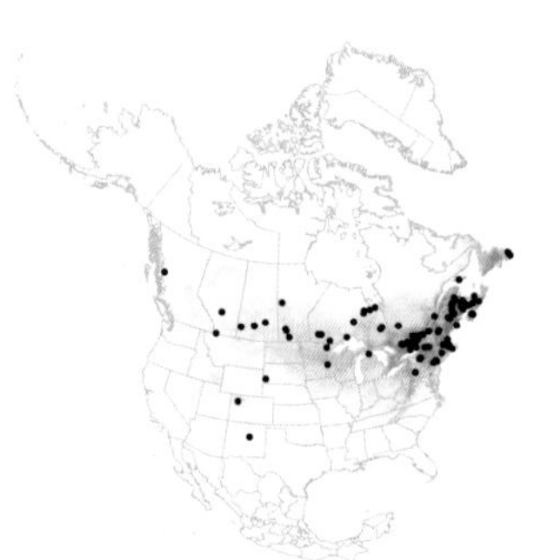

# *Orthonevra anniae*

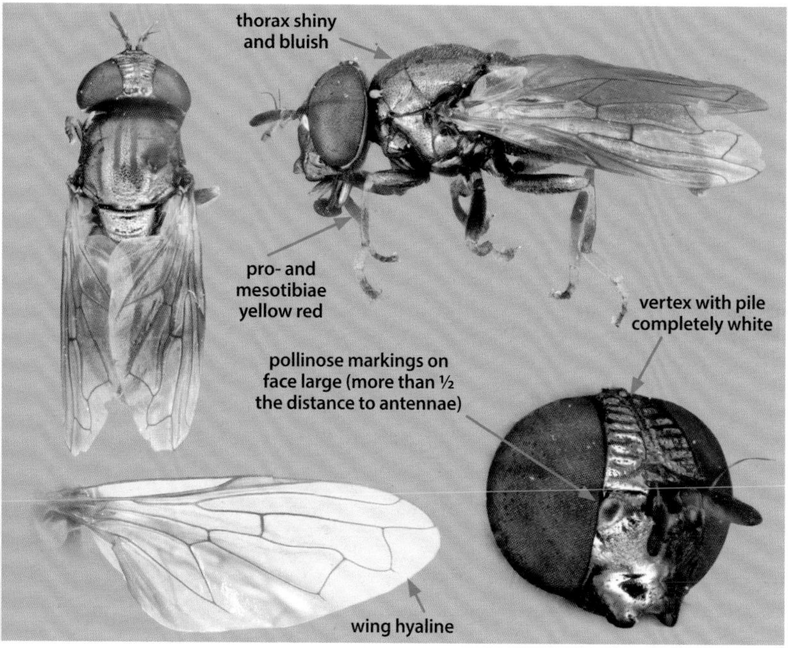

thorax shiny and bluish

pro- and mesotibiae yellow red

vertex with pile completely white

pollinose markings on face large (more than ½ the distance to antennae)

wing hyaline

# *Orthonevra pulchella*

pro- and mesotibiae black medially

pollinose markings on face small (less than ½ the distance to the antennae)

wing hyaline

## *Orthonevra pictipennis* Dusky-veined Mucksucker

**SIZE RANGE:** 4.5–6.0 mm

**IDENTIFICATION:** This species has the eye with a single medial band, wings that are darkened around the crossveins (sometimes only faintly), white thoracic pile, and tibiae that are black medially. **ABUNDANCE:** Uncommon. **FLIGHT TIMES:** Mid-May to early October (early March to early November in the south). **NOTES:** Flowers visited include *Caltha*, *Pastinaca*, *Primula*, *Pyrus*, and *Sedum*.

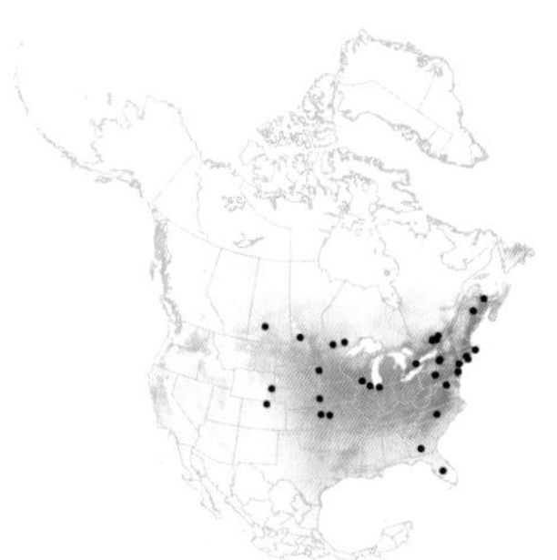

## *Orthonevra weemsi* Weems's Mucksucker

**SIZE RANGE:** 4.8–5.8 mm

**IDENTIFICATION:** This species has the eye with a single medial band, darkened crossveins on the wings, brown thoracic pile, and yellowish tibiae. **ABUNDANCE:** Rare. **FLIGHT TIMES:** Early May and mid-July (from early March in the south). **NOTES:** This species has been collected from bogs and lake edges. Flowers visited include *Crataegus* and *Photinia*.

## *Orthonevra pictipennis*

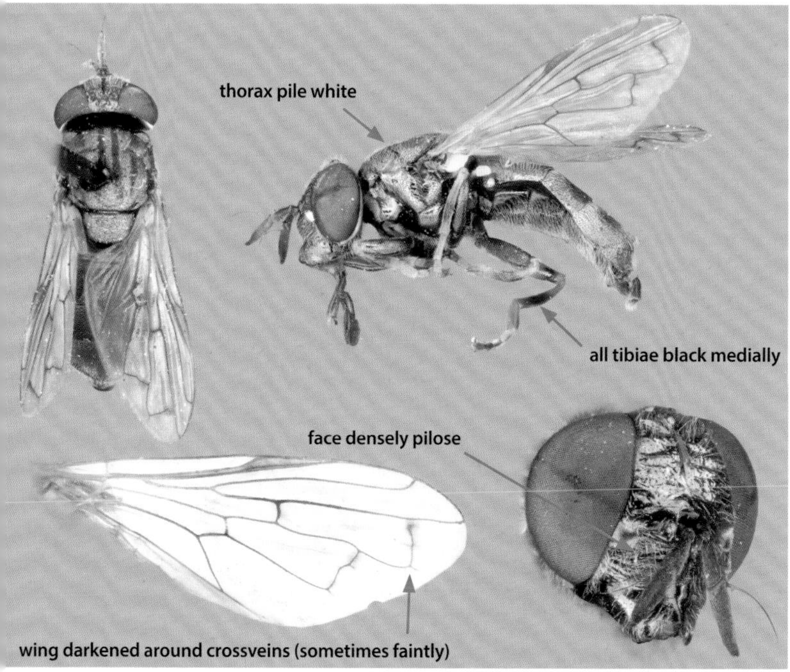

thorax pile white

all tibiae black medially

face densely pilose

wing darkened around crossveins (sometimes faintly)

## *Orthonevra weemsi*

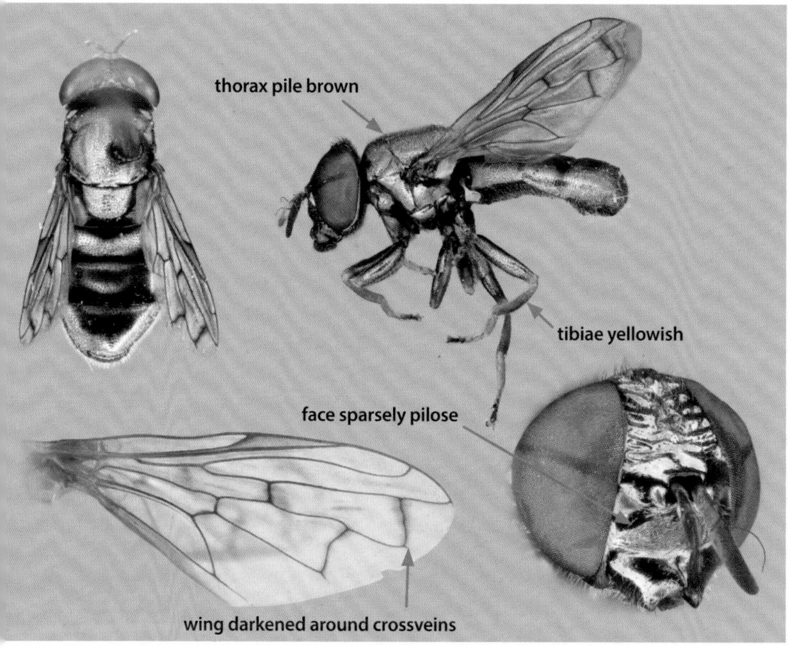

thorax pile brown

tibiae yellowish

face sparsely pilose

wing darkened around crossveins

## *Orthonevra robusta* Short-horned Mucksucker

**SIZE RANGE:** 6.0–6.1 mm
**IDENTIFICATION:** The lack of eye markings on the two species on this page may mislead you to think that they are *Chrysogaster* or *Chrysosyrphus* species. Compare the wing venation though. The recurrent $M_1$ vein is distinctly different from that in *Chrysogaster* and *Chrysosyrphus*. *Orthonevra robusta* is readily distinguished from the species below by the all black legs and shorter antennal flagellum (flagellum shorter than scape and pedicel combined). **ABUNDANCE:** Rare. **FLIGHT TIMES:** Mid-May to early July. **NOTES:** Work is needed to find out if *O. robusta* is a complex of species. The northern population is considerably disjunct from the western range where the species was first described. Specimens have been collected in fens and around springs and ponds.

## *Orthonevra* undescribed species 1
Fee's Mucksucker

**SIZE RANGE:** 5.4–6.5 mm
**IDENTIFICATION:** This and the preceding species lack eye markings typical of *Orthonevra*. This species is readily distinguished from *O. robusta* by the bicolored legs and long antennal flagellum (longer than scape and pedicel combined). **ABUNDANCE:** Rare and local. **FLIGHT TIMES:** Early to mid-June. **NOTES:** Known only from Scott Bog, New Hampshire, where all known specimens were collected by Frank Fee. Specimens were found nectaring on *Cornus*, *Fragaria*, and *Taraxacum*.

## *Orthonevra robusta*

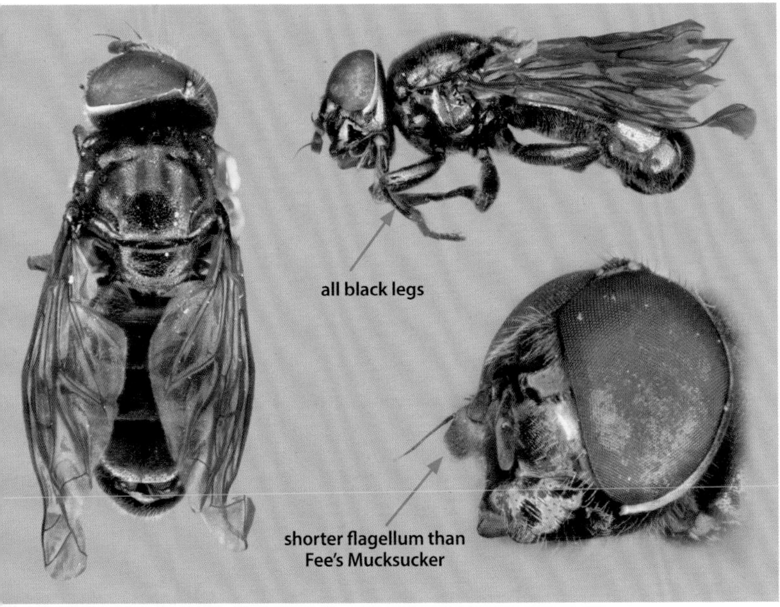

all black legs

shorter flagellum than Fee's Mucksucker

## *Orthonevra* undescribed species 1

$M_1$ vein recurrent (angled back toward wing base) in this species and *O. robusta*. Compare with *Chrysogaster* and *Chrysosyrphus* on following pages

bicolored legs

longer flagellum than *O. robusta*

# Chrysogaster

These small black flies are most similar to *Cheilosia*, *Hiatomyia*, *Orthonevra*, and *Chrysosyrphus*. *Hiatomyia* have plumose antennae and *Cheilosia* have a strong facial tubercle and scutellar bristles. *Orthonevra* have wing vein $M_1$ recurrent (see previous page for most similar species). *Chrysosyrphus* and *Chrysogaster* have smooth faces, no scutellar bristles, $M_1$ angled toward the wing tip, and females have a distinctly wrinkled frons. *Chrysogaster* lack the subscutellar fringe of pile found in *Chrysosyrphus*. The antennae of *Chrysogaster* are inserted at the mid- to lower ⅓ of the head while they are at the upper ⅓ in *Chrysosyrphus*. *Chrysogaster* also have short pedicel bristles (long in *Chrysosyrphus*). Female facial pollinosity of *Chrysogaster* is usually concentrated below antennal insertions and extends to the edge of the eye. There are 38 world species of *Chrysogaster* but only two in North America. Larvae are aquatic in accumulations of decaying vegetation and mud in pools, ponds, and slow-moving streams.

## Chrysogaster antitheus

### Short-haired Wrinklehead

**SIZE RANGE:** 5.0–7.0 mm
**IDENTIFICATION:** Male sternite 2 with sparse, short pile (equal length or shorter than that of the sternite 1). Female sternite 1 with a few short pili.
**ABUNDANCE:** Fairly common. **FLIGHT TIMES:** Mid-May to late July (from late March in the south).
**NOTES:** Found on flowers of *Aronia*, *Aruncus*, *Castanea*, *Ceanothus*, *Physocarpus*, and *Viburnum* in sedge meadows, bogs, and forests.

## Chrysogaster inflatifrons

### Long-haired Wrinklehead

**SIZE RANGE:** 6.8–9.1 mm
**IDENTIFICATION:** Male sternite 2 with abundant, erect, long pile, longer than on sternite 1. Female sternite 1 with abundant, long, erect pile.
**ABUNDANCE:** Uncommon. **FLIGHT TIMES:** Late May to early July (from mid-February in the south).
**NOTES:** The only floral record is from *Malus pumila*.

## *Chrysogaster* – compare with *Chrysosyrphus* (page 256)

short pedicel bristles
female facial pollinosity forms band
antennae insert at or below midpoint of head

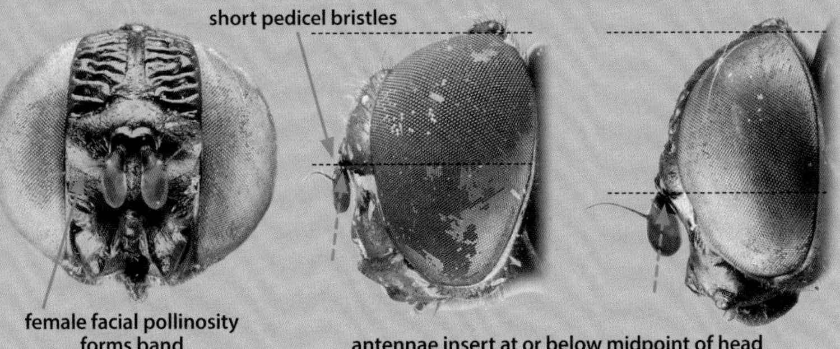

| *Chrysogaster antitheus* | *Chrysogaster inflatifrons* |
| --- | --- |

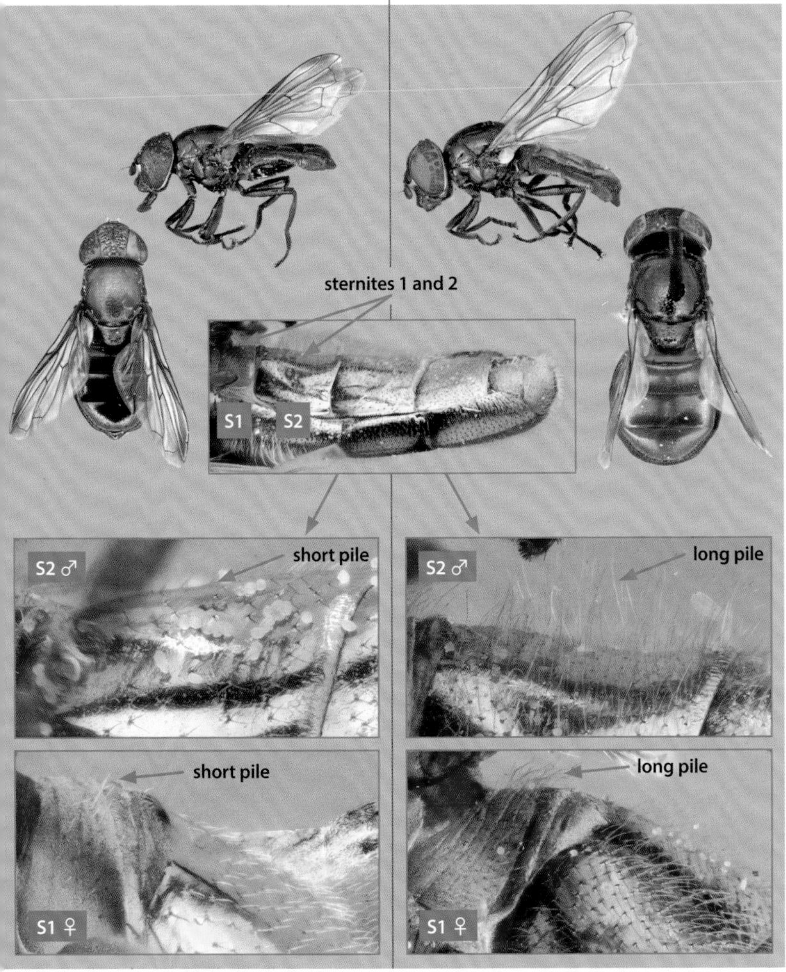

sternites 1 and 2
S1
S2
S2 ♂
short pile
S1 ♀
short pile
S2 ♂
long pile
S1 ♀
long pile

# Chrysosyrphus

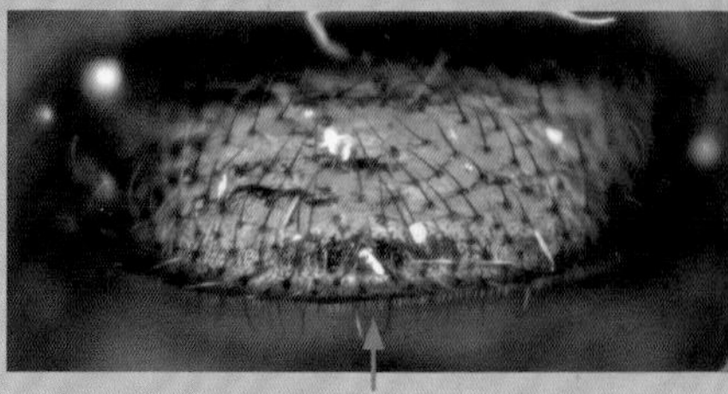


This genus has sometimes been treated as part of *Chrysogaster* but molecular data show that it should be treated separately. Adults are most easily distinguished from *Chrysogaster* by the presence of a subscutellar fringe of pile. The antennae are also distinctive as they are inserted on the upper ⅓ of the head. The pedicel has long distinct bristles and the female facial pollinosity is weak, forming an inverted triangle widely separated from the eyes. There are six species of *Chrysosyrphus*, five of them Nearctic and one found in our region. Larvae are unknown but are likely aquatic like their relatives. There is no workable key to Nearctic *Chrysosyrphus*.

## *Chrysosyrphus latus* Variable Wrinklehead

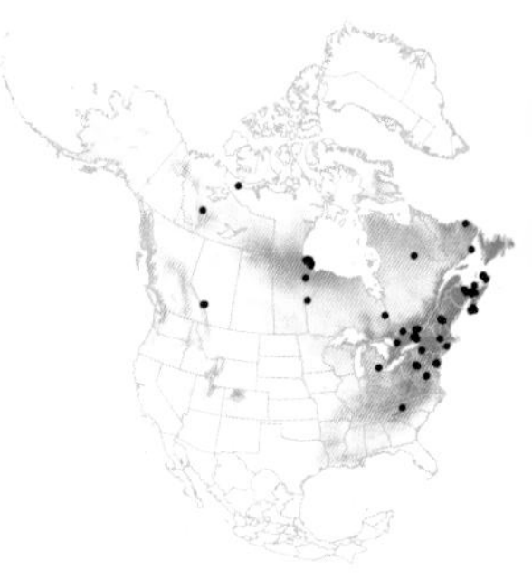

**SIZE RANGE:** 5.5–6.5 mm

**IDENTIFICATION:** Males of our eastern *Chrysosyrphus* can easily be distinguished from all other *Chrysosyrphus* species (not included in this guide) by the presence of a pile fringe on the frons. The thorax and abdomen of this species are covered in long, yellow or black pile (the species is extremely variable for this trait). Male pile is erect while female pile is appressed. **ABUNDANCE:** Rare. **FLIGHT TIMES:** Early May to early August. **NOTES:** *Chrysosyrphus frontosus* was formerly considered to be widespread, but we now recognize it as a Mexican endemic. Three former synonyms of *C. frontosus* are actually synonyms of *C. latus* (*C. bigelowi*, *C. ithaca*, and *C. versipellis*). Flowers visited include *Chamaedaphne*, *Crataegus*, *Fragaria*, and *Prunus*. Records with habitat data are all from bogs.

## *Chrysosyrphus* – compare with *Chrysosgaster (page 254)*

long pedicel bristles
female facial pollinosity forming inverted triangle
antennae insert on upper ⅓ of head

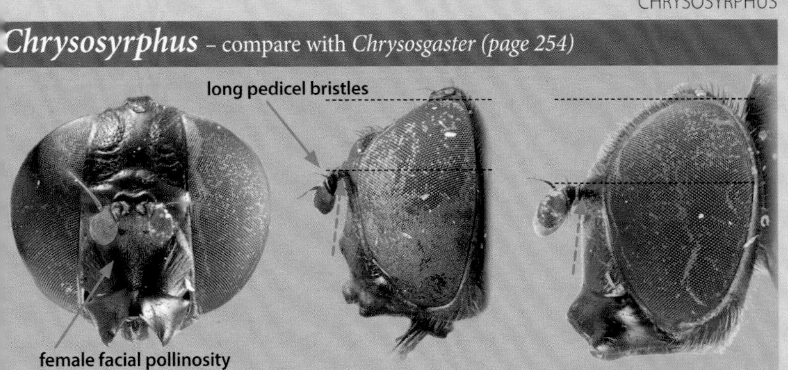

## *Chrysosyrphus latus*

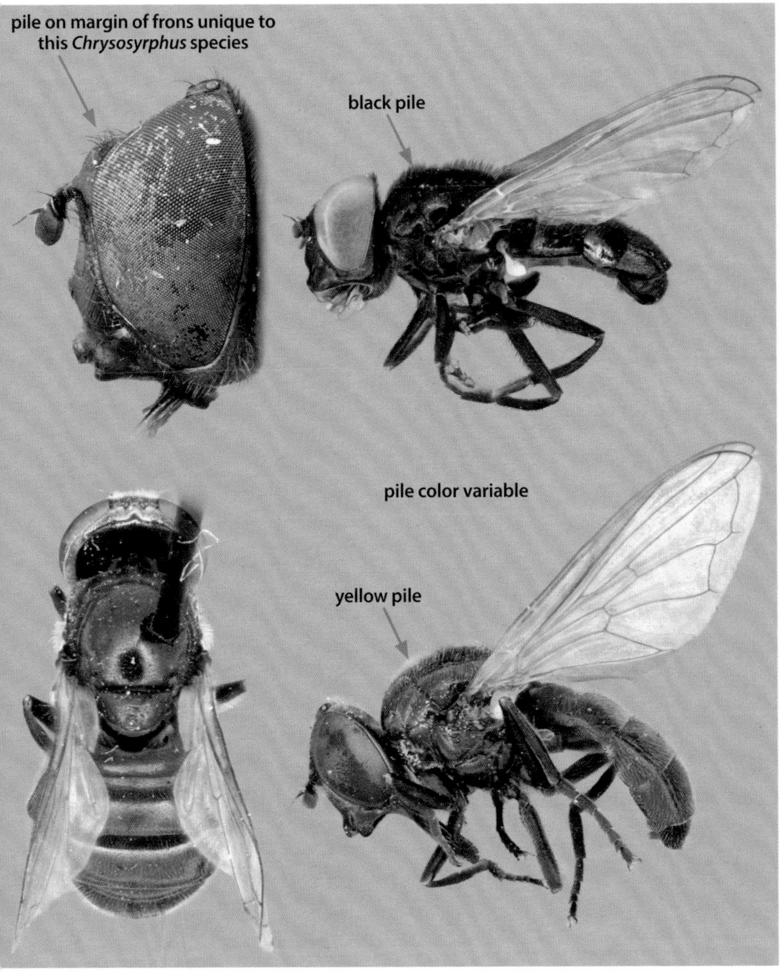

pile on margin of frons unique to this *Chrysosyrphus* species
black pile
pile color variable
yellow pile

# Lejota

Two of the world's seven species of *Lejota* occur in North America. Our two species are easy to distinguish from each other, but not so easy to distinguish from other black syrphids. *Lejota* species are completely black syrphids with a distinctly produced frontal prominence, straight face below the frontal prominence, vein $R_{4+5}$ + $M_1$ not longer than crossvein h, and vein $M_1$ directed toward the apex. Larvae of this genus occur in decaying heartwood of well-decayed tree stumps, and adults are best searched for around fallen timber. Fluke and Weems (1956) were the last to treat these species and provide a key.

## *Lejota aerea* Golden Trunksitter

**SIZE RANGE:** 7.0–9.8 mm
**IDENTIFICATION:** Partly orange legs; basal two tarsomeres, apex of femur, and base of tibia are all orange. Body covered in golden pile.
**ABUNDANCE:** Uncommon. **FLIGHT TIMES:** Early May to mid-June (from early March in the south).
**NOTES:** This is a hardwood forest species and has been found associated with bare, dry, weathered logs. There are no floral records for this species.

## *Lejota cyanea* Cobalt Trunksitter

**SIZE RANGE:** 6.8–8.4 mm
**IDENTIFICATION:** All black legs. Body with a bluish sheen, covered in white to light yellow pile.
**ABUNDANCE:** Uncommon. **FLIGHT TIMES:** Late April to mid-June. **NOTES:** This species has been recorded from hardwood forest, spruce forest, and the edge of a bog. Flowers visited include *Fragaria* and *Prunus*.

# Lejota

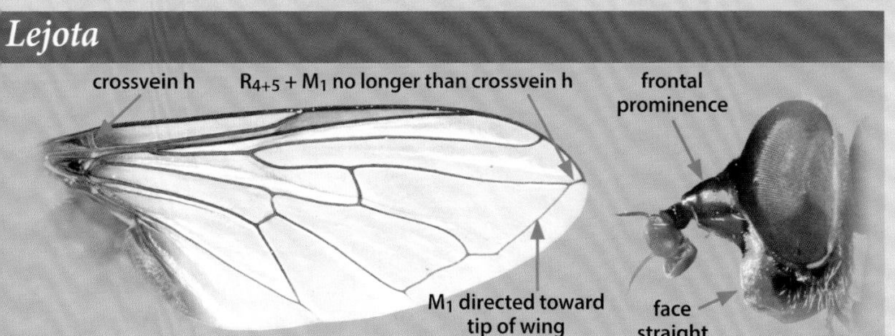


## Lejota aerea

## Lejota cyanea

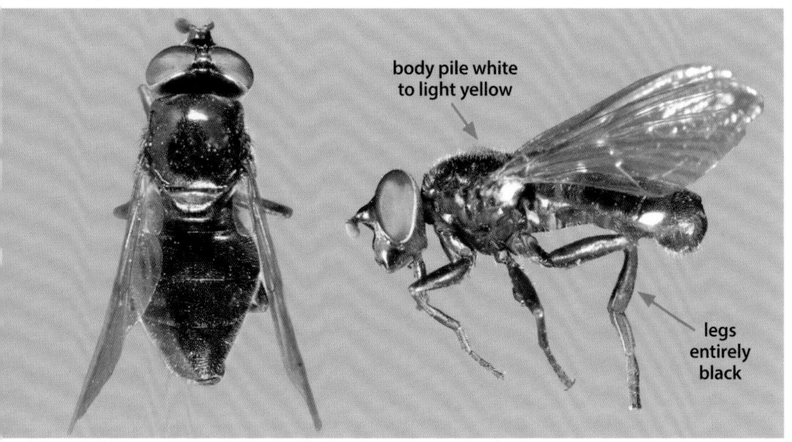

# *Cynorhinella*

Species in this genus have a distinctive, black, elongate face with a central tubercle. There is only one other species of *Cynorhinella*, *C. bella*, which occurs in the western Nearctic and is larger in size. The larvae are unknown.

## *Cynorhinella longinasus* Eastern Longnose

**SIZE RANGE:** 5.6–7.2 mm
**IDENTIFICATION:** The face of this species is tuberculate and produced forward. These flies are dark with a shiny body and bare eyes and arista. **ABUNDANCE:** Rare. **FLIGHT TIMES:** Early April to late June. **NOTES:** This species is very rarely encountered away from hilltops. On hilltops, it sits on vertical surfaces (tree trunks and towers). Larvae may be in fallen trees (adults have been collected perching on fallen tree trunks). Flowers visited include *Acer spicatum* (great for small syrphids), *Amelanchier*, and *Prunus*.

# *Hiatomyia*

*Hiatomyia* is an endemic Nearctic genus with 21 species, all but one of which occur in western North America. Although they look superficially like *Cheilosia*, they can quickly be identified by a combination of their bare eyes and plumose arista. As with *Cheilosia*, the face is tuberculate. For a key to species, see Hull and Fluke (1950). Larval habits are unknown.

## *Hiatomyia cyanescens* Cobalt Deltawing

**SIZE RANGE:** 5.9–9.2 mm
**IDENTIFICATION:** This species has a plumose arista, a bare eye, and a dark, shiny body. **ABUNDANCE:** Uncommon. **FLIGHT TIMES:** Mid-May to mid-July. **NOTES:** In the field, *Hiatomyia* species, often hold their wings at a 45-degree angle when at rest. Other black syrphids rarely hold their wings like this. Flowers visited include *Acer*, *Aruncus*, *Physocarpus*, and *Viburnum*.

## Cynorhinella longinasus

long, black
tuberculate face

## Hiatomyia cyanescens

plumose
arista

bare eye

tuberculate
face

# *Cheilosia*

*Cheilosia* species are small dark syrphids with a distinctive facial tubercle and usually bare antennae. This is the most diverse genus of syrphids in the world, with 496 currently recognized species. Eighty-two of these occur in the Nearctic and 25 are within the range of this guide. One eastern species of uncertain status, *C. tantalus*, is known only from the missing holotype and is not included in this guide. Only five species are currently considered to be Holarctic in distribution, but this is certainly an underestimate as the faunas have never been thoroughly compared. *Cheilosia* is arguably the most taxonomically intractable genus of syrphids in North America and is in dire need of revision. The eastern species are much better resolved than the western species but the taxonomy presented here must be viewed as tentative. Despite this caution, species concepts presented here have been formulated by referring to DNA barcode data, reexamining morphology, and checking primary types. For the most recent keys to the Nearctic specimens, see Fluke and Hull (1945, 1946) and Hull and Fluke (1950).

To help with identification, *Cheilosia* is divided into groups below based on combinations of characters. Examples of these important characters are given on the next page. Female *Cheilosia* are difficult to distinguish and are often not identifiable. Most *Cheilosia* larvae are quite host specific and feed in (miners) or on living plant tissue (roots, stems, and/or leaves). A few species feed on basidiomycete mushrooms (such as *Amanita*, *Boletus*, *Leccinum*, and *Suillus*), one species feeds on truffles (ascomycete mushrooms in the genus *Tuber*), at least one species feeds on rotting plant tissues, and one specialized European species lives in resin outflows from bark beetle (Scolytidae) larval holes on spruce trees. Although *Cheilosia* species may be locally abundant on flowers in our region, they do not come close to having the abundance and local diversity that they have in Europe and Asia. There they dominate in meadows, with multiple species found per location.

# *Cheilosia* species groups

Eyes bare, face bare, antennal pits joined

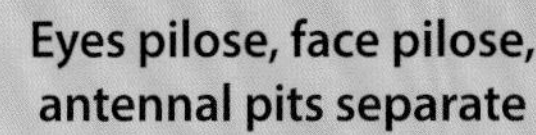

Eyes pilose, face pilose, antennal pits separate

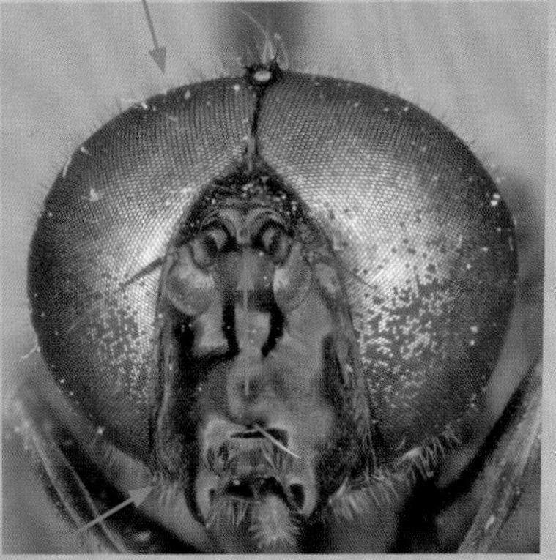

Katepisternal pile separated

Katepisternal pile joined

Scutellar bristles absent

Scutellar bristles present

## Eye and face pilose

### *Cheilosia primoveris* Pale-haired Blacklet

**SIZE RANGE:** 8.0–9.6 mm **IDENTIFICATION:** Eye pilose, face pilose, thorax covered in yellow pile, legs partly yellow at joint between the femur and tibia, antenna yellow brown. Katepisternal pile separated. Antennal pits separated by a flat chitinous extension. **ABUNDANCE:** Rare. **FLIGHT TIMES:** Mid-April to early June. **NOTES:** Flowers visited include *Caltha* and *Prunus*.

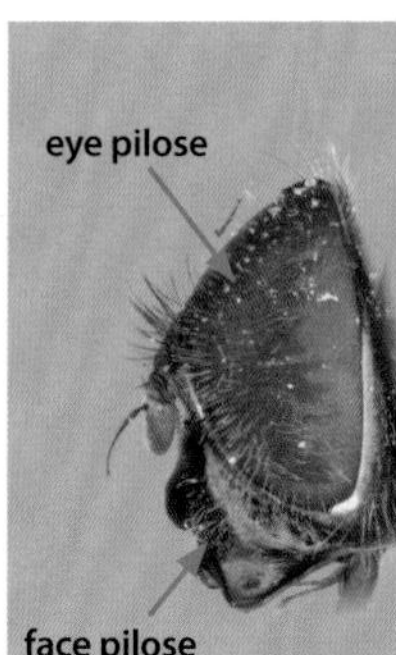


### *Cheilosia pontiaca* Ebony Blacklet

**SIZE RANGE:** 7.8–9.7 mm **IDENTIFICATION:** Eye pilose, face pilose, thorax covered in black pile, legs almost entirely black, antenna orange. Katepisternal pile separated. Antennal pits separated. **ABUNDANCE:** Rare. **FLIGHT TIMES:** Mid-May to late June. **NOTES:** This is also a deciduous and mixed forest species. Most known specimens were collected on the summit of Mont Rigaud, Quebec. Known to visit flowers of *Cornus canadensis*.

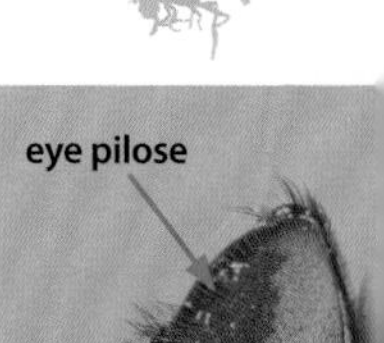


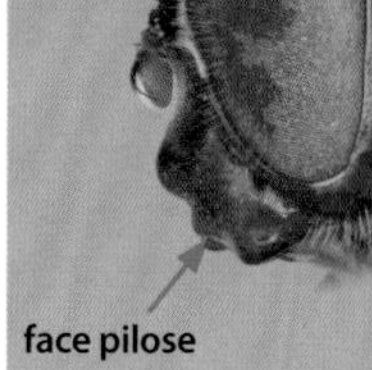


### *Cheilosia yukonensis* Yukon Blacklet

**SIZE RANGE:** 8.5–10.2 mm **IDENTIFICATION:** Eye pilose, face pilose, thorax covered in black pile, legs almost entirely black, antenna orange. Katepisternal pile joined. Antennal pits separated. **ABUNDANCE:** Uncommon. **FLIGHT TIMES:** Mid-June to early September. **NOTES:** Nothing is known about this species.

## *Cheilosia primoveris*

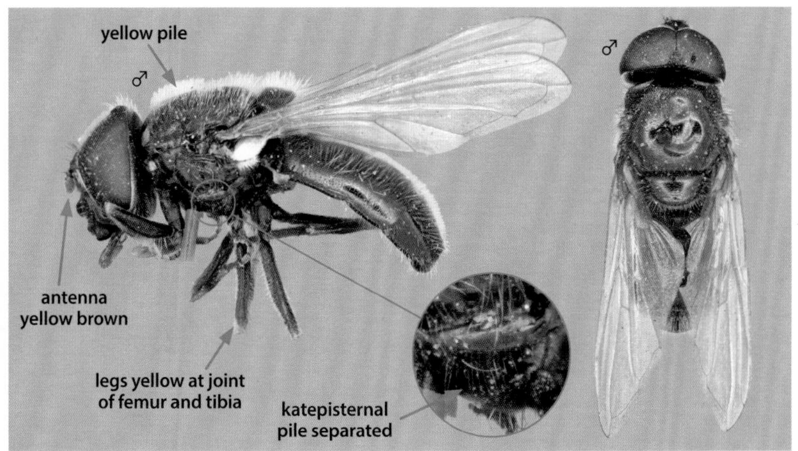


## *Cheilosia pontiaca*

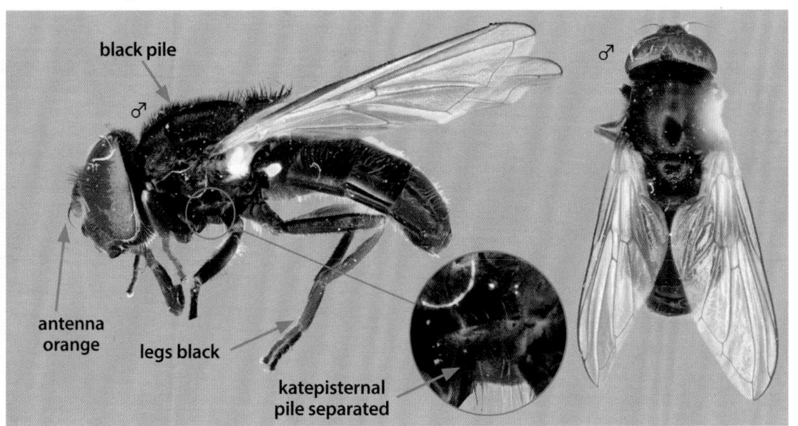


## *Cheilosia yukonensis*

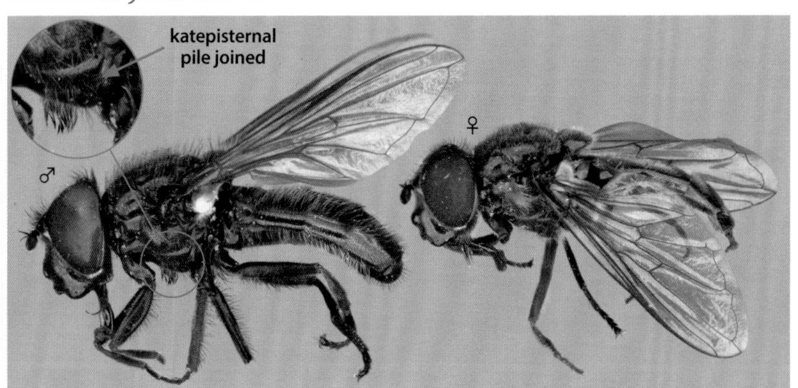

## Eye pilose, face bare, scutellar bristles present

### *Cheilosia subchalybea* Ultramarine Blacklet

**SIZE RANGE:** 7.1–7.6 mm

**IDENTIFICATION:** Eye pilose, face bare. Head oval from frontal view, scutellum with bristles, body blue black, face shiny, antenna orange, pile on scutum and scutellum black. Katepisternal pile separated. **ABUNDANCE:** Rare. **FLIGHT TIMES:** Mid-May to mid-June. **NOTES:** One record is from mixed forest. Nothing else is known about this species.

### *Cheilosia albitarsis* Buttercup Blacklet

**SIZE RANGE:** 9.6–10.0 mm

**IDENTIFICATION:** Eye of male pilose, face bare. Head oval from frontal view, scutellum with bristles, body black, sometimes with bluish sheen; antenna dark. Legs entirely black except sometimes tarsi 2–4 on pro- and/or mesolegs. Female with bare eye, antennal pits separated. Katepisternal pile separated. **ABUNDANCE:** Common. **FLIGHT TIMES:** Late May to late July. **NOTES:** This species is found in a variety of habitats from forests to agricultural fields to montane forest meadows in Europe, but it is a synanthropic species, favored by current farming practices. Adults are found along forest tracks and the edges of clearings, along hedges, and in pastures. Males may be found hovering at heights up to 5 m in these locations but they also regularly settle on foliage of shrubs or low-growing plants. Flowers visited include *Ajuga*, *Allium*, *Caltha*, *Crataegus*, *Matricaria*, *Potentilla*, *Ranunculus*, *Sorbus*, and *Stellaria*, but it is most reliably found on *Ranunculus* in our region. Larvae may feed on roots of *Ranunculus*, but this needs to be confirmed. *Cheilosia albitarsis* is sometimes considered a synonym of *C. bardus*, but the type specimen of the latter is missing and the species concept is thus uncertain. Until the genus is completely revised, we prefer to use *C. albitarsis* for this species. The distribution suggests that this species is introduced in the Nearctic.

# *Cheilosia subchalybea*

katepisternal
pile separated
body blue black
♀
black pile
♂
antenna
orange

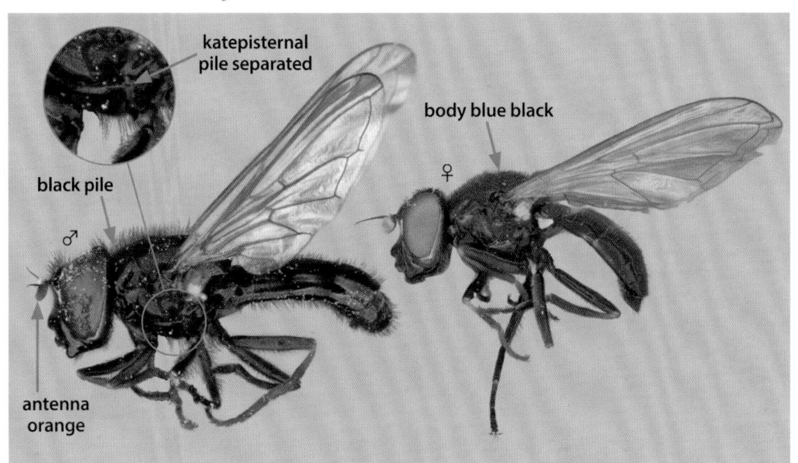

# *Cheilosia albitarsis*

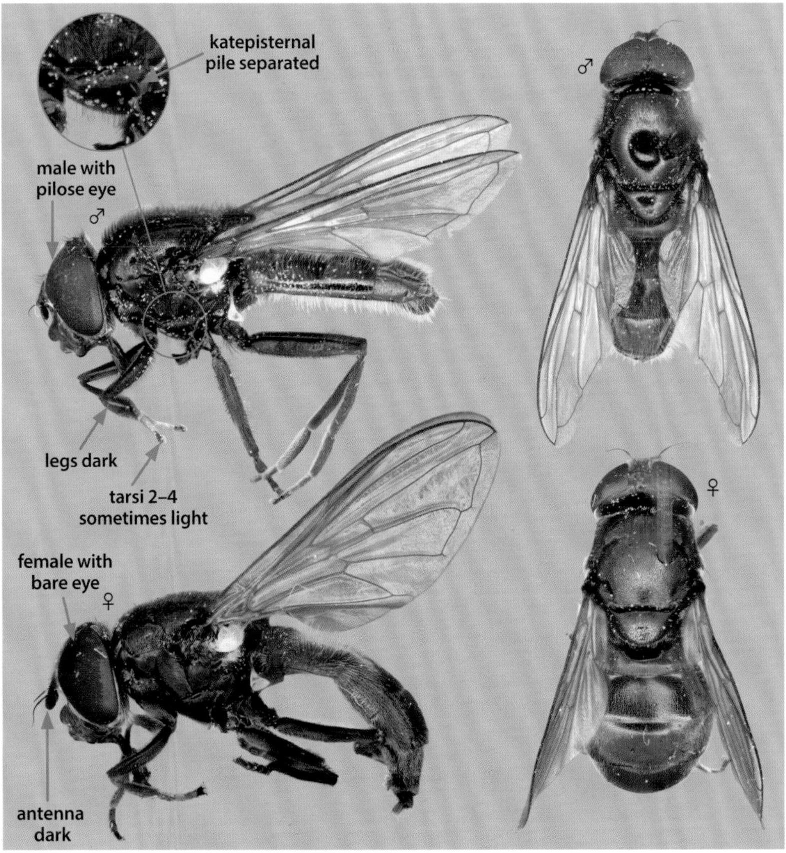

katepisternal
pile separated
♂
male with
pilose eye
♂
legs dark
tarsi 2–4
sometimes light
♀
female with
bare eye
♀
antenna
dark

## *Cheilosia hunteri* Hunter's Blacklet

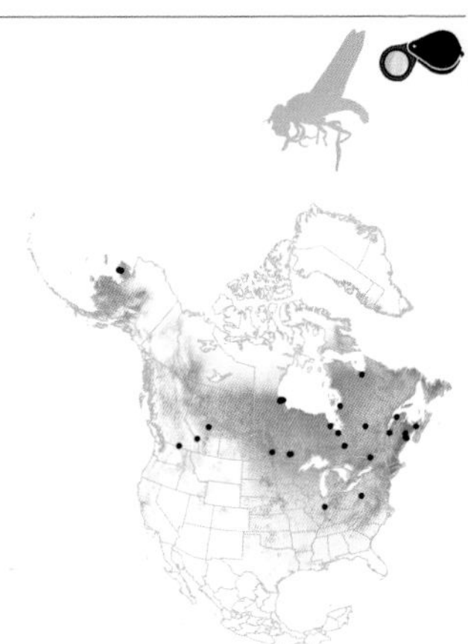

**SIZE RANGE:** 9.4–11.2 mm
**IDENTIFICATION:** Eye pilose, face bare. Head oval from frontal view, scutellum with bristles; legs partly yellow. Katepisternal pile separated.
Male with tergite 4 extensively pale pilose, pleura pale pilose. Female with face produced forward, body pale pilose, usually yellowish.
**ABUNDANCE:** Uncommon. **FLIGHT TIMES:** Mid-May to late July. **NOTES:** There is a single floral record from *Heracleum*. *Cheilosia nigrofasciata* is treated here as a new synonym of *C. hunteri*.

## *Cheilosia orilliaensis* Black-backed Blacklet

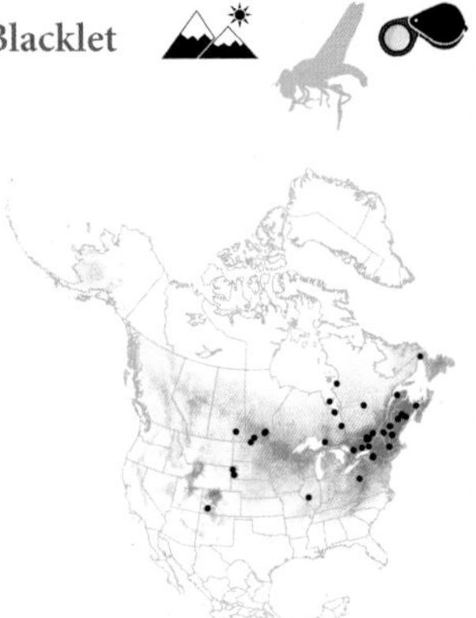

**SIZE RANGE:** 7.7–9.9 mm
**IDENTIFICATION:** Eye pilose, face bare. Head oval from frontal view, scutellum with bristles; legs partly yellow. Katepisternal pile joined. Male with abundant black pile on at least posterior ½ of tergite 4, pleura black pilose. Female with face not produced forward, body pale pilose, usually whitish. **ABUNDANCE:** Common. **FLIGHT TIMES:** Early May to early August. **NOTES:** We treat *C. consentiens* (known only from the holotype female) as a new synonym of *C. orilliaensis*. A single floral record is from *Trillium*.

# *Cheilosia hunteri*

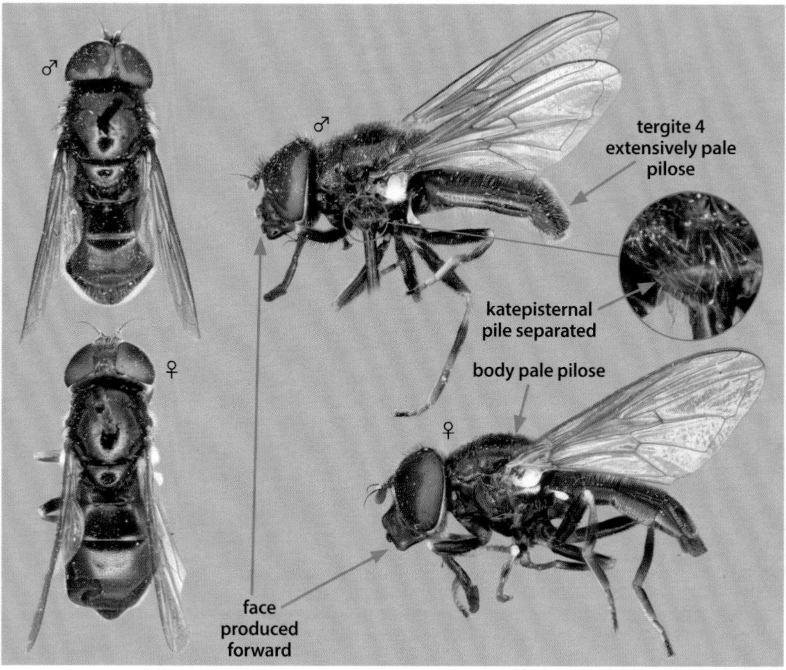

♂

♂

tergite 4 extensively pale pilose

katepisternal pile separated

♀

body pale pilose

♀

face produced forward

# *Cheilosia orilliaensis*

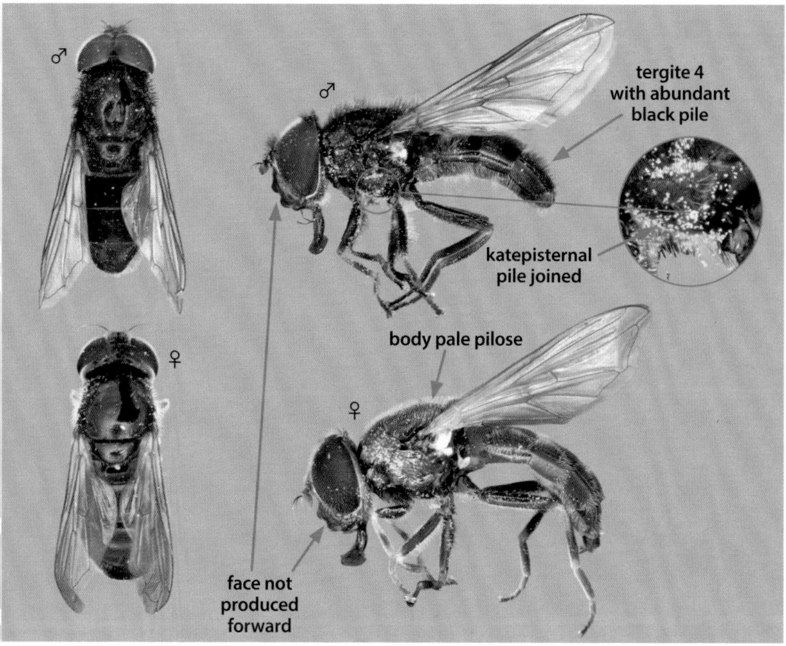

♂

♂

tergite 4 with abundant black pile

katepisternal pile joined

body pale pilose

♀

♀

face not produced forward

## Eye pilose, face bare, scutellar bristles absent

# *Cheilosia lasiophthalma*

### Amber-haired Blacklet

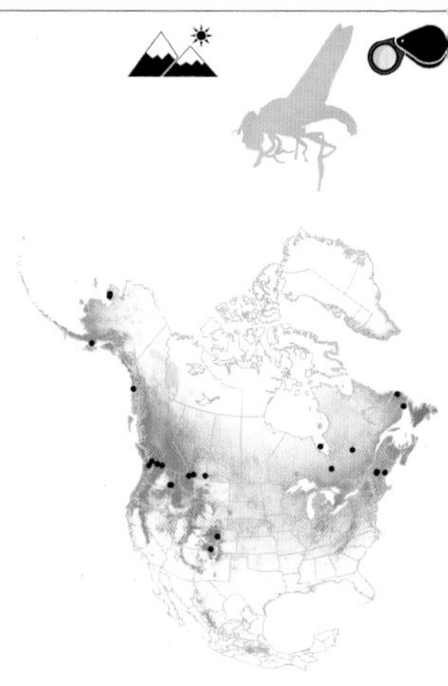

**SIZE RANGE:** 11.7–12.8 mm
**IDENTIFICATION:** Eye pilose, face bare. Head triangular from frontal view, large (greater than 10 mm), body densely yellow pilose, anepimeron yellow to brown pilose. Katepisternal pile joined.
**ABUNDANCE:** Uncommon. **FLIGHT TIMES:** Late May to mid-July (from early April in the west).
**NOTES:** *Cheilosia browni* and *C. nigroapicata* are both treated here as new synonyms of *C. lasiophthalma*. Flowers visited include *Crataegus* and *Prunus*.

# *Cheilosia bigelowi* Bigelow's Blacklet

**SIZE RANGE:** 8.6–9.1 mm
**IDENTIFICATION:** Eye pilose, face bare. Head oval from frontal view, body not unusually pilose, procoxa with a small spur on lateral edge. Scutellum without black bristles. Katepisternal pile joined. Male frontal pile black. Tergite 4 white pilose on male. **ABUNDANCE:** Rare. **FLIGHT TIMES:** Early to late June. **NOTES:** Nothing is known about the ecology of this species.

## *Cheilosia lasiophthalma*

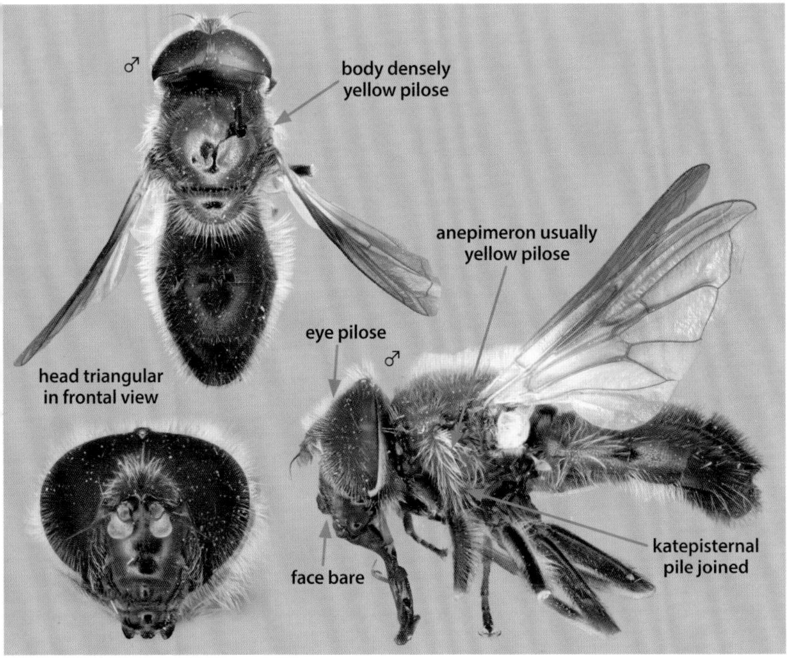

♂
body densely yellow pilose
anepimeron usually yellow pilose
eye pilose
♂
head triangular in frontal view
katepisternal pile joined
face bare

## *Cheilosia bigelowi*

♂
♂
tergite 4 white pilose
male frons black pilose
head oval in frontal view
small spur on lateral edge of procoxa

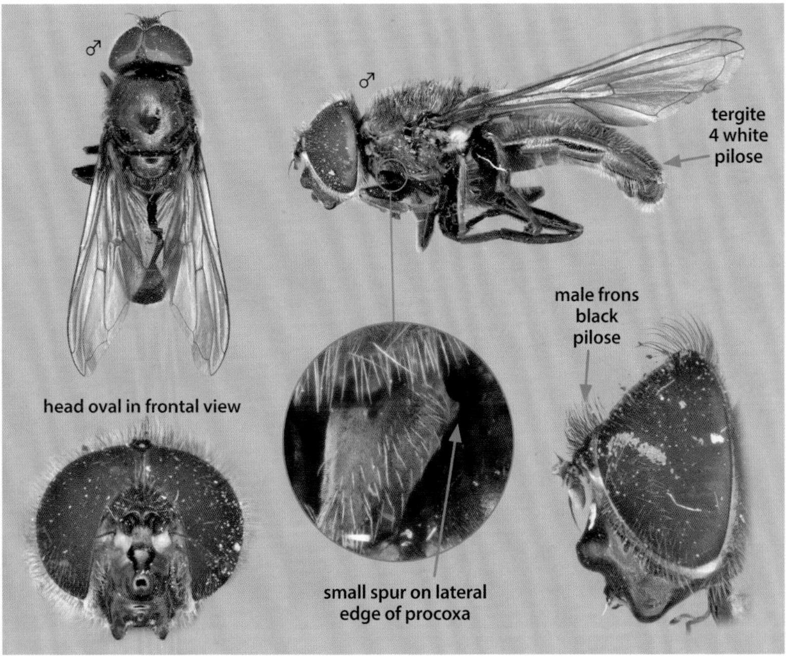

## *Cheilosia cynoprosopa* White-fronted Blacklet

**SIZE RANGE:** 8.7–9.7 mm

**IDENTIFICATION:** Eye pilose, face bare. Head oval from frontal view, body not unusually pilose, scutellum without bristles, front and thorax largely white pilose. Katepisternal pile joined. Tergite 4 white pilose on male. Female metatarsus with entirely pale setae. **ABUNDANCE:** Rare. **FLIGHT TIMES:** Early May to late July. **NOTES:** One specimen was collected from *Solidago* flowers.

## *Cheilosia swannanoa* Autumn Blacklet

**SIZE RANGE:** 7.9–9.5 mm

**IDENTIFICATION:** Eye pilose, face bare. Katepisternal pile joined. Posterior ½ of tergites 2–4 extensively black pilose on male. Frons black pilose or white with some black pile. **ABUNDANCE:** Rare. **FLIGHT TIMES:** Mid-August to late September. **NOTES:** Flowers visited include *Eupatorium* and *Solidago*. Many specimens were found around moist soil or puddles while others were found hovering along the edge of woods.

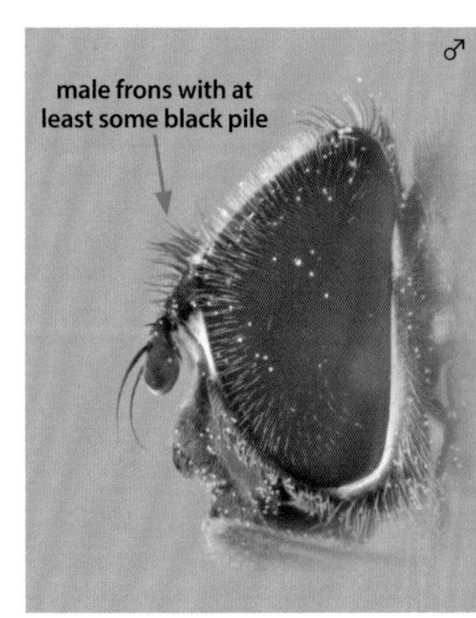

# *Cheilosia cynoprosopa*

♂
frons and thorax covered in light pile
♂
scutellum without bristles
♀
metatarsus with entirely pale setae
♀
tergite 4 white pilose
♂

# *Cheilosia swannanoa*

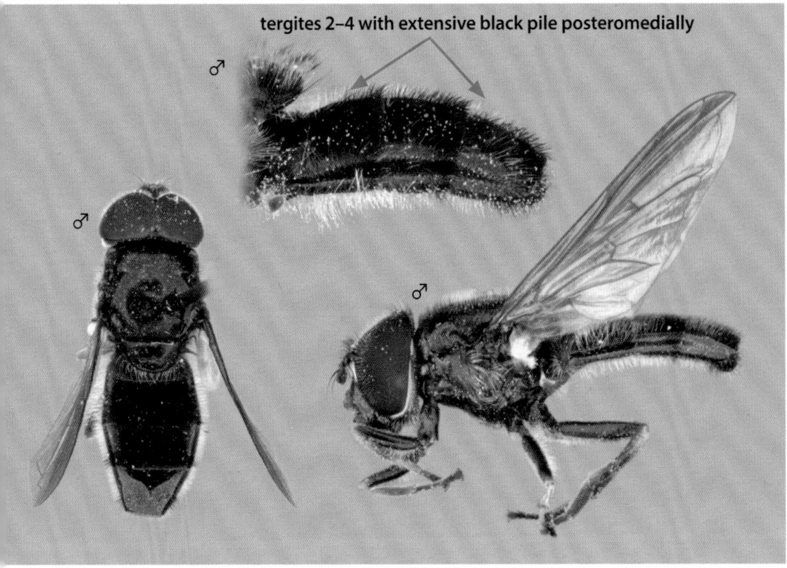

tergites 2–4 with extensive black pile posteromedially
♂
♂
♂

## Eye bare and face pilose

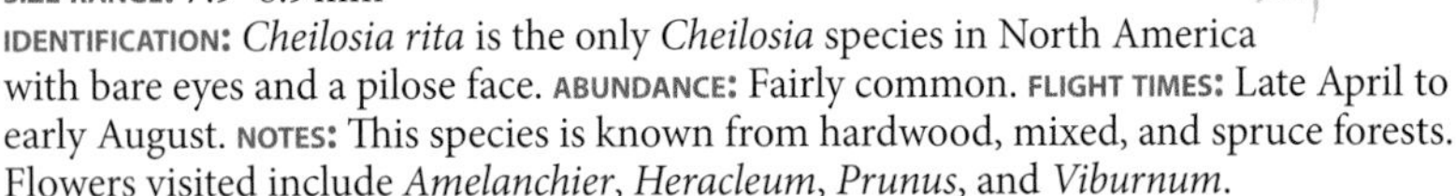

# *Cheilosia rita* Inky Blacklet

**SIZE RANGE:** 7.9–8.9 mm
**IDENTIFICATION:** *Cheilosia rita* is the only *Cheilosia* species in North America with bare eyes and a pilose face. **ABUNDANCE:** Fairly common. **FLIGHT TIMES:** Late April to early August. **NOTES:** This species is known from hardwood, mixed, and spruce forests. Flowers visited include *Amelanchier*, *Heracleum*, *Prunus*, and *Viburnum*.

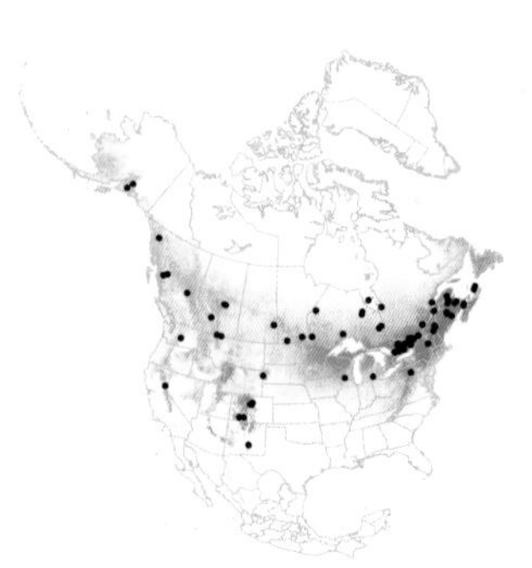

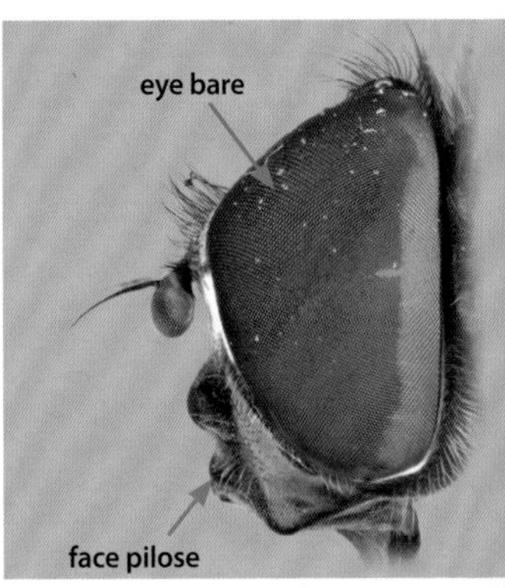


## Eye bare, face bare, antennal pits separate

# *Cheilosia wisconsinensis* Golden-legged Blacklet

**SIZE RANGE:** 8.0–9.2 mm
**IDENTIFICATION:** Eye bare, face bare, bristles on scutellum, antennal pits separated by chitinous extension. Female legs yellow. **ABUNDANCE:** Rare. **FLIGHT TIMES:** Mid-May to late August. **NOTES:** Nothing is known about this species. *Cheilosia albitarsis* females will key here but can be distinguished by dark antennae and legs.

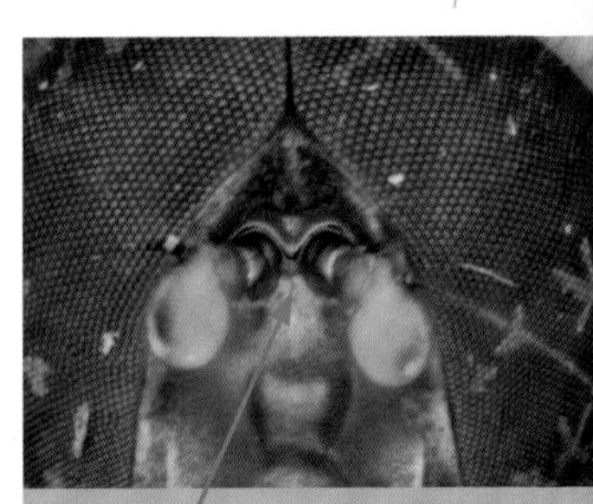

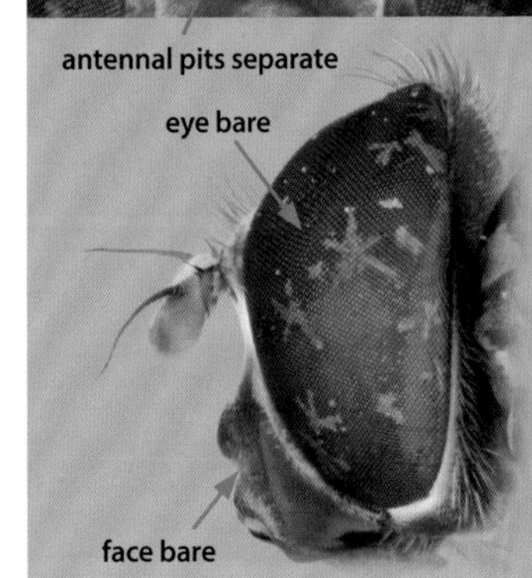

## *Cheilosia rita*

## *Cheilosia wisconsinensis*

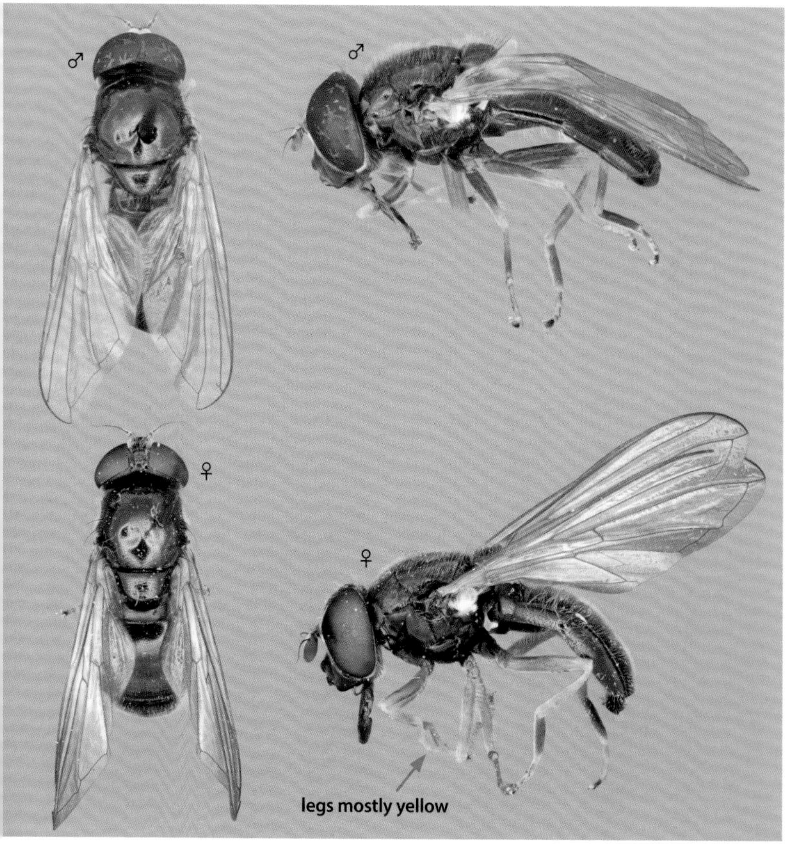

♂

♂

♀

♀

legs mostly yellow

**Eye bare, face bare, antennal pits joined, scutellar bristles absent**

## *Cheilosia capillata* Scar-horned Blacklet

**SIZE RANGE:** 10.1–10.4 mm
**IDENTIFICATION:** Eye bare, face bare, no bristles on scutellum, katepisternal pile patches joined. Male frontal pile yellow white, basal edge of costa with long, black bristles, tibiae mostly yellow. Female flagellum with long sensory pit on inner side. **ABUNDANCE:** Uncommon. **FLIGHT TIMES:** Early April to late June. **NOTES:** *Aruncus* is the only flower noted as visited.

## *Cheilosia comosa* Jack Pine Blacklet

**SIZE RANGE:** 7.3–8.9 mm
**IDENTIFICATION:** Eye bare, face bare, no bristles on scutellum, katepisternal pile patches broadly separated. Male frontal pile yellow white or mixed, not entirely black, basal edge of costa with very short pale bristles, some short black ones mixed in, tibia yellow with distinctive black band medially. Female metatibia mostly black, metabasitarsus orange to reddish brown, not contrasting with other tarsomeres; flagellum small, less than 1/2 as long as frontal width. **ABUNDANCE:** Rare. **FLIGHT TIMES:** Mid-April to early June. **NOTES:** Habitat notes indicate it can be found in jack pine and sandy spruce communities. *Cheilosia caltha* and *C. sensua* are treated as new synonyms of *C. comosa*.

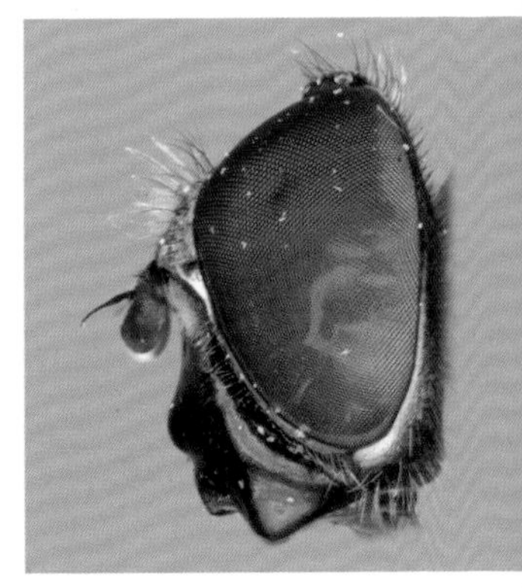

## Cheilosia capillata

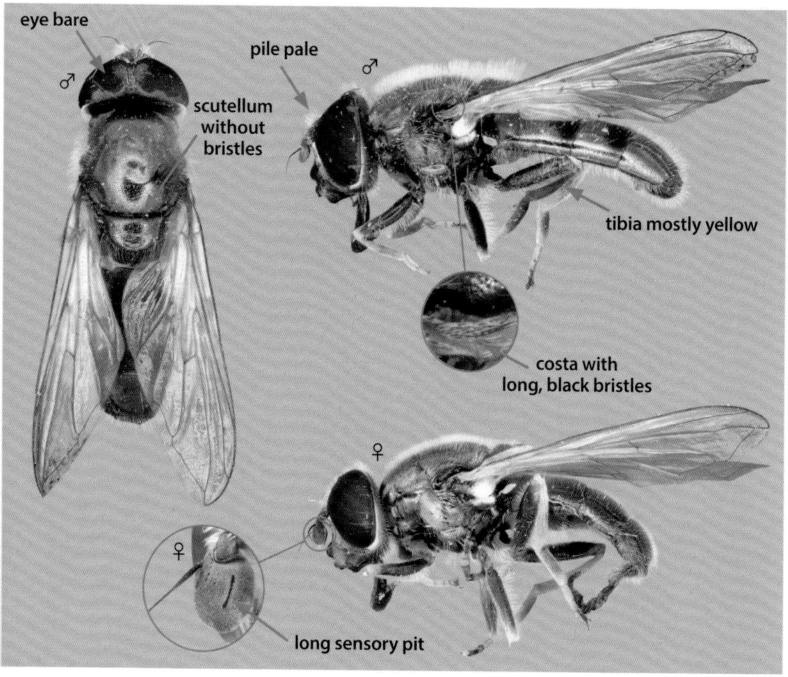

eye bare
♂
pile pale
♂
scutellum without bristles
tibia mostly yellow
costa with long, black bristles
♀
♀
long sensory pit

## Cheilosia comosa

♂
frontal pile at least partly yellow white
♂
tibia with black band
♀
♀

## *Cheilosia prima* Swarthy Blacklet

**SIZE RANGE:** 7.8–10.4 mm

**IDENTIFICATION:** Eye bare, face bare, no bristles on scutellum, katepisternal pile patches broadly separated. Male frontal pile black, frontal triangle small, shorter than eye contiguity, pro- and metatarsi with tarsomeres 2–4 orange; mesotarsus with tarsomeres 1–4 orange. Female metatibia yellow, metabasitarsus brown, strongly contrasting with other tarsomeres; flagellum large, about 2/3 as long as frontal width. **ABUNDANCE:** Common. **FLIGHT TIMES:** Late March to late June (late February to early September in the south). **NOTES:** This is a hardwood and mixed forest species. Flowers visited include *Prunus* and *Tussilago*.

**Eye bare, face bare, antennal pits joined, scutellar bristles present**

## *Cheilosia leucoparea* Dusky Blacklet

**SIZE RANGE:** 7.3–9.0 mm

**IDENTIFICATION:** Eye bare, face bare, bristles on scutellum, antennal pits confluent. Antenna yellow, face black, postpronotum black. Mesonotal pile short, subappressed, black. Anepisternum nonpollinose. At least two apical tarsomeres black on pro- and mesoleg. Female frons narrow and flagellum large, so that frons is narrower than flagellum is long. **ABUNDANCE:** Rare. **FLIGHT TIMES:** Late June to late September. **NOTES:** Flowers visited include *Cicuta* and *Symphyotrichum*.

## Cheilosia prima

♂

frontal pile black

♂

♀

katepisternal pile separated

pro- and metatarsi with tarsomeres 2–4 orange

♀

flagellum large

metatibia yellow

metabasitarsus brown

## Cheilosia leucoparea

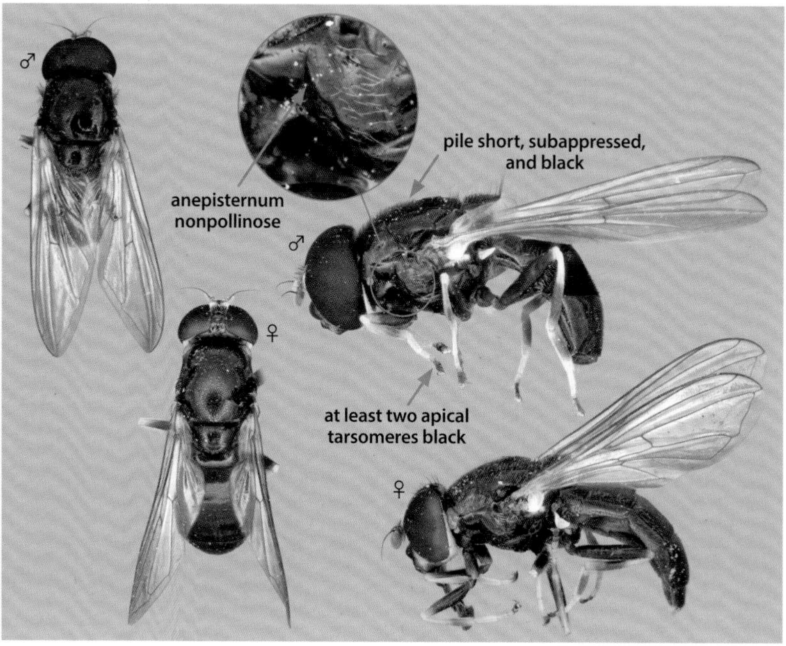

♂

pile short, subappressed, and black

anepisternum nonpollinose

♂

♀

at least two apical tarsomeres black

♀

## *Cheilosia* undescribed species 17-1 Arctic Blacklet

**SIZE RANGE:** 7.5–9.1 mm

**IDENTIFICATION:** Eye bare, face bare, bristles on scutellum, antennal pits confluent. Antenna black, face black, postpronotum black, femora, tibiae, and tarsi entirely black (joints rarely yellow.) Male black pilose on mesonotum, posterior corner of tergite 3 black pilose. Female pale pilose, metabasitarsus swollen, variable in size, but always thicker than hind tibia. **ABUNDANCE:** Uncommon. **FLIGHT TIMES:** Late May to late August. **NOTES:** A Holarctic species restricted to taiga ecoregions. The variable thickness of the female metabasitarsus along with preliminary DNA data suggests that this may be a complex of closely related species.

## *Cheilosia laevis* Jet Blacklet

**SIZE RANGE:** 7.4–8.9 mm

**IDENTIFICATION:** Eye bare, face bare, bristles on scutellum, antennal pits confluent. Antenna black, face black, postpronotum black, femora, tibiae, and tarsi entirely black (joints rarely yellow.) Male admixed black and yellow pilose on mesonotum, posterior corner of tergite 3 pale pilose. Female pale pilose, metabasitarsus never thicker than hind tibia. **ABUNDANCE:** Rare. **FLIGHT TIMES:** Late June to mid-September (outliers from early May to mid-October). **NOTES:** Flowers visited include *Solidago* and *Symphyotrichum*. Records are from hardwood forests and fens.

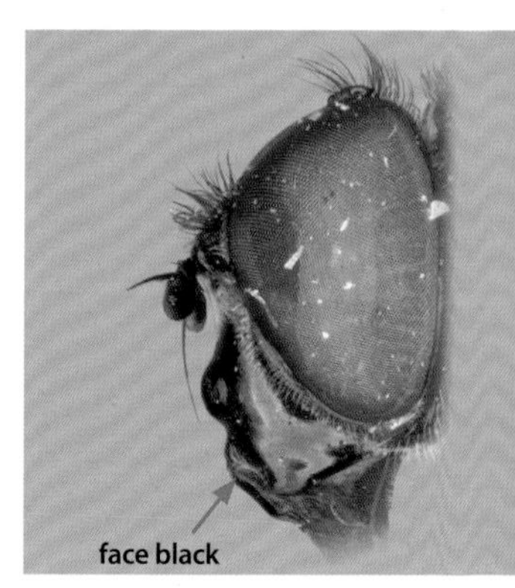

## *Cheilosia* undescribed species 17-1

♂

black pilose

♂

corner of tergite 3 black pilose

legs black

♀

## *Cheilosia laevis*

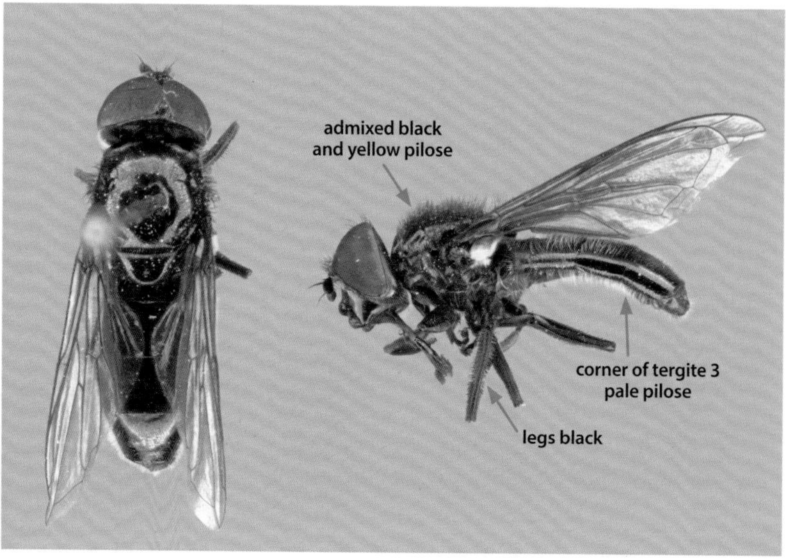

admixed black and yellow pilose

corner of tergite 3 pale pilose

legs black

## *Cheilosia latrans* Steely Blacklet

**SIZE RANGE:** 7.3–9.1 mm

**IDENTIFICATION:** Eye bare, face bare, bristles on scutellum, antennal pits confluent. Antenna orange, sometimes darkly so, postpronotum variable but never bright yellow, femora and tibiae variable in color but never completely black, with at least two apical tarsomeres black on pro- and mesolegs. Male black pilose on mesonotum, with face black and posterior corner of tergite 3 black pilose; tergites 3 and 4 long pilose. Female with face yellow, white pilose, metabasitarsus never swollen. **ABUNDANCE:** Very common. **FLIGHT TIMES:** Mid-May to mid-October (to early November in the south). **NOTES:** This species is found in mixed and coniferous forests, including bogs. Recorded from flowers of *Eupatorium*, *Eutrochium*, *Heracleum*, *Saxifraga*, *Senecio*, *Spiraea*, and *Viburnum*. *Cheilosia tristis* is a synonym.

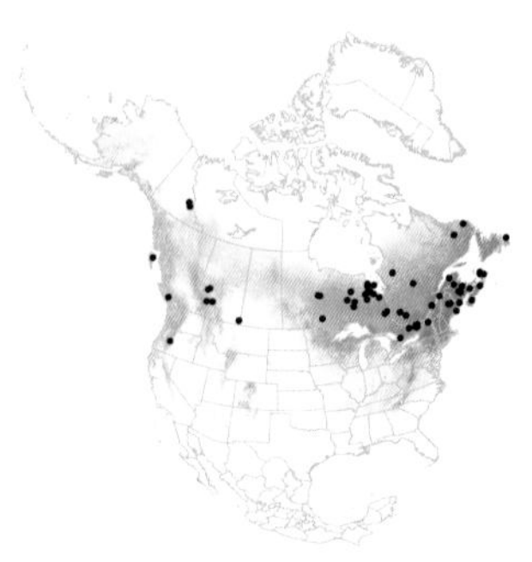

## *Cheilosia shannoni* Shannon's Blacklet

**SIZE RANGE:** 7.3–9.1 mm

**IDENTIFICATION:** Eyes bare, face bare, bristles on scutellum, antennal pits confluent. Antenna orange, sometimes darkly so, face and postpronotum black, femora and tibiae variable in color but never completely black, with at least two apical tarsomeres black on pro- and mesolegs. Male black pilose on mesonotum, with posterior corner of tergite 3 black pilose; tergites 3 and 4 short pilose. Female metabasitarsus conspicuously swollen. **ABUNDANCE:** Uncommon. **FLIGHT TIMES:** Late June to mid-September (outliers from early May to mid-October). **NOTES:** Flowers visited include *Solidago* and *Symphyotrichum*. Records are from hardwood forests and fens.

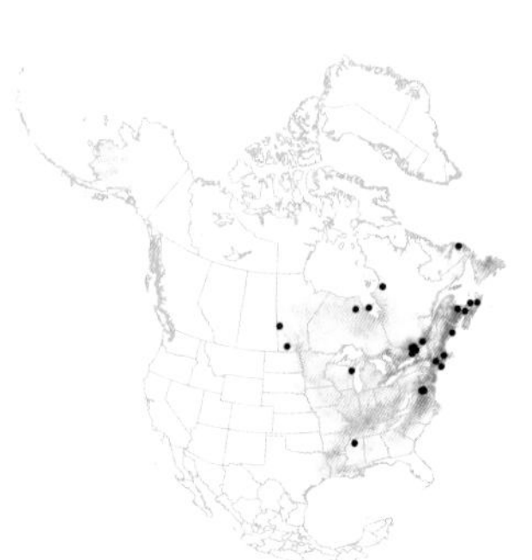

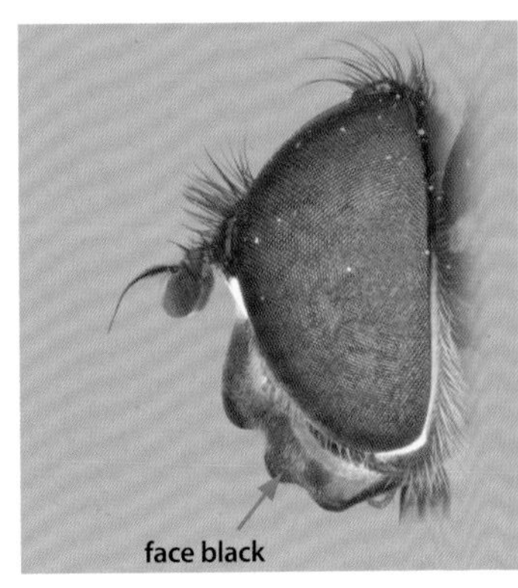

## Cheilosia latrans

♂

black pilose

♂

♂

tergites 3 and 4 long pilose

at least two apical tarsomeres black

♀

corner of tergite 3 black pilose

## Cheilosia shannoni

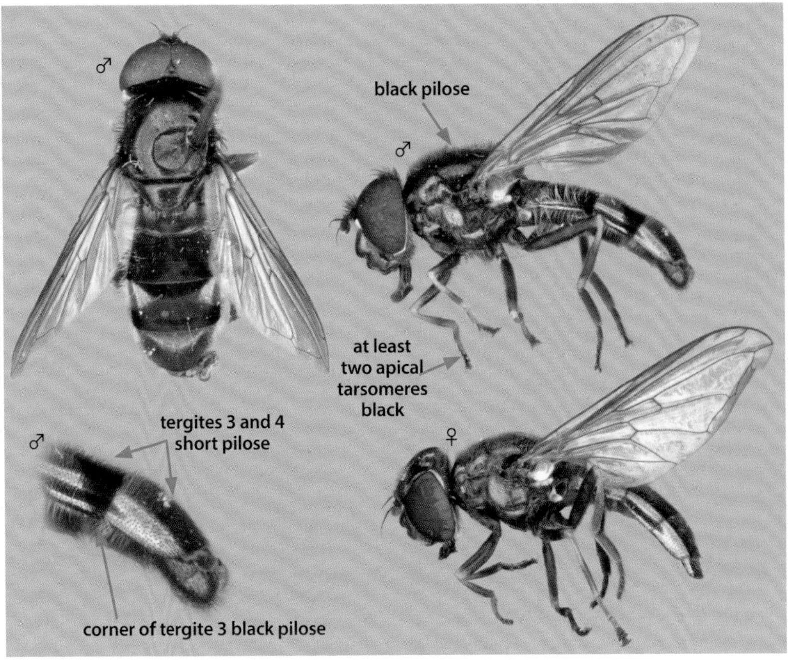

♂

black pilose

♂

at least two apical tarsomeres black

♂

tergites 3 and 4 short pilose

♀

corner of tergite 3 black pilose

## *Cheilosia pallipes* Yellow-shouldered Blacklet

**SIZE RANGE:** 7.3–9.0 mm

**IDENTIFICATION:** Eyes bare, face bare, bristles on scutellum, antennal pits confluent. Antenna orange. Face partly yellow. Femora and tibiae variable in color but never completely black. Yellow pilose on mesonotum, postpronotum yellow. Male pro- and mesoleg with only apical tarsomere black and posterior corner of tergite 3 pale pilose. Female same as above except fifth tarsomere variable. Females are indistinguishable from those of *Cheilosia* undescribed species 76-1. **ABUNDANCE:** Common. **FLIGHT TIMES:** Early June to mid-September (from mid-May in the south). **NOTES:** Flowers visited include *Cryptotaenia*, *Physocarpus*, and *Sambucus*.

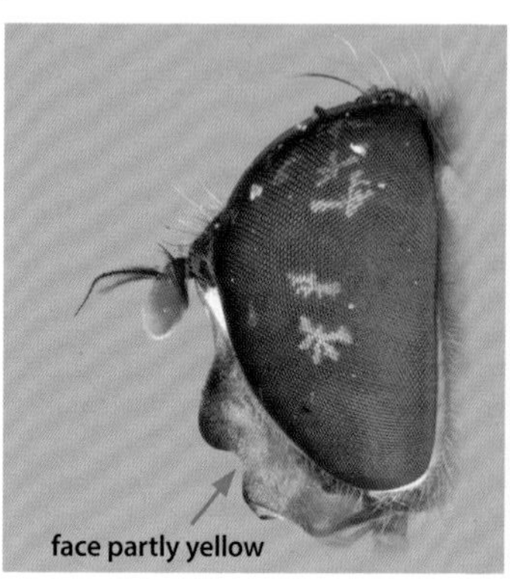


## *Cheilosia* undescribed species 76-1
### Bicolored Blacklet

**SIZE RANGE:** 7.2–9.0 mm

**IDENTIFICATION:** Eye bare, face bare, bristles on scutellum, antennal pits confluent. Antenna orange, sometimes darkly so. Femora and tibiae variable in color but never completely black, with only apical tarsomere black on pro- and mesoleg. Male admixed black and yellow pilose on mesonotum, face black, postpronotum black, pro- and mesoleg with only fifth tarsomere black and posterior corner of tergite 3 pale pilose. Female white pilose with face yellow and postpronotum yellow. Females are indistinguishable from those of *C. pallipes* (see *C. pallipes* for photos). **ABUNDANCE:** Common. **FLIGHT TIMES:** Late May to mid-August. **NOTES:** Recorded from flowers of *Cicuta*, *Heracleum*, *Ilex*, and *Viburnum*.

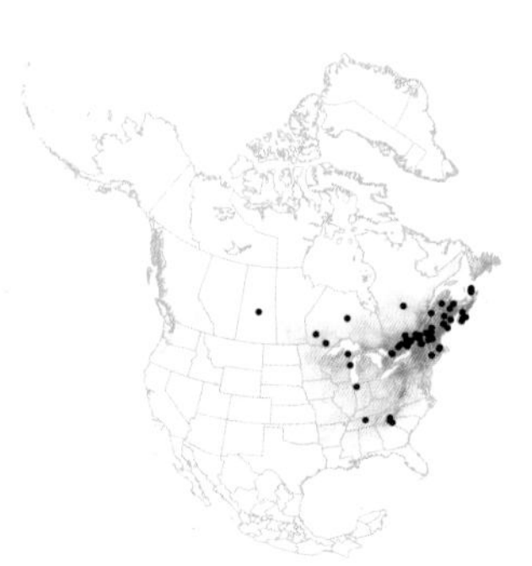

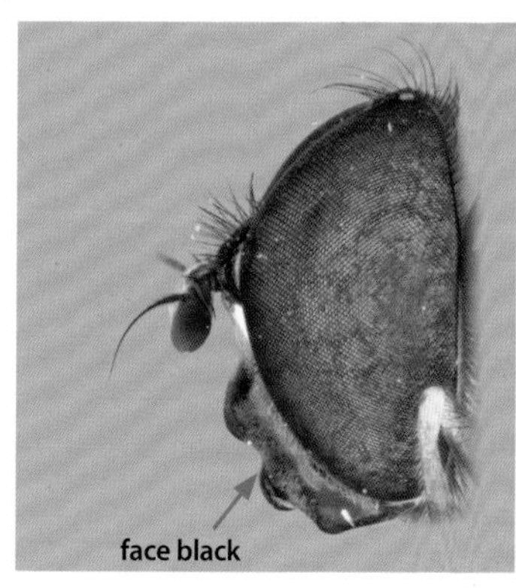

## *Cheilosia pallipes*

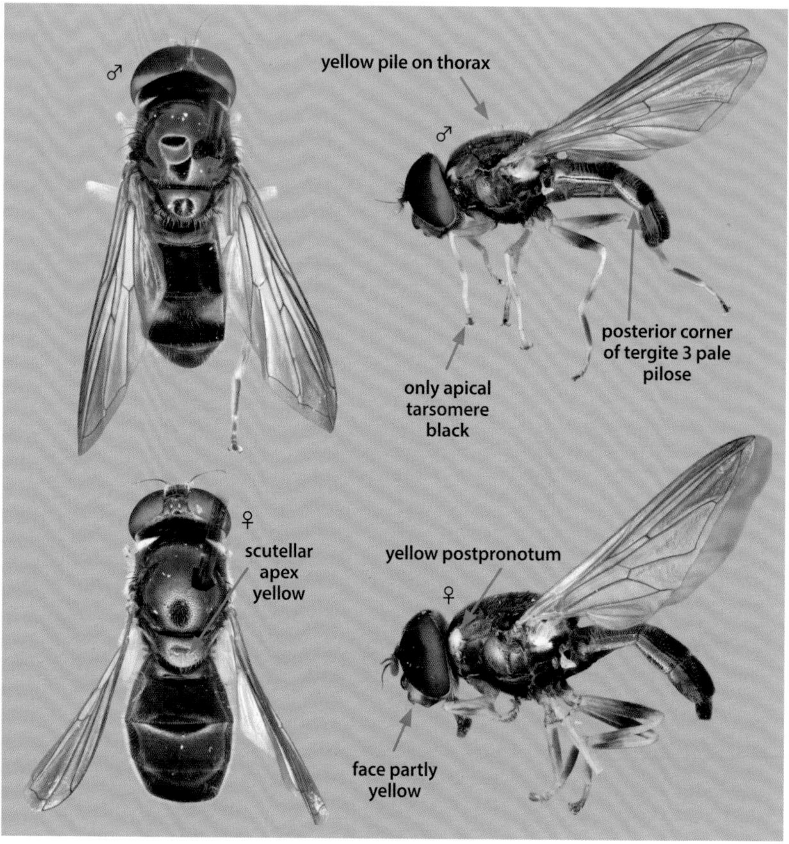

♂
yellow pile on thorax
♂
posterior corner of tergite 3 pale pilose
only apical tarsomere black
♀
scutellar apex yellow
yellow postpronotum
♀
face partly yellow

## *Cheilosia* undescribed species 76-1

♂
admixed black and yellow pilose
♂
only apical tarsomere black
posterior corner of tergite 3 pale pilose

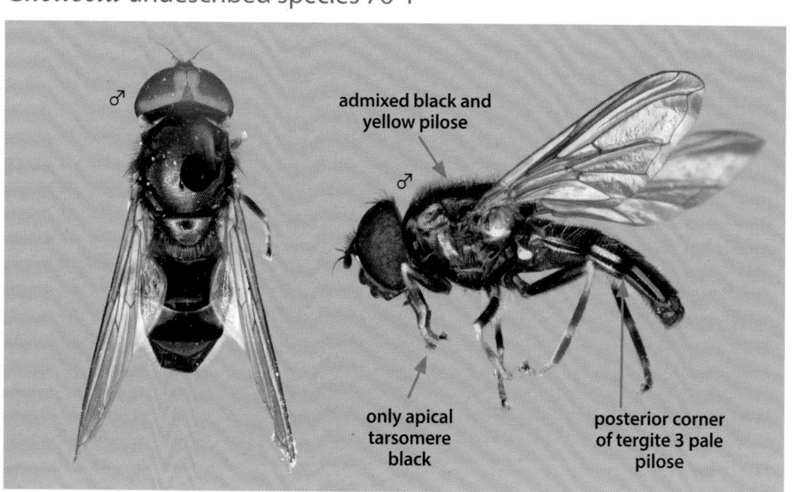

## *Cheilosia* species 17-2 Stygian Blacklet

**SIZE RANGE:** 7.4–8.9 mm
**IDENTIFICATION:** Eye bare, face bare, bristles on scutellum, antennal pits confluent. Antenna orange, sometimes darkly so. Femora and tibiae variable in color but never completely black, with at least two apical tarsomeres black on pro- and mesoleg. Male black pilose on mesonotum, with face black, postpronotum black, and posterior corner of tergite 3 pale pilose. Female with face yellow, postpronotum yellow, white pilose, metabasitarsus never swollen. **ABUNDANCE:** Rare. **FLIGHT TIMES:** Late June to late July. **NOTES:** This may or may not be a new species. Until a revision is completed it should be treated with this morphospecies number.

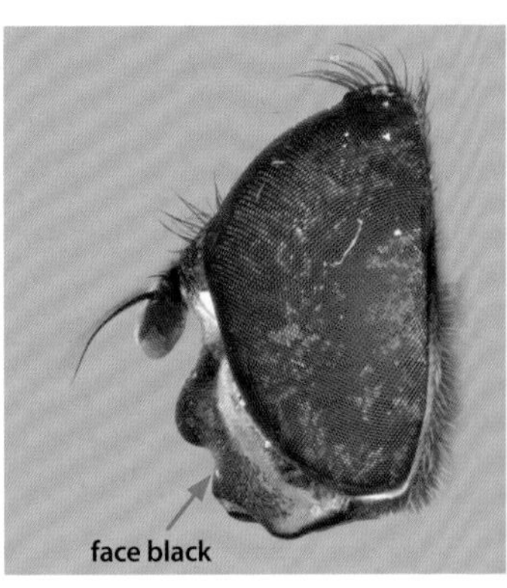


## *Cheilosia* undescribed species 17-3
### Black-faced Blacklet

**SIZE RANGE:** 7.2–9.0 mm
**IDENTIFICATION:** Eye bare, face bare, bristles on scutellum, antennal pits confluent. Antenna orange, sometimes darkly so. Femora and tibiae variable in color but never completely black, with only fifth tarsomere black on pro- and mesoleg. Male pale pilose on mesonotum, with face black, postpronotum black, and posterior corner of tergite 3 pale pilose. Female with face black, postpronotum yellow, white pilose on mesonotum, metabasitarsus never swollen. **ABUNDANCE:** Rare. **FLIGHT TIMES:** Late May to early July. **NOTES:** Has been found on flowers of *Aruncus* and *Saxifraga*.

## *Cheilosia* species 17-2

## *Cheilosia* undescribed species 17-3

# Psilota

*Psilota* species are dark, pilose flies. Their eyes and face are densely pilose and their body is covered in long pile. They have no tubercle on the face (see *Cheilosia* for a tuberculate face), the oral margin projects anteriorly, and the antennal flagellum is elongate. They are most likely to be confused with Pipizinae. To distinguish from these, look at the oral margin (see introductory section on pages 34–35). It is notched anteriorly in *Psilota* and rounded in the pipizines. There are three described Nearctic species, two of which occur in our region. No key exists for the Nearctic species. DNA data suggest that there are many species and that the current species concepts are meaningless. We thus pool the pale-pilose species here and illustrate a black-pilose species that we have found. Careful examination of male genitalia combined with DNA data will be the only way forward to working out this taxonomically challenging group. Larvae appear to be associated with large living or recently fallen trees, where they may occur in accumulations of decaying sap under bark. Adults of European species are often found drinking from puddles and streams in hot weather, and some species are thought to be mostly arboreal.

## *Psilota buccata*, *flavidipennis*, and related

Haireyes

**SIZE RANGE:** 6.3–9.0 mm

**IDENTIFICATION:** Members of this species group all have pale pile. **ABUNDANCE:** Rare. **FLIGHT TIMES:** Early March to late August (mid-February to early October in the extreme south). **NOTES:** There is no way to currently identify species within the yellow-pilose group. It clearly includes many species. Flowers visited include *Cornus*, *Crataegus*, *Pastinaca*, *Photinia*, *Physocarpus*, *Prunus* (most records), *Pyrus*, *Salix*, and *Viburnum*.

## *Psilota* undescribed species 17-1 Black Haireye

**SIZE RANGE:** 5.8–7.0 mm

**IDENTIFICATION:** At least one species (possibly more) with thick black pile is found in the east. It is similar to the western *P. thatuna* but we believe it is a different species. **ABUNDANCE:** Rare. **FLIGHT TIMES:** Late May to mid-June. **NOTES:** Two specimens were found nectaring on *Pastinaca*. Others were found in mixed woods at the edge of a clearing, in an area of pines and blueberries, and on an open hilltop in mixed forest.

## *Psilota*

## *Psilota buccata*, *P. flavidipennis*, and related

## *Psilota* undescribed species 17-1

## *Heringia*

*Heringia* species are small, black, pilose syrphids. Like other pipizines, they have a rounded oral margin and the oral suture has been reduced to a pit (see pages 34–35). The eye, face, and body are densely pilose. They can be distinguished from most other pipizines by the combination of a bare anterior anepisternum, pilose katepimeron, elongate flagellum, and the lack of a projection on the mesocoxa of the males. Until recently, *Heringia* was treated as two subgenera, *Heringia* and *Neocnemodon*. Based on DNA data, these are now treated as genera. In general, the antennae of *Heringia* are longer and the legs are simple. *Neocnemodon* species typically have shorter antennae (but see *H. intensica*), and males of all species have modifications on their mesocoxae and most have modified metatrochanters. *Neocnemodon* mesotibiae are also expanded and have a groove on the underside. Females of both genera are unidentifiable using morphology, and only males are treated here. For the most recent key, see Curran (1921). Known larvae are predators of gall-making or leaf-curling aphids, adelgids, or psyllids. There are 11 world species and five Nearctic species. Three species occur within the range of the field guide.

### *Heringia intensica* Dusky Smoothleg

SIZE RANGE: 7.0 mm
IDENTIFICATION: Males with metafemur slightly enlarged apically and with small apicoventral triangular projection. Flagellum very short compared to other *Heringia* species. Meso- and metacoxae simple. Metatrochanter simple. Sternites simple. ABUNDANCE: Rare. FLIGHT TIMES: Late April to late August. NOTES: The Nova Scotia specimen was collected in a *Corema* barren.

## *Heringia* and *Neocnemodon*

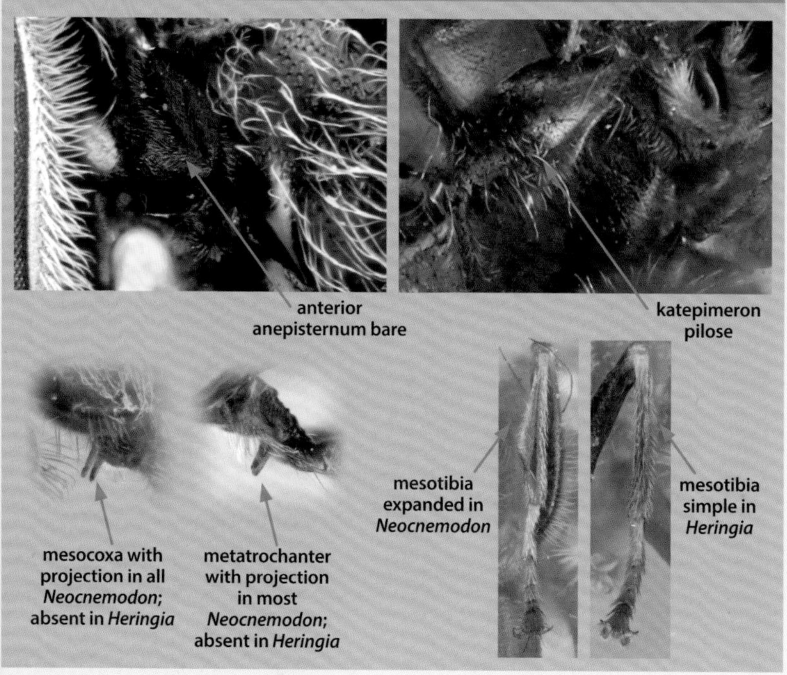

## *Heringia intensica*

## *Heringia canadensis* Canadian Smoothleg

**SIZE RANGE:** 5.7–9.0 mm
**IDENTIFICATION:** Males with metafemur of uniform width and without apical ventral triangular projection. Flagellum elongate. Thorax white pilose. Meso- and metacoxae simple. Metatrochanter simple. Sternites simple. **ABUNDANCE:** Fairly common. **FLIGHT TIMES:** Early May to late August (late March to early November in the south). **NOTES:** Found in mixed mesic forest.

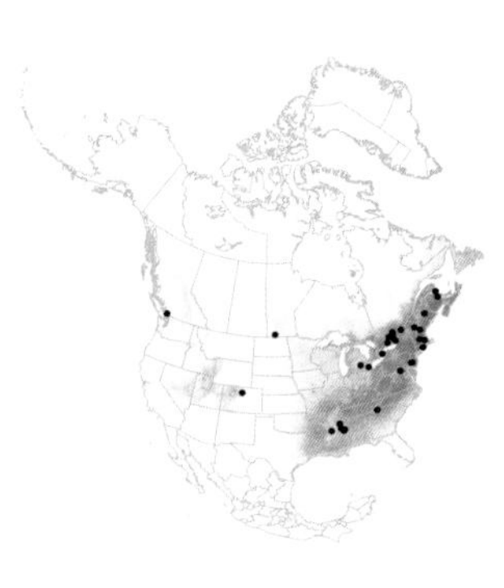

## *Heringia salax* Eastern Smoothleg

**SIZE RANGE:** 8.0–8.5 mm
**IDENTIFICATION:** Males with metafemur of uniform width and without apical ventral triangular projection. Flagellum elongate. Pile of thorax yellow brown. Meso- and metacoxae simple. Metatrochanter simple. Sternites simple. **ABUNDANCE:** Fairly common. **FLIGHT TIMES:** Late April to mid-September. **NOTES:** Typically found in hardwood forests but also noted from fens, and sand barrens. Larvae are predators of the aphid *Eriosoma lanigerum* and the phylloxerid *Daktulosphaira vitifoliae*.

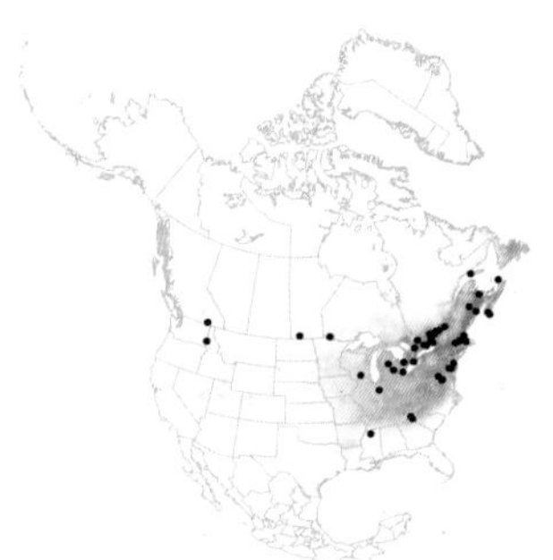

## *Heringia canadensis*

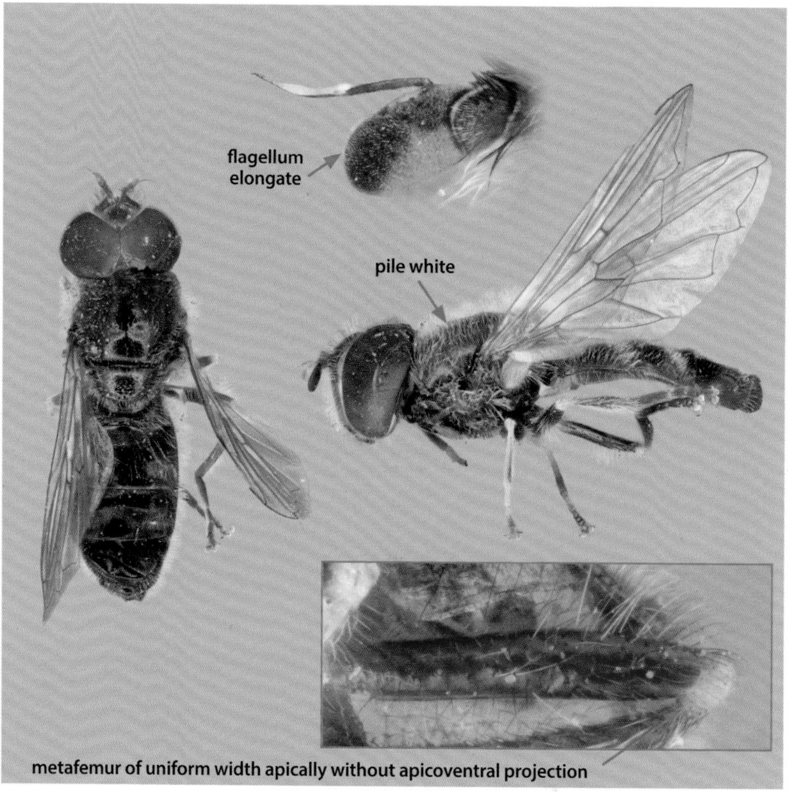
flagellum elongate
pile white
metafemur of uniform width apically without apicoventral projection

## *Heringia salax*

flagellum elongate
pile yellow brown

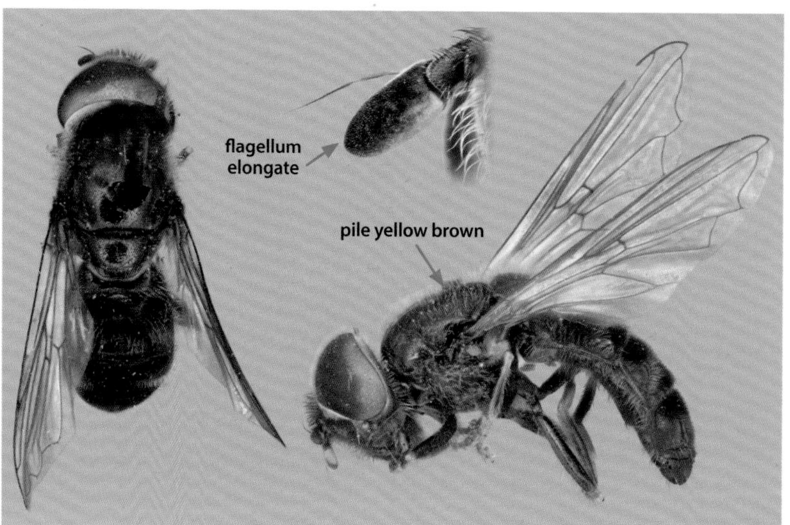

# Neocnemodon

See *Heringia* above for notes on how to separate *Neocnemodon*. Unlike *Heringia* males, males of all *Neocnemodon* have processes on the meso- and, usually, metalegs. As with *Heringia*, female *Neocnemodon* specimens are not identifiable using morphology, and only males are treated here. For the most recent key, see Curran (1921). There are 35 world and 22 Nearctic species of *Neocnemodon*. *Neocnemodon latitarsis* and *N. pubescens* are European species that have been intercepted in New Brunswick, but neither appears to be established, so they are not treated here. We have 15 species in the northeast. The group is badly in need of revision. Larvae are mostly predators of aphids and adelgids but also have been recorded feeding on beetle and moth larvae.

**mesocoxal process (present in males of all *Neocnemodon* species, absent in *Heringia*)**

## *Neocnemodon squamulae* Dingy Spikeleg

SIZE RANGE: 5.3–7.1 mm
IDENTIFICATION: Similar to *N. unicolor*, *N. intermedia*, and *N. longiseta*. Best differentiated from them by the paler calypters. Mesocoxa with process. Metacoxa and trochanter simple. Sternites simple. ABUNDANCE: Rare. FLIGHT TIMES: Mid-May to early August. NOTES: There are no data on habitat or flower usage for this species.

## *Neocnemodon unicolor* Unadorned Spikeleg

SIZE RANGE: 6.0 mm
IDENTIFICATION: Metatrochanter simple as in *N. squamulae*, *N. intermedia*, and *N. longiseta*. Sternites simple. ABUNDANCE: Known only from the holotype specimen from Guelph, Ontario. FLIGHT TIMES: The single specimen is from 22 June 1913. NOTES: In the original description and keys this species is noted to have a simple mesocoxa as in *Heringia*. Examination of the holotype showed that this is not the case, and it has a mesocoxal process like all *Neocnemodon* species. It is likely conspecific with *N. intermedia*, but genitalia and DNA should be examined as part of a complete revision of the genus.

# Neocnemodon squamulae

calypter grayish with pale fringe of pile

metatrochanter simple as in next three species

# Neocnemodon unicolor

calypter brownish with brownish fringe of pile as in following two species

legs darker than in following two species

mesocoxa with process as in all *Neocnemodon*

metatrochanter simple

## *Neocnemodon intermedia* Midnight Spikeleg

SIZE RANGE: 5.6–7.2 mm
IDENTIFICATION: Most similar to *N. squamulae*, *N. unicolor*, and *N. longiseta*. Darker calypter than in *N. squamulae*. Best differentiated from *N. longiseta* by smaller size and from *N. unicolor* by yellower legs. Mesocoxa with process. Metacoxa and metatrochanter simple. Sternites simple. ABUNDANCE: Rare. FLIGHT TIMES: Mid-May to mid-July in the east (western specimen from late September). NOTES: Species concepts in *Neocnemodon* need to be revised. Similar species may be conspecific.

## *Neocnemodon longiseta* Long-horned Spikeleg

SIZE RANGE: 8.0–8.5 mm
IDENTIFICATION: Similar to *N. squamulae*, *N. unicolor*, and *N. intermedia*. Darker calypter than in *N. squamulae*. Best differentiated from *N. unicolor* and *N. intermedia* by larger size. Mesocoxa with process. Metacoxa and metatrochanter simple. Sternites simple. ABUNDANCE: Rare. FLIGHT TIMES: Late May to early August. NOTES: Several specimens were collected in open habitat.

## Neocnemodon intermedia

## Neocnemodon longiseta

## *Neocnemodon ontarioensis*

Smooth-bellied Spikeleg

**SIZE RANGE:** 5.2–7.6 mm
**IDENTIFICATION:** Mesocoxa and metatrochanter with processes. Metacoxa simple. Sternite 3 smooth along knife-like central ridge. Sternite 4 with pilose protuberance. **ABUNDANCE:** Rare. **FLIGHT TIMES:** Mid-April to mid-August. **NOTES:** Two specimens were collected at a bleeding elm tree.

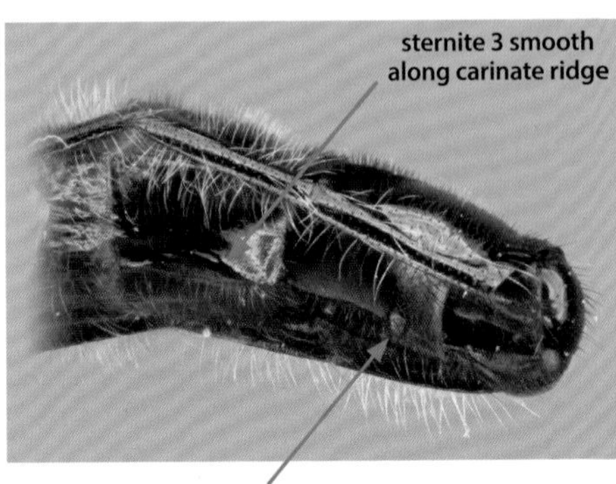


sternite 4 with protuberance in this and following three species, tubercle-like in this and *N. trochanterata*; protuberance on sternite 4 more pilose than in *N. trochanterata*

## *Neocnemodon trochanterata* Rough-bellied Spikeleg

**SIZE RANGE:** 7.0 mm
**IDENTIFICATION:** Mesocoxa and metatrochanter with processes. Metacoxa simple. Sternite 3 rugose along knife-like central ridge. Sternite 4 with less pilose protuberance. **ABUNDANCE:** Very rare. **FLIGHT TIMES:** Mid-May. **NOTES:** This species is known only from the type series. Revision of *Neocnemodon* is needed to test the species concepts of this and similar species. It is possible that *N. ontarioensis* is simply a variant of this species.

protuberance on sternite 4 less pilose than in *N. ontarioensis*

# *Neocnemodon ontarioensis*

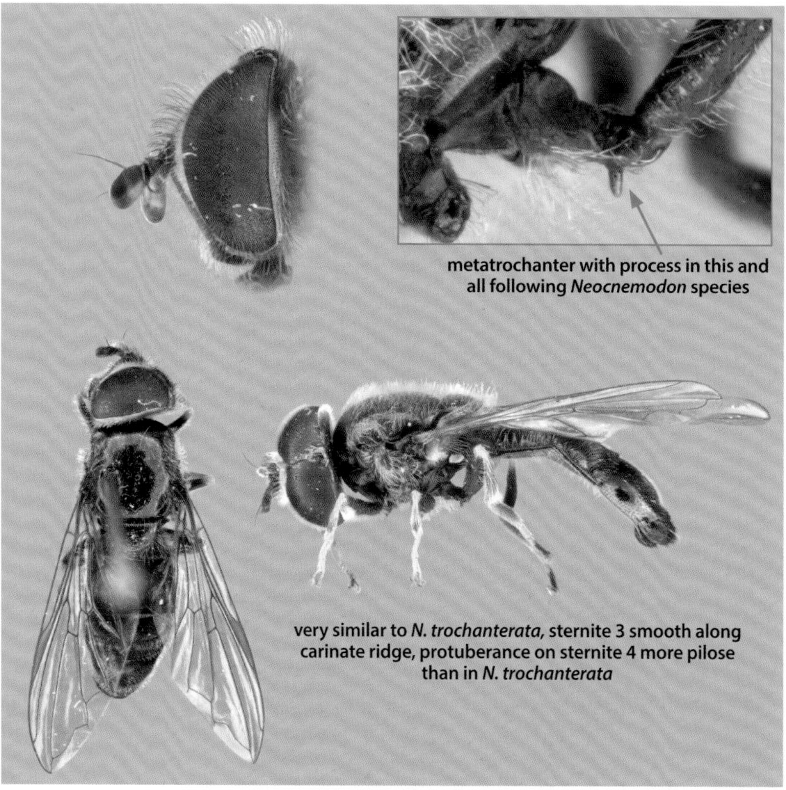
metatrochanter with process in this and all following *Neocnemodon* species
very similar to *N. trochanterata*, sternite 3 smooth along carinate ridge, protuberance on sternite 4 more pilose than in *N. trochanterata*

# *Neocnemodon trochanterata*

metatrochanter with process

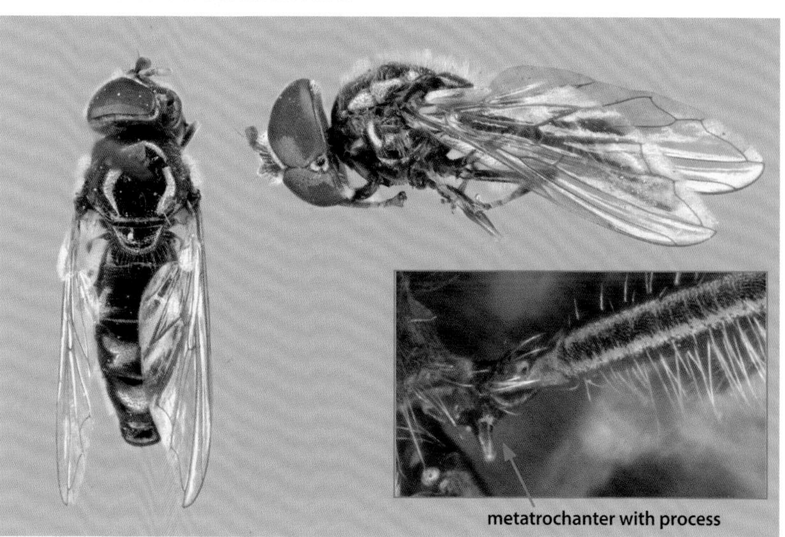

## *Neocnemodon venteris* Dark-winged Spikeleg

**SIZE RANGE:** 6.5–7.5 mm
**IDENTIFICATION:** Mesocoxa and metatrochanter with processes. Metacoxa simple. Sternite 3 with protuberance. Sternite 4 with spur-like protuberance. Wing evenly infuscated. **ABUNDANCE:** Rare. **FLIGHT TIMES:** Late April to mid-June. **NOTES:** Most specimens have been collected around moist soil or puddles in the forest.

## *Neocnemodon myerma* Smoke-tipped Spikeleg

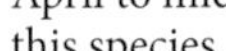

**SIZE RANGE:** 8.2–8.6 mm
**IDENTIFICATION:** Mesocoxa and metatrochanter with processes. Metacoxa simple. Sternite 3 with protuberance. Sternite 4 with spur-like protuberance. **ABUNDANCE:** Rare. **FLIGHT TIMES:** Late April to mid-June. **NOTES:** Nothing is known about this species.

## *Neocnemodon venteris*

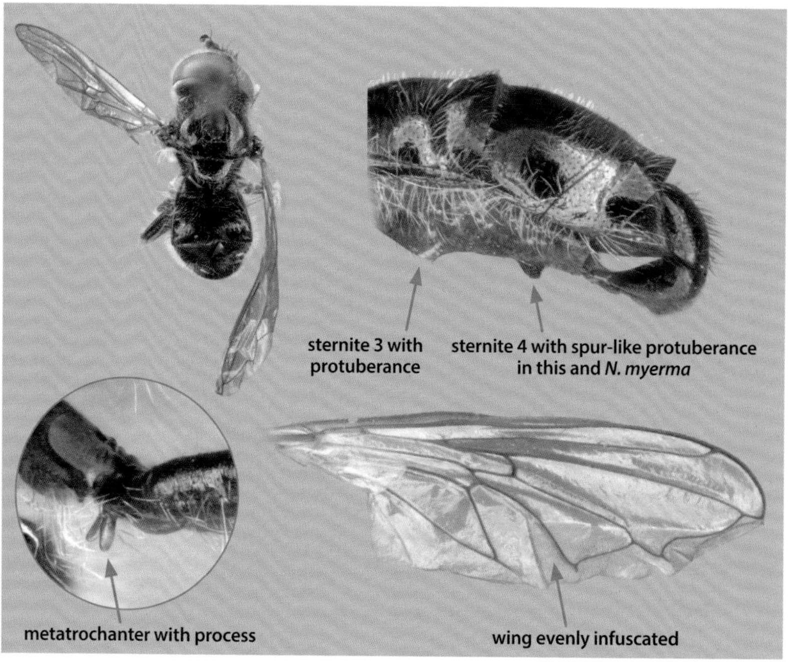

sternite 3 with protuberance

sternite 4 with spur-like protuberance in this and *N. myerma*

metatrochanter with process

wing evenly infuscated

## *Neocnemodon myerma*

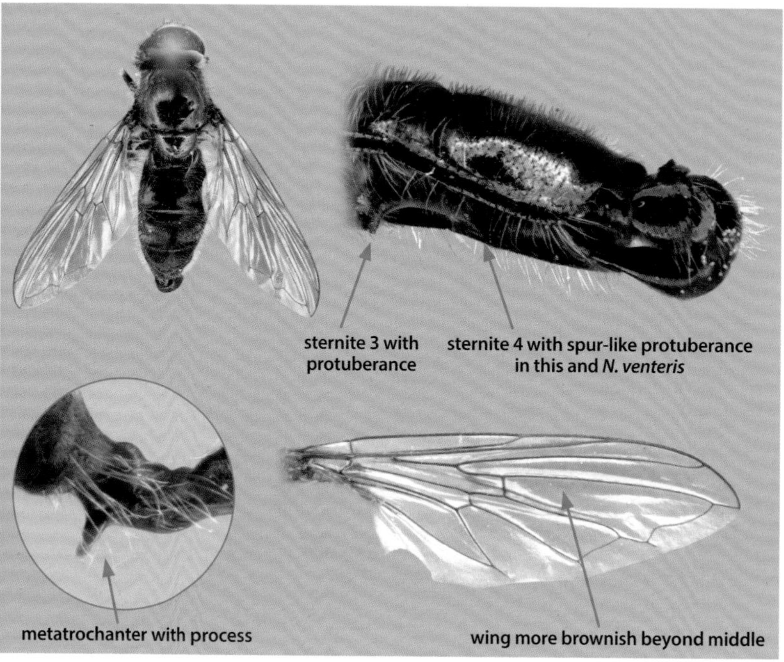

sternite 3 with protuberance

sternite 4 with spur-like protuberance in this and *N. venteris*

metatrochanter with process

wing more brownish beyond middle

## *Neocnemodon rita* Black-faced Spikeleg

**SIZE RANGE:** 5.5–6.5 mm

**IDENTIFICATION:** Face entirely black pilose. Mesocoxa and metatrochanter with processes. Metacoxa with short but conspicuous spur directed outward. Sternites 3 and 4 simple. **ABUNDANCE:** Common. **FLIGHT TIMES:** Late May to late August (from late March in the west). **NOTES:** This species is found in mixed woods. Flowers visited include *Fragaria*, *Ilex*, *Prunus*, *Rubus*, and *Vaccinium*. Larvae have been reared from *Adelges piceae* (Balsam Woolly Adelgid, a serious invasive pest), *Eriosoma lanigerum* (Woolly Apple Aphid), and according to one specimen label, *Paralobesia piceana* (a tortricid moth).

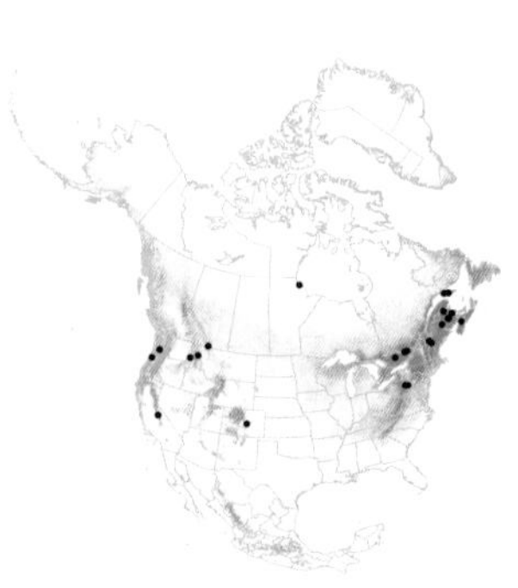

## *Neocnemodon coxalis* White-faced Spikeleg

**SIZE RANGE:** 6.5–7.5 mm

**IDENTIFICATION:** Face partly white pilose. Mesocoxa and metatrochanter with processes. Metacoxa with short but conspicuous spurs directed outward. Sternites 3 and 4 simple. **ABUNDANCE:** Common. **FLIGHT TIMES:** Late March to early October. **NOTES:** They have been recorded in several habitats including mixed woods, cranberry fields, orchards, forest edges, mesic ravines, and open areas. The single floral record is from *Malus*. Larvae are known predators of *Adelges piceae* (Balsam Woolly Adelgid).

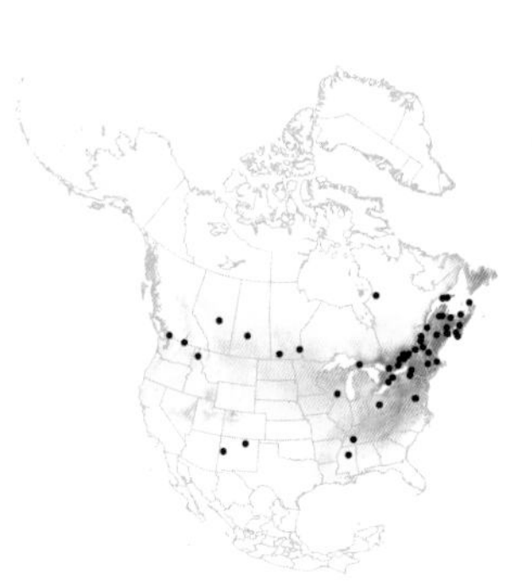

## *Neocnemodon rita*

metacoxa with short spur
metatrochanter with process

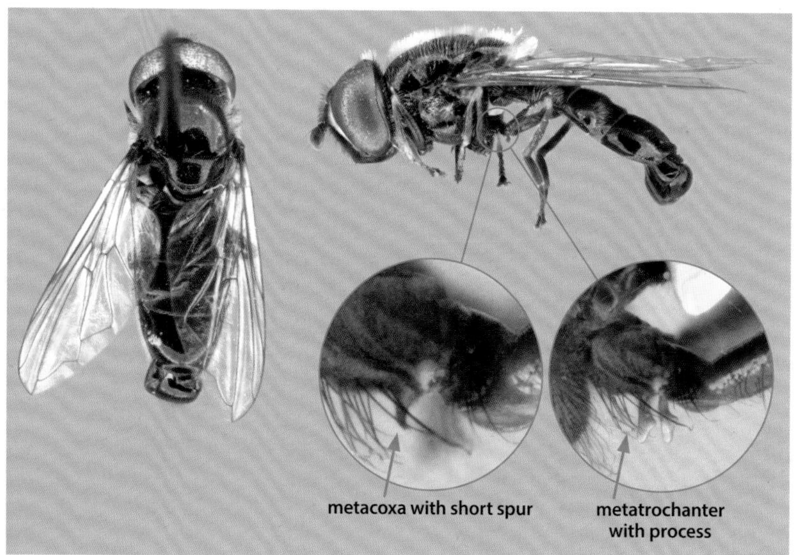

## *Neocnemodon coxalis*

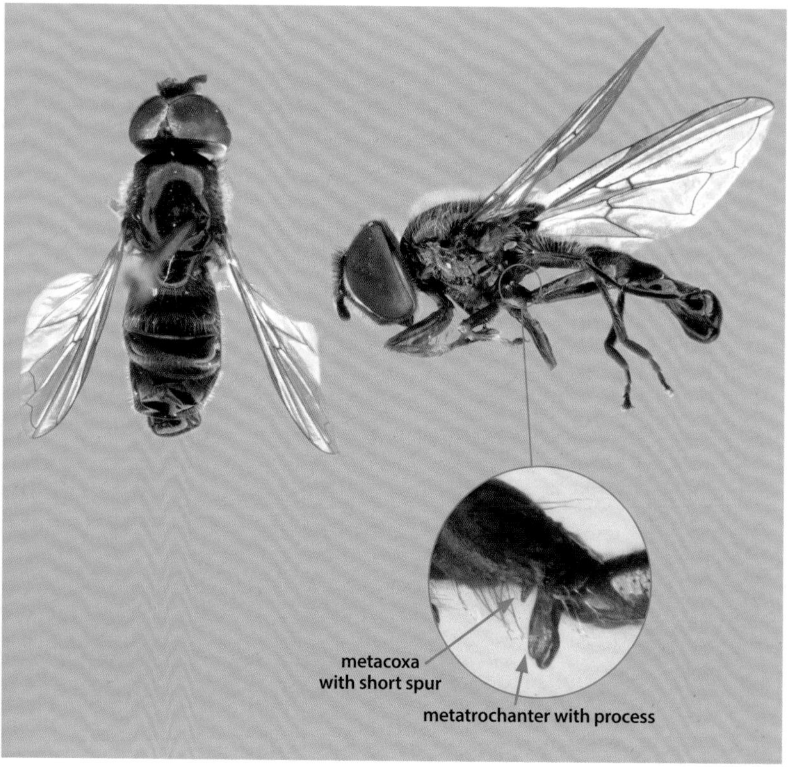

metacoxa with short spur
metatrochanter with process

## *Neocnemodon carinata* Carinate Spikeleg

**SIZE RANGE:** 6.3–7.9 mm
**IDENTIFICATION:** Mesocoxa and metatrochanter with processes. Metacoxa simple. Sternite 3 with sharp central ridge; ridge continues onto the base of sternite 4. **ABUNDANCE:** Rare. **FLIGHT TIMES:** Early June to late July. **NOTES:** Nothing is known about this species.

## *Neocnemodon pisticoides* Long-spined Spikeleg

**SIZE RANGE:** 5.8–7.9 mm
**IDENTIFICATION:** Mesocoxa with process. Metacoxa simple. Metatrochanter with long process. Sternite 3 with sharp central ridge. Sternite 4 simple. **ABUNDANCE:** Rare. **FLIGHT TIMES:** Late May to early November. **NOTES:** Some specimens were collected in mixed woods while others were in gardens and orchards. One specimen was reared from *Eriosoma lanigerum* (Woolly Apple Aphid).

## Neocnemodon carinata

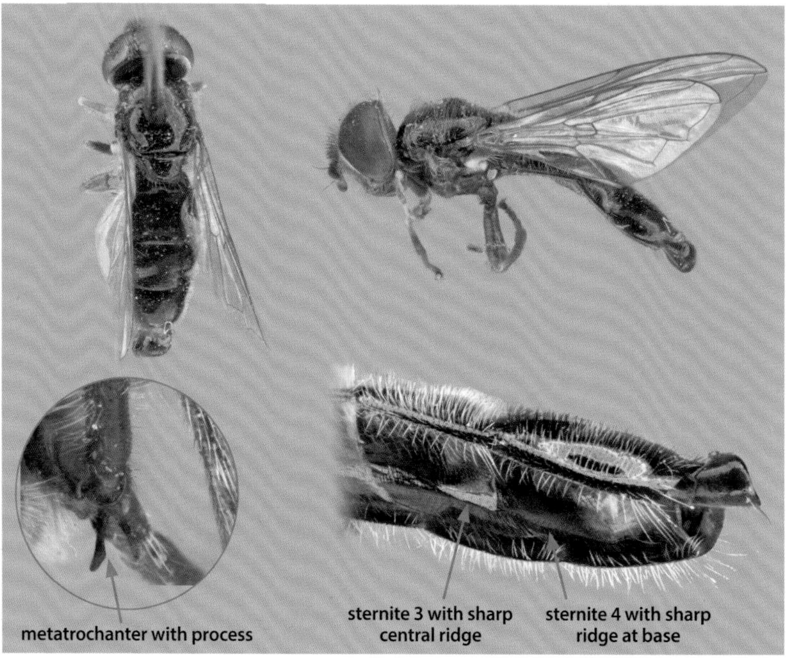

metatrochanter with process

sternite 3 with sharp central ridge

sternite 4 with sharp ridge at base

## Neocnemodon pisticoides

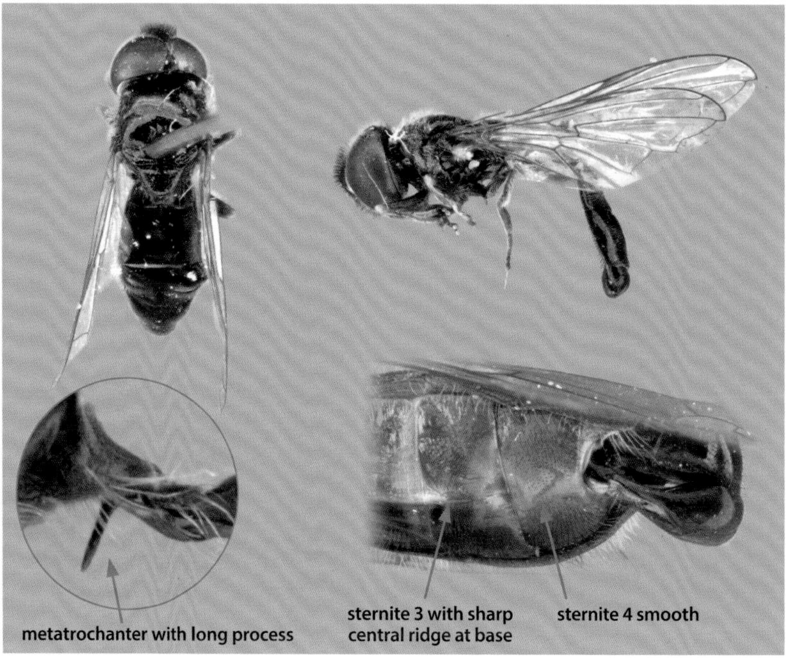

metatrochanter with long process

sternite 3 with sharp central ridge at base

sternite 4 smooth

## *Neocnemodon cevelata* Thick-horned Spikeleg

**SIZE RANGE:** 5.2–5.9 mm
**IDENTIFICATION:** Flagellum broader than in all other *Neocnemodon*. Mesocoxa with process. Metacoxa simple. Metatrochanter with process shorter than in all other species. Sternites simple. **ABUNDANCE:** Rare. **FLIGHT TIMES:** Early May to early August. **NOTES:** Females also have a broad flagellum and may be identified to species on this basis (flagellum broader than long in female).

## *Neocnemodon elongata* Elongate Spikeleg

**SIZE RANGE:** 4.8–8.8 mm
**IDENTIFICATION:** Facial pile completely black. Mesocoxa with process. Metacoxa simple. Metatrochanter with long process. Sternites simple. **ABUNDANCE:** Uncommon. **FLIGHT TIMES:** Early May to mid-September. **NOTES:** Flowers visited include *Alliaria*, *Daucus*, *Ranunculus*, *Rhododendron*, and *Spiraea*. Reared from *Eriosoma lanigerum*. Often found drinking from moist soil.

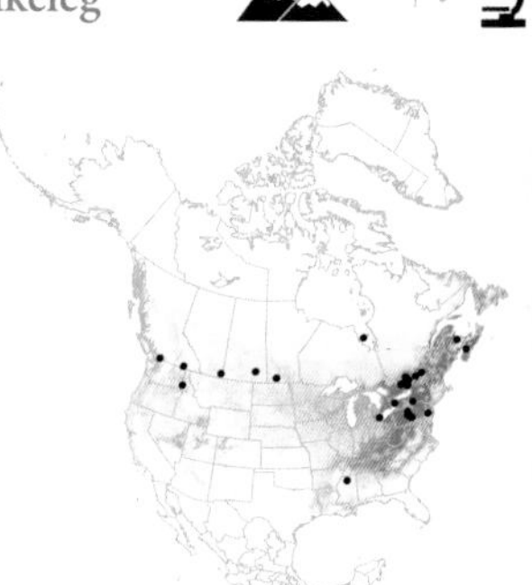

## *Neocnemodon calcarata* Opaque Spikeleg

**SIZE RANGE:** 5.5–7.0 mm
**IDENTIFICATION:** Like *N. elongata* except facial pile partly white. **ABUNDANCE:** Uncommon. **FLIGHT TIMES:** Early June to mid-August. **NOTES:** The habitat was recorded as mixed woods for some specimens.

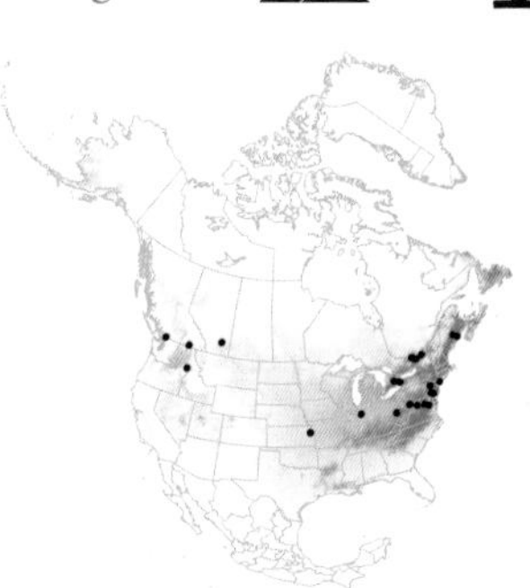

## *Neocnemodon cevelata*

## *Neocnemodon elongata*

## *Neocnemodon calcarata*

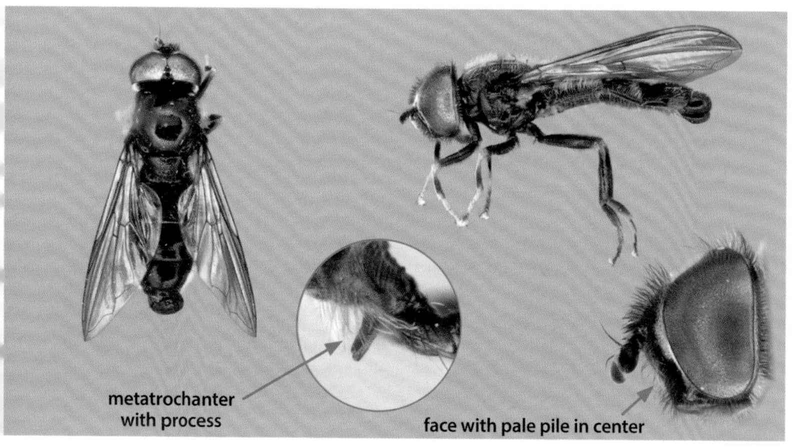

# *Pipiza*

*Pipiza* are small black syrphids that vary from having all black abdomens to having paired yellow spots on tergite 2 and sometimes also tergite 3. They can be mistaken for *Heringia* and *Trichopsomyia* and so should be checked for a bare anterior anepisternum and katepimeron. There are 52 world species; 11 in the Nearctic and seven from the northeast. A recent revision in Europe (Vujić *et al.* 2013) turned much of the original taxonomy on its head and illustrated how difficult this group is. Despite recent work by Coovert (1996) in the Nearctic, taxonomic concepts need to be reevaluated incorporating genetic data. Many problems with current concepts exist but cannot be solved without complete revision. We thus follow Coovert here with the caveat that changes are needed. *Pipiza* species are often found flying through herbaceous vegetation or around shrubs. Known larvae are predators of aphids and phylloxera (mostly gall-making or leaf-rolling aphids that create waxy secretions). Characters illustrated below generally work, but male genitalia should be checked for confirmation (see Coovert [1996] for illustrations).

## *Pipiza quadrimaculata* Four-spotted Pithead

**SIZE RANGE:** 5.0–8.0 mm

**IDENTIFICATION:** This species is easily identified by the two pairs of yellow-orange abdominal spots; all other *Pipiza* species have one pair or are entirely black. The spots can be obscure, especially the second pair. If in doubt, this is the only species with the antennae inserted at or below the middle of the head. **ABUNDANCE:** Fairly common. **FLIGHT TIMES:** Mid-May to early August. **NOTES:** This species is widespread in North America and Eurasia. North American specimens are genetically similar to Old World specimens but have accrued a few differences. Treating them as one Holarctic species seems merited. Adults may be found along forest tracks, in clearings, in treed fens, and in open, mature coniferous or mixed forest. Flowers visited include *Alliaria*, *Allium*, *Caltha*, *Cardamine*, *Cornus*, *Euphorbia*, *Fragaria*, *Malus*, *Pastinaca*, *Potentilla*, *Ranunculus*, *Rhododendron*, *Rubus*, *Salix*, *Sambucus*, and *Sorbus*. Larvae are known to feed on *Cinara* aphids and have been documented overwintering in leaf litter in Europe.

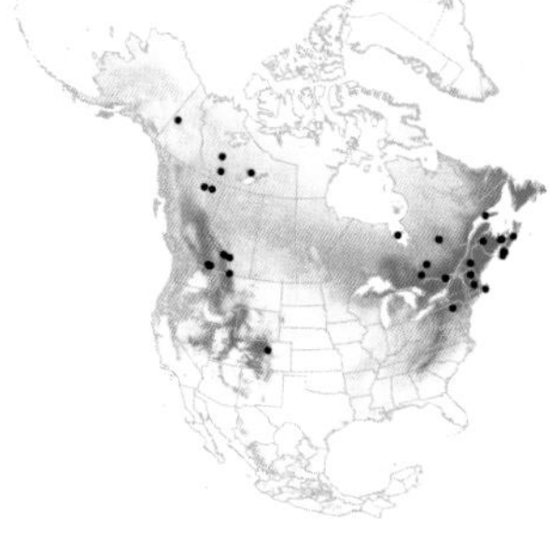

# Pipiza

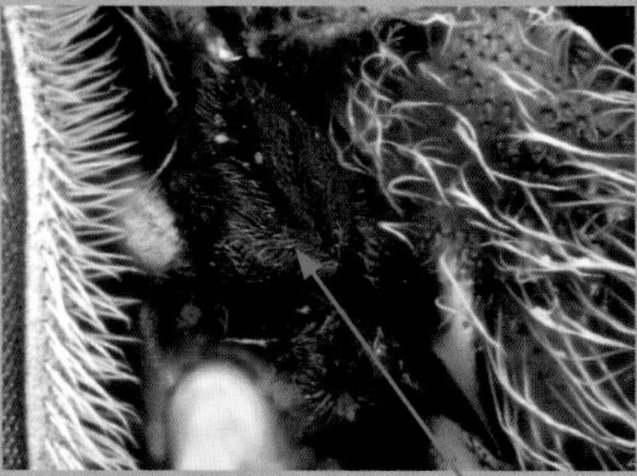

anterior anepisternum bare

katepimeron bare

# Pipiza quadrimaculata

♂

♀

two pairs of yellow-orange spots

♂

antennae inserted at or below middle of head

## *Pipiza atrata* Ebony Pithead

**SIZE RANGE:** 7.0–10.0 mm
**IDENTIFICATION:** Both sexes with a completely black abdomen and a slender metafemur without a denticulate ridge. Male with completely black-pilose scutum and metabasitarsus with black semiappressed pile anterodorsally. Female tergite 5 at least twice as wide as long. **ABUNDANCE:** Rare. **FLIGHT TIMES:** Late May to late July. **NOTES:** Eastern specimens are dubious. Any specimens from east of the Rocky Mountains should be carefully checked and vouchered. Flowers visited include *Rhododendron* and *Rosa*.

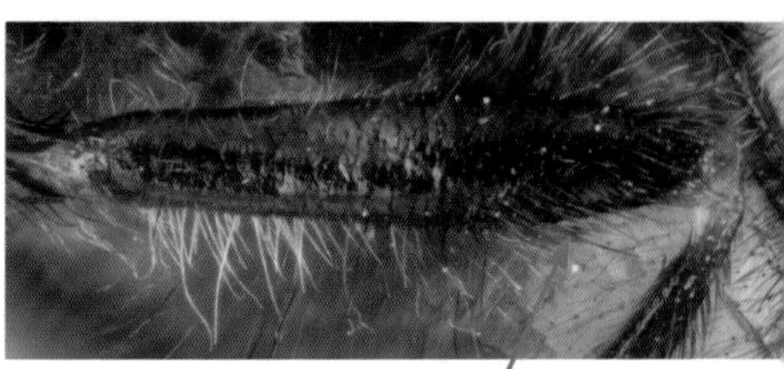

metafemur slender, without a distinct apicoventral ridge

## *Pipiza nigripilosa* Pale-haired Pithead

**SIZE RANGE:** 6.0–8.5 mm
**IDENTIFICATION:** Male abdomen black. Female abdomen with one pair of yellow-orange spots on tergite 2. Males with a slender metafemur without a denticulate ridge and usually a white-yellow pilose scutum. If scutum black pilose, metabasitarsus with pale semiappressed pile anterodorsally. Male cell bm bare posterobasally and flagellum rounded apically, pale ventrally. Female with tergite 5 with at least some black pile and subquadrate, only slightly shorter than wide.The occasional male of *P. nigripilosa* with an entirely black-pilose thorax can be distinguished from *P. atrata* by the smaller size and pale-pilose metabasitarsus. **ABUNDANCE:** Common. **FLIGHT TIMES:** Early May to mid-July (from early April in the south). **NOTES:** Flowers visited include *Crataegus*, *Malus*, *Prunus*, *Ranunculus*, and *Rubus*.

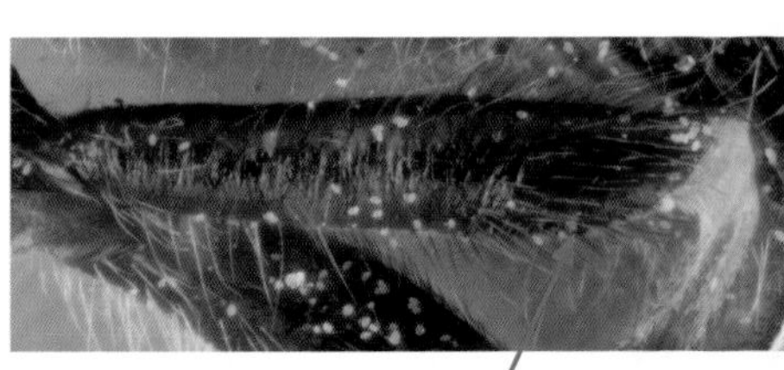

metafemur slender, without a distinct apicoventral ridge

# *Pipiza atrata*

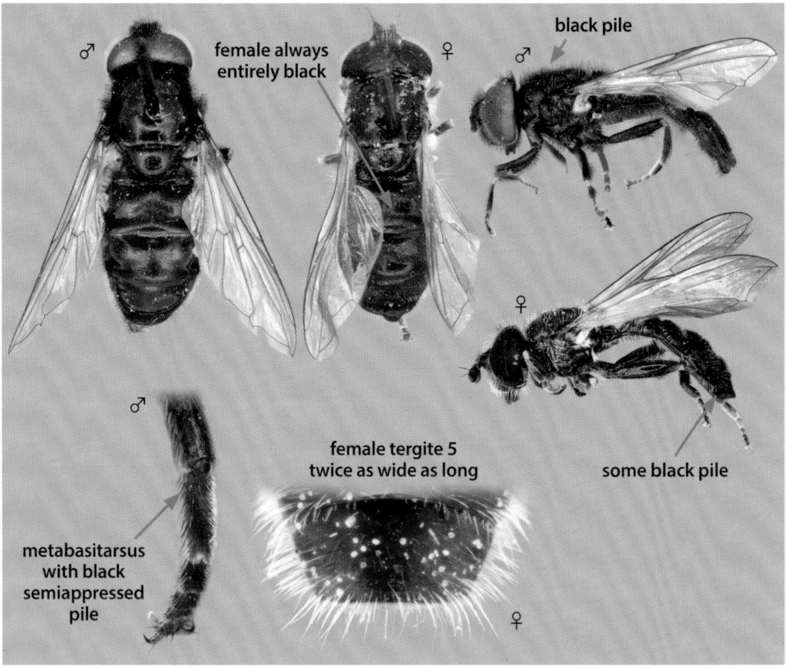

♂
female always entirely black
♀
black pile
♂
♀
♂
female tergite 5 twice as wide as long
some black pile
metabasitarsus with black semiappressed pile
♀

# *Pipiza nigripilosa*

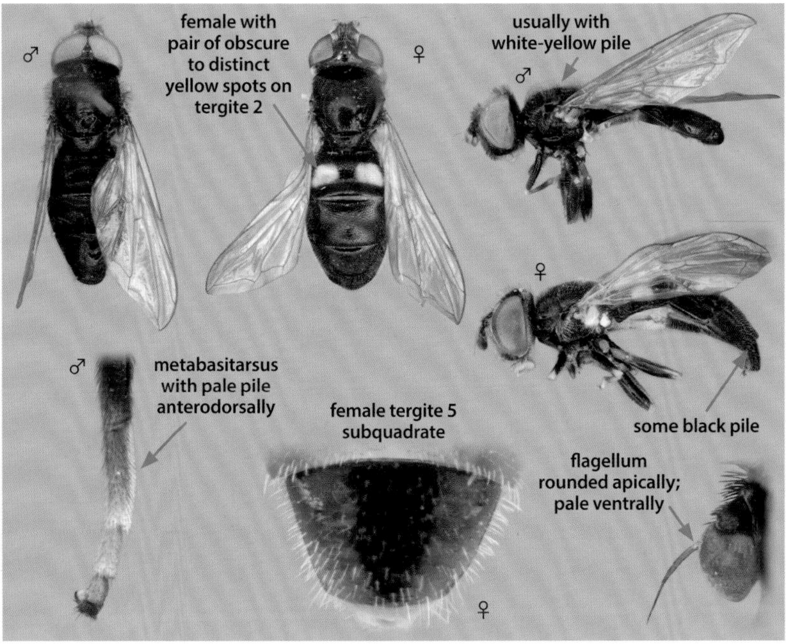

♂
female with pair of obscure to distinct yellow spots on tergite 2
♀
usually with white-yellow pile
♂
♀
♂
metabasitarsus with pale pile anterodorsally
female tergite 5 subquadrate
some black pile
flagellum rounded apically; pale ventrally
♀

## *Pipiza macrofemoralis* Large-legged Pithead

**SIZE RANGE:** 5.5–8.5 mm

**IDENTIFICATION:** Male abdomen black. Female abdomen usually with pair of yellow-orange spots on tergite 2. Male metafemur usually swollen with a distinct rounded apicoventral denticulate ridge. This species can otherwise be recognized by the male cell bm (cell mostly microtrichose) and flagellum (obliquely truncate apically, entirely black, about as long as wide). Scutum white-yellow pilose. Female tergite 5 with at least some black pile and at least twice as wide as long (similar to *P. atrata*). **ABUNDANCE:** Fairly common. **FLIGHT TIMES:** Late May to mid-August. **NOTES:** Genetic data place this species within the *Pipiza noctiluca* species complex and it is likely synonymous with one of the European species (*P. noctiluca* or *P. notata*). It seems to be confined to the coniferous forest region. There is a single floral record from *Heracleum*.

metafemur swollen with low, rounded apicoventral toothed ridge, with black pile apically

## *Pipiza puella* Sumac Gall Pithead

**SIZE RANGE:** 7.5–11.0 mm

**IDENTIFICATION:** Male abdomen black. Female abdomen with pair of yellow-orange spots on tergite 2. Male metafemur strongly swollen with a distinct apicoventral denticulate ridge, which is angulate apically; black pilose apically. Male flagellum distinctly longer than wide. Female tergite 5 with at least some black pile and at least twice as wide as long (similar to *P. atrata*). **ABUNDANCE:** Common. **FLIGHT TIMES:** Early May to late September (from early April in the south). **NOTES:** Flowers visited include *Alliaria*, *Brassica*, *Crataegus*, *Eutrochium*, *Heracleum*, *Hydrangea*, *Oxypolis*, *Pastinaca*, *Rhexia*, *Rubus*, and *Sambucus*. Larvae have been associated with aphids in galls on *Rhus typhina* on multiple occasions (presumably these are Staghorn Sumac Aphids, *Melaphis rhois*, a type of woolly aphid).

metafemur swollen with distinct angulate apicoventral toothed ridge and black pile apically

## Pipiza macrofemoralis

♂

tergite 2 black or with a pair of small to medium rounded spots

♀

♂

flagellum obliquely truncate apically; entirely black; about as long as wide

♀

female tergite 5 with some black pile; twice as wide as long

white-yellow pile

♂

♀

some black pile

## Pipiza puella

♂

tergite 2 black or with a pair of small to medium rounded spots

♀

♂

male flagellum longer than wide

♂

♀

some black pile

## *Pipiza cribbeni* Yellow-haired Pithead

**SIZE RANGE:** 7.0–9.5 mm

**IDENTIFICATION:** Male abdomen usually entirely black. Female tergite 2 with large, distinctly subrectangular spots narrowly divided to confluent medially. Metafemur swollen with a rounded, apicoventral denticulate ridge; entirely pale pilose apically. Scutum golden-yellow pilose in both sexes. Female with tergite 5 pale pilose, about two times as wide as long; scape and pedicel pale pubescent dorsally. **ABUNDANCE:** Fairly common. **FLIGHT TIMES:** Mid-April to mid-May. **NOTES:** Flowers visited include *Malus* and *Prunus*. Several specimens have been collected around moist soil.

## *Pipiza femoralis* White-haired Pithead

**SIZE RANGE:** 6.5–9.5 mm

**IDENTIFICATION:** Male tergite 2 with yellow bands. Female tergite 2 with large, distinctly subrectangular spots narrowly divided to confluent medially. Metafemur swollen with a distinct, angulate apicoventral denticulate ridge; entirely pale pilose apically. Scutum white pilose in males, white to yellowish in females. Female with tergite 5 pale pilose, about two times as wide as long; scape and pedicel black pubescent dorsally. *Pipiza femoralis* is closest to *P. cribbeni* but averages narrower and slightly smaller and is usually distinctly white pilose. The black pubescence of the scape and pedicel will distinguish females from *P. cribbeni*. **ABUNDANCE:** Very common. **FLIGHT TIMES:** Early April to late August. **NOTES:** Typically found landing on foliage or hovering in deciduous forests. Flowers visited include *Alliaria*, *Crataegus*, *Prunus*, and *Ribes*. Larvae are predators of *Colopha*, *Eriosoma*, *Georgiaphis*, *Pemphigus* (aphids), and *Daktulosphaira* (phylloxera).

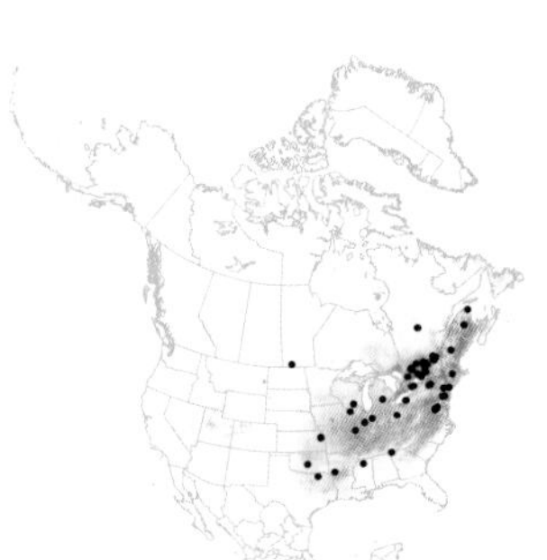

## *Pipiza cribbeni*

## *Pipiza femoralis*

# Trichopsomyia

There are 29 world species of *Trichopsomyia*. Twelve of these are Nearctic and four occur in our region (plus one undescribed species). The previous key to Nearctic species was by Curran (1921) (as *Pipizella*). As with all other Pipizinae, revision of this genus is badly needed. Known larvae are predators of gall-making Psyllidae (jumping plant lice), aphids, and phylloxera; however, an Australian species has gone off on a tangent and is predatory in arboreal ant nests. *Trichopsomyia* are difficult to distinguish from other pipizines. They are our only genus of pipizines with a pilose anterior anepisternum. They often show a greenish metallic sheen.

## Trichopsomyia apisaon

**Black-haired Psyllid-killer**

**SIZE RANGE:** 5.1–7.1 mm
**IDENTIFICATION:** Males with eyes touching for a distance of at least ½ the length of the vertical triangle; frontal triangle entirely black pilose. Female with cell dm microtrichose at base, sometimes with small bare area along anterior and posterior margins; cell $r_{4+5}$ mostly bare at base; scutellar pile short. **ABUNDANCE:** Common. **FLIGHT TIMES:** Late May to late September (from late March in the south). **NOTES:** This species has been found in a variety of habitats but seems to prefer open areas such as meadows and bogs. One floral record is from *Cornus canadensis*. Larvae are predators of *Eriosoma lanigerum* and *Daktulosphaira vitifoliae*.

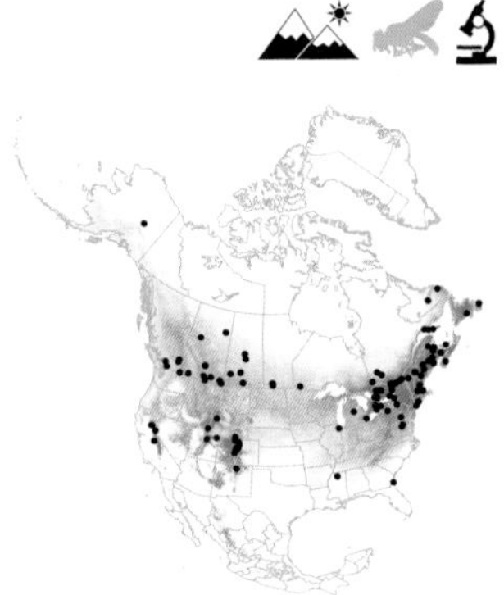

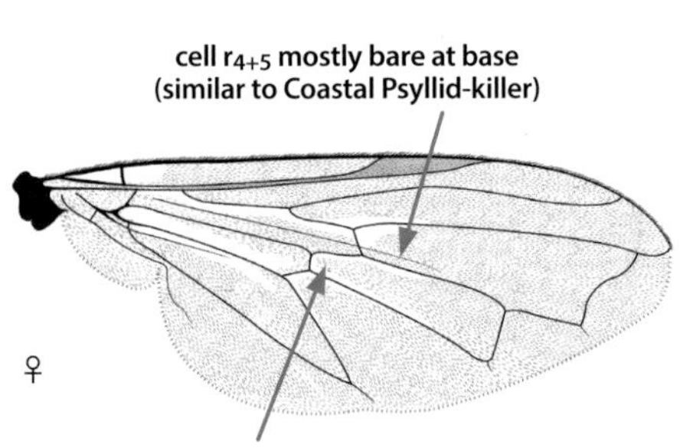


male with eyes touching in this and the following two species

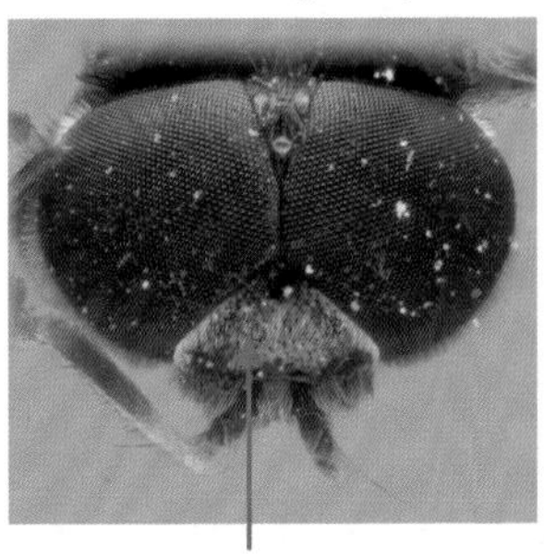

## *Trichopsomyia*

anterior anepisternum pilose

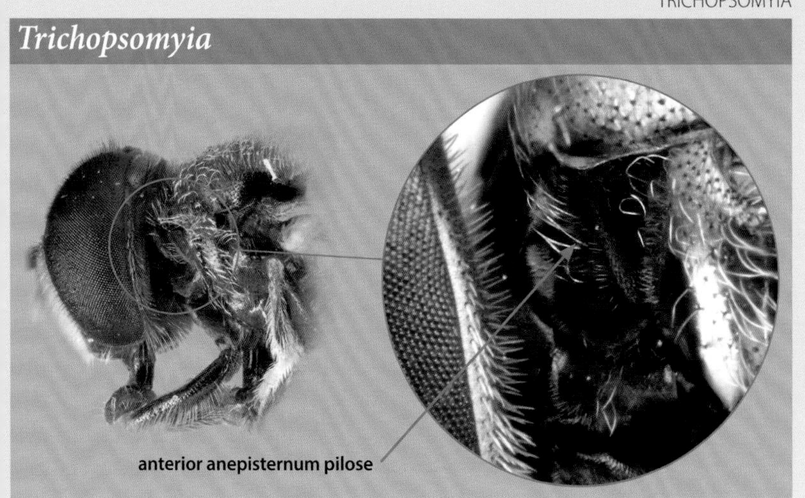

## *Trichopsomyia apisaon*

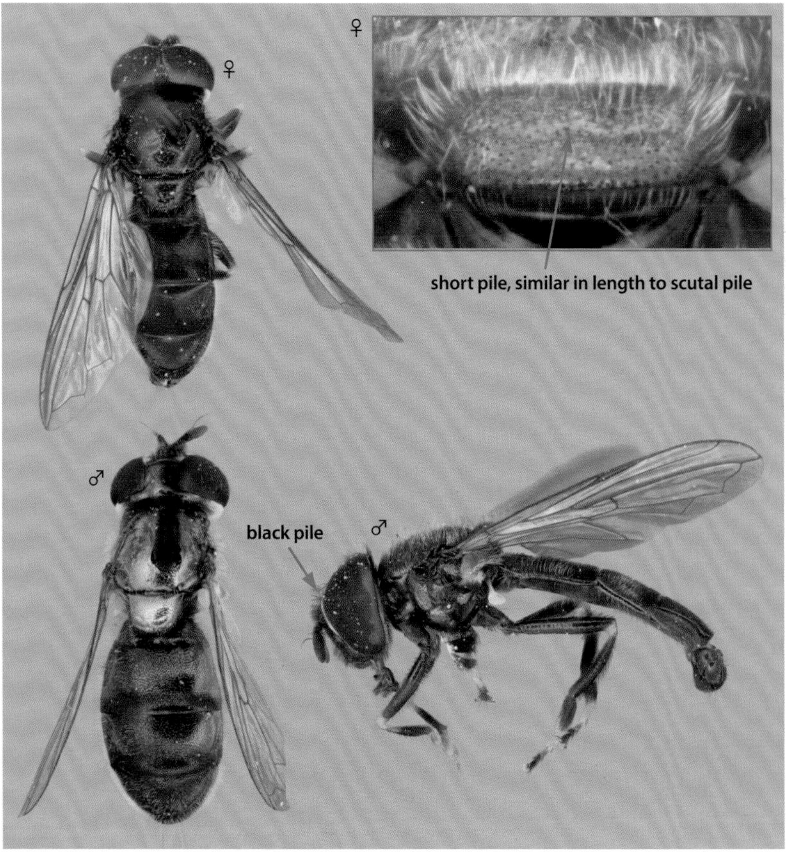

♀
♀
short pile, similar in length to scutal pile
♂
black pile
♂

## *Trichopsomyia banksi* White-faced Psyllid-killer

**SIZE RANGE:** 4.3–5.5 mm
**IDENTIFICATION:** Males with eyes touching for a distance of at least ½ the length of the vertical triangle; frontal triangle mostly white pilose. Both sexes with metabasitarsus with longer, anterior marginal pile black. Female with cell dm bare at base. **ABUNDANCE:** Rare. **FLIGHT TIMES:** Early May to late August (early March to early November in the south). **NOTES:** Flowers visited include *Baccharis*, *Euphorbia*, *Fragaria*, and *Polygonum*. One record is from a sphagnum bog.

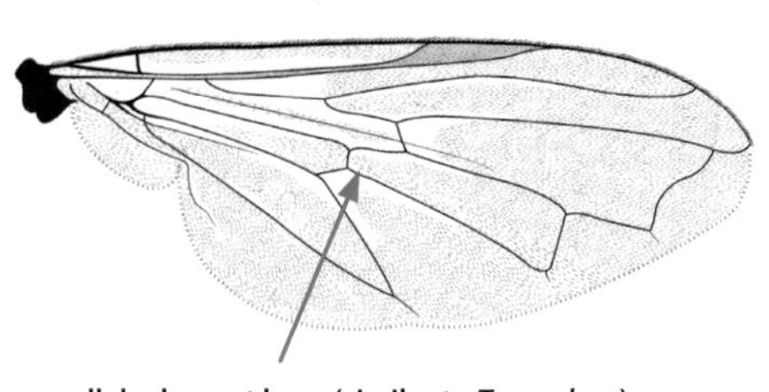

cell dm bare at base (similar to *T. recedens*)

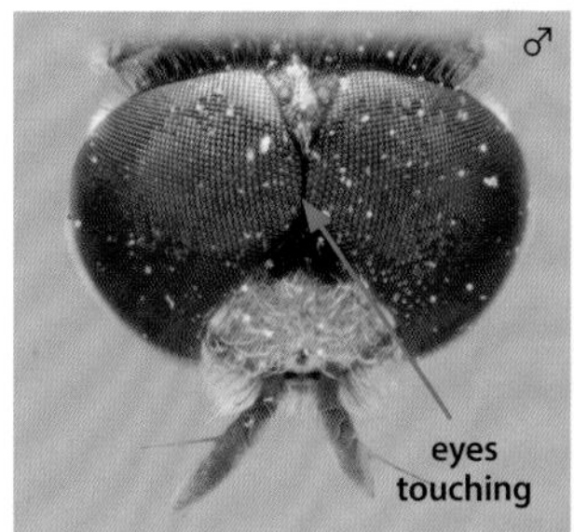


## *Trichopsomyia recedens* Shadowy Psyllid-killer

**SIZE RANGE:** 4.6–6.8 mm
**IDENTIFICATION:** Males with eyes touching for a distance of at least ½ the length of the vertical triangle; frontal triangle mostly white pilose. Both sexes with metabasitarsus with anterior marginal pile white. Female with cell dm bare at base. **ABUNDANCE:** Uncommon. **FLIGHT TIMES:** Mid-May to late August. **NOTES:** There is a single floral record from *Thalictrum*. Specimens with habitat data were collected in old quarries and regenerating fields. One record is from a hilltop.

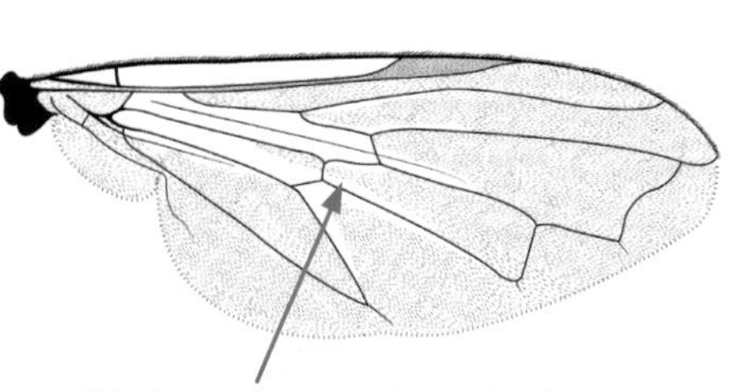

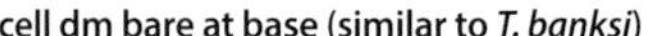

cell dm bare at base (similar to *T. banksi*)

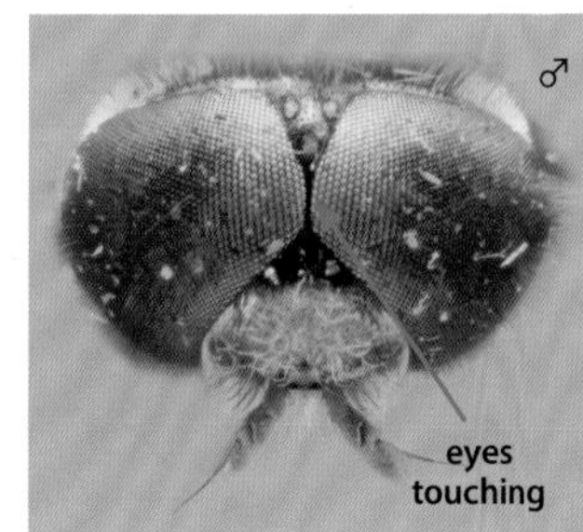

# Trichopsomyia banksi

♂ ♀

white pile

♂

♀

metatibia with black pile at apex

metabasitarsus with black pile

# Trichopsomyia recedens

♂ ♀

white pile

♂

♀

metatibia with white pile at apex

metabasitarsus with white pile

## *Trichopsomyia* undescribed species 1

### Coastal Psyllid-killer

**SIZE RANGE:** 5.5–6.0 mm
**IDENTIFICATION:** Males with eyes separated by about ½ the width of the anterior ocellus, nearly in contact for only a short distance; metatarsus with white anterodorsal pile. Female with cell dm microtrichose at base, sometimes with small bare areas along anterior and posterior margins; scutellar pile long. Both sexes with cell $r_{4+5}$ bare at base. **ABUNDANCE:** Rare. **FLIGHT TIMES:** Early June to late September. **NOTES:** Found along the coast.

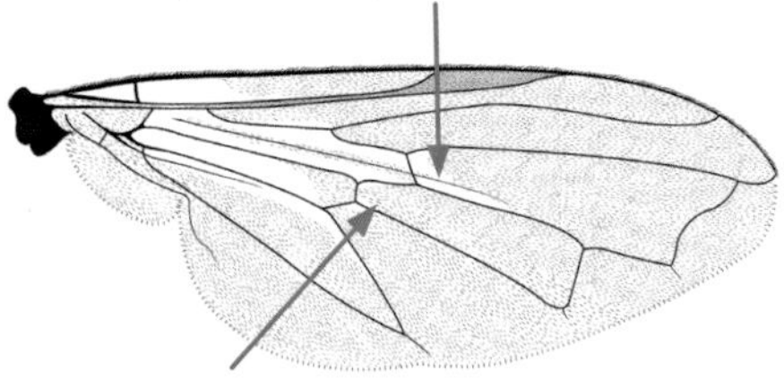

## *Trichopsomyia pubescens* Wide-eyed Psyllid-killer

**SIZE RANGE:** 4.5–6.5 mm
**IDENTIFICATION:** Males with eyes separated by about ½ the width of the anterior ocellus, nearly in contact for only a short distance; metatarsus with brown black anterodorsal pile; cell $r_{4+5}$ microtrichose. Female unknown, wing presumably similar to male. **ABUNDANCE:** Rare. **FLIGHT TIMES:** Late January to early September (early June to early July in the north). **NOTES:** Nothing is known about this species.

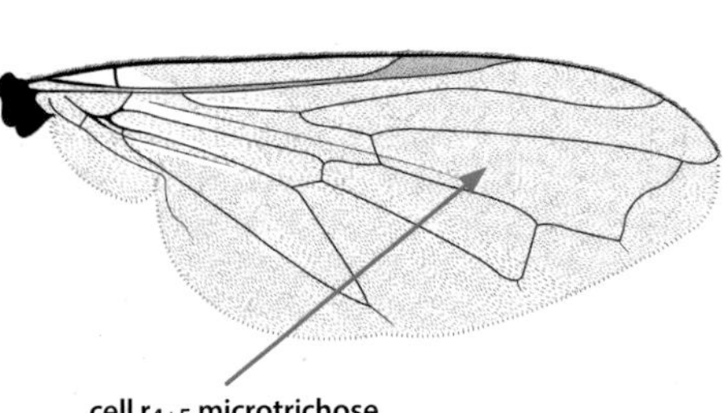

## *Trichopsomyia* undescribed species 1

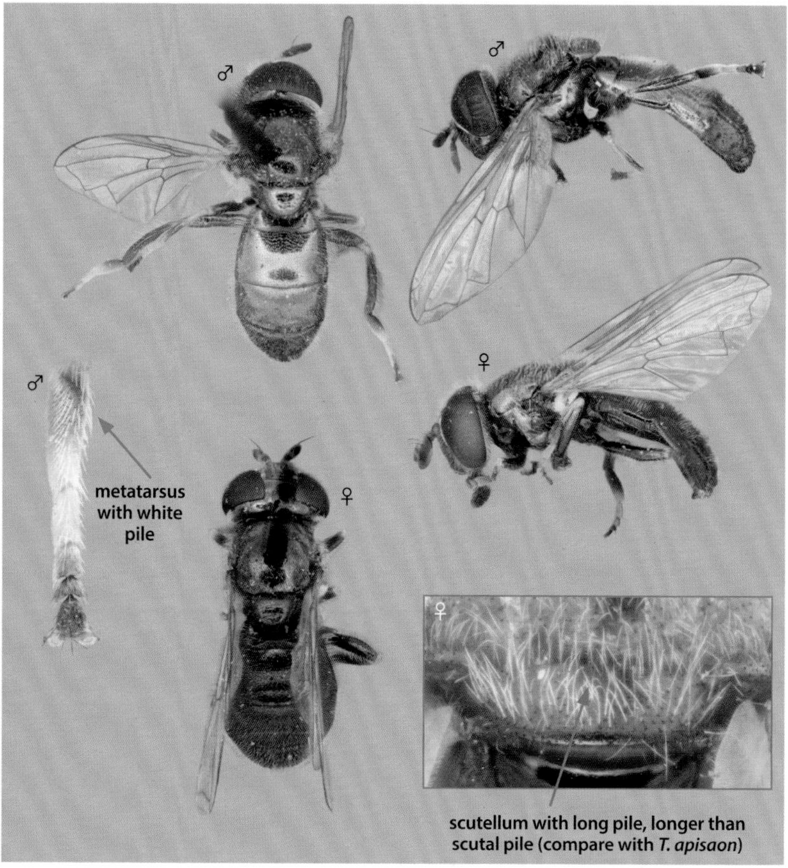
♂
♂
♂
metatarsus with white pile
♀
♀
♀
scutellum with long pile, longer than scutal pile (compare with *T. apisaon*)

## *Trichopsomyia pubescens*

♂
metatarsus with anterodorsal pile brown black
♂
♂

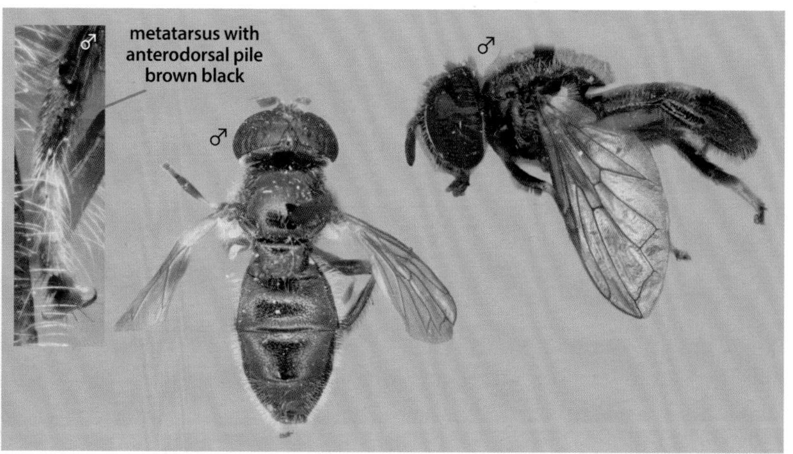

# SYRPHINAE

## *Paragus*

*Paragus* species are small, punctate, yellow-faced syrphids that have black or black and orange abdomens. Unlike in other syrphines, tergite 1 is well developed and is typically half as long as tergite 2. There are two subgenera. *Paragus (Pandasyopthalmus)* is easily recognizable. *P. (Paragus)* is difficult and only males are identifiable by their genitalia. Only eight of the 104 species are Nearctic (six in our area). Adults are often found among low vegetation and prefer hot, dry areas. They appear to have many generations per year. For a key to the New World species, see Vockeroth (1986b). Genitalia drawings are from Vockeroth (1992). Larvae feed on aphids and other soft-bodied insects.

### *Paragus (Pandasyopthalmus)* — *Paragus (Paragus)* spp.

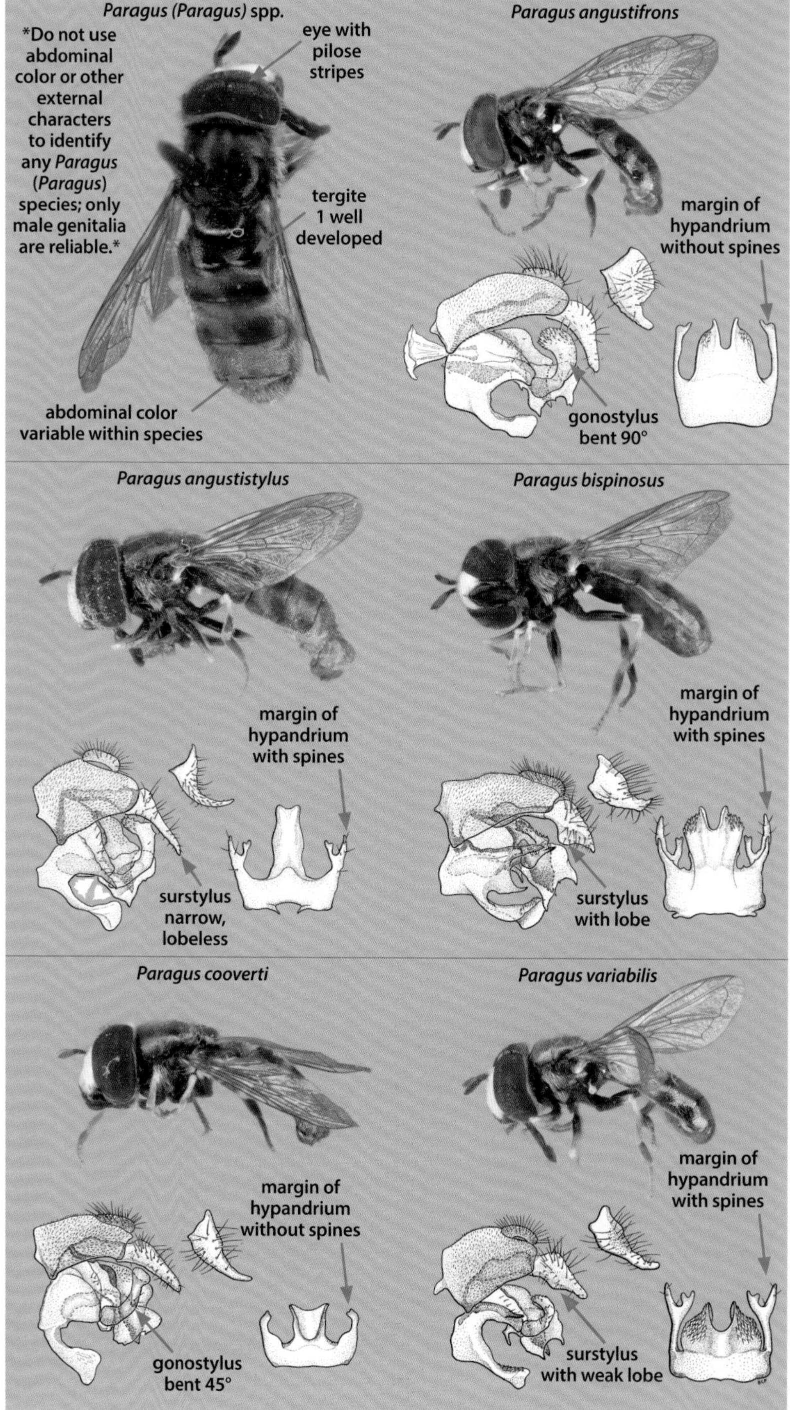
*Paragus (Paragus)* spp.
*Do not use abdominal color or other external characters to identify any *Paragus* (*Paragus*) species; only male genitalia are reliable.*
eye with pilose stripes
tergite 1 well developed
abdominal color variable within species
*Paragus angustifrons*
margin of hypandrium without spines
gonostylus bent 90°
*Paragus angustistylus*
margin of hypandrium with spines
surstylus narrow, lobeless
*Paragus bispinosus*
margin of hypandrium with spines
surstylus with lobe
*Paragus cooverti*
margin of hypandrium without spines
gonostylus bent 45°
*Paragus variabilis*
margin of hypandrium with spines
surstylus with weak lobe

## *Paragus (Pandasyopthalmus) haemorrhous*

### Black-backed Grass Skimmer

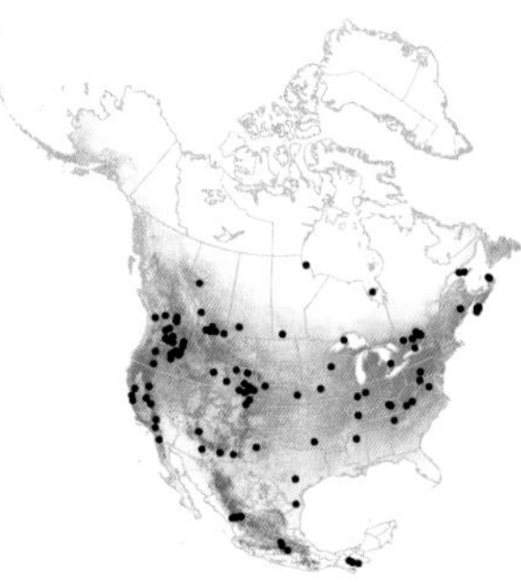

**SIZE RANGE:** 4.3–5.9 mm
**IDENTIFICATION:** This is our only *Paragus* species that can be recognized with external characters. The scutellum is entirely black and the eye is uniformly pilose. The face always has a black medial stripe and the terminal abdominal segments are often orange. **ABUNDANCE:** Uncommon. **FLIGHT TIMES:** Mid-April to early October (mid-March to early November in the south). **NOTES:** This species is Holarctic, Afrotropical, and Neotropical south to Costa Rica. Adults are known from forest, field, meadow, prairie, desert, and agricultural habitats. They have been recorded from Apiaceae, Asteraceae, Rosaceae, and Verbenaceae flowers. Larvae feed on ground-dwelling aphids that feed on a variety of plants, including cultivated crops. The large range and considerable variation within this species suggests that it is a complex of species.

## *Paragus (Paragus) angustifrons*

### Narrow-faced Grass Skimmer

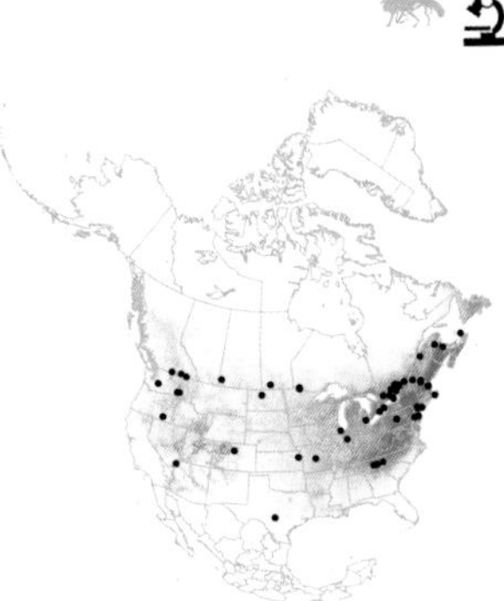

**SIZE RANGE:** 4.8–6.1 mm
**IDENTIFICATION:** Identification by male genitalia. Margin of hypandrium without spines; gonostylus bent 90 degrees, with apex enlarged and denticulate. **ABUNDANCE:** Uncommon. **FLIGHT TIMES:** Early May to late September (as early as late March in other parts of range). **NOTES:** Adults are known from mixed forest habitats. Larvae unknown.

## *Paragus (Paragus) angustistylus*

### Thin-spined Grass Skimmer

**SIZE RANGE:** 4.7–6.9 mm
**IDENTIFICATION:** Identification by male genitalia. Margin of hypandrium with spines; surstylus narrow, without lobes. **ABUNDANCE:** Uncommon. **FLIGHT TIMES:** Early May to mid-September (to early November in the south). **NOTES:** Larvae unknown.

## *Paragus (Paragus) bispinosus*

Two-spined Grass Skimmer

**SIZE RANGE:** 5.5–6.6 mm
**IDENTIFICATION:** Identification by male genitalia. Margin of hypandrium with spines; surstylus broad, with strong medial lobe. **ABUNDANCE:** Rare. **FLIGHT TIMES:** Mid-June to late August. **NOTES:** Adults have been collected from *Rumex* and are likely to be present on other flowers. Larvae unknown.

## *Paragus (Paragus) cooverti*

Coovert's Grass Skimmer

**SIZE RANGE:** 5.0–6.4 mm
**IDENTIFICATION:** Identification by male genitalia. Margin of hypandrium without spines; gonostylus bent 45 degrees and with apex enlarged and smooth. **ABUNDANCE:** Rare. **FLIGHT TIMES:** Early June to early July. **NOTES:** Larvae unknown but this and all others treated here presumably feed on aphids as others in this genus do.

## *Paragus (Paragus) variabilis*

Variable Grass Skimmer

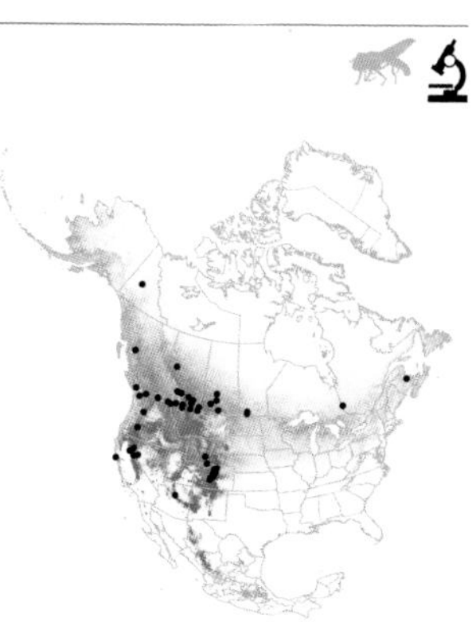

**SIZE RANGE:** 4.8–6.5 mm
**IDENTIFICATION:** Identification by male genitalia. Margin of hypandrium with spines; surstylus broad, with weak medial lobe. **ABUNDANCE:** Uncommon in the west, very rare in the east. **FLIGHT TIMES:** Mid-June to late August (from early May in the west). **NOTES:** Larvae unknown. There is considerable variation within this species and it is potentially a complex.

# *Melanostoma*

The single Nearctic member of this genus is one of the most abundant syrphids in our region. There are 57 world species in *Melanostoma* but their taxonomy is challenging and revision of most world species is needed. Our species is undoubtedly part of a complex. Standard DNA barcoding (using the gene COI) will not help to resolve taxonomic issues in *Melanostoma*, and faster evolving markers such as ITS are needed. See Haarto and Ståhls (2014) for a discussion of how they at least partly resolved *Melanostoma* taxonomy in Europe.

*Platycheirus* is the most likely genus to be confused with *Melanostoma*. They are both slender flies with black, tuberculate faces, black scutums, and black abdomens with yellow-orange markings. The way to definitively distinguish these genera is to look at the metasternum. In *Melanostoma* it is reduced to a small diamond. Several supporting characters can often be used to distinguish between the two genera. All male *Platycheirus* with quadrate markings on the tergites also have an expanded protibia and/or protarsomeres, while *Melanostoma* have unmodified prolegs. All female *Platycheirus* with triangular markings on the tergites have a tuft of setae at the base of the profemur, while female *Melanostoma* have triangular markings but no such tuft. However, melanistic *Melanostoma* with no pale markings are occasionally collected and those must be identified positively using the metasternal character.

*Melanostoma* larvae feed on aphids. For a genus and species diagnosis and description, see Vockeroth (1992).

## *Melanostoma mellinum* Variable Duskyface

**SIZE RANGE:** 4.8–10.0 mm

**IDENTIFICATION:** The abdominal markings on tergites 2–4 are diagnostic for both sexes. Learning this makes it easy to separate them from *Platycheirus* in the field. Males typically have quadrate markings, while females have triangular-oval markings. Prolegs simple. **ABUNDANCE:** Abundant. **FLIGHT TIMES:** Late April to early October (from late January to mid-November in the west). **NOTES:** This species is common in mixed woodlands and damp meadows, where it can often be found feeding on grass pollen and other flowers. Adults have been collected at flowers of Apiaceae, *Cornus*, *Elymus*, *Equisetum*, *Heracleum*, *Kalmia*, *Myosotis*, *Oryzopsis*, *Rhododendron*, *Rubus*, *Sonchus*, *Symphyotrichum*, *Tragopogon*, *Trillium*, and *Vaccinium*. They are commonly collected by sweeping sedges and grasses.

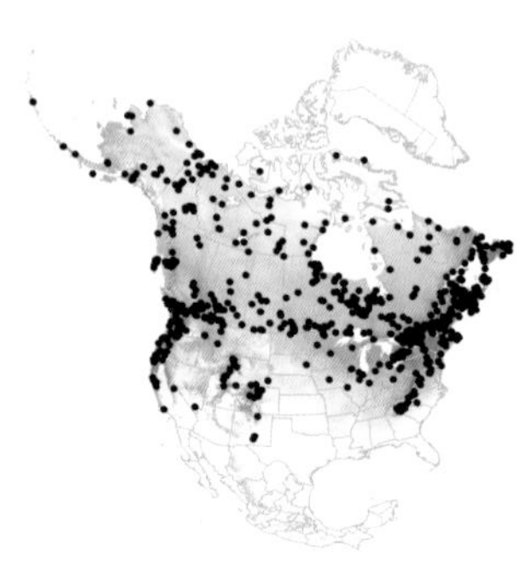

# *Melanostoma mellinum*

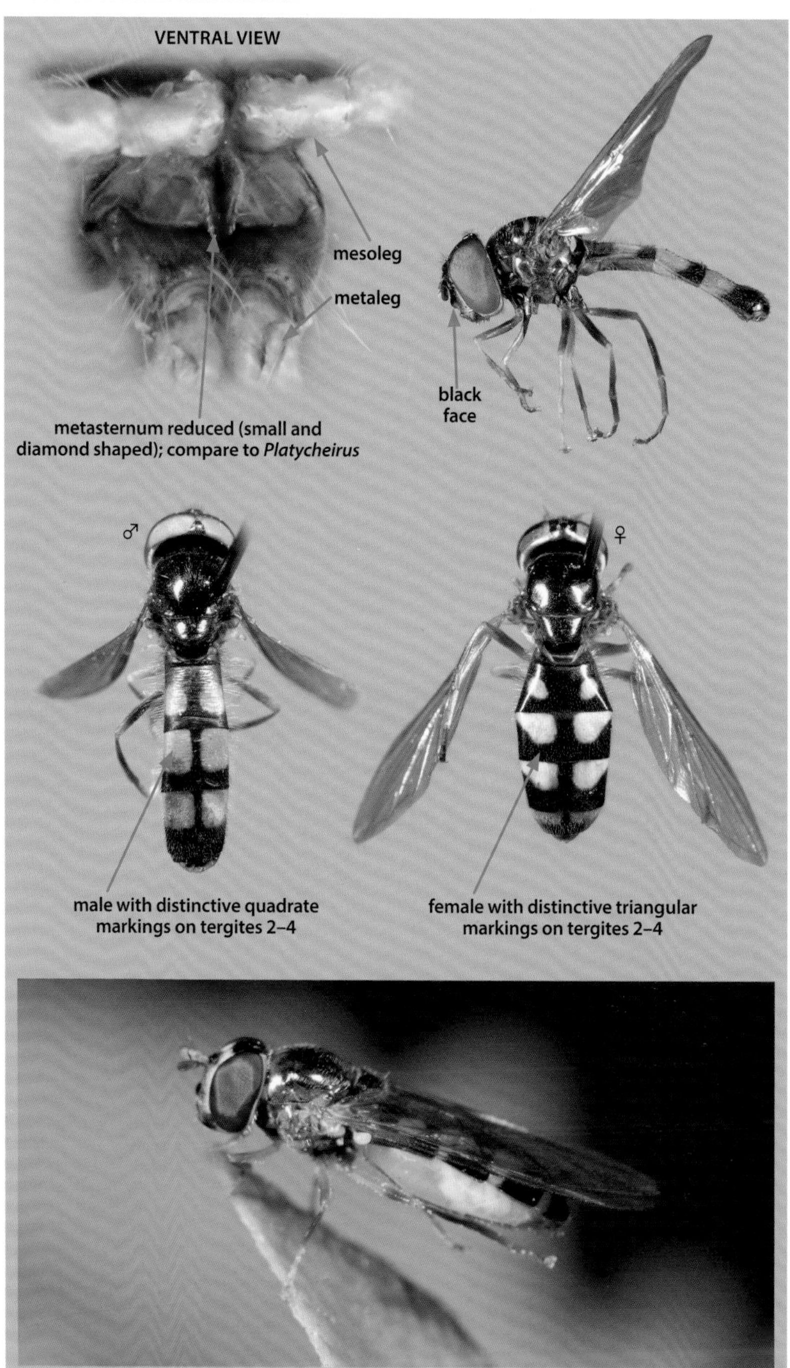

# *Platycheirus*

This is the most diverse syrphine genus in the Nearctic region, with approximately 220 species worldwide, 75 in the Nearctic, and 44 species present in the northeast. The genus is diagnosed by an entirely bare postpronotum (like all Nearctic Syrphinae), a black scutum, scutellum, and face, and an unmodified metasternum. Depending on the species, *Platycheirus* adults are often found in boreal/mixed woods, humid meadows and woodlands, or in wetland areas. Many species are associated with sedges and the adults may feed on their pollen and that of other wind-pollinated plants. *Platycheirus* larvae, where they are known, feed on aphids and other soft-bodied insects. The most recent species descriptions and key to species can be found in Young *et al.* (2016b). The genus can be divided into eight species groups, which are based largely on the morphology of the male proleg (shown on the right). ***P. stegnus*** **group** DIAGNOSIS: long, posterior black setae on the pro- and mesotibiae and facial pollinosity arranged in either oblique ripples or punctures in both sexes. ***P. granditarsis*** **group** DIAGNOSIS: large, irregular orange markings on the tergites; probasitarsus expanded in *P. granditarsis* only. ***P. albimanus*** **group** (illustrated with *P. quadratus*) DIAGNOSIS: expanded probasitarsus in combination with the tibia expanded over its entire length. ***P. ambiguus*** **group** (illustrated with *P. coerulescens*) DIAGNOSIS: long, curled seta at the apex of the profemur. ***P. manicatus*** **group** DIAGNOSIS: expanded probasitarsus in combination with the unexpanded tibia. ***P. chilosia*** **group** DIAGNOSIS: long, posterior, curled setae on the probasitarsus. ***P. pictipes*** **group** DIAGNOSIS (in the northeast): unmodified proleg. ***P. peltatus*** **group** (illustrated with *P. naso*) DIAGNOSIS: expanded probasitarsus in combination with the tibia expanded only at the apex.

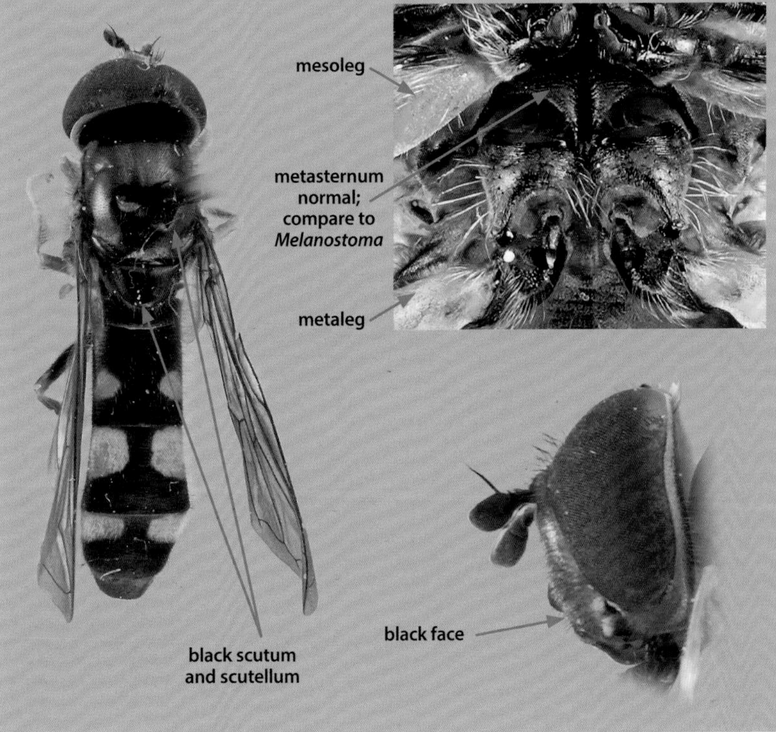

# *Platycheirus* species groups

## *P. stegnus* group (*P. stegnus*)

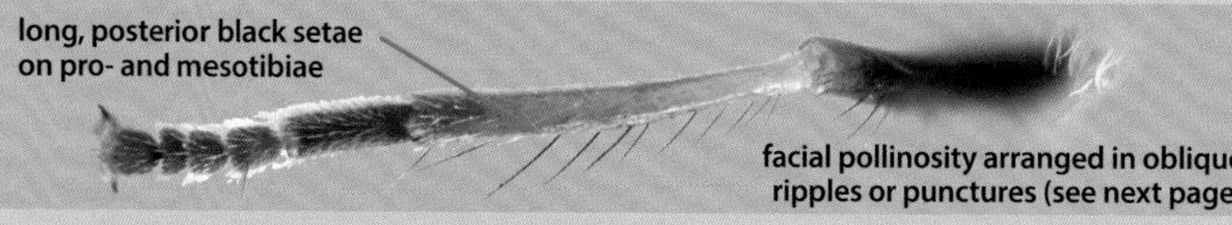


## *P. granditarsis* group (*P. granditarsis*)

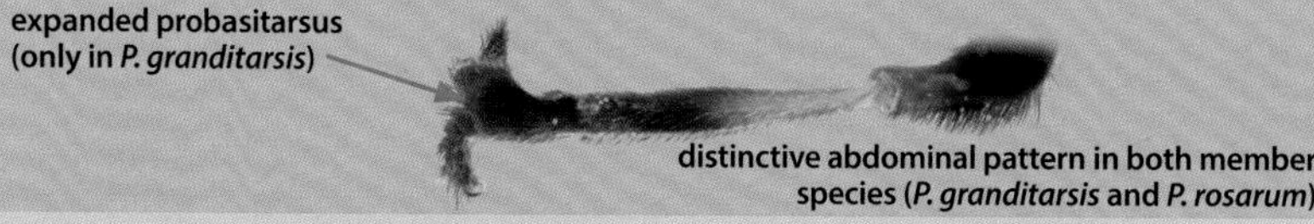


## *P. albimanus* group (*P. quadratus*)

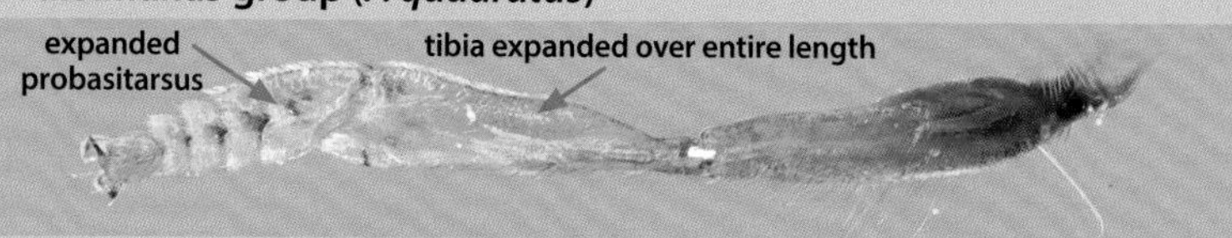


## *P. ambiguus* group (*P. coerulescens*)

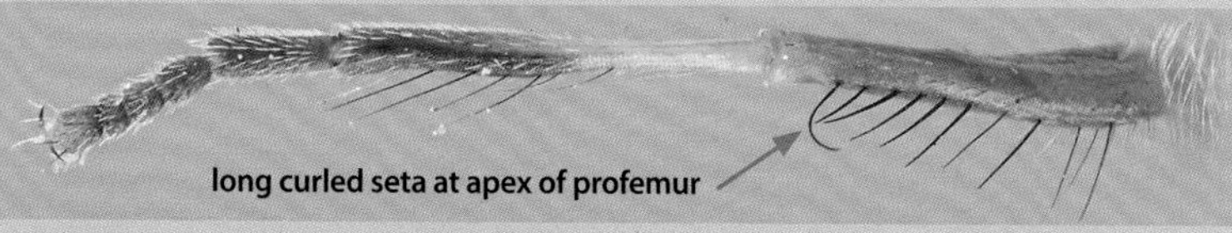


## *P. manicatus* group (*P. manicatus*)

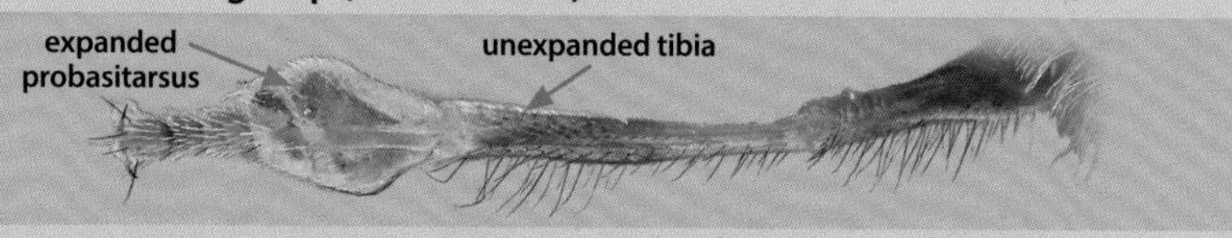


## *P. chilosia* group (*P. chilosia*)

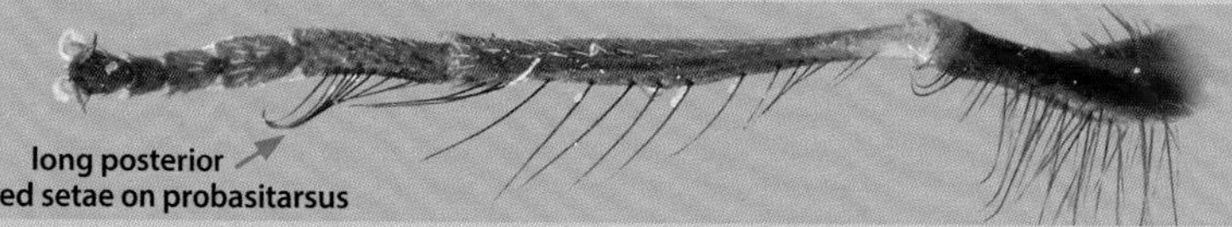


## *P. pictipes* group (*P. pictipes*)

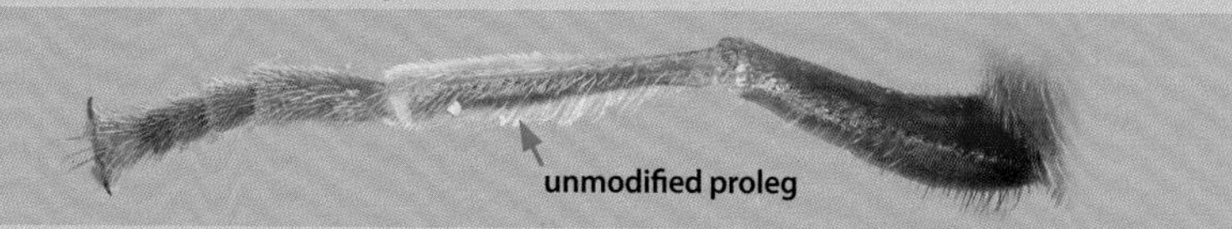


## *P. peltatus* group (*P. naso*)

## *stegnus* group

### *Platycheirus obscurus* Eastern Forest Sedgesitter

**SIZE RANGE:** 6.8–9.4 mm

**IDENTIFICATION:** This species has a distinctly ventrally produced face, with the anterior oral margin extending as far as the tubercle, but not beyond (see *P. trichopus* below). Laterally, the face has distinct oblique ripples present in the grayish pollinosity. The male pro- and mesotibiae have a row of weak posterior setae on apical ½ to ⅔. The male metatibia has a posterior row of irregular setae, the longest of which are approximately 2.5–3 times as long as the width of the tibia. **ABUNDANCE:** Common. **FLIGHT TIMES:** Mid-April to late October (from early March in the south). **NOTES:** Larvae have been reared on four species of aphids as well as rotting *Stellaria*. Males are frequently found hovering in mixed woodlands, fens, bogs, and swamps. Flowers visited include *Hieracium*, *Primula*, *Solidago*, and *Taraxacum*.

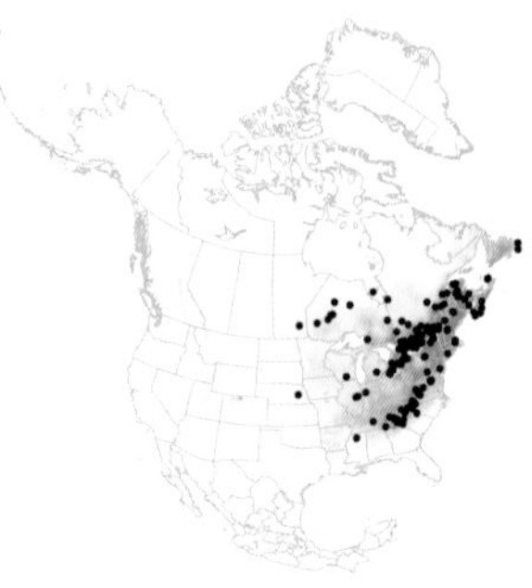

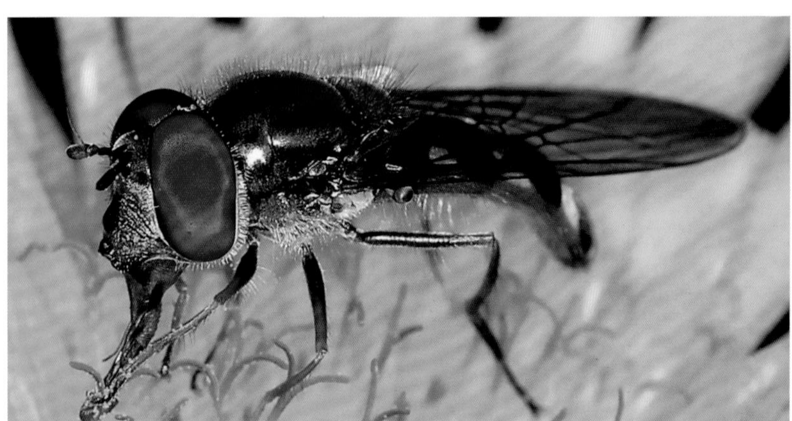

### *Platycheirus confusus* Confusing Sedgesitter

**SIZE RANGE:** 6.0–8.2 mm

**IDENTIFICATION:** This species is very similar to *P. obscurus* and differs in that the anterior oral margin is never produced as far forward as the tubercle. **ABUNDANCE:** Common. **FLIGHT TIMES:** Late April to late July (to early September in the west). **NOTES:** Usually collected in humid meadows, marshes, sphagnum bogs, and surrounding mixed forest. Flowers visited include *Euthamia*, *Primula*, *Prunus*, and *Ranunculus*.

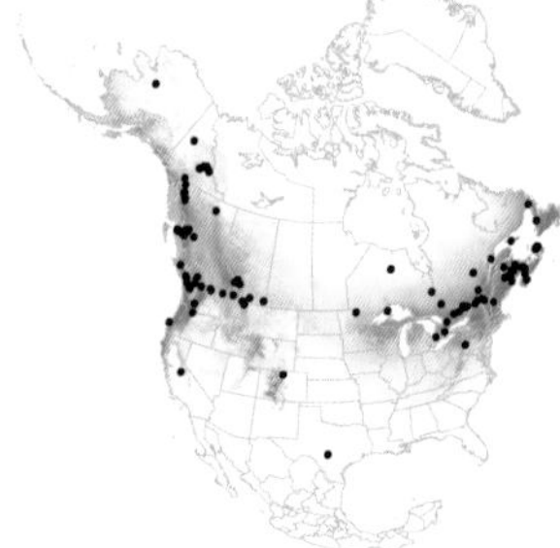

## *Platycheirus obscurus*

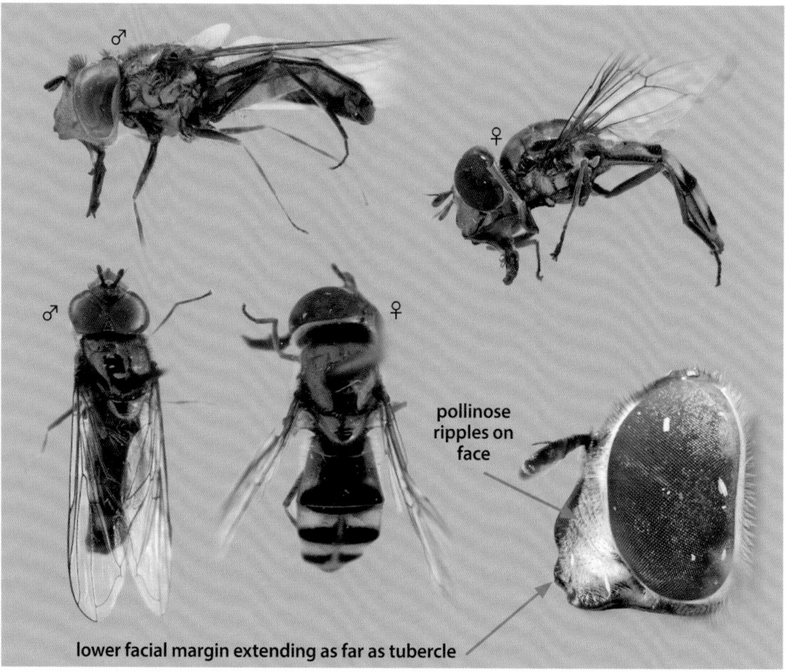

## *Platycheirus confusus*

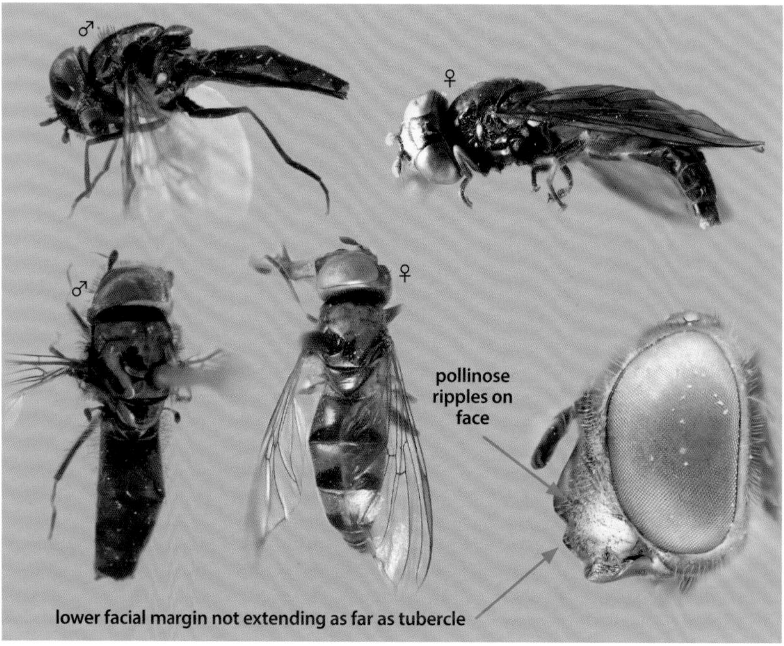

## *Platycheirus trichopus* Western Forest Sedgesitter

**SIZE RANGE:** 7.0–9.6 mm
**IDENTIFICATION:** This species has a face that is distinctly produced ventrally, with the anterior oral margin extending beyond the facial tubercle. Laterally, the face has distinct oblique ripples present in the grayish pollinosity. The male pro- and mesotibiae have a row of weak posterior setae on apical ½ to ⅔. The male metatibia has a posterior row of irregular setae, the longest of which are approximately 3.5–4 times as long as the width of the tibia. **ABUNDANCE:** Common in the west. **FLIGHT TIMES:** August (late January to early November in the west). **NOTES:** It has been collected in pine forests. The taxonomy of this species has been confused with *P. obscurus* historically.

### *granditarsis* group

## *Platycheirus granditarsis* Hornhand Sedgesitter

**SIZE RANGE:** 7.7–10.5 mm
**IDENTIFICATION:** Males of this species are easily identifiable due to the probasitarsus, which has a large, anterior triangular process on the apical ½. The first four tarsomeres of the mesoleg are strongly flattened, each with a broad, apically rounded process; these are progressively shorter from the first to third tarsomere. Females are also easily diagnosed by overall gestalt, as the abdomen is relatively broad, with irregularly shaped medially confluent orange spots on tergites 2–4. In addition, the face is bare and shiny, with silver pollinosity present only at the lateral edges. **ABUNDANCE:** Common. **FLIGHT TIMES:** Late May to late September (early March to early November in the south). **NOTES:** Found in humid open grasslands, marshes, fens, and edges of raised bogs. Flowers visited: *Alisma*, white Apiaceae, *Bidens*, *Leontodon*, *Lycopus*, *Polygonum*, *Ranunculus*, and *Senecio*.

## *Platycheirus trichopus*

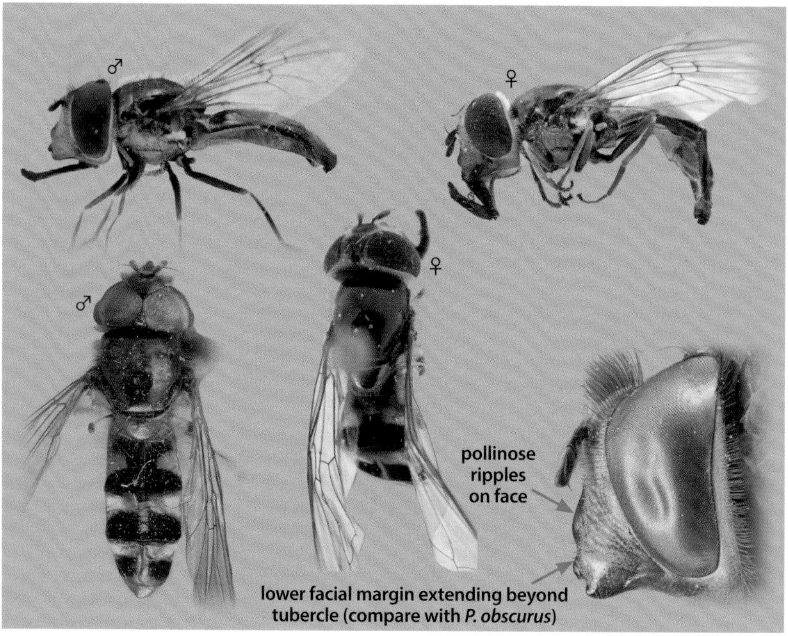

## *Platycheirus granditarsis*

## *Platycheirus rosarum* Fourspot Sedgesitter

**SIZE RANGE:** 7.3–9.1 mm
**IDENTIFICATION:** This species has unmodified legs and a narrow, thinly pollinose face. Tergites 3 and 4 each have a pair of pale yellow spots reaching the anterior and lateral margins of the tergites, with spots of tergite 3 somewhat longer than spots of tergite 4. These spots are sometimes confluent medially in the male and often in the female. The abdomen widens toward the apex in both sexes, but it is more noticeable in the female. **ABUNDANCE:** Common. **FLIGHT TIMES:** Early April to mid-September. **NOTES:** Found in wetlands, along pond, stream, and river margins with tall herbaceous vegetation, around the periphery of raised bogs, and in humid, seasonally flooded grasslands. Flowers visited: *Caltha*, *Knautia*, *Lythrum*, *Nymphaea*, *Potentilla*, *Ranunculus*, and *Scirpus*.

### *albimanus* group

## *Platycheirus quadratus* Meadow Sedgesitter

**SIZE RANGE:** 7.1–9.1 mm
**IDENTIFICATION:** Males of this species have a posterior, subbasal tuft of two to three long, white, wavy setae on the profemur and a protibia broadened from base to apex. The tarsomeres are widened, with the first almost as wide as the tibia and the rest progressively narrower. This species can be distinguished from close relatives by the mesotibia, which is broadened on the anterior ¾ and has a dense ventral brush of many mixed black and yellow setae. The female is indistinguishable from *immarginatus*, *neoperpallidus*, and *perpallidus*, but can be placed in this group by the profemur, which has a posterior tuft of setae similar to the male. In addition, pollinosity on the frons above the antennal insertions forms two lateral triangles, and the spots of their tergites are slightly wider than long. **ABUNDANCE:** Common. **FLIGHT TIMES:** Early April to late October (early March to early November in the south). **NOTES:** Found in wetlands, *Carex* marshes, fens, and humid grasslands. Flies and sits among tall waterside and emergent vegetation. Larvae have been found overwintering in seedheads of *Hibiscus laevis*. There is a single floral record from *Zizia*.

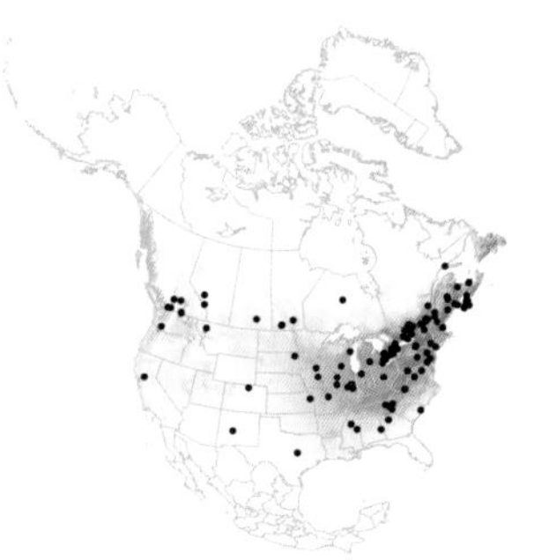

# Platycheirus rosarum

# Platycheirus quadratus

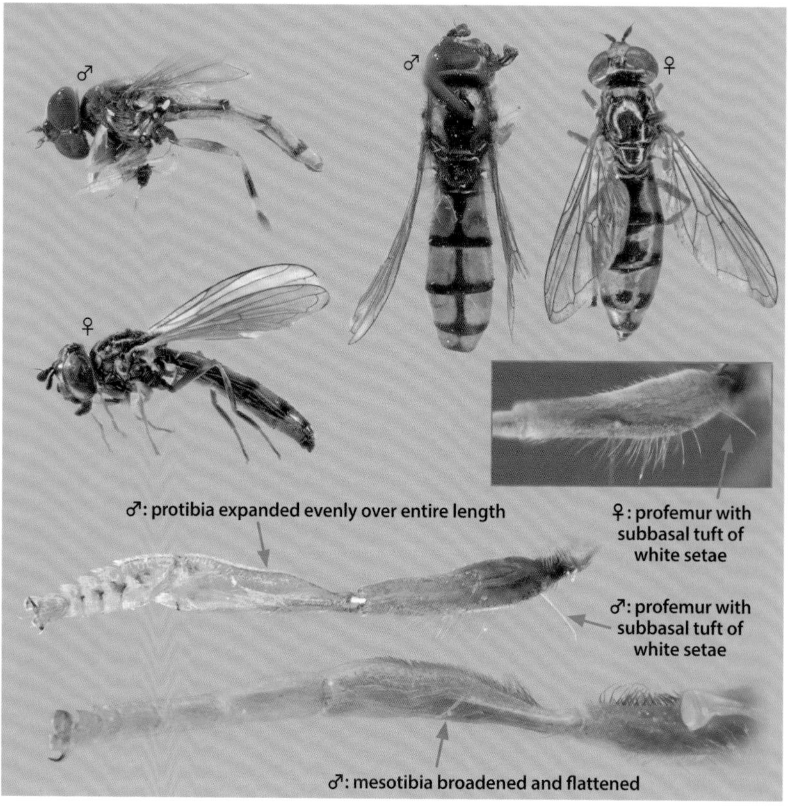

♂

♂

♀

♀

♂: protibia expanded evenly over entire length

♀: profemur with subbasal tuft of white setae

♂: profemur with subbasal tuft of white setae

♂: mesotibia broadened and flattened

## *Platycheirus immarginatus*

### Comb-legged Sedgesitter

**SIZE RANGE:** 6.5–9.6 mm
**IDENTIFICATION:** Males of this species possess a proleg very similar to *P. quadratus*, except the subbasal white tuft of the profemur is followed by a regularly spaced row of 4–5 posterior, long, black setae. The mesofemur has an anteroventral row of 10–22 short, stiff, black setulae on the apical ⅔, this row usually ending in 1 or 2 longer setae that are strongly curved toward base of femur. The mesofemur also has 3–6 ventral, black or yellow setae on basal ½, these setae approximately twice as long as the femoral diameter. Female: see *Platycheirus quadratus*. **ABUNDANCE:** Common. **FLIGHT TIMES:** Early May to mid-September (mid-April in the southwest). **NOTES:** Found in wetlands including *Carex* marsh, taiga wetlands, fens, and freshwater coastal marshes. Flies and sits among tall waterside and emergent vegetation.

## *Platycheirus neoperpallidus*

### Yellowcomb Sedgesitter

**SIZE RANGE:** 5.7–9.1 mm
**IDENTIFICATION:** Males of this species possess a proleg very similar to *P. quadratus*. It can be distinguished from related species by the mesotibia, which has subappressed, wavy yellow pile on basal ⅔ of the anteroventral surface. Female: see *Platycheirus quadratus*. **ABUNDANCE:** Uncommon. **FLIGHT TIMES:** Early June to late August. **NOTES:** This species has been collected from *Carex*, *Equisetum*, and grasses around lakes and marshes.

## Platycheirus immarginatus

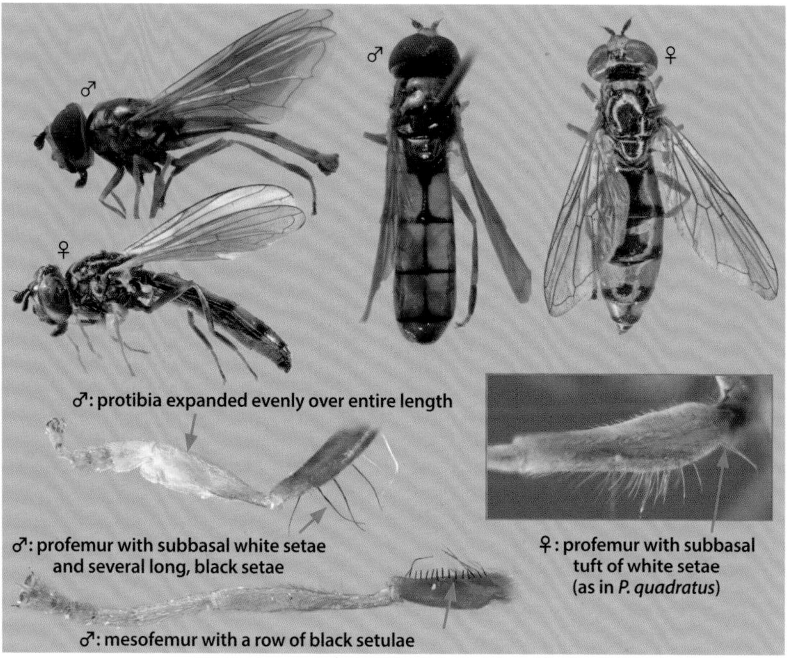

♂

♂

♀

♀

♂: protibia expanded evenly over entire length

♂: profemur with subbasal white setae and several long, black setae

♀: profemur with subbasal tuft of white setae (as in *P. quadratus*)

♂: mesofemur with a row of black setulae

## Platycheirus neoperpallidus

♂

♂

♀

♀

♂: profemur with subbasal tuft of white setae (as in *P. quadratus*)

♂: mesotibia with ventral, subappressed yellow pile

♀: profemur with subbasal tuft of white setae (as in *P. quadratus*)

## *Platycheirus perpallidus* Perplexing Sedgesitter

**SIZE RANGE:** 5.7–9.1 mm

**IDENTIFICATION:** Males of this species possess a proleg very similar to *P. quadratus*. It can be distinguished from related species by the mesotibia, which has erect, dense, wavy black pile on basal ⅔ of the anteroventral surface. Female: see *Platycheirus quadratus*. **ABUNDANCE:** Uncommon. **FLIGHT TIMES:** Mid-June to mid-July (early May to early August in the west). **NOTES:** Found in wetland habitats and along river and lake margins, among tall emergent waterside vegetation. This species has been recorded nectaring and pollen feeding on Cyperaceae, *Juncus*, *Rubus*, and *Salix*.

## *Platycheirus angustatus* Delicate Sedgesitter

**SIZE RANGE:** 5.7–7.9 mm

**IDENTIFICATION:** Males of this species possess a profemur similar to *P. quadratus*. The protibia is broadened from base to apex, with the posteroapical angle distinctly produced into a point. This species can be diagnosed by the probasitarsus, which is narrowed posteriorly on basal ⅓, parallel sided on apical ⅔, and has a V-shaped incision on the underside. Females of this species have a vertical face, with the bottom of oral margin rounded, not produced forward. The abdomen is parallel sided and narrow with tergite 2 distinctly longer than wide. Pale spots on the tergites are longer than wide, with those of tergites 3 and 4 meeting the anterior and lateral margins of the tergites. Tergites 5 and 6 entirely dark. **ABUNDANCE:** Uncommon. **FLIGHT TIMES:** Late April to early September. **NOTES:** This species has been recorded from wetlands, marshy meadows, mixed forests, and tundra habitats. It has been recorded nectaring and pollen feeding at *Aegopodium*, Cyperaceae, *Leontodon*, *Lycopus*, Poaceae, *Polygonum*, *Ranunculus*, and *Rubus*.

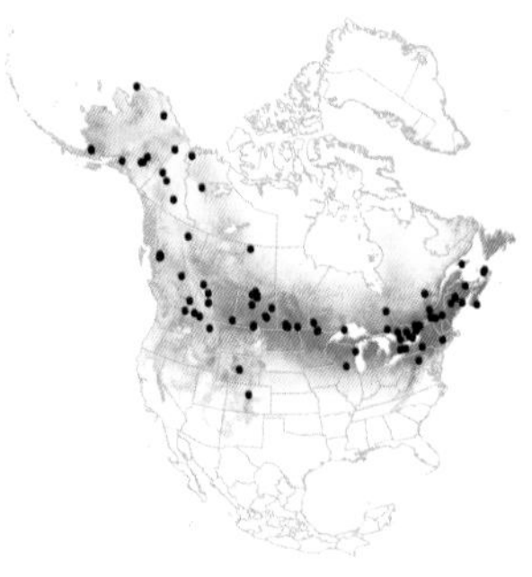

## *Platycheirus perpallidus*

♂

♂

♀

♀

♂: profemur with subbasal tuft of white setae (as in *P. quadratus*)

♀: profemur with subbasal tuft of white setae (as in *P. quadratus*)

♂: mesotibia with ventral, erect black pile

## *Platycheirus angustatus*

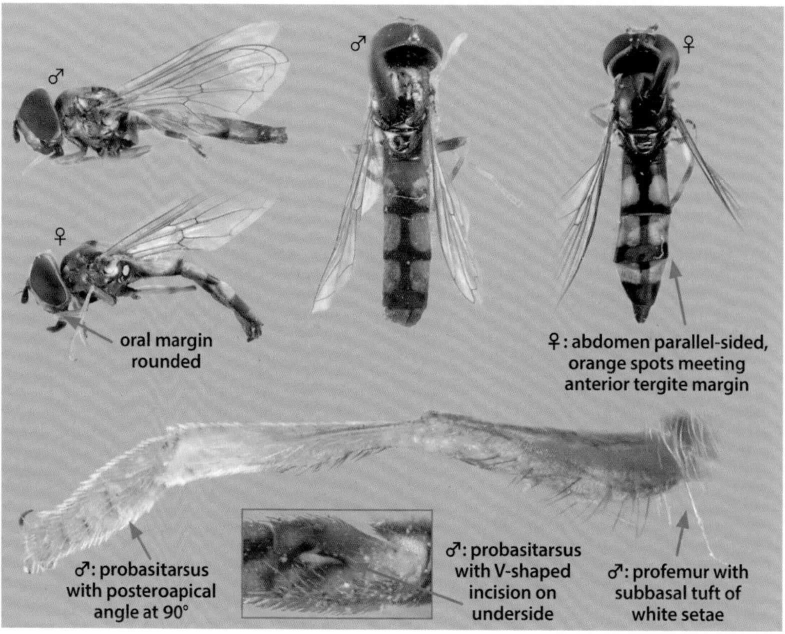

♂

♂

♀

♀

oral margin rounded

♀: abdomen parallel-sided, orange spots meeting anterior tergite margin

♂: probasitarsus with posteroapical angle at 90°

♂: probasitarsus with V-shaped incision on underside

♂: profemur with subbasal tuft of white setae

## *Platycheirus clypeatus* Smoky-winged Sedgesitter

**SIZE RANGE:** 6.0–8.8 mm

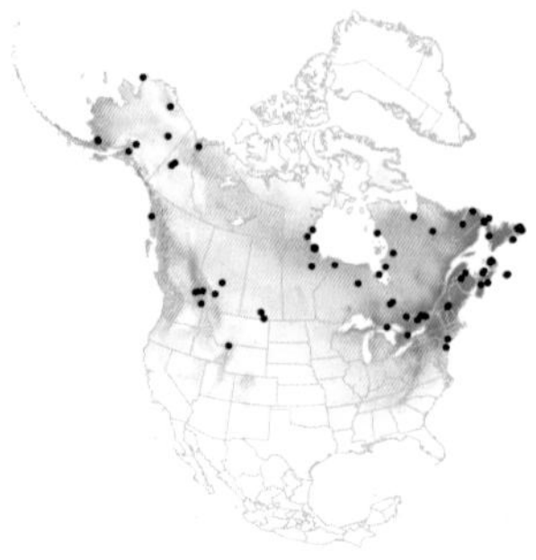

**IDENTIFICATION:** Males of this species have a proleg very similar to *P. angustatus*, except that the probasitarsus has a shallow, pale groove on the underside. Remaining protarsomeres slightly narrower than first and unmodified. Spots of tergites are yellow or orange, sometimes with a faint silvery pollinose overlay. Females of this species have a vertical face, with the bottom of the oral margin rounded and not produced forward. The abdomen is narrowly oval. Spots of tergites are yellow, sometimes with a faint to strong silvery pollinose overlay. Spots of tergite 2 are small, triangular, and well separated from anterior margin of the tergite. **ABUNDANCE:** Uncommon. **FLIGHT TIMES:** Early May to early October. **NOTES:** This species is found in humid meadows, wetlands, fens, and along the margins of rivers, streams, lakes, and ponds sitting in emergent vegetation. It has been recorded nectaring and pollen feeding on white Apiaceae, *Caltha*, Cyperaceae, *Luzula*, *Plantago*, Poaceae, *Polygonum*, *Ranunculus*, and *Salix*.

## *Platycheirus hyperboreus* Pearly Sedgesitter

**SIZE RANGE:** 5.3–8.7 mm

**IDENTIFICATION:** Males of this species have a proleg very similar to *P. angustatus*, except that the probasitarsus has the anterior margin straight and posterior margin rounded, so margins are not parallel on the apical ⅔. This species is diagnosed by the spots on the abdomen that range from yellow to dark brown or even absent, but always with a silvery pollinose overlay. Females of this species are almost indistinguishable from *P. clypeatus*, with strong silvery pollinose spots on the tergites. **ABUNDANCE:** Common. **FLIGHT TIMES:** Late March to early October. **NOTES:** This species is found in wetlands, marshes, fens, palsa mires, swampy woodlands, tundra, and along water margins. It has been recorded nectaring and pollen feeding on Apiaceae, *Carex*, *Parnassia*, and *Ranunculus*.

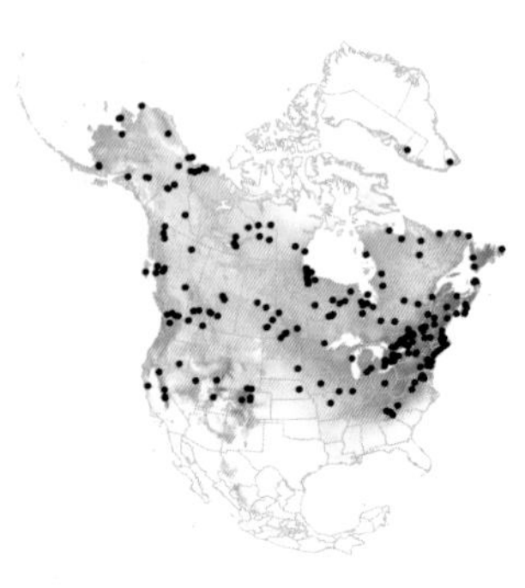

# Platycheirus clypeatus

# Platycheirus hyperboreus

## *Platycheirus aeratus* Coquillett's Sedgesitter

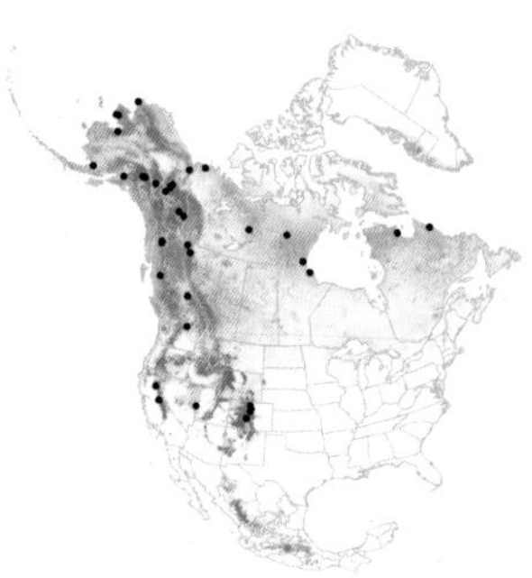

**SIZE RANGE:** 5.3–7.3 mm
**IDENTIFICATION:** Males of this species possess a proleg very similar to *P. quadratus*, except that the protibia is slightly less broadened and has wavy posterior black setae. This species can be distinguished by its dark legs and the mesotibia, which has short, dense, wavy anteroventral pile. Additionally, the abdomen has silvery pollinose spots, often dark otherwise. Females of this species have a vertical face, with the bottom of oral margin rounded. Females can be distinguished from related species because all femora are dark on the basal ⅔ to ¾. **ABUNDANCE:** Uncommon. **FLIGHT TIMES:** Mid-June to mid-August. **NOTES:** This species has been recorded from tidal flats, tundra, and marshy water margins. It has been recorded feeding at *Carex* and *Eriophorum*.

## *Platycheirus podagratus* Variable Sedgesitter

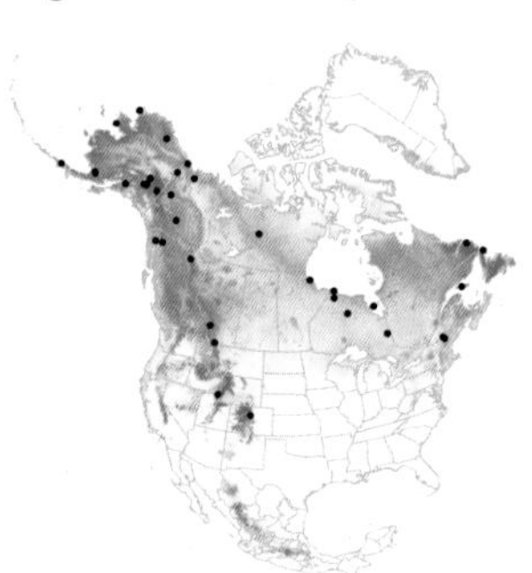

**SIZE RANGE:** 5.7–7.7 mm
**IDENTIFICATION:** Males of this species are the only *Platycheirus* in the northeast with a protibia that is strongly and abruptly broadened on the apical ⅖ and slightly narrowed at the apex. Likewise, females of this species have uniquely marked tergites, with small triangular or rounded spots that are usually separated from the lateral margin. When in doubt, check the color of the metafemur and whether the oral margin is rounded or produced forward and compare to *P. thompsoni* and *P. flabella*. **ABUNDANCE:** Uncommon. **FLIGHT TIMES:** Late May to early July (to mid-August in the west). **NOTES:** This species is found flying low to the ground in wetlands, mires, acid fens, taiga, tundra, alpine grasslands, and along water margins of oligotrophic lakes, rivers, and streams. It has been recorded feeding at *Carex*.

## *Platycheirus aeratus*

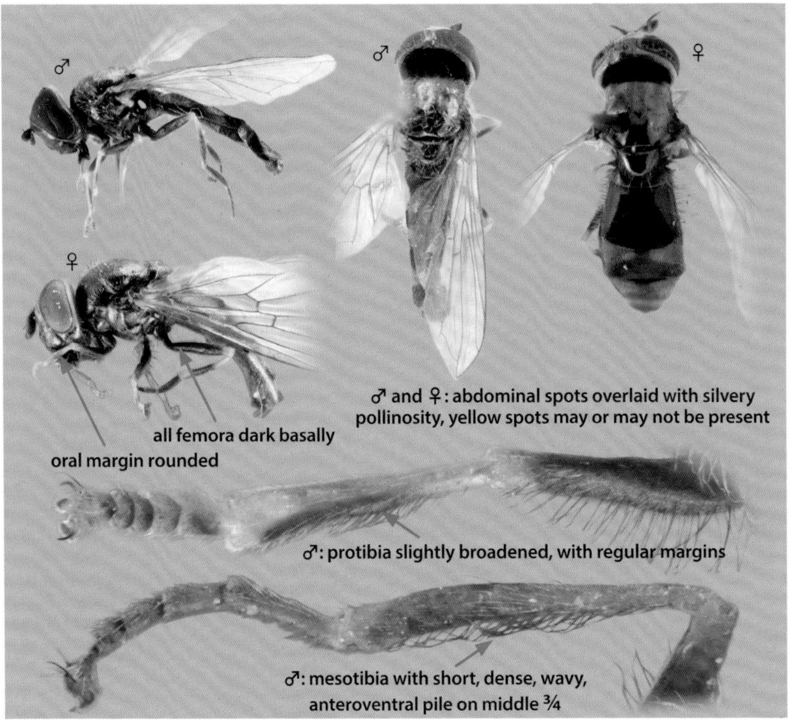

♂

♂

♀

♀

♂ and ♀: abdominal spots overlaid with silvery pollinosity, yellow spots may or may not be present

all femora dark basally

oral margin rounded

♂: protibia slightly broadened, with regular margins

♂: mesotibia with short, dense, wavy, anteroventral pile on middle ¾

## *Platycheirus podagratus*

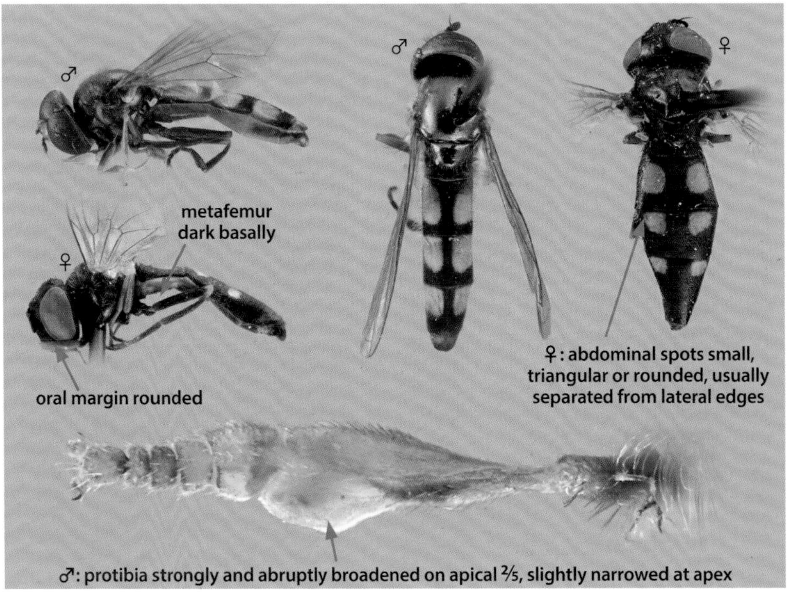

♂

♂

♀

metafemur dark basally

♀

oral margin rounded

♀: abdominal spots small, triangular or rounded, usually separated from lateral edges

♂: protibia strongly and abruptly broadened on apical ⅖, slightly narrowed at apex

## *Platycheirus normae* Paddlearm Sedgesitter

**SIZE RANGE:** 7.1–8.4 mm
**IDENTIFICATION:** Males of this species are the only *Platycheirus* with a strongly broadened, paddle-shaped protibia with a longitudinal keel, followed by a long, narrow probasitarsus. Similarly, females of this species are the only ones with a yellow protrochanter that has a blunt, ventral triangular process on it. The yellow spots on the tergites of both sexes are also large and usually medially confluent. **ABUNDANCE:** Very rare. **FLIGHT TIMES:** Early July to late August. **NOTES:** This species has been recorded nectaring on *Sagittaria*.

## *Platycheirus scambus* Blackspine Sedgesitter

**SIZE RANGE:** 6.5–9.6 mm
**IDENTIFICATION:** Males of this species have no subbasal tuft of white setae on the profemur, but have a posterior row of four to five long, black setae. They also have a protibia that is broadened from base to apex. The tarsomeres are widened, with the first almost as wide as the tibia. On the mesoleg, the mesofemur has a row of six to eight black setulae and the mesotibia has appressed, ventral black pile. Females of this species similarly have no subbasal white tuft on the profemur and can be identified by the yellow-pollinose face, with the area below the facial tubercle entirely pollinose and the yellow spots of the tergites with a sinuous posterior margin. **ABUNDANCE:** Common. **FLIGHT TIMES:** Early May to early September (from mid-April in the west). **NOTES:** This species can be found in dry forest, mixed woodlands, wetlands, marshes, water margins, and coastal salt-marshes. Often found in emergent vegetation at water margins, or hovering up to 3 m off the ground (males). It has been recorded feeding at *Carex*, *Ranunculus*, *Schoenoplectus*, *Scirpus*, *Spartina*, and *Urtica*.

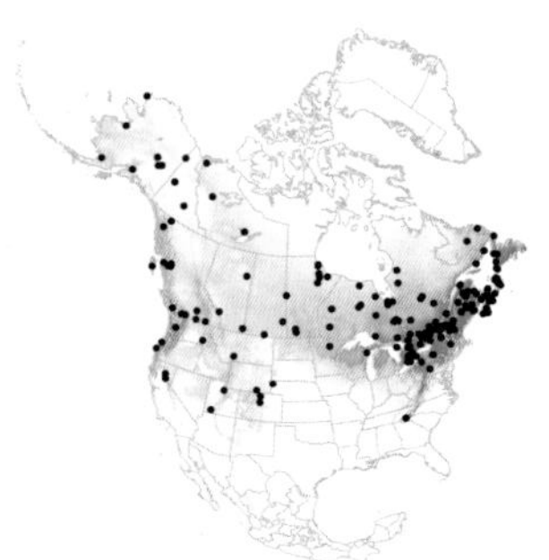

## Platycheirus normae

## Platycheirus scambus

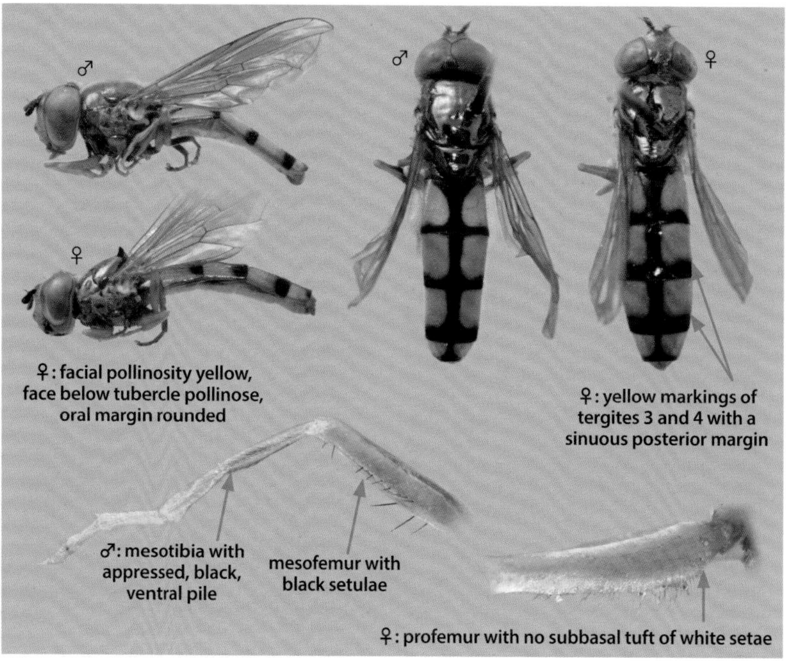

## *Platycheirus scamboides* Yellowspine Sedgesitter

**SIZE RANGE:** 8.2–9.1 mm
**IDENTIFICATION:** The male of this species is very similar to *P. scambus*, differing in that the appressed ventral pile on the mesotibia is yellow instead of black and the setulae on the mesofemur are yellow. The female of this species is unknown, but may be indistinguishable from *P. scambus*. **ABUNDANCE:** Rare. **FLIGHT TIMES:** Late May to early September. **NOTES:** This species has been collected in bogs and surrounding mixed woodlands.

## *Platycheirus modestus* Yellow Sedgesitter

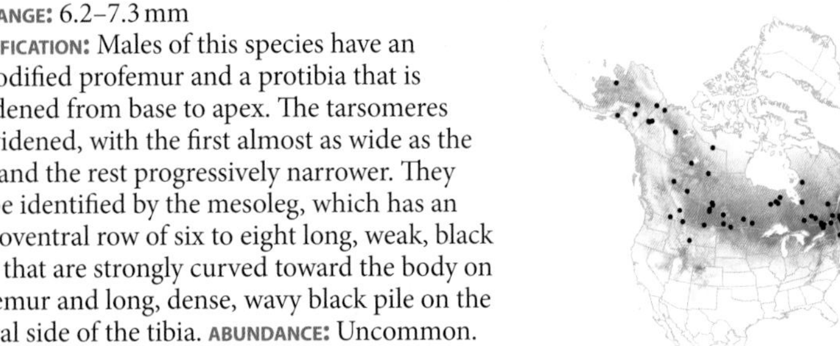

**SIZE RANGE:** 6.2–7.3 mm
**IDENTIFICATION:** Males of this species have an unmodified profemur and a protibia that is broadened from base to apex. The tarsomeres are widened, with the first almost as wide as the tibia and the rest progressively narrower. They can be identified by the mesoleg, which has an anteroventral row of six to eight long, weak, black setae that are strongly curved toward the body on the femur and long, dense, wavy black pile on the ventral side of the tibia. **ABUNDANCE:** Uncommon. **FLIGHT TIMES:** Mid-May to mid-August. **NOTES:** This species can be found in marshes, fens, wetlands, and tundra, usually in association with *Carex*.

## *Platycheirus scamboides*

♂
♂
♂: mesotibia with appressed, yellow, ventral pile
mesofemur with yellow setulae

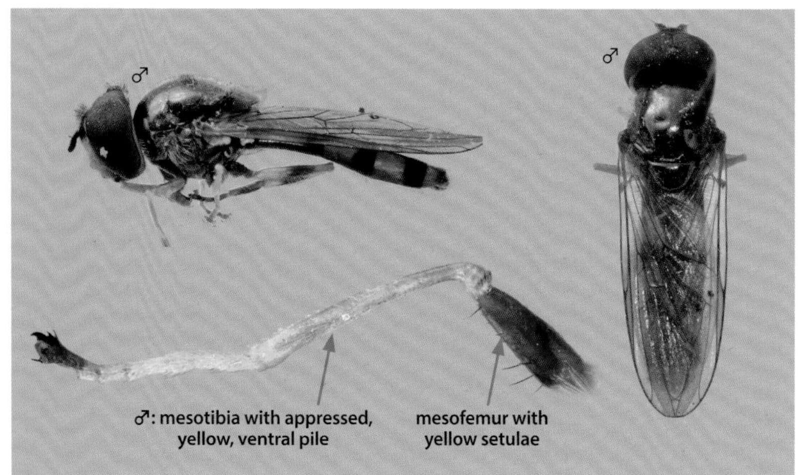

## *Platycheirus modestus*

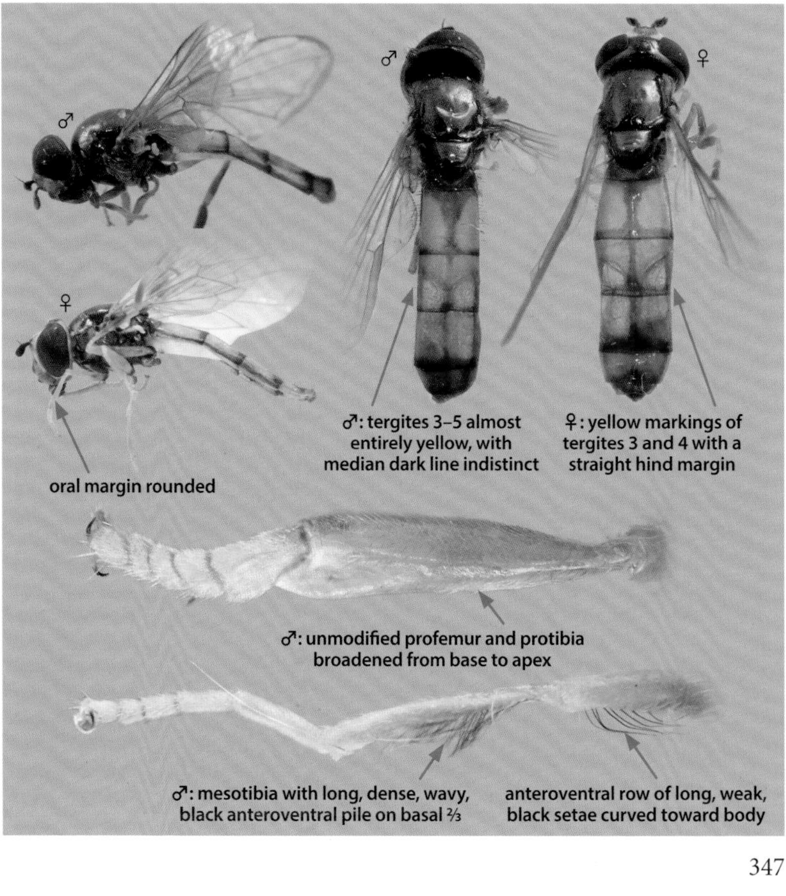

♂
♂
♀
♀
oral margin rounded
♂: tergites 3–5 almost entirely yellow, with median dark line indistinct
♀: yellow markings of tergites 3 and 4 with a straight hind margin
♂: unmodified profemur and protibia broadened from base to apex
♂: mesotibia with long, dense, wavy, black anteroventral pile on basal ⅔
anteroventral row of long, weak, black setae curved toward body

## *Platycheirus orarius* Salt Marsh Sedgesitter

**SIZE RANGE:** 7.9–9.6 mm

**IDENTIFICATION:** Males of this species have a proleg similar to *P. scambus*. They can be identified by the dense, wavy, erect, mostly pale pile on the ventral surface of the mesotibia. Females of this species can be identified by the combination of the rounded oral margin and the pollinosity of the frons being uniform, not forming two triangles. **ABUNDANCE:** Locally common. **FLIGHT TIMES:** Early June to mid-August. **NOTES:** This species is known from salt marshes and tidal flats and is often around small ponds in these habitats.

## *Platycheirus varipes* Silver Sedgesitter

**SIZE RANGE:** 6.2–9.6 mm

**IDENTIFICATION:** Males of this species have a proleg similar to *P. scambus*. They can be identified by the entirely black tergites with only silvery pollinose spots. Females have the oral margin produced forward into a point, the tergites entirely dark with only silvery pollinosity, and the profemur with no outstanding setae. **ABUNDANCE:** Rare. **FLIGHT TIMES:** Early June to early August. **NOTES:** This species is known from boreal and subarctic forest, tundra, and mires. It has been recorded visiting small, white Brassicaceae and *Ranunculus*.

## *Platycheirus orarius*

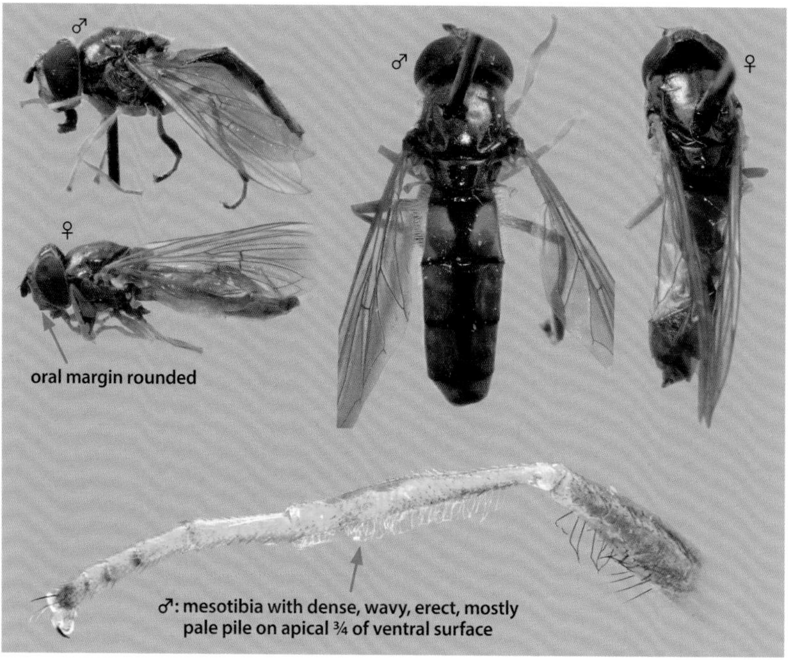

♂

♂

♀

♀

oral margin rounded

♂: mesotibia with dense, wavy, erect, mostly pale pile on apical ¾ of ventral surface

## *Platycheirus varipes*

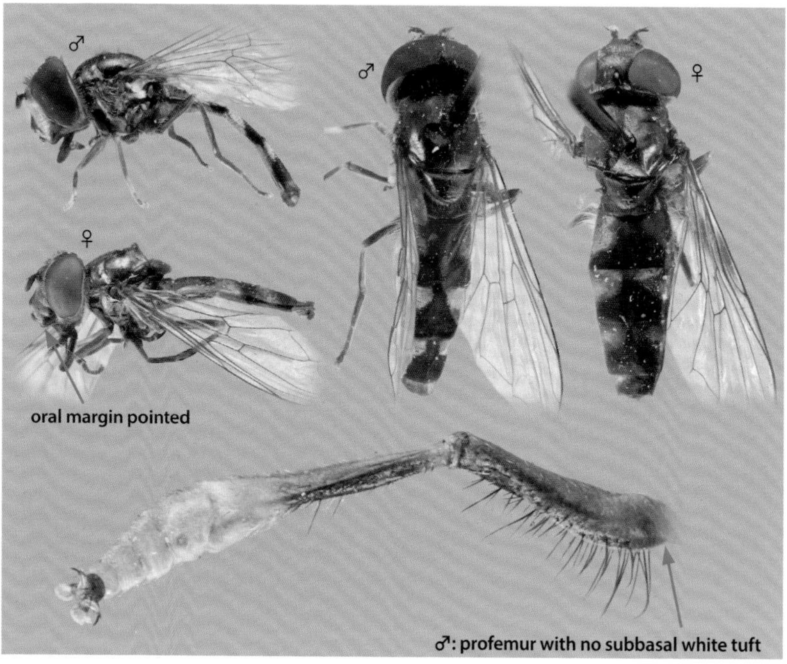

♂

♂

♀

♀

oral margin pointed

♂: profemur with no subbasal white tuft

## *Platycheirus nodosus* Twospear Sedgesitter

**SIZE RANGE:** 5.3–8.7 mm
**IDENTIFICATION:** Males of this species are instantly identifiable in the northeast by the two subbasal tufts of setae with broadened apices on the profemur. Similarly, the females are identifiable by the triangular spots on the tergites and two subbasal tufts of white setae on the profemur. **ABUNDANCE:** Uncommon. **FLIGHT TIMES:** Late May to late July (to mid-August in the west). **NOTES:** They can be found along water margins of rivers, streams, and lakes and have been collected from emergent vegetation in these habitats.

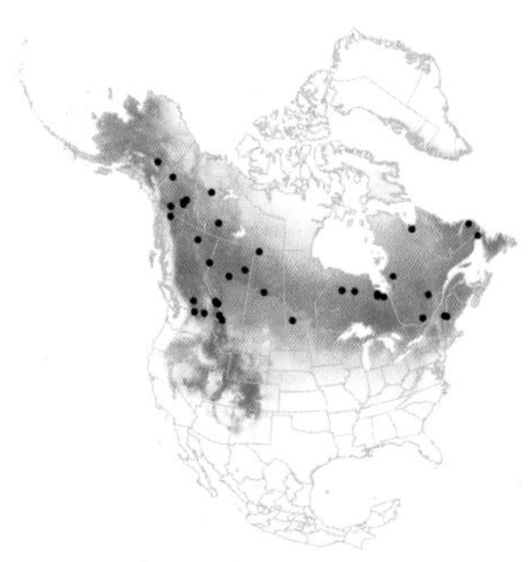

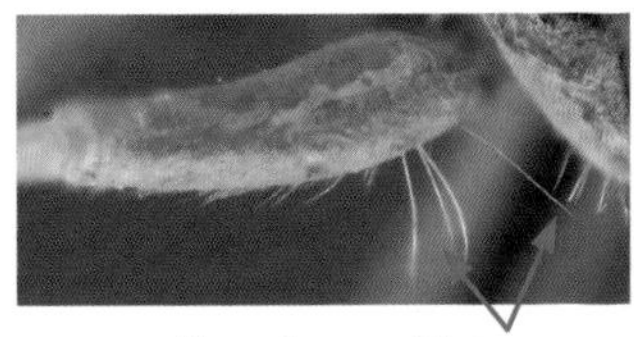

♀: profemur with two subbasal tufts of white setae

## *Platycheirus thompsoni* Thompson's Sedgesitter

**SIZE RANGE:** 7.8–9.0 mm
**IDENTIFICATION:** Males of this species are very similar to *P. nodosus*, but there is only one subbasal tuft of setae with broadened apices on the profemur. Females of this species have triangular markings on the tergites that are usually smaller than those of *P. nodosus* and a single subbasal tuft of setae on the profemur. **ABUNDANCE:** Uncommon. **FLIGHT TIMES:** Early May to late July. **NOTES:** This species has been collected in marshes, sphagnum bogs, and mixed woodlands, usually in association with *Carex* and/or *Salix*.

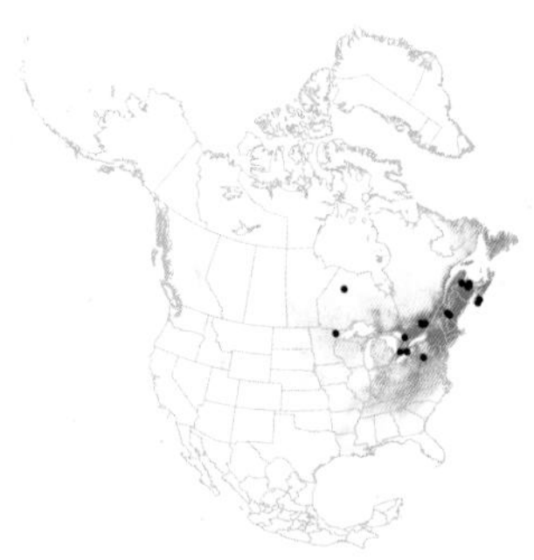

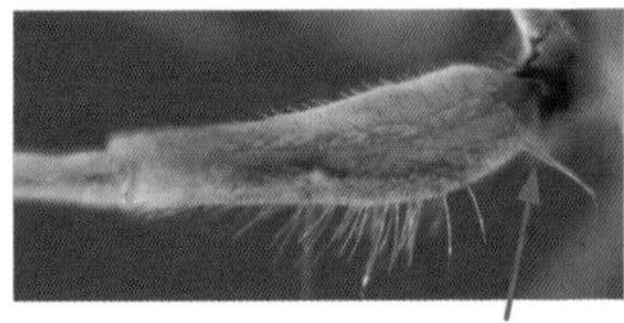

♀: profemur with subbasal tuft of white setae (as in *P. quadratus*)

## Platycheirus nodosus

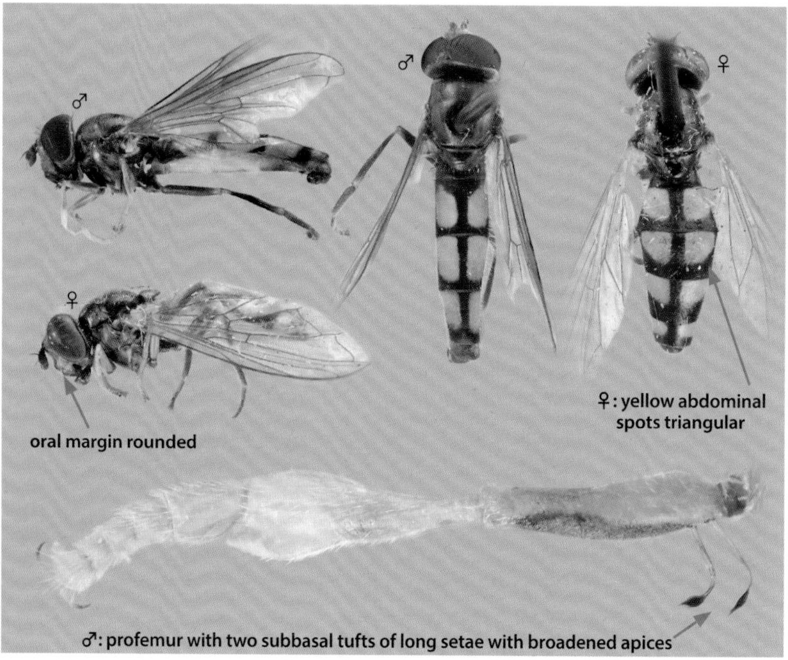

♂

♂

♀

♀

♀: yellow abdominal spots triangular

oral margin rounded

♂: profemur with two subbasal tufts of long setae with broadened apices

## Platycheirus thompsoni

♂

♂

♀

♀

♀: yellow abdominal spots triangular

oral margin rounded

♂: profemur with one subbasal tuft of long setae with broadened apices, sometimes preceded by a single long ,slender seta

## *Platycheirus scutatus* Many-tufted Sedgesitter

**SIZE RANGE:** 6.8–8.7 mm

**IDENTIFICATION:** Males of this species are the only *Platycheirus* in the northeast with a ventral spur on the mesocoxa. Additionally, they have a profemur with a subbasal tuft of white setae followed by two larger tufts of black setae. The protibia is smoothly broadened from base to apex, with the posteroapical angle broadly rounded. Females have orange spots with a silvery overlay that rarely meet the anterior margin of the tergites, a metabasitarsus that is swollen, approximately 50% thicker medially than at either end, and arista that are microtrichose, with individual microtrichia at least ½ the width of the arista on the basal ⅓. **ABUNDANCE:** Uncommon. **FLIGHT TIMES:** Late May to mid-September. **NOTES:** This species is found in most types of deciduous forest habitat. It has been recorded feeding at flowers of *Achillea*, *Berberis*, *Campanula*, *Euphorbia*, *Geranium*, *Leontodon*, *Ranunculus*, *Rosa*, *Salix*, *Silene*, *Stellaria*, *Symphyotrichum*, *Taraxacum*, and *Tripleurospermum*.

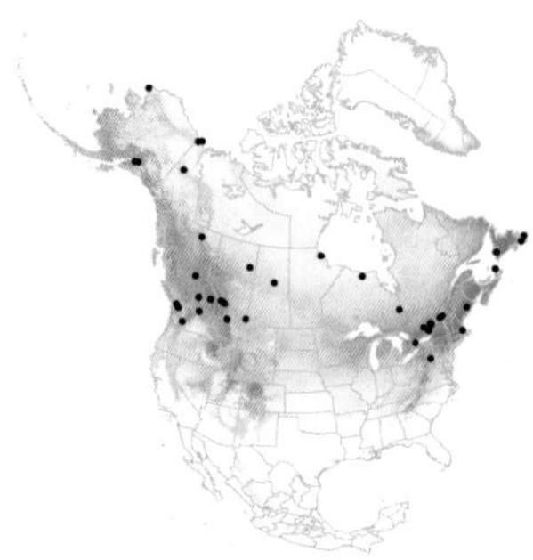

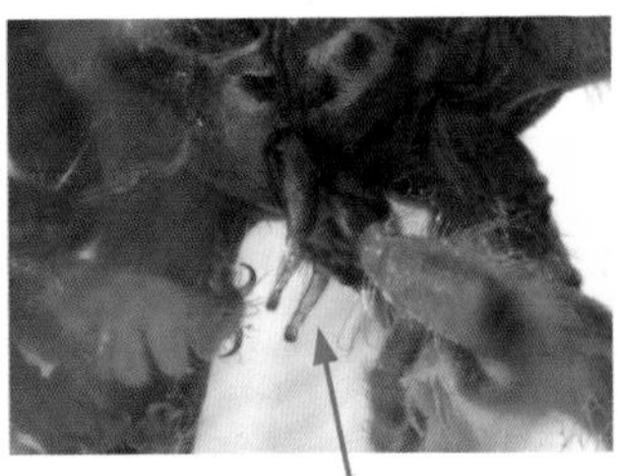

♂: mesocoxa with ventral spur

## *Platycheirus albimanus* Three-tufted Sedgesitter

**SIZE RANGE:** 6.2–9.6 mm

**IDENTIFICATION:** Males of this species have a profemur with a subbasal tuft of white setae followed by two larger tufts of black setae. The protibia is strongly broadened on the apical ⅓ and the posteroapical angle is acute. Females have the anterior oral margin produced forward into a point when viewed laterally, silvery pollinose spots on the abdomen, sometimes with dull orange markings underneath, a subbasal tuft of white setae on the profemur, and a metafemur and tibia that are mostly orange. **ABUNDANCE:** Uncommon. **FLIGHT TIMES:** Early June to mid-August (early January to late October in the west). **NOTES:** This species is found in deciduous forest and tundra. It may also be found in farmlands, urban parks, and gardens.

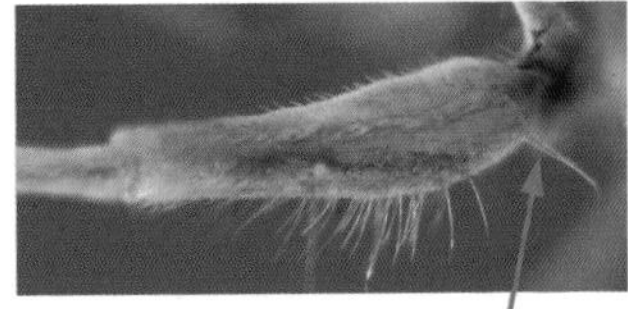

♀: profemur with subbasal tuft of white setae (as in *P. quadratus*)

## *Platycheirus scutatus*

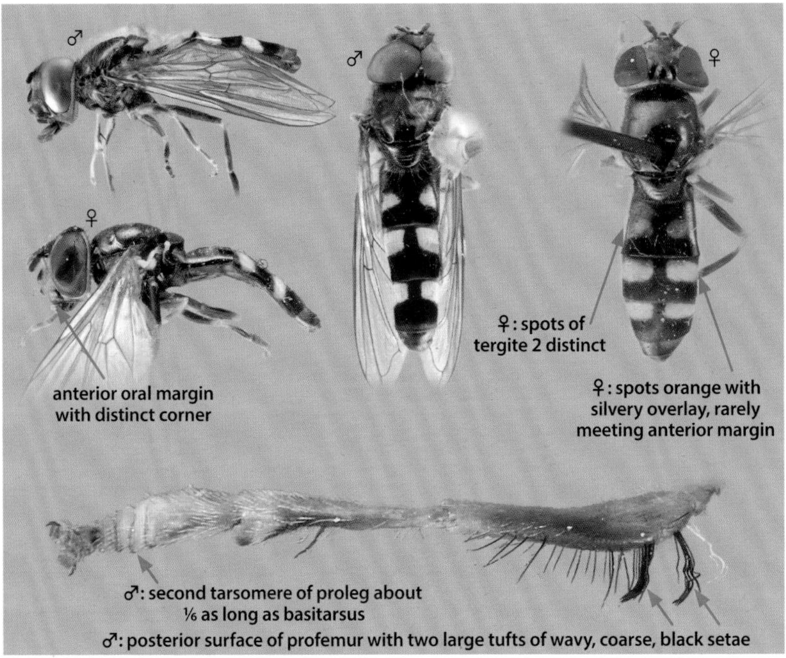

♂

♂

♀

♀

♀: spots of tergite 2 distinct

anterior oral margin with distinct corner

♀: spots orange with silvery overlay, rarely meeting anterior margin

♂: second tarsomere of proleg about ⅙ as long as basitarsus

♂: posterior surface of profemur with two large tufts of wavy, coarse, black setae

## *Platycheirus albimanus*

♂

♂

♀

♀

♀: metafemur and tibia mostly orange, with posterior brown stripes

♀: spots either orange with silvery overlay or entirely silvery, rarely meeting anterior margin

♂: protibia strongly broadened on apical ⅓ and with posteroapical angle acute

♂: posterior surface of profemur with two large tufts of wavy, coarse, black setae

## *Platycheirus nigrofemoratus*

### Black-legged Sedgesitter

**SIZE RANGE:** 6.2–7.3 mm
**IDENTIFICATION:** Males of this species have a profemur with a subbasal tuft of white setae followed by two larger tufts of black setae, similar to *P. albimanus* and *P. scutatus*. However, the protibia is uniformly broadened from base to tip and the posteroapical angle is slightly rounded. Females of this species have the anterior oral margin produced forward into a point when viewed laterally, only silvery pollinose spots on the abdomen, a subbasal tuft of white setae on the profemur, and a dark brown metaleg.
**ABUNDANCE:** Rare. **FLIGHT TIMES:** Late June to late July (to early August in the west). **NOTES:** *Platycheirus nigrofemoratus* has been collected on open ground and near riverbanks in tundra habitats.

## *ambiguus* group

## *Platycheirus coerulescens* Hooked Sedgesitter

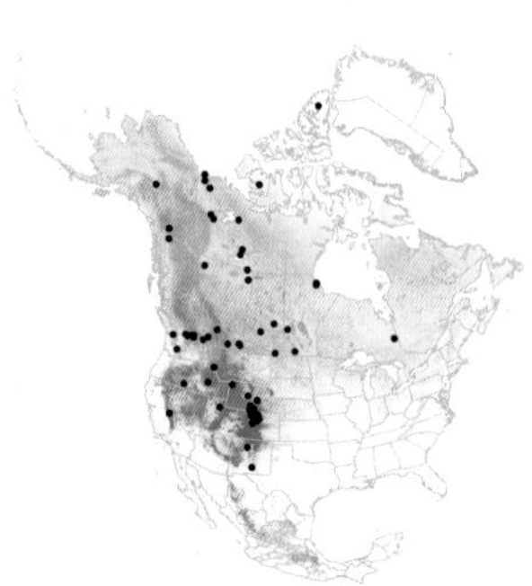

**SIZE RANGE:** 5.7–9.1 mm
**IDENTIFICATION:** Males of this species have a profemur that is almost entirely orange, with a posterior row of long, strong setae ending in a longer seta with a curled apex. Females have the anterior oral margin projected forward into a point, rhomboid orange spots on the tergites that do not reach the anterior margins of the tergite and are overlaid with strong, silvery pollinosity, and a regular row of weak, white setae on the posterior surface of the profemur.
**ABUNDANCE:** Uncommon. **FLIGHT TIMES:** Late April to early September. **NOTES:** This species is known from boreal, montane, prairie, and tundra habitats.

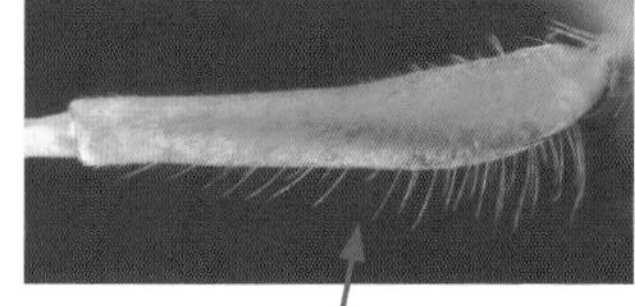

♀: profemur with an evenly spaced row of very weak, white setae posteriorly

# *Platycheirus nigrofemoratus*

♂ ♂ ♀

♀

anterior oral margin with distinct corner

♀: legs brown

♀: abdomen with only silvery pollinose spots

♂: protibia uniformly broadened from base to tip, posteroapical angle slightly rounded

♂: posterior surface of profemur with two large tufts of wavy, coarse, black setae

# *Platycheirus coerulescens*

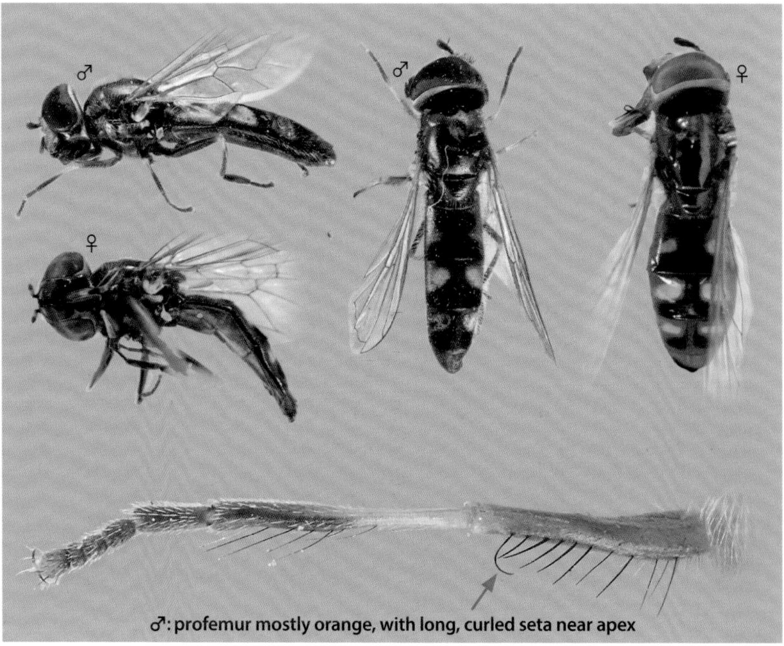

♂ ♂ ♀

♀

♂: profemur mostly orange, with long, curled seta near apex

## *Platycheirus lundbecki* Lundbeck's Sedgesitter

**SIZE RANGE:** 5.3–6.5 mm
**IDENTIFICATION:** Males of this species are very similar to *P. coerulescens*, differing in that the profemur is mostly brown and the tergites are often somewhat duller. Females are very similar to those of *P. coerulescens* as well, differing in that the row of weak setae on the profemur is black instead of white. **ABUNDANCE:** Very rare. **FLIGHT TIMES:** Late June to early August. **NOTES:** This species can be found in boreal forest, open ground, and tundra/taiga habitats, often near water.

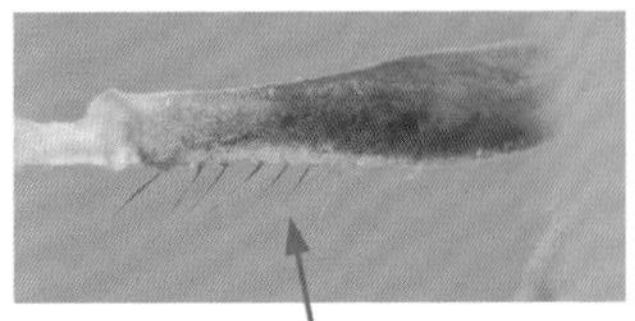

♀: profemur with an evenly spaced row of long, dark setae posteriorly

## *Platycheirus kelloggi* Broad-bodied Sedgesitter

**SIZE RANGE:** 7.7–10.1 mm
**IDENTIFICATION:** Males of this species have a profemur with two to three long, thin, black setae with curly apices; these setae contrast strongly with a preceding row of short, dense, wavy, pale pile. Tergite 2 has a pair of orange spots and tergites 3 and 4 have orange spots widely separated from the lateral margins. Females have an extremely broad abdomen, with large orange spots that have no trace of silvery pollinosity. The spots of tergite 2 are triangular, not reaching the anterior or lateral margins of the tergite. The spots of tergites 3 and 4 are rectangular, meeting the anterior edge but separated laterally. The pro- and mesofemora are entirely orange and the pro- and mesotibiae are almost entirely orange with only the tips obscurely brown. **ABUNDANCE:** Rare. **FLIGHT TIMES:** Mid- to late July (late June to early August in the west). **NOTES:** This species has been collected in marshy meadows and tundra.

# Platycheirus lundbecki

♂ ♂ ♀

♀

anterior oral margin with distinct corner

♂: profemur mostly brown, with long, curled seta near apex

# Platycheirus kelloggi

♂ ♂ ♀

♀

♀: with large orange spots on tergites

♂: profemur with long, curled seta near apex and short, dense, wavy pale pile closer to body

## *manicatus* group

### *Platycheirus flabella* Smallspot Sedgesitter

**SIZE RANGE:** 6.8–9.1 mm
**IDENTIFICATION:** Males of this species have a probasitarsus that is flattened, slightly expanded apically, and approximately twice as long as broad. The mesotibia has a posteroventral row of weak setae but no anteroventral long, wavy pile. Females have a face that is produced forward on its lower ½, pro- and mesolegs that are mostly dark, and small subquadrate pale spots on the tergites that are well separated from the anterior margin. **ABUNDANCE:** Rare. **FLIGHT TIMES:** Early May to late July (to mid-August in the west). **NOTES:** This species has been collected in boreal, montane, and tundra habitats.

### *Platycheirus thylax* Yellow-legged Sedgesitter

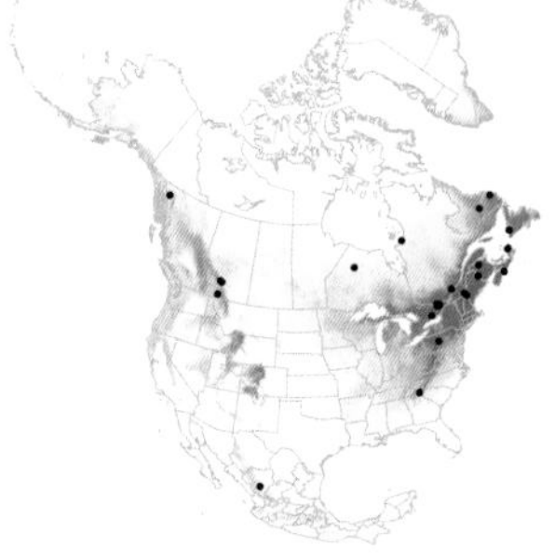

**SIZE RANGE:** 5.6–6.8 mm
**IDENTIFICATION:** Males of this species are very similar to *P. flabella*, except that the probasitarsus is flattened, expanded apically to approximately twice the width of the protibia, and as long as broad. Like *P. flabella*, the mesotibia has no anteroventral patch of long, wavy pile. The protrochanter also has many black, ventral setae. Females are likewise similar to female *P. flabella*, except that the pro- and mesolegs are mostly pale. **ABUNDANCE:** Uncommon. **FLIGHT TIMES:** Late April to late June (to mid-July in the south and west). **NOTES:** This species has been collected in open sphagnum fens, in mixed upland woods, and in rich deciduous floodplain forest. It has been recorded feeding at *Salix* flowers.

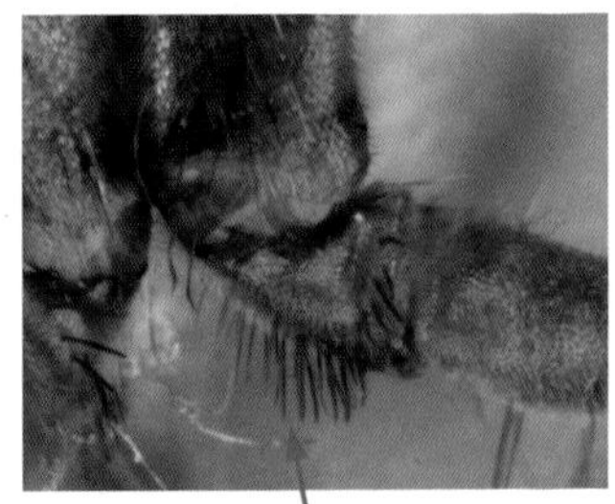

♂: protrochanter with black ventral setae

## *Platycheirus flabella*

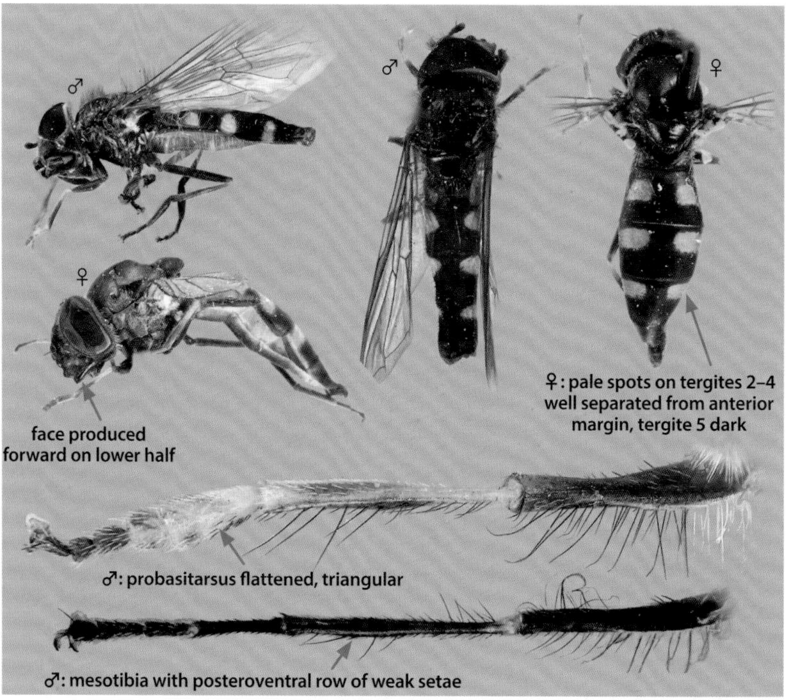

♂

♂

♀

♀

♀: pale spots on tergites 2–4 well separated from anterior margin, tergite 5 dark

face produced forward on lower half

♂: probasitarsus flattened, triangular

♂: mesotibia with posteroventral row of weak setae

## *Platycheirus thylax*

♀: pollinosity of frons uniform (arrow above)

♂

♂

♀

♀

anterior oral margin produced into a point

♀: pro- and mesolegs mostly pale

♂: probasitarsus flattened, triangular

## *Platycheirus discimanus* Yellowfoot Sedgesitter

**SIZE RANGE:** 6.2–6.8 mm
**IDENTIFICATION:** Males of this species are instantly identifiable by the bright yellow, laterally compressed first two mesotarsomeres. Females have a dark abdomen with faint pollinose spots on the tergites, the anterior oral margin produced into a point, and a bare frons, free of pollinosity. **ABUNDANCE:** Very rare. **FLIGHT TIMES:** Late April to early June. **NOTES:** This species has been collected in cool, humid woodlands and upland spruce forests. It has been recorded nectaring at *Salix*.

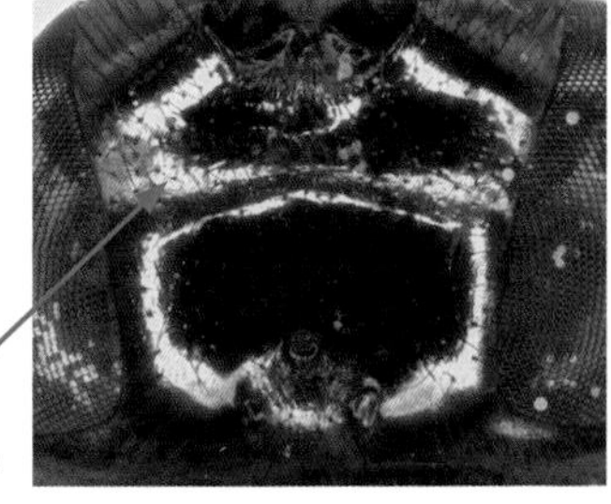

♀: frons shiny, not pollinose

## *Platycheirus groenlandicus* Arctic Sedgesitter

**SIZE RANGE:** 5.7–8.2 mm
**IDENTIFICATION:** Males of this species are very similar to *P. flabella*, except that the probasitarsus is flattened and expanded apically to approximately twice the width of the protibia and is as long as broad. Additionally, the mesotibia has a tuft of long, wavy pile on the basal ⅓ of the anteroventral surface. Females are very similar in appearance to *P. discimanus*, except that the frons is thinly and uniformly pollinose. Unlike females of *P. thylax*, the pro-and mesolegs are dark. **ABUNDANCE:** Uncommon. **FLIGHT TIMES:** Mid-June to late July (to early August in the west). **NOTES:** This species has been recorded from subalpine forest and tundra in association with *Betula* and *Salix*.

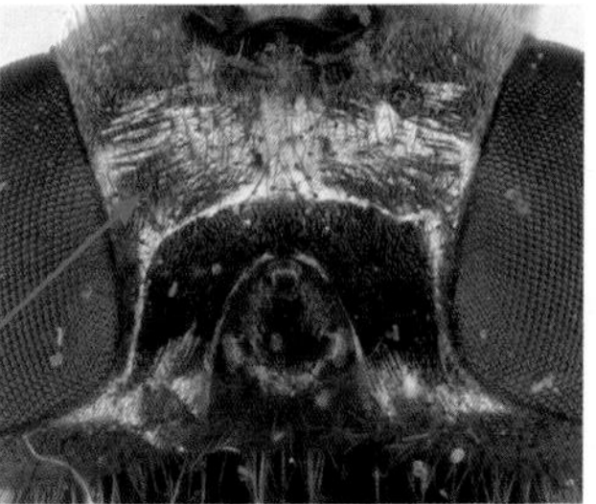

♀: frons thinly and uniformly pollinose

## Platycheirus discimanus

♂

♂

♀

anterior oral margin with distinct corner

♀

♂: first two mesotarsomeres laterally compressed, bright yellow

## Platycheirus groenlandicus

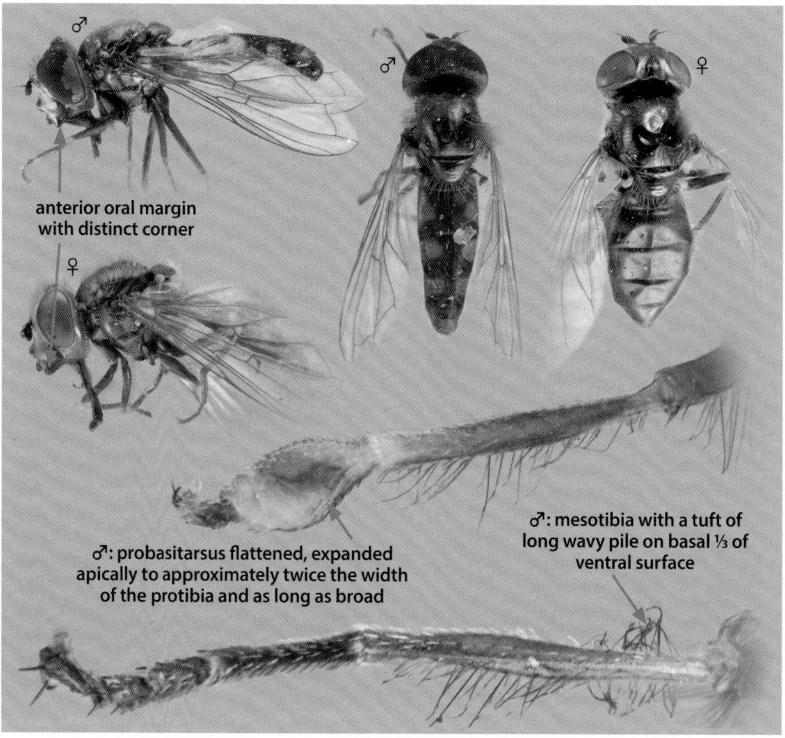

♂

♂

♀

anterior oral margin with distinct corner

♀

♂: mesotibia with a tuft of long wavy pile on basal ⅓ of ventral surface

♂: probasitarsus flattened, expanded apically to approximately twice the width of the protibia and as long as broad

## *chilosia* group

### *Platycheirus chilosia* Bristlehand Sedgesitter

**SIZE RANGE:** 4.8–7.6 mm
**IDENTIFICATION:** Males of this species are the only *Platycheirus* in the northeast with long, weak, posterior setae with curled apices on the probasitarsus. Females have a dark abdomen, sometimes with faint pollinose spots on the tergites, the anterior oral margin produced into a point, uniform pollinosity on the frons, and an irregular, posterior row of thin setae on the protibia. **ABUNDANCE:** Uncommon. **FLIGHT TIMES:** Late June to late July. **NOTES:** This species has been collected in high boreal and Arctic habitats as well as from barren, rocky slopes in montane regions. Southern records are from relatively high altitudes (1600 m). *Platycheirus carinata* is a commonly used synonym.

## *pictipes* group

### *Platycheirus luteipennis*
Coppery Sedgesitter

**SIZE RANGE:** 8.7–10.1 mm
**IDENTIFICATION:** Males of this species have a sparsely pollinose face with a median keel between the antennal bases, unmodified legs, and overall coppery color. Females have extremely short pile on the thorax, no longer than the length of the scape, and share the male's coppery color. **ABUNDANCE:** Rare. **FLIGHT TIMES:** Late June to early August. **NOTES:** Little is known about the habitat of this rare species.

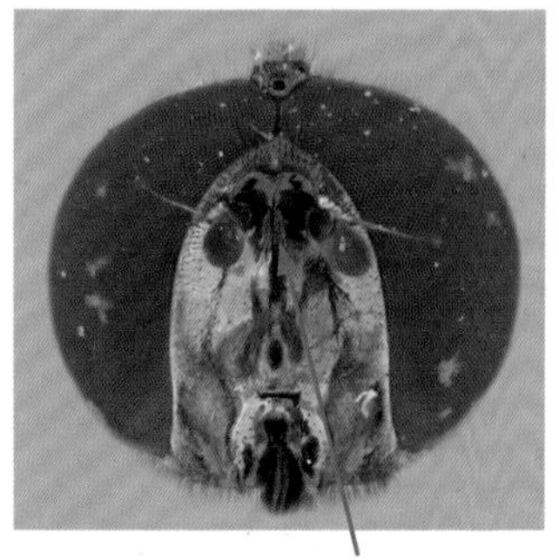

♂: face sparsely pollinose, upper part of face with median keel

## Platycheirus chilosia

♂

♀: pollinosity of frons uniform

♂

♀

anterior oral margin produced into a point

♀

♀: protibia with an irregular row of slightly thickened setae

♂: probasitarsus with five to seven long, weak, posterior setae, curled at apices

## Platycheirus luteipennis

♂

♀

♂

♀

♂: legs unmodified, body coppery

♀: pile of scutum, scutellum, and pleura short, no longer than length of scape; body coppery

## *Platycheirus striatus* Mundane Sedgesitter

**SIZE RANGE:** 7.3–11.0 mm
**IDENTIFICATION:** Males of this species have a densely pollinose face with a median keel or grooves between the antennal bases, unmodified legs, and an overall gray color. The female is unknown. **ABUNDANCE:** Rare. **FLIGHT TIMES:** Early June to late July (from early May in the west). **NOTES:** Little is known about the habitat of this rare species.

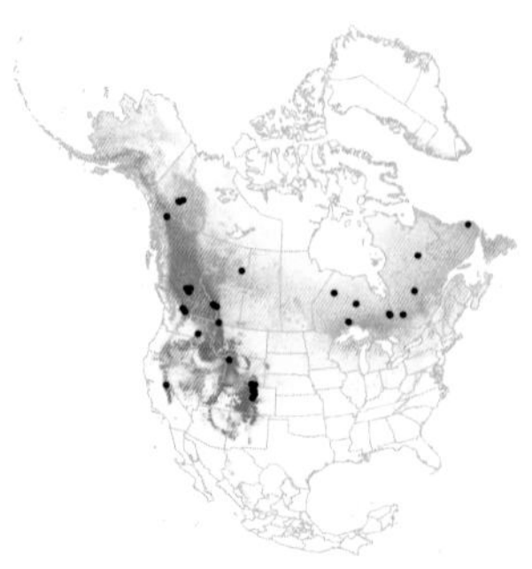

**♂: face densely pollinose, upper part of face either with keel or shallow grooves**

## *Platycheirus pictipes* Cobalt Sedgesitter

**SIZE RANGE:** 7.7–10.5 mm
**IDENTIFICATION:** Males of this species are identifiable by their unmodified legs and metallic blue color. Females are identifiable by their short thoracic pile (no longer than the length of the scape) and their metallic blue color. **ABUNDANCE:** Common. **FLIGHT TIMES:** Late May to mid-August (to late September in the west). **NOTES:** This species is found in mixed and coniferous forests.

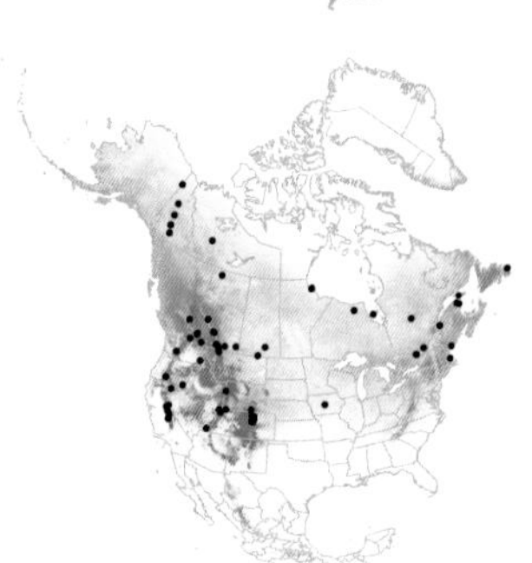

## *Platycheirus striatus*

♂
♂
♂: legs unmodified, body gray

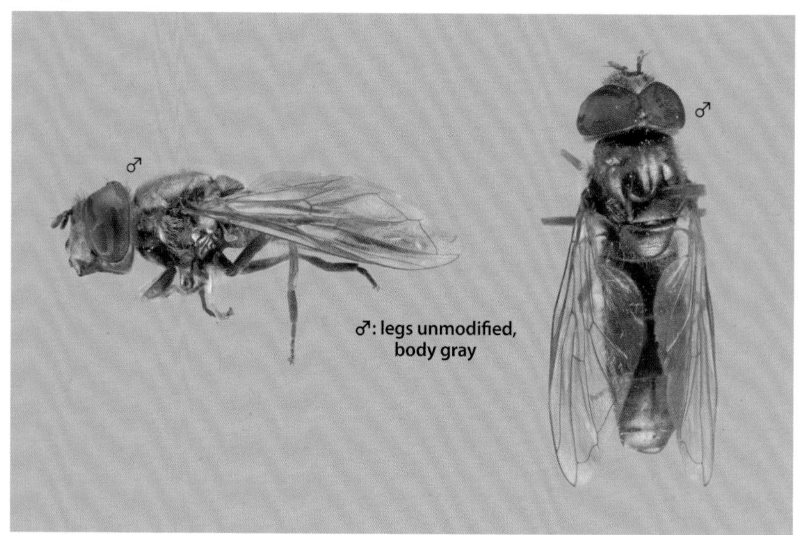

## *Platycheirus pictipes*

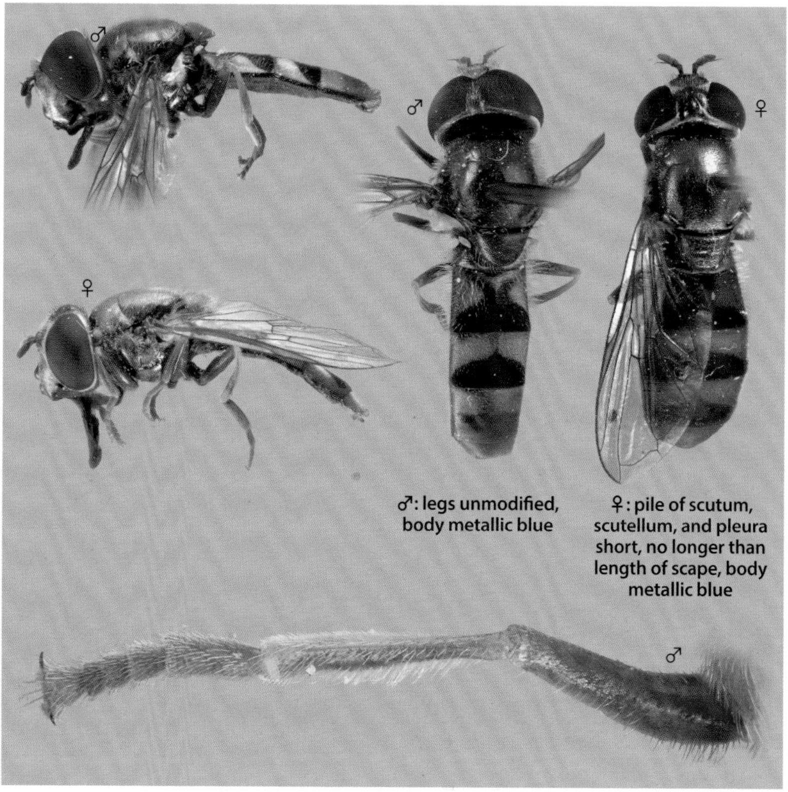
♂
♂
♀
♀
♂: legs unmodified, body metallic blue
♀: pile of scutum, scutellum, and pleura short, no longer than length of scape, body metallic blue
♂

## *peltatus* group

### *Platycheirus parmatus*

Shieldhand Sedgesitter

**SIZE RANGE:** 9.1–10.1 mm
**IDENTIFICATION:** Males of *P. parmatus* and *P. jaerensis* are the only species in the *peltatus* group with many thin, unmodified posterior setae on the profemur. Males also have protibia with many posterior setae longer than the width of the protibia, a protibia expanded only apically, and a probasitarsus that is flattened with an anterior projection. Females have a face that is produced forward, pale, rectangular spots on the abdominal tergites that are separated from the anterior and lateral edges of the tergites, and tergite 5 with distinct spots. **ABUNDANCE:** Rare. **FLIGHT TIMES:** Mid-May to mid-June (to early August in the west). **NOTES:** This species is found in mixed and coniferous forests and males may be found hilltopping near their preferred habitat. It has been recorded feeding at *Alliaria*, *Allium*, *Anemone*, *Ranunculus*, *Salix*, *Stellaria*, and *Vaccinium*.

### *Platycheirus jaerensis* Palehorn Sedgesitter

**SIZE RANGE:** 9.2–10.1 mm
**IDENTIFICATION:** Males of this species have many long, thin, unmodified posterior setae on the profemur (also in *P. parmatus*) but only short posterior setae on the protibia. The protibia is expanded only apically and the probasitarsus is flattened with no anterior projection. Females have a face that is distinctly produced and sparsely pollinose with the shiny black ground color clearly visible through the pollinosity. The female flagellum is bright orange and the pale spots on the tergites are large. **ABUNDANCE:** Very rare. **FLIGHT TIMES:** Late May to late June. **NOTES:** This species is found in boreal pine forests, mixed spruce-pine-birch forests, and raised bogs. It has been recorded nectaring at *Geranium*, *Ranunculus*, *Taraxacum*, and *Vaccinium*.

## *Platycheirus parmatus*

♂

♂

♀

face distinctly produced below

♀

♂: yellow spots of tergites 3 and 4 slightly wider than long

♀: tergites with spots narrowly separated from anterior and lateral edges, tergite 5 with distinct spots

♂: probasitarsus flattened with an anterior projection

♂: protibia with many posterior setae longer than tibial width

## *Platycheirus jaerensis*

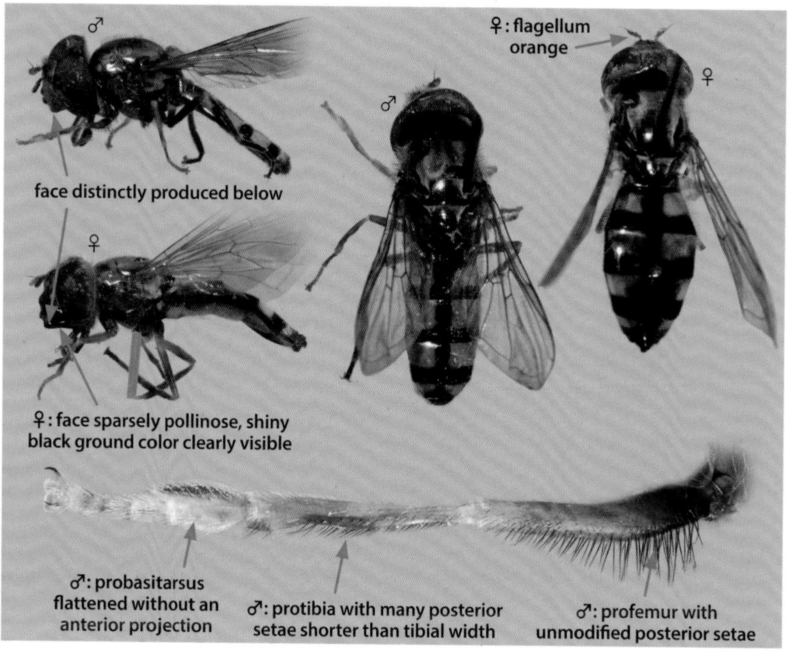

♂

♀: flagellum orange

♀

♂

face distinctly produced below

♀

♀: face sparsely pollinose, shiny black ground color clearly visible

♂: probasitarsus flattened without an anterior projection

♂: protibia with many posterior setae shorter than tibial width

♂: profemur with unmodified posterior setae

## *Platycheirus latitarsis* Flathand Sedgesitter

**SIZE RANGE:** 7.1–10.1 mm
**IDENTIFICATION:** Males of this species have many long, flattened, posterior setae on the profemur. The protibia is expanded only apically and the probasitarsus is flattened, with no distinct keels or projections. The female is unknown.
**ABUNDANCE:** Rare. **FLIGHT TIMES:** Late July (from early June in the west). **NOTES:** Nothing is known about the natural history of this rare fly.

## *Platycheirus naso* Tufted Sedgesitter

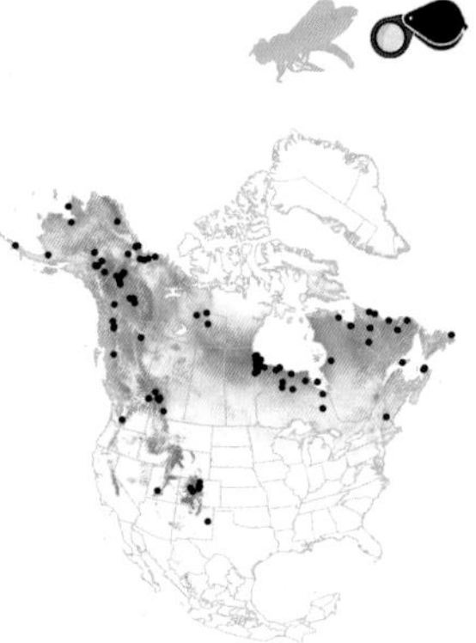

**SIZE RANGE:** 7.1–10.1 mm
**IDENTIFICATION:** Males of this species have many long, flattened, posterior setae on the profemur. The protibia is expanded only apically and the probasitarsus has a distinct longitudinal keel running its entire length. The mesotibia has long, wavy, ventral, black or yellow pile on the basal ⅓, up to three times the tibial width. Females have a face that is strongly produced forward ventrally, with the anterior oral margin usually projected as far forward as the tubercle. Pollinosity of the frons is in the shape of two distinct lateral triangles above the antennal bases. Pale spots on tergite 2 are lunulate and do not reach the anterior or lateral margins of the tergite. The upper ½ of the anepimeron has a tuft of dense, pale pile in both sexes; this tuft is denser in males, with insertions of individual pili impossible to discern.
**ABUNDANCE:** Common. **FLIGHT TIMES:** Early June to early August (from mid-May in the west).
**NOTES:** This species is found in taiga, subalpine, and boreal areas and has been collected at flowers of *Epilobium*, *Geranium*, *Ranunculus*, *Rubus*, *Tanacetum*, and *Taraxacum*.

♂: another view to show probasitarsus with distinct keel over entire length

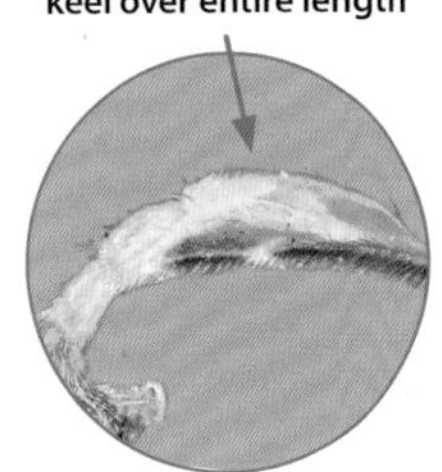

## *Platycheirus latitarsis*

## *Platycheirus naso*

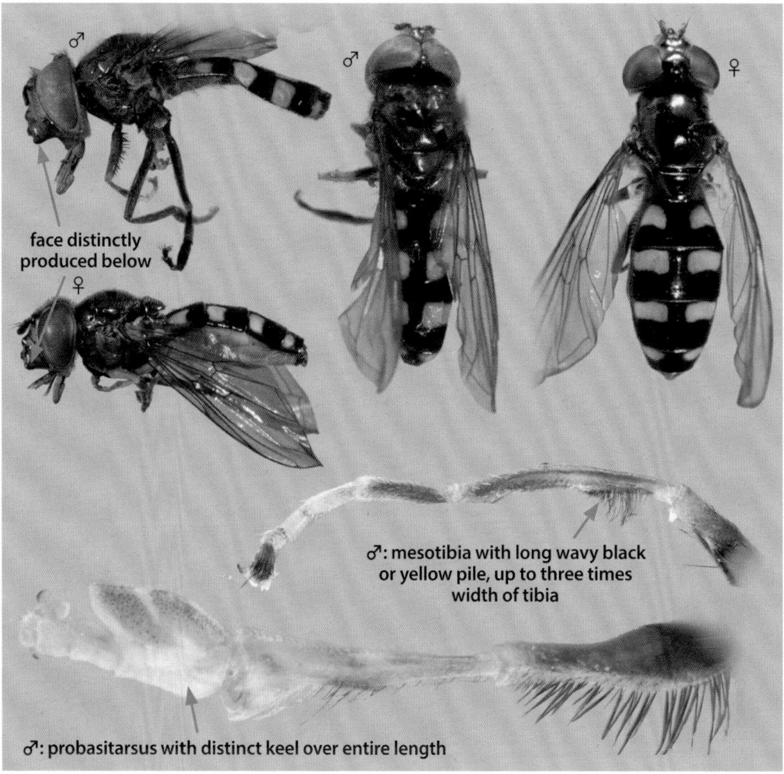

♂
♂
♀
face distinctly produced below
♀
♂: mesotibia with long wavy black or yellow pile, up to three times width of tibia
♂: probasitarsus with distinct keel over entire length

# *Platycheirus nearcticus*

## Nearctic Broadhand Sedgesitter

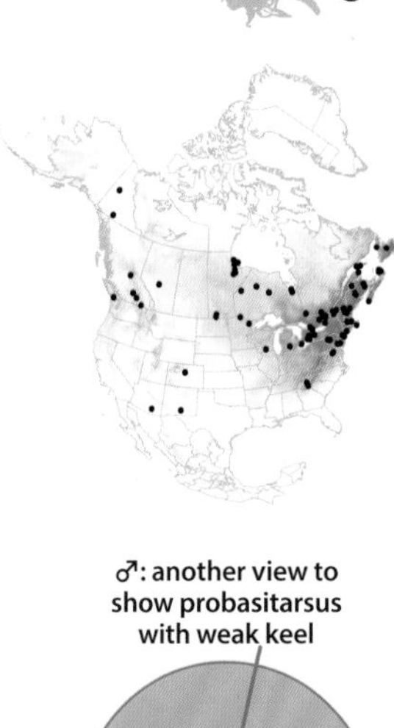

**SIZE RANGE:** 8.4–10.5 mm
**IDENTIFICATION:** Males of this species have many long, flattened, posterior setae on the profemur. The protibia is expanded only apically and the probasitarsus has a weak, dorsal keel on the apical ½ only. The mesotibia has short, straight, ventral, black or yellow pile on the basal ⅓, approximately 1.5 times the tibial width. Females have a face that is distinctly produced on the ventral ½ and the pale spots on tergite 2 are large, subquadrate, and reach the anterior and/or lateral margins of the tergite. Pollinosity of the frons is in the shape of two distinct lateral triangles above the antennal bases. Abdominal spots never have a pollinose overlay. **ABUNDANCE:** Common. **FLIGHT TIMES:** Early April to late September. **NOTES:** This species has been collected in various forest types, including spruce-aspen-birch and maple-elm forests. It can also be found in open, humid areas near forest, in sphagnum fens, sedge-dominated bogs, floodplains, on roadside flowers near ditches, and in other humid, disturbed areas. It has been recorded nectaring at *Ranunculus*.

♂: another view to show probasitarsus with weak keel

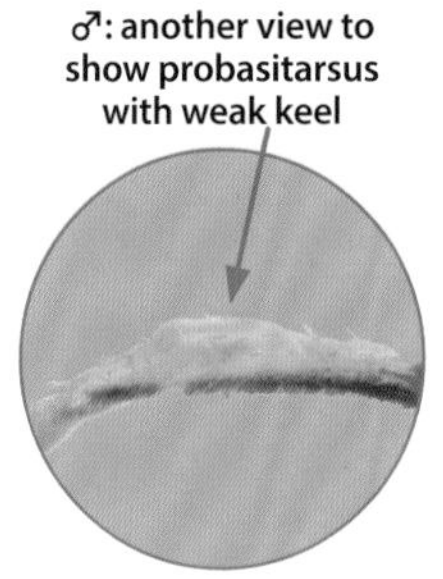

# *Platycheirus nielseni* Nielsen's Sedgesitter

**SIZE RANGE:** 8.4–9.6 mm
**IDENTIFICATION:** Males of this species are very similar to *P. nearcticus*, differing in that the mesotibia has the apex strongly swollen and curved downward slightly. Females have a face that is distinctly produced ventrally, the apex of the mesotibia is slightly swollen, and the pale spots of the tergites are overlaid with strong, silvery pollinosity. Pollinosity of the frons is in the shape of two distinct lateral triangles above the antennal bases. **ABUNDANCE:** Rare. **FLIGHT TIMES:** Late June to late July (to late August in the west). **NOTES:** Usually found along brooks in open conifer forest and birch woodlands.

# Platycheirus nearcticus

♂

♂

♀

face distinctly produced below

♀

♂: mesotibia with short, straight pile approximately 1.5 times the width of the tibia

♂: probasitarsus with weak dorsal keel on apical ½ only

# Platycheirus nielseni

♂

♂

♀

face distinctly produced below

♀

♂: mesotibia with apex strongly swollen

# *Platycheirus amplus*

## Northern Broadhand Sedgesitter

**SIZE RANGE:** 7.2–8.7 mm
**IDENTIFICATION:** Males of this species are very similar to *P. nearcticus*, except that the mesotibia has long, wavy, ventral, black or yellow pile on the basal ⅓, with these pili approximately three times the tibial width. Likewise, females are very similar to *P. nearcticus*, except that the pollinosity of the frons is more diffuse, with two lateral triangles visible above antennal insertions, but also with scattered pollinosity covering the rest of the frons more thinly. **ABUNDANCE:** Rare. **FLIGHT TIMES:** Early June to late July (to late August in the west). **NOTES:** This species is found in spruce forest, fens, bogs, and other areas with slow-moving water. Usually collected flying among sedges and similar vegetation. Males hover 1 m above the ground. This species has been collected on Apiaceae, *Polygonum*, *Ranunculus*, *Salix*, and *Taraxacum*.

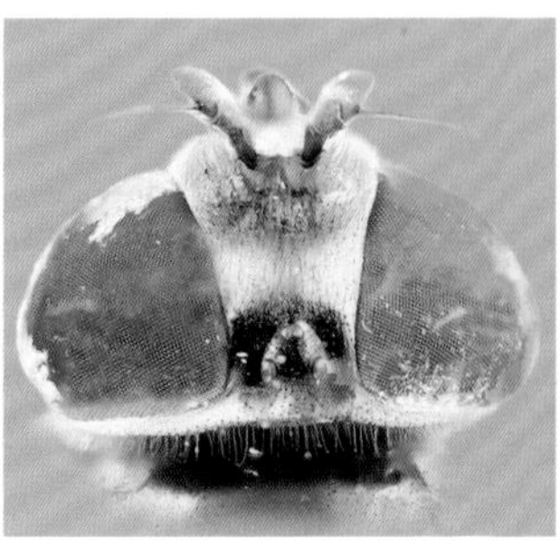

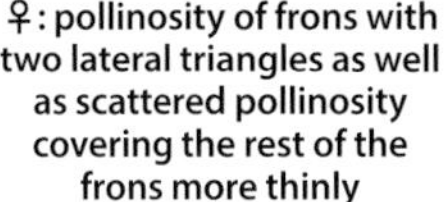

**♀: pollinosity of frons with two lateral triangles as well as scattered pollinosity covering the rest of the frons more thinly**

# *Platycheirus inversus* Knobfoot Sedgesitter

**SIZE RANGE:** 8.7–9.6 mm
**IDENTIFICATION:** Males of this species are very similar to *P. nearcticus*, except that the mesotibia has long, wavy, ventral, black or yellow pile on the basal ⅓ that is approximately three times the tibial width (as in *P. amplus*) and the basitarsus of the metaleg is strongly swollen basally before being abruptly constricted at midlength. Females are similar to *P. nearcticus*, except that pale spots of the abdominal tergites are overlaid with weak, silvery pollinosity. **ABUNDANCE:** Rare. **FLIGHT TIMES:** Late May to late July. **NOTES:** This species has been collected in moist, deciduous woods, mixed forests, and in bogs. The single floral record is from *Cornus canadensis*.

**♀: pollinosity of frons concentrated into two lateral triangles**

# Platycheirus amplus

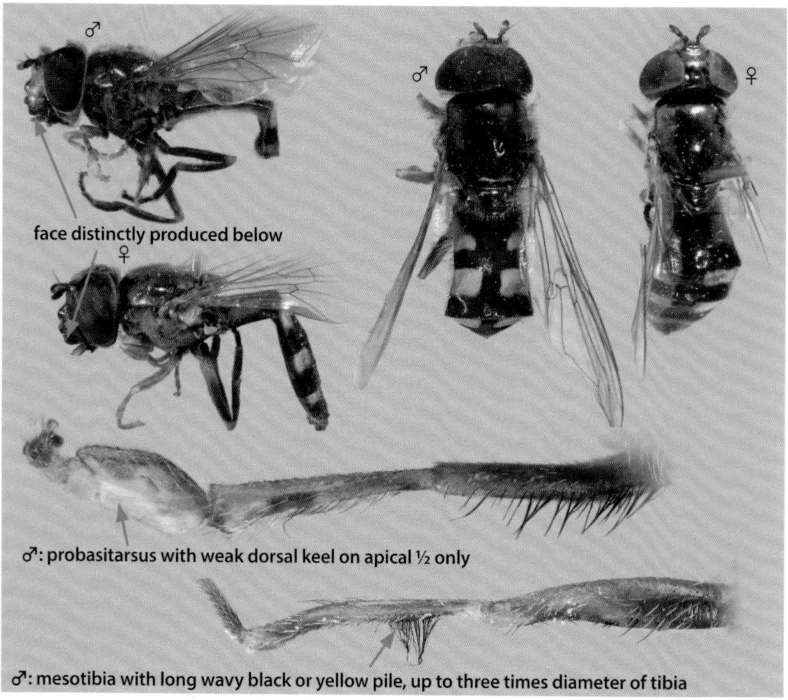

♂

♂ ♀

face distinctly produced below

♀

♂: probasitarsus with weak dorsal keel on apical ½ only

♂: mesotibia with long wavy black or yellow pile, up to three times diameter of tibia

# Platycheirus inversus

♂

♂ ♀

face distinctly produced below

♀

♂: metabasitarsus constricted at midlength

# Baccha

Two species of *Baccha* occur in North America and there are 16 species worldwide, with most of the diversity in the Palearctic and Indomalayan regions. These flies are small and slender, with elongate, petiolate abdomens. They are most similar to species of *Ocyptamus*, *Pelecinobaccha*, and *Pseudodoros*, but are smaller and more fragile in appearance than these flies. Larvae are predators of ground-dwelling aphids.

## *Baccha cognata* American Dainty

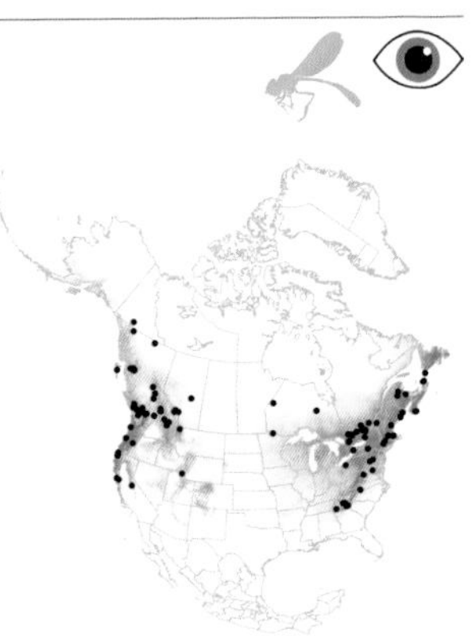

**SIZE RANGE:** 7.2–10.2 mm
**IDENTIFICATION:** Readily identified by the narrow abdomen and unmarked wings. The face is black with a small tubercle, the oral margin is not produced, and the scutellum is black. The female ocellar triangle is pollinose. **ABUNDANCE:** Common. **FLIGHT TIMES:** Early May to early October (from late March in California). **NOTES:** *Baccha cognata* is resurrected from synonymy with *B. elongata*. They are restricted to the Nearctic while *B. elongata* occurs in Alaska, Yukon, the Northwest Territories, and the Old World. The species are genetically distinct and females of *B. elongata* have a shiny ocellar triangle.

# Pseudodoros

Of three species of *Pseudodoros*, one makes it into North America. It is most similar to species of *Baccha*, *Ocyptamus*, and *Pelecinobaccha* in overall body shape and gestalt. Wing pattern and scutellum color differentiate it from these genera. Larvae feed on aphids (Belliure and Michaud 2001).

## *Pseudodoros clavatus* Fourspot

**SIZE RANGE:** 7.0–12.0 mm
**IDENTIFICATION:** Easily identifiable by the yellow scutellum with a brown medial band (black in similar species). The wing has a slim brown leading edge and brown band through $r_1$ and $r_{2+3}$. The face is yellow with a black medial stripe and produced oral margin. **ABUNDANCE:** Common in the south, rare in our region. **FLIGHT TIMES:** Late March to late December. **NOTES:** This widespread tropical species is a vagrant to our area. The larvae are aphidophagous and are important predators of aphids on *Citrus*.

## *Baccha cognata*

petiolate abdomen

wing without brown markings

## *Pseudodoros clavatus*

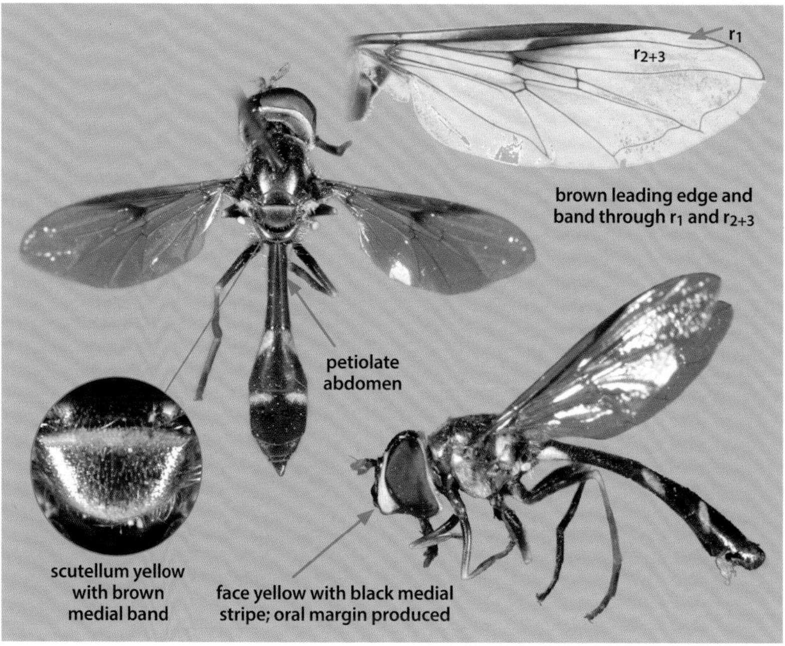

$r_1$

$r_{2+3}$

brown leading edge and band through $r_1$ and $r_{2+3}$

petiolate abdomen

scutellum yellow with brown medial band

face yellow with black medial stripe; oral margin produced

# *Ocyptamus*

*Ocyptamus* is a large, mostly Neotropical group of slender flies. Formerly treated in a broader sense, this group had 273 species. Miranda (2011), Mengual *et al.* (2012), and Miranda *et al.* (2016) highlight the paraphyly of this genus, and the doctoral thesis by Miranda (2011) suggests that several genera be resurrected and new genera be described. Some of these changes were made formal in Miranda *et al.* (2014) and many resurrected generic names were used by Miranda in subsequent publications (2017a, b). We follow his authority on this and thus currently recognize 128 species of *Ocyptamus*, but note that further revision at the generic level is still expected that will ultimately restrict *Ocyptamus* to a group of about 25 species. Only two species in this redefined genus occur in our region. One of these (*O. fascipennis*) is expected to be placed into a new genus by Miranda (following Miranda 2011) in the future. Larvae of *Ocyptamus* in the broad sense are known to feed on aphids, scales, mealybugs, and other soft-bodied insects.

*Ocyptamus fuscipennis*

*Ocyptamus fascipennis*

## *Ocyptamus fascipennis*

### Eastern Band-winged Hover Fly

**SIZE RANGE:** 9.0–13.3 mm

**IDENTIFICATION:** This slender fly's wing has only a medial band darkened. Its face is entirely yellow. **ABUNDANCE:** Common. **FLIGHT TIMES:** Early June to late September (late March to late October in the south). **NOTES:** Adults are found in forest, bog, and marsh habitats and visit flowers of *Cephalanthus*, *Cornus*, *Persicaria*, *Solidago*, and *Spiraea*. Larvae are known to feed on mealybugs on *Chrysanthemum*.

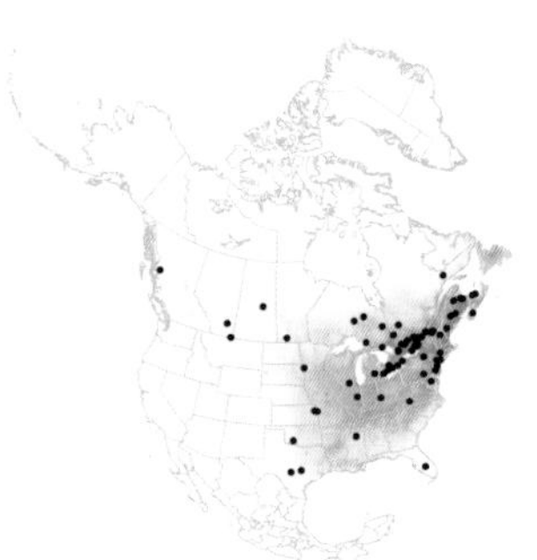

# *Ocyptamus*

***Ocyptamus* in the broad sense is an extremely variable and colorful group of flies.**

*"Ocyptamus" parvicornis* (Florida)

*Ocyptamus (Pseudoscaeva) diversifasciatus* (Colorado)

*Ocyptamus cylindricus* group (Texas)

*Ocyptamus funebris* (Costa Rica)

*"Ocyptamus" lepidus* group (Costa Rica)

## *Ocyptamus fascipennis*

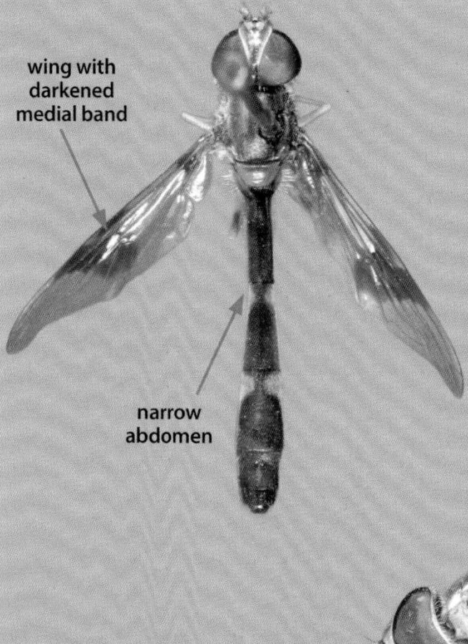

wing with darkened medial band

narrow abdomen

face entirely yellow

## *Ocyptamus fuscipennis*

### Dusky-winged Hover Fly

**SIZE RANGE:** 6.8–11.3 mm
**IDENTIFICATION:** The wing on this species is variably pigmented, but the leading edge and at least part of cell dm are always darkened. Its face is yellow. **ABUNDANCE:** Common. **FLIGHT TIMES:** Early June to early October (year round in the south). **NOTES:** Adults are found in forests and visit a variety of flowers. Larvae are known to feed on aphids on *Citrus* and other species.

dm

leading edge of wing always dark, cell dm always at least partially dark

# *Pelecinobaccha*

*Pelecinobaccha* is another mostly Neotropical genus of slender flies. There are 53 species, mostly in Central and South America, with one species occurring north of Mexico. This genus was recently revised and split from *Ocyptamus* by Miranda *et al.* (2014). Larvae are known to feed on coccids, but little else is known about the ecology of this group.

## *Pelecinobaccha costata* Cobalt Hover Fly

**SIZE RANGE:** 8.7–11.5 mm
**IDENTIFICATION:** This slender fly's wing has a darkened anterior edge, but cell dm is always clear. Its face has a black medial stripe. **ABUNDANCE:** Uncommon. **FLIGHT TIMES:** Mid-June to early October (year round in the south). **NOTES:** Adults are found in forest habitats where they have been recorded visiting Apiaceae and *Euonymus*. Larvae feed on arboreal scale insects (Coccoidea) and possibly aphids.

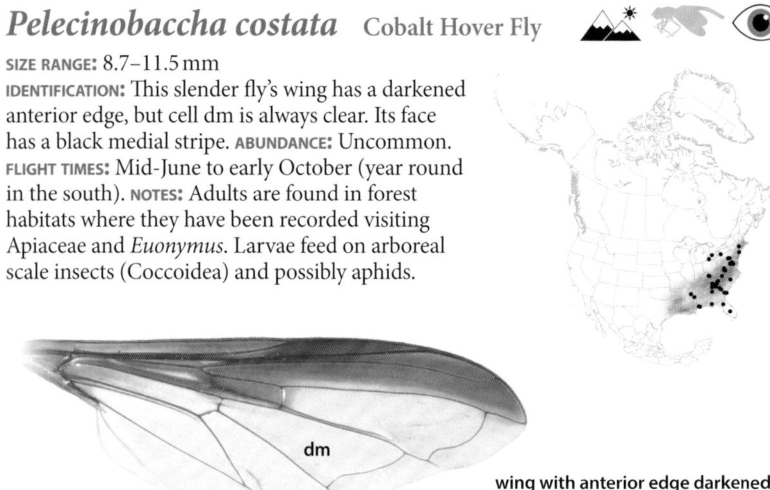


wing with anterior edge darkened, cell dm never darkened

## *Ocyptamus fuscipennis*

narrow abdomen
face yellow

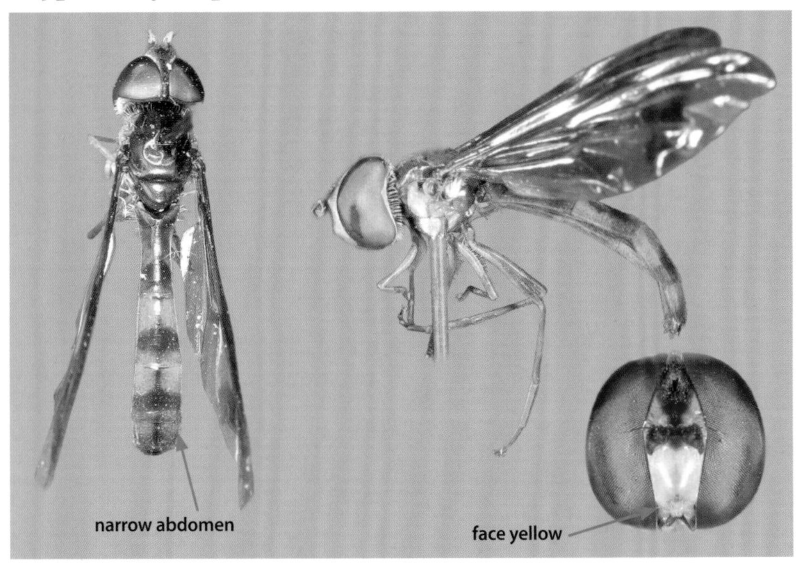

## *Pelecinobaccha costata*

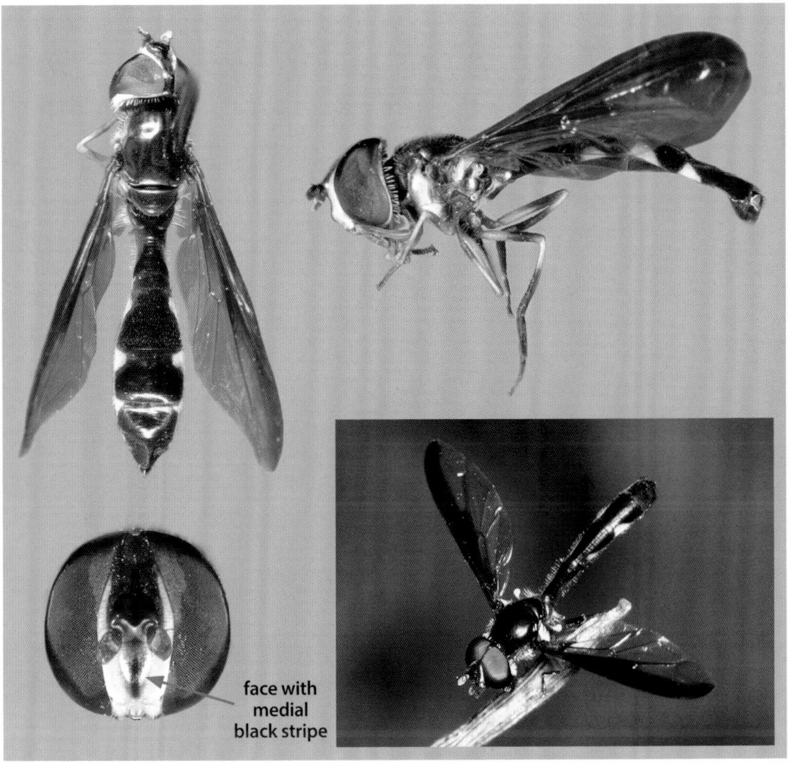
face with
medial
black stripe

# Toxomerus

*Toxomerus* is a New World genus with 140 species. Seven species occur in our region, but only two are broadly distributed. Recently *T. floralis* has been introduced to the Afrotropics. These small flies are commonly encountered at flowers and are immediately recognizable by their distinctive abdominal markings. If in doubt, check the diagnostic triangular emargination (arrow, right) on the posterior margin of the eye, which is at or above the height of the antennal insertion. Most larvae of this genus are predatory, feeding on aphids and other soft-bodied insects. A few species, however, have developed other feeding strategies. Some larvae are known to be phytophagous, feeding on pollen, and one species is known to scavenge on dead prey trapped in sundews. For a key to species (under the genus name *Mesogramma*), see Hull (1943).

## *Toxomerus marginatus*

### Margined Calligrapher

**SIZE RANGE:** 4.9–5.7 mm
**IDENTIFICATION:** This species has an entirely yellow abdominal margin. The margin may appear yellow in other species where the black pigment has been reduced, but normal specimens have alternating black and yellow margins. **ABUNDANCE:** Abundant. **FLIGHT TIMES:** Mid-March to early October (year round in the south). **NOTES:** Adults can be found in diverse habitats such as forests, meadows, fields, savannas, prairies, marshes, bogs, fens, deserts, stream margins, and alpine areas. They are highly adaptable and can occur in very disturbed habitats. Flowers visited include *Abies*, *Anemone*, Apiaceae, *Arabis*, *Aruncus*, *Brassica*, *Cichorium*, *Echium*, *Fraxinus*, *Glyceria*, *Ipomoea*, *Medicago*, *Physocarpus*, *Picea*, *Pinus*, Poaceae, *Rubus*, Solanaceae, *Trifolium*, *Vaccinium*, *Valerianella*, and *Zizia*. Larvae feed on a wide variety of aphids and other soft-bodied insects such as thrips, caterpillars, and mealybugs. This is the most common *Toxomerus* species in the northeast and is one of the most abundant of all the syrphids in the area covered in this guide.

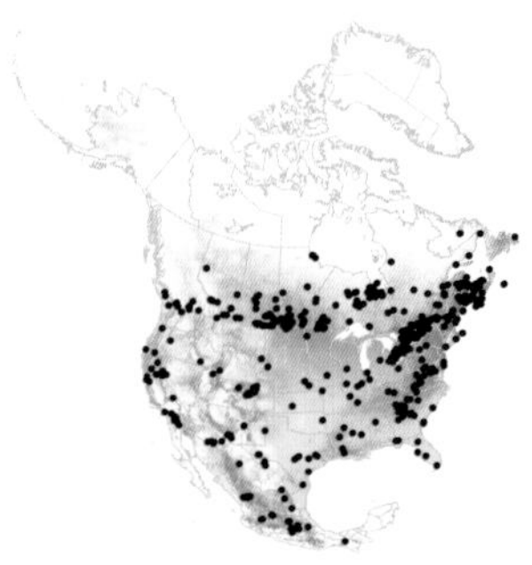

# Toxomerus marginatus

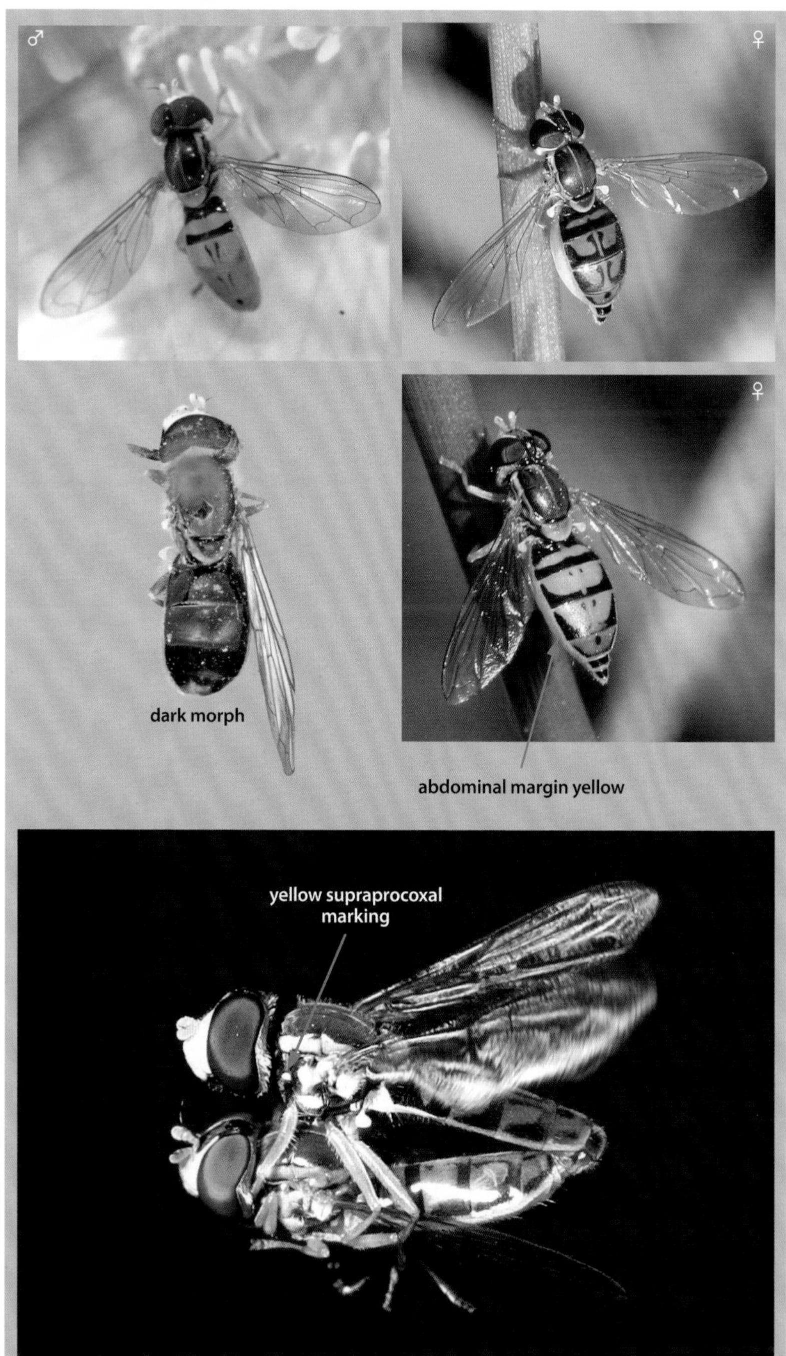

## *Toxomerus verticalis* Banded Calligrapher

**SIZE RANGE:** 4.7–6.0 mm
**IDENTIFICATION:** This species is recognized by the simple, transverse pairs of yellow abdominal markings on tergites 2–5, which are rounded and do not meet medially. **ABUNDANCE:** Vagrant. **FLIGHT TIMES:** Late March to mid-July. **NOTES:** This species is typically found in Florida and the Antilles. It has occasionally been found in other southeastern US states, with one outlying specimen from Minnesota that we did not examine. Larvae are not known. Recorded visiting *Spermacoce*.

## *Toxomerus politus* Maize Calligrapher

**SIZE RANGE:** 7.0–9.0 mm
**IDENTIFICATION:** Distinctive abdominal markings make this species easily recognizable. There is a yellow band along the anterior edge of tergites 3–5 and a pair of yellow medial markings. **ABUNDANCE:** Uncommon. **FLIGHT TIMES:** Late July to early October (late March to mid-December in the south). **NOTES:** Adults are known from agricultural habitats and have been recorded from *Plantago* flowers. Larvae are known to feed on *Zea mays* (corn) pollen and sap. They are not known to be a major pest of corn.

## Toxomerus verticalis

## Toxomerus politus

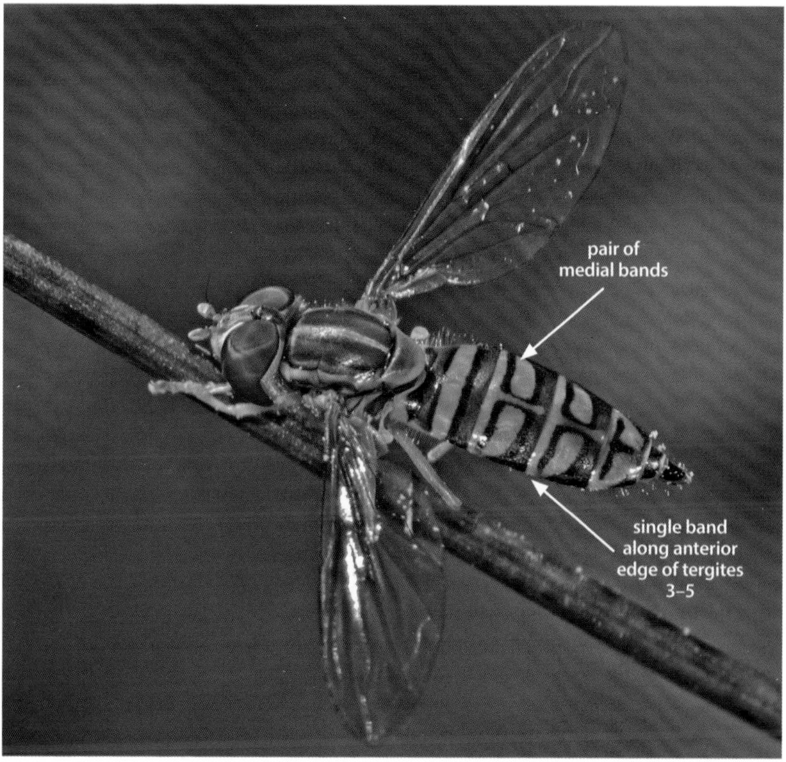

pair of
medial bands

single band
along anterior
edge of tergites
3–5

## *Toxomerus geminatus* Eastern Calligrapher

**SIZE RANGE:** 6.1–7.6 mm
**IDENTIFICATION:** Males have an arcuate metafemur. Both sexes have a black scutellum with yellow posterior margin and lack the yellow supraprocoxal marking found on all of our *Toxomerus* species except this and *T. boscii* (see. *T. corbis* for a close-up of this yellow spot). To separate *T. geminatus* from *T. boscii* look at the amount of black on tergite 3. In *geminatus* most or all of the posterior edge is black. **ABUNDANCE:** Common. **FLIGHT TIMES:** Late April to late October (from mid-March in the south). **NOTES:** Adults can be found in diverse habitats such as forests, meadows, fields, savannas, marshes, bogs, and fens. Flowers visited include *Eupatorium*, *Euthamia*, *Heracleum*, *Physocarpus*, Poaceae, *Potentilla*, *Rhexia*, *Rubus*, *Solidago*, *Tanacetum*, and *Vaccinium*. Larvae have been recorded feeding on a variety of aphids and mites.

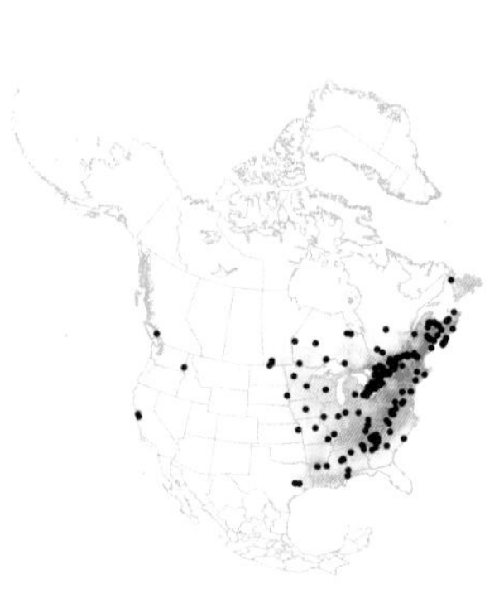

## *Toxomerus boscii* Thin-lined Calligrapher

**SIZE RANGE:** 4.7–6.8 mm
**IDENTIFICATION:** This species does not have a yellow supraprocoxal marking (only *T. geminatus* shares this character state). To separate this species from *T. geminatus* look at the amount of black on tergite 3. In this species it is restricted and most or all of the posterior edge is yellow. **ABUNDANCE:** Uncommon. **FLIGHT TIMES:** Early August (early May to mid-December in the south). **NOTES:** Adults have been recorded visiting *Phyla nodiflora*. Larvae have been recorded feeding on aphids on Fabaceae.

## *Toxomerus geminatus*

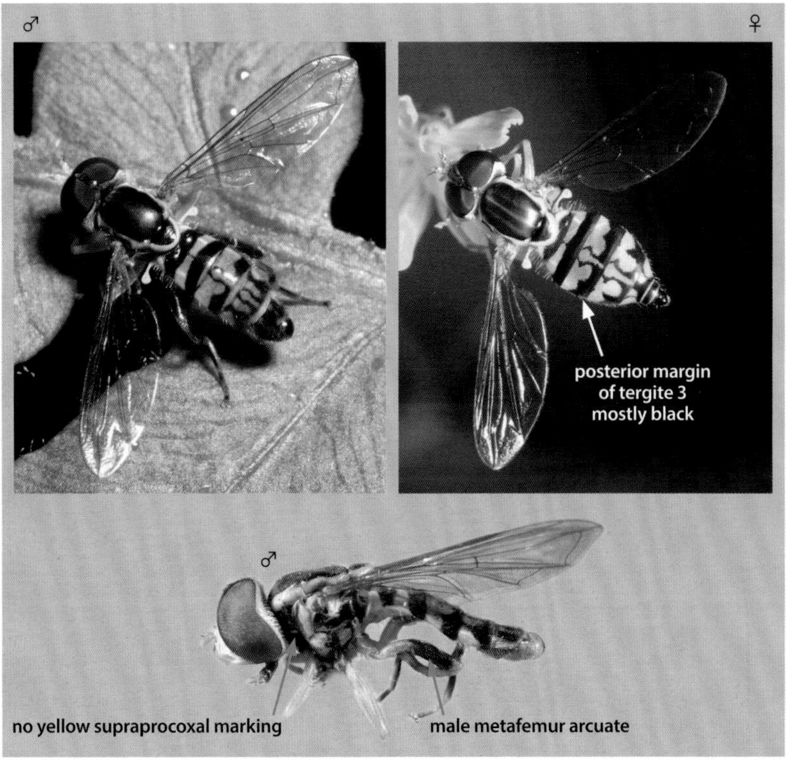

♂
♀
posterior margin of tergite 3 mostly black
♂
no yellow supraprocoxal marking
male metafemur arcuate

## *Toxomerus boscii*

♂
♂
♀
posterior edge of tergite 3 mostly yellow

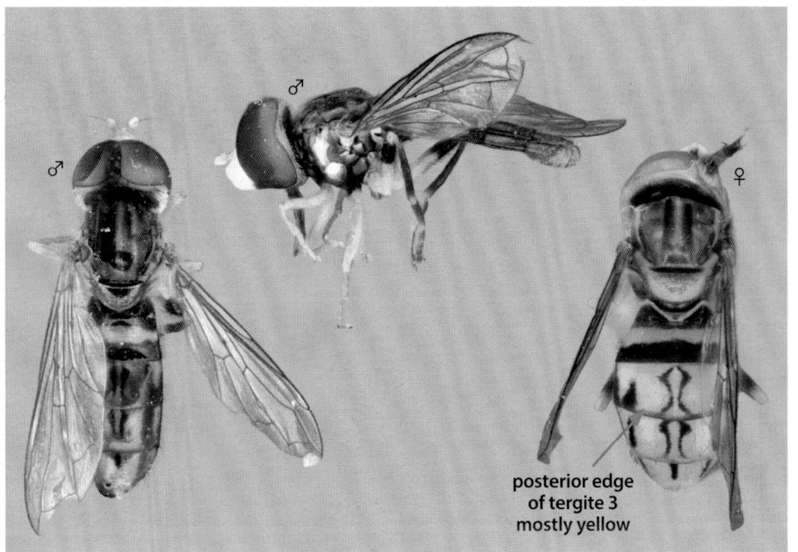

## *Toxomerus corbis* Black-sided Calligrapher

**SIZE RANGE:** 5.6–6.8 mm
**IDENTIFICATION:** This species has a yellow supraprocoxal marking, the abdominal margin is black and yellow, and the anepimeron is black (it is partly yellow in *T. jussiaeae*). **ABUNDANCE:** Vagrant. **FLIGHT TIMES:** Late March to late December over range. **NOTES:** Collected from *Baccharis* and *Symphyotrichum*. Larvae have been recorded feeding on mites on *Gossypium*. We have examined specimens from North Carolina to Florida but there is a historical record from Virginia.

## *Toxomerus jussiaeae* Orange-backed Calligrapher

**SIZE RANGE:** 5.1–6.2 mm
**IDENTIFICATION:** Males with arcuate metafemur (shared only with *T. geminatus*). Scutellum bright yellow orange, anepimeron partly yellow, and tergite 2 has two yellow markings that do not meet medially (most other species have a yellow band). **ABUNDANCE:** Uncommon. **FLIGHT TIMES:** Early June to late August. **NOTES:** Adults are known from habitats around lakes, ponds, and streams. They have been recorded visiting *Ludwigia* (formerly *Jussiaea*, hence the name) and *Nuphar*. Larvae are unknown.

# Toxomerus corbis

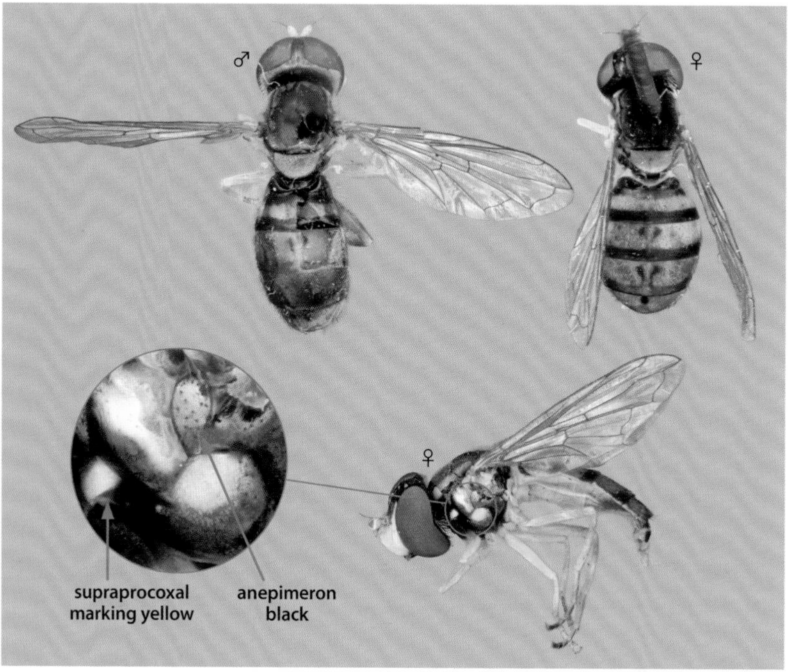

♂

♀

♀

supraprocoxal marking yellow

anepimeron black

# Toxomerus jussiaeae

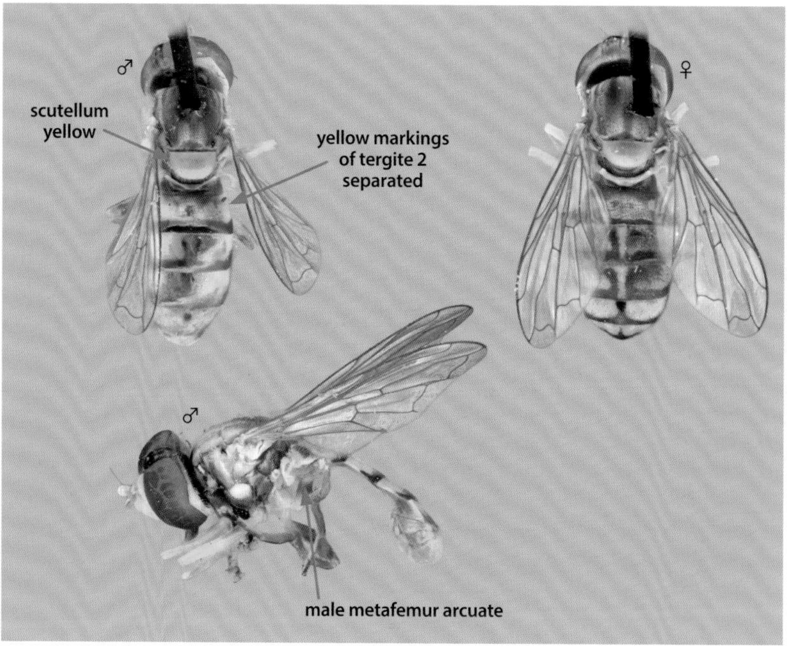

♂

♀

scutellum yellow

yellow markings of tergite 2 separated

♂

male metafemur arcuate

# *Allograpta*

*Allograpta* species are small, black and yellow syrphines with parallel-sided abdomens, lateral yellow stripes on their scutum, and a mainly yellow face. Species of our region are easy to spot because of their distinctive oblique, apical abdominal markings. The genus is distributed throughout the world (74 species) but is absent from most of the Palearctic region. It is most diverse in the Neotropical and Australasian regions. Four species occur in the Nearctic. *Allograpta (A.) radiata* occurs in Florida and *Allograpta (Fazia) micrura* is western. Larvae are typically predatory, feeding on soft-bodied insects, particularly aphids and other Hemiptera. Some species have been shown to feed on pollen, mine leaves, or bore in stems. For a key to species, see Curran (1932).

## *Allograpta (Allograpta) obliqua*

### Oblique Stripetail

**SIZE RANGE:** 5.4–9.0 mm
**IDENTIFICATION:** This species is similar to *A. exotica* but can be distinguished by the extra yellow fascia on the anterior margin of tergite 2 and the yellow katepimeron. **ABUNDANCE:** Abundant. **FLIGHT TIMES:** Mid-April to late September (mid-January to early November in the south). **NOTES:** Adults are known from a wide variety of habitats and have been recorded visiting a wide variety of flowers. Larvae have been recorded feeding on many types of prey, including mites, Lepidoptera, psyllids, mealybugs, whiteflies, and at least 50 species of aphids.

## *Allograpta (Allograpta) exotica*

### Exotic Stripetail

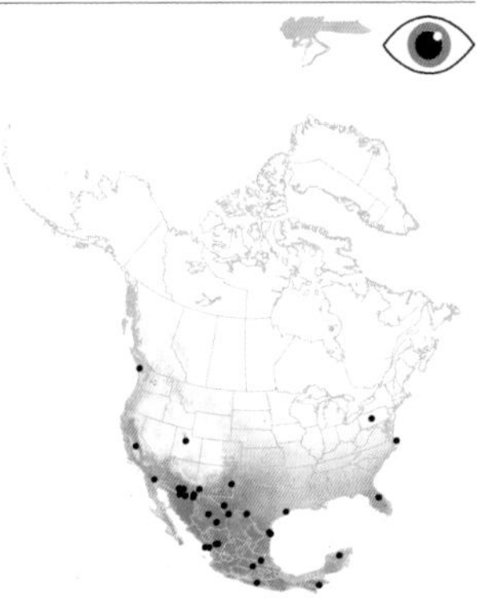

**SIZE RANGE:** 4.7–8.5 mm
**IDENTIFICATION:** This species is similar to *A. (A.) obliqua* but can be distinguished by the single yellow band on tergite 2 (*Allograpta (A.) obliqua* has a second anterior band that is broken medially) and the black katepimeron. **ABUNDANCE:** Vagrant in our region, common in the south. **FLIGHT TIMES:** Mid- to late August (mid-March to late September in the south). **NOTES:** Adults are known from agricultural habitats where they have been recorded visiting *Spiraea* and thistle. Larvae have been found to feed on 18 species of aphids from a variety of crops and other plants, such as *Brassica oleracea*, *Bromus*, *Capsicum annuum*, *Citrus*, *Hordeum vulgare*, *Phalaris*, *Prunus*, *Rosa*, *Triticum aestivum*, and *Zea mays*.

## *Allograpta (Allograpta) obliqua*

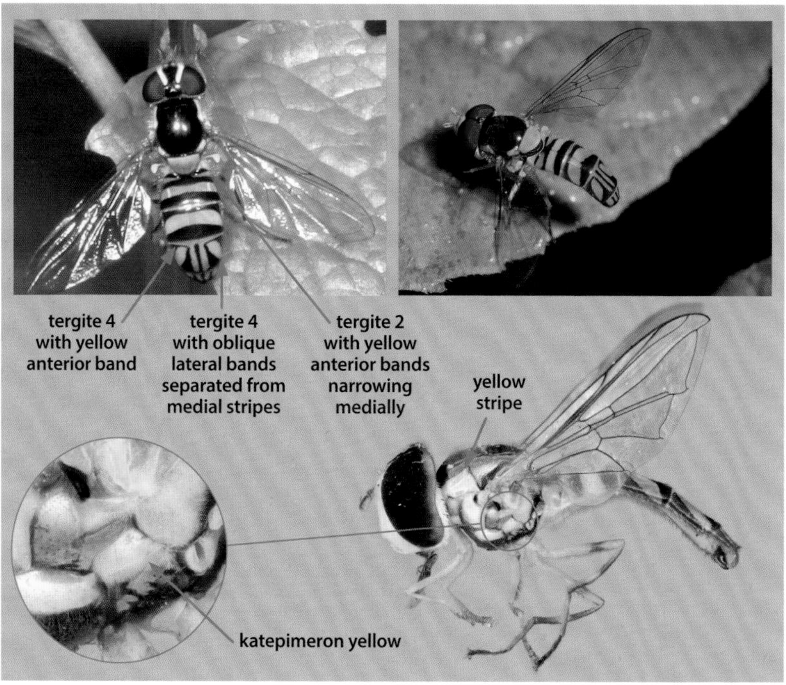

tergite 4 with yellow anterior band

tergite 4 with oblique lateral bands separated from medial stripes

tergite 2 with yellow anterior bands narrowing medially

yellow stripe

katepimeron yellow

## *Allograpta (Allograpta) exotica*

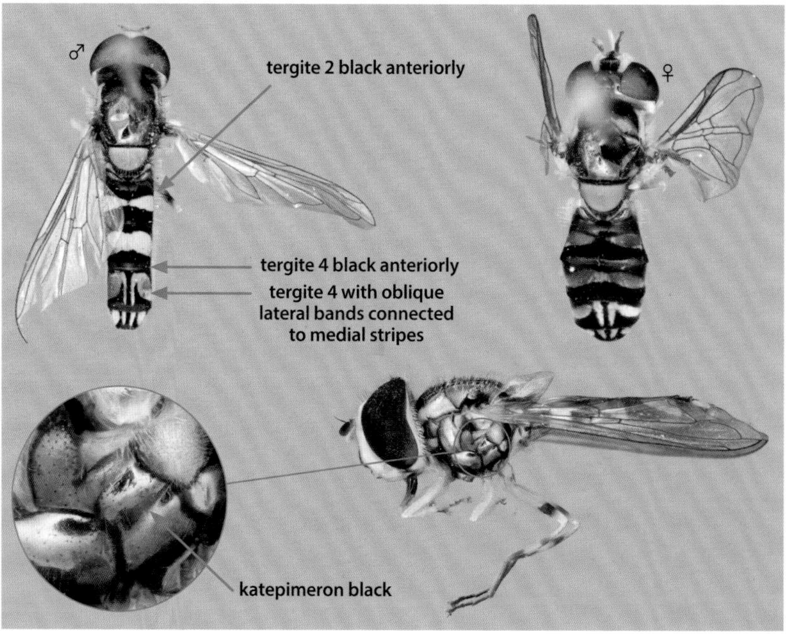

♂

tergite 2 black anteriorly

♀

tergite 4 black anteriorly

tergite 4 with oblique lateral bands connected to medial stripes

katepimeron black

# *Sphaerophoria*

*Sphaerophoria* species are small, narrow, black and yellow syrphines. They have long, slender, black, parallel-sided abdomens with simple yellow abdominal markings, either yellow bands or sometimes pairs of yellow markings on each tergite. They have lateral yellow stripes on the scutum that may stop at the transverse suture or continue to the postalar callus. The scutellum is without a ventral scutellar fringe, or nearly so. Males are easily recognized by the large, globose genitalia at the apex of the abdomen. They can often be found in low vegetation. Larvae are aphidophagous, often feeding on ground-layer aphids. For a key to the species of North America, see Knutson (1973).

♂

male genitalia
large, globose

## *Sphaerophoria novaeangliae*

**Black-striped Globetail**

**SIZE RANGE:** 6.8–8.2 mm
**IDENTIFICATION:** This species is easily recognized. It is our only *Sphaerophoria* to have a broad, medial, black stripe on its face (other species may have a faint stripe, but not as broad and dark as in *S. novaeangliae*). The lateral yellow stripes on the scutum end at the transverse suture, and the anteroventral anepimeron is black.
**ABUNDANCE:** Uncommon. **FLIGHT TIMES:** Early May to early August. **NOTES:** Adults are found in forest habitats. They have been recorded from *Rubus*. Larvae are unknown but likely feed on aphids as others in this genus do.

# *Sphaerophoria*

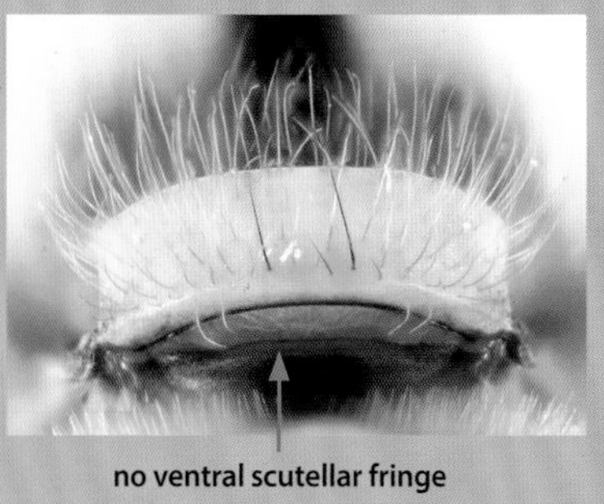

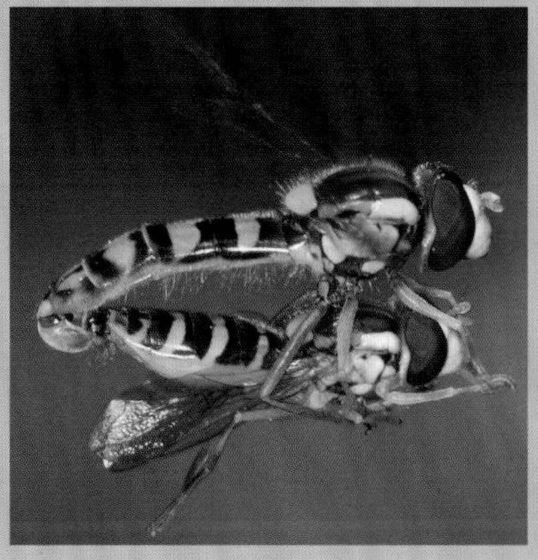

no ventral scutellar fringe

## *Sphaerophoria novaeangliae*

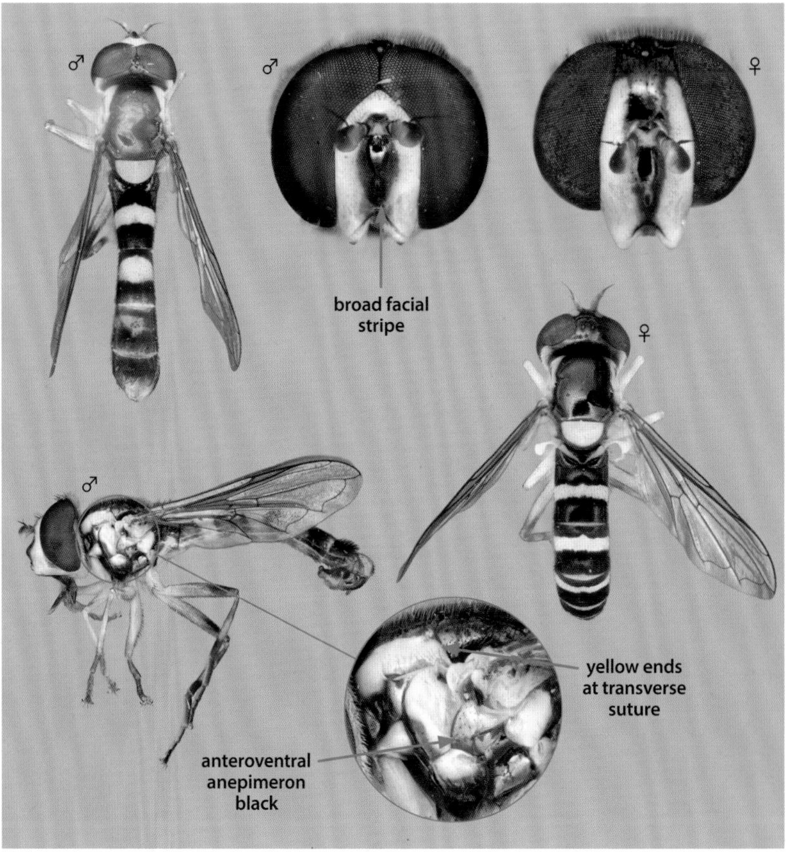

♂
♂
♀
broad facial stripe
♀
♂
yellow ends at transverse suture
anteroventral anepimeron black

## *Sphaerophoria contigua* Tufted Globetail

**SIZE RANGE:** 6.5–8.8 mm
**IDENTIFICATION:** The lateral yellow stripes on the scutum end at the transverse suture, the anteroventral anepimeron is yellow, and the tergites have undivided bands. Male with dorsal lobe of surstylus with apical tuft of anteromedially directed pile. Female with frontal stripe without metallic purple sheen and slightly broadened anteriorly. **ABUNDANCE:** Common. **FLIGHT TIMES:** Mid-April to early October (mid-February to late October in the south). **NOTES:** Adults are found in forest, meadow, bog, marsh, beach, and garden habitats. Flowers visited include *Eupatorium*, *Hedyotis*, *Solidago*, and *Tanacetum*. Larvae have been recorded feeding primarily on aphids but also mites, thrips, and Lepidoptera on a variety of plants (mainly crops of the families Adoxaceae, Anacardiaceae, Asteraceae, Brassicaceae, Celastraceae, Convolvulaceae, Cornaceae, Cucurbitaceae, Fabaceae, Fagaceae, Grossulariaceae, Juglandaceae, Malvaceae, Oleaceae, Onagraceae, Poaceae, Polygonaceae, Rosaceae, Rutaceae, Salicaceae, Solanaceae, and Vitaceae).

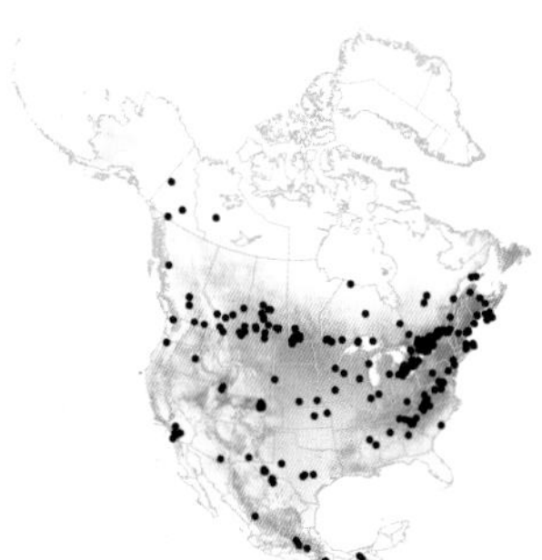

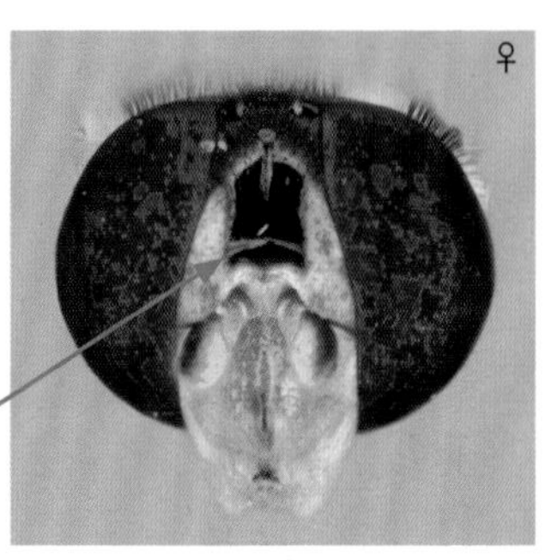


## *Sphaerophoria pyrrhina* Violaceous Globetail

**SIZE RANGE:** 5.6–6.9 mm
**IDENTIFICATION:** Lateral yellow stripes on the scutum end at the transverse suture, and the anteroventral anepimeron is black. Male with dorsal lobe of surstylus not longer than broad, obtuse apically; outer surface not shiny and with abundant long pile. Female with frontal stripe narrowed anteriorly and with metallic purple bronze sheen. **ABUNDANCE:** Uncommon. **FLIGHT TIMES:** Early June to late July (late March to late November in the south). **NOTES:** This species is associated with agricultural habitats. Flowers visited include *Apium*, *Fragaria*, *Leucanthemum*, *Solidago*, *Spiraea*, and *Zizia*. Larvae feed on aphids and have been collected from *Capsicum*.

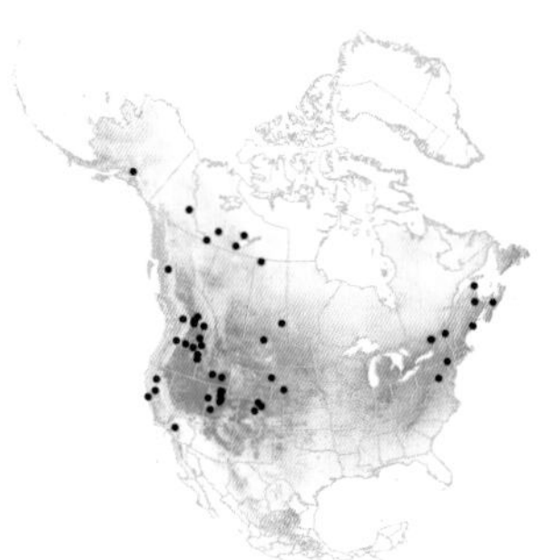

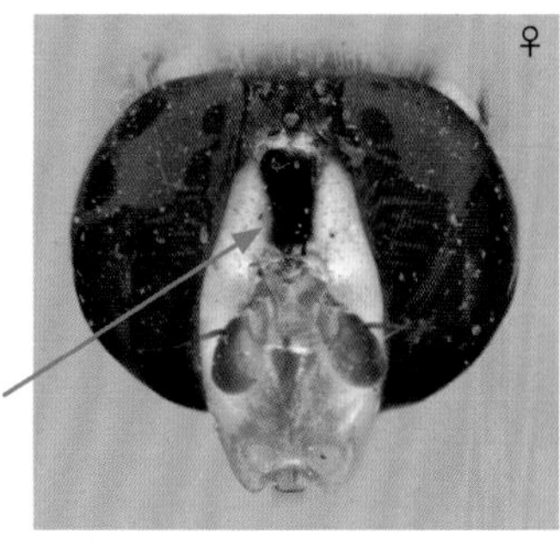

# *Sphaerophoria contigua*

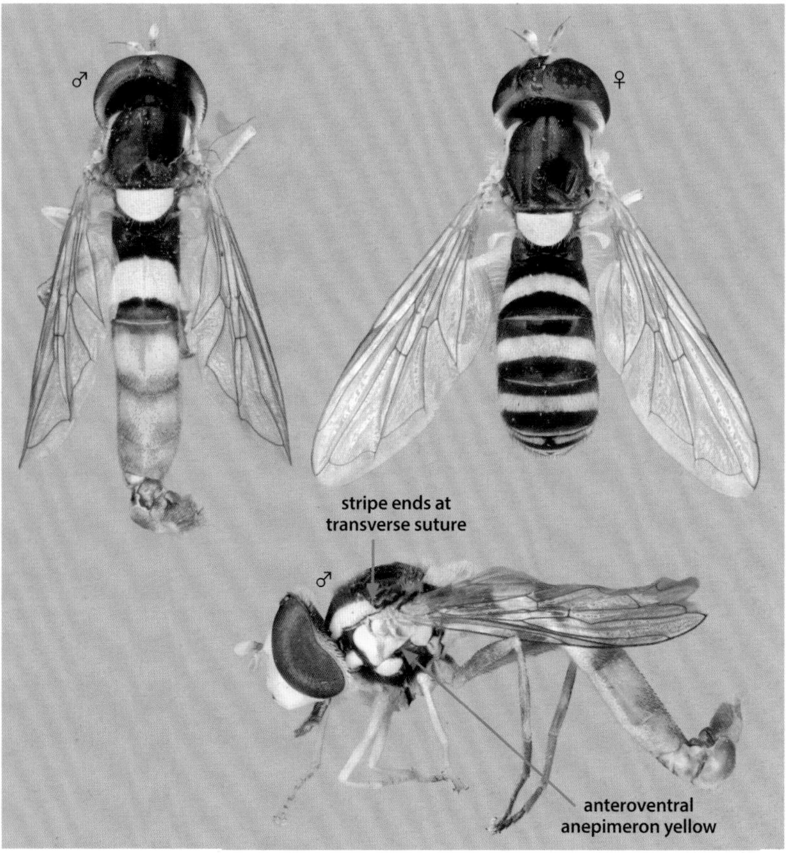

♂
♀
stripe ends at transverse suture
♂
anteroventral anepimeron yellow

# *Sphaerophoria pyrrhina*

♀
stripe ends at transverse suture
♂
anteroventral anepimeron black

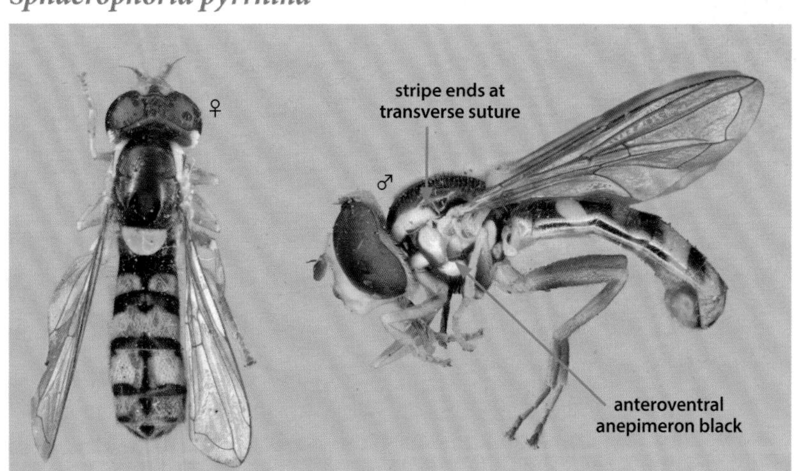

## *Sphaerophoria scripta* Greenland Globetail

**SIZE RANGE:** 9.0–12.0 mm

**IDENTIFICATION:** Lateral yellow stripes on the scutum extend past the transverse suture, onto the postalar callus. Male with abdomen often longer than wings, metafemur with dense black setulae posteroventrally on apical ¾; dorsal lobe of surstylus elongate with broad, triangular, bare, flattened area at base. Female metafemur with longer, stronger pile on posteroventral surface. **ABUNDANCE:** Uncommon. **FLIGHT TIMES:** Late May to late July (February to November in Palearctic) **NOTES:** Found in Greenland and throughout the northern Palearctic. Adults are known from forest, meadow, grassland, and agricultural habitats. Adults have been recorded from a wide variety of flowers. Larvae are known to feed on a wide variety of aphids, Lepidoptera, and psyllids on herbaceous plants, including many crop species.

## *Sphaerophoria bifurcata* Forked Globetail

**SIZE RANGE:** 7.1–8.0 mm

**IDENTIFICATION:** Lateral yellow stripes on the scutum extend past the transverse suture, onto the postalar callus. Male with apical tarsomere of pro- and mesolegs paler than basal tarsomere or tarsus uniformly yellow to pale brown; pile on sternite 2 longer than width of metafemur; ventral lobe of surstylus with ventral margin nearly straight and with two apical processes differing slightly in length. Female with black frontal stripe forked anteriorly. **ABUNDANCE:** Rare. **FLIGHT TIMES:** Mid-May to late August. **NOTES:** Adults are known from forest and open habitats. They have been recorded from *Fragaria*. Larvae presumably feed on aphids on herbaceous plants as others in this genus do.

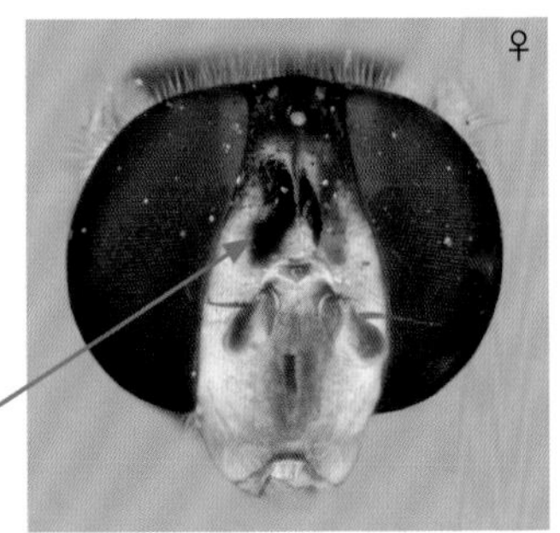

# *Sphaerophoria scripta*

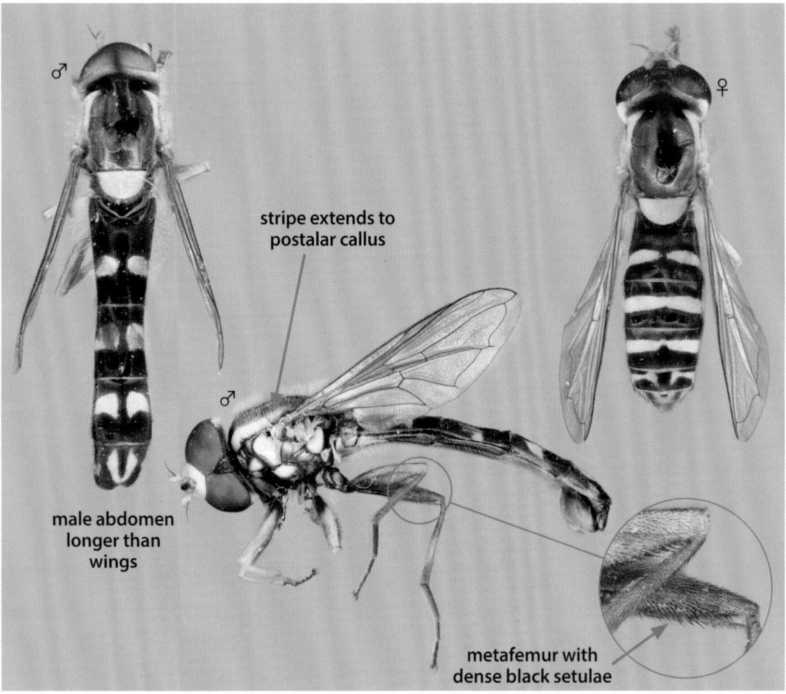

♂

♀

stripe extends to postalar callus

♂

male abdomen longer than wings

metafemur with dense black setulae

# *Sphaerophoria bifurcata*

♂

♀

stripe extends to postalar callus

♂

# *Sphaerophoria brevipilosa*

## Broad-striped Globetail

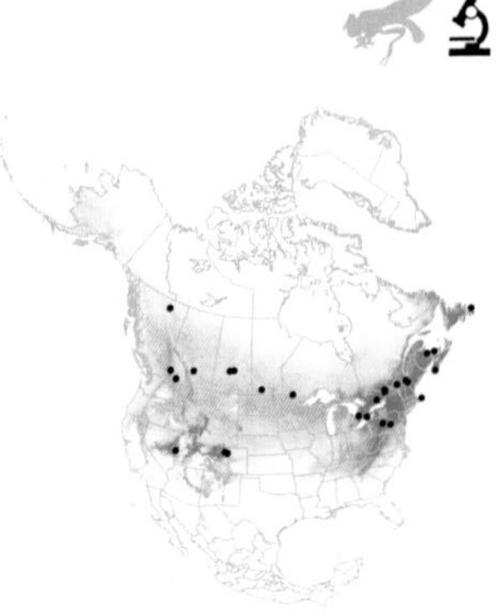

**SIZE RANGE:** 6.3–8.1 mm
**IDENTIFICATION:** Lateral yellow stripes on the scutum extend past the transverse suture onto the postalar callus, bright yellow behind transverse suture (in males). Male with apical tarsomere of pro- and mesolegs paler than basitarsus or tarsus uniformly yellow to pale brown; dorsal lobe of surstylus with fringe of short sparse pile; ventral lobe with apex obliquely truncate. Pile on sternite 2 shorter than width of metafemur in both sexes. Females cannot be distinguished from *S. longipilosa*. **ABUNDANCE:** Rare. **FLIGHT TIMES:** Early May to mid-August. **NOTES:** This species has been recorded from salt marshes, *Corema* barrens, and a small homestead clearing in mixed forest. Flowers visited include *Fragaria*, *Ranunculus*, and *Rhododendron*. Larvae are known to feed on aphids on *Prunus avium*.

# *Sphaerophoria longipilosa*

## Narrow-striped Globetail

**SIZE RANGE:** 7.5–8.2 mm
**IDENTIFICATION:** Lateral yellow stripes on the scutum extend past the transverse suture onto the postalar callus, narrowed and darkened beyond transverse suture (in males). Male with apical tarsomere of pro- and mesolegs paler than basitarsus or tarsus uniformly yellow to pale brown; dorsal lobe of surstylus with fringe of long dense pile; ventral lobe with stout finger-like apical process. Pile on sternite 2 shorter than width of metafemur in both sexes. Females cannot be distinguished from *S. brevipilosa*. **ABUNDANCE:** Rare. **FLIGHT TIMES:** Early May to mid-June. **NOTES:** Adults are found in forest and fen habitats. They have been recorded from *Crataegus*, *Fragaria*, *Prunus*, and *Salix*. Larvae presumably feed on aphids on herbaceous plants as others in this genus do.

## *Sphaerophoria brevipilosa* male

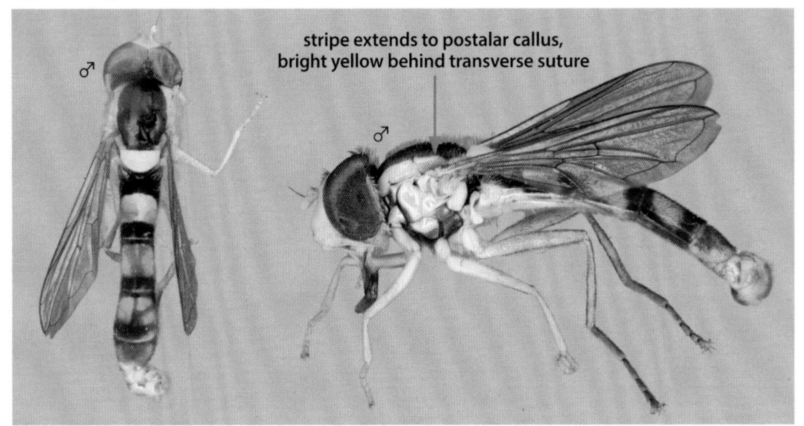


## *Sphaerophoria longipilosa* male

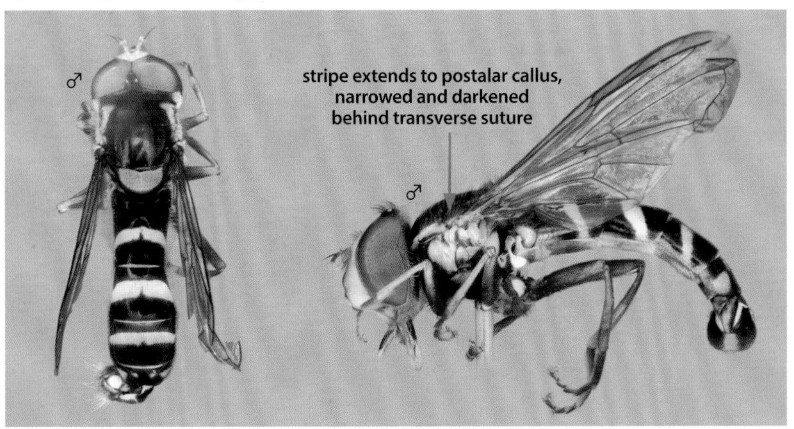


## *Sphaerophoria brevipilosa* or *longipilosa* female

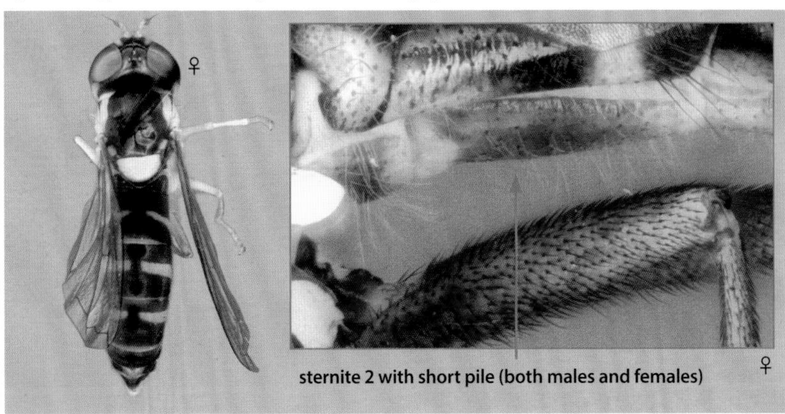

# *Sphaerophoria philanthus*

## Black-footed Globetail

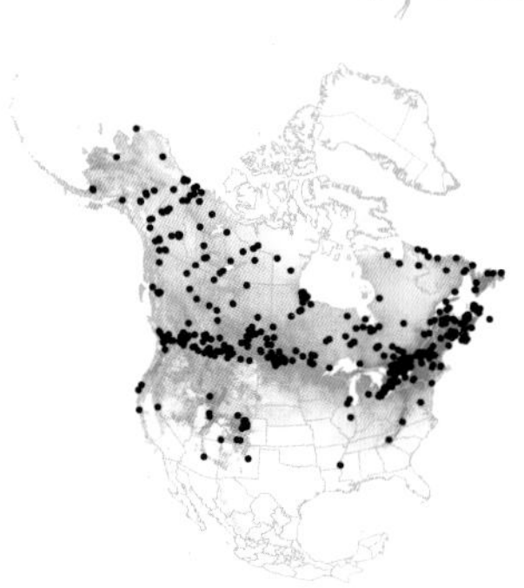

**SIZE RANGE:** 7.6–9.6 mm
**IDENTIFICATION:** Male with apical tarsomere of pro- and meso legs brown black, distinctly darker than basitarsus or all tarsomeres black; ventral lobe of surstylus with flattened preapical process on dorsal margin; inner lobe of surstylus present, acute apically. **ABUNDANCE:** Abundant. **FLIGHT TIMES:** Mid-April to late October. **NOTES:** Adults are known from forest, meadow, field, bog, marsh, swamp, and tundra habitats. They have been recorded visiting Asteraceae, *Avena*, *Coreopsis rosea*, *Crataegus*, *Medicago*, Poaceae, *Rubus*, *Salix*, *Solidago*, *Thlaspi*, *Tiarella*, and *Trifolium*. Larvae feed on aphids and Lepidoptera that are mainly on species of cultivated crops and some herbaceous plants. This species is Holarctic.

# *Sphaerophoria asymmetrica*

## Asymmetric Globetail

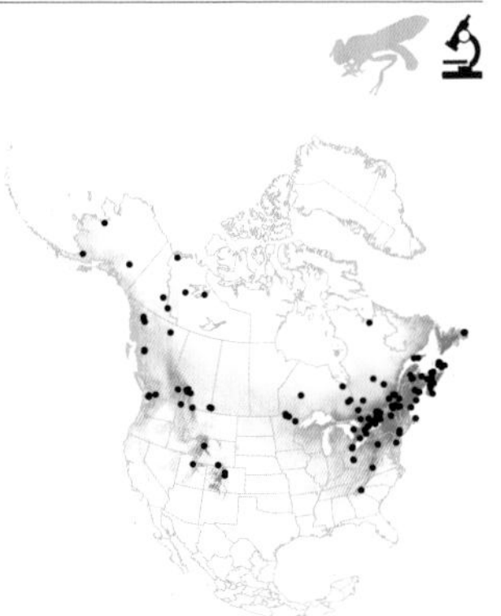

**SIZE RANGE:** 6.8–9.0 mm
**IDENTIFICATION:** Male with apical tarsomere of pro- and mesolegs paler than basitarsus or tarsus uniformly yellow to pale brown; anteroventral lobes of surstylus asymmetrical. **ABUNDANCE:** Common. **FLIGHT TIMES:** Late April to late October. **NOTES:** Adults are known from forest, meadow, field, marsh, and tundra habitats and have been recorded on flowers of *Lythrum*, *Spiraea*, and *Symphyotrichum*. Larvae feed on aphids on crops and herbaceous plants such as *Asparagus*, *Beta vulgaris*, and *Medicago sativa*.

# *Sphaerophoria abbreviata* Variable Globetail

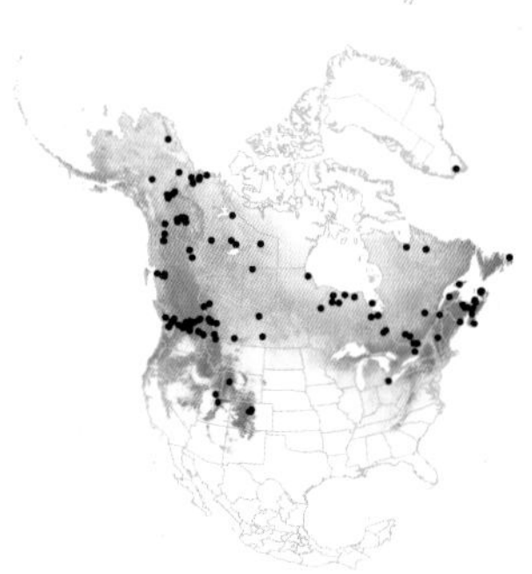

**SIZE RANGE:** 6.3–10.2 mm
**IDENTIFICATION:** Male with apical tarsomere of pro- and mesolegs paler than basitarsus or tarsus uniformly yellow to pale brown; ventral lobe of surstylus with ventral margin curved and with apical processes markedly different in length. **ABUNDANCE:** Common. **FLIGHT TIMES:** Late May to mid-September (from mid-April in the west). **NOTES:** Adults are known from forest, bog, tundra, and open grassy habitats. Flowers visited include *Caltha*, *Ranunculus*, *Rubus*, and *Taraxacum*. Larvae feed on aphids on crops and herbaceous plants, such as *Secale cereale*. This species is Holarctic.

# *Sphaerophoria philanthus, asymmetrica,* and *abbreviata*

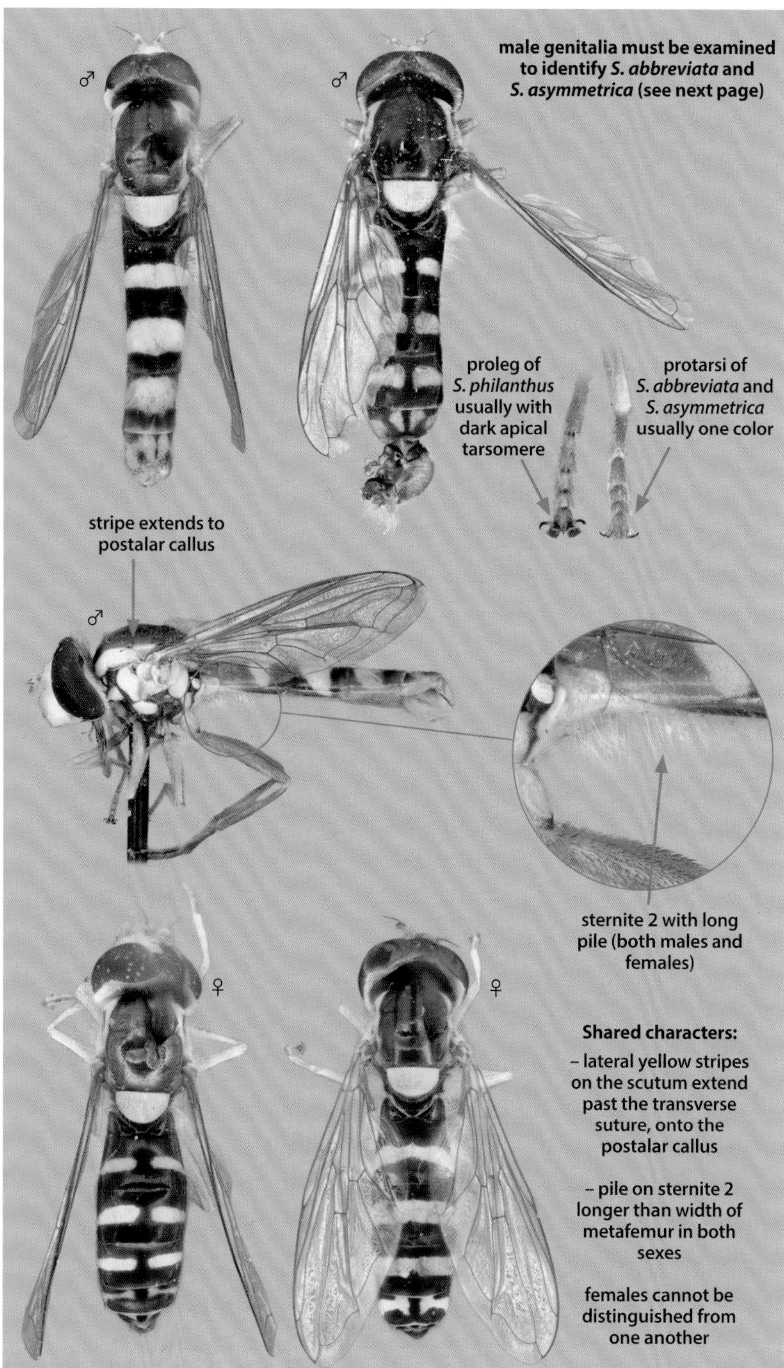

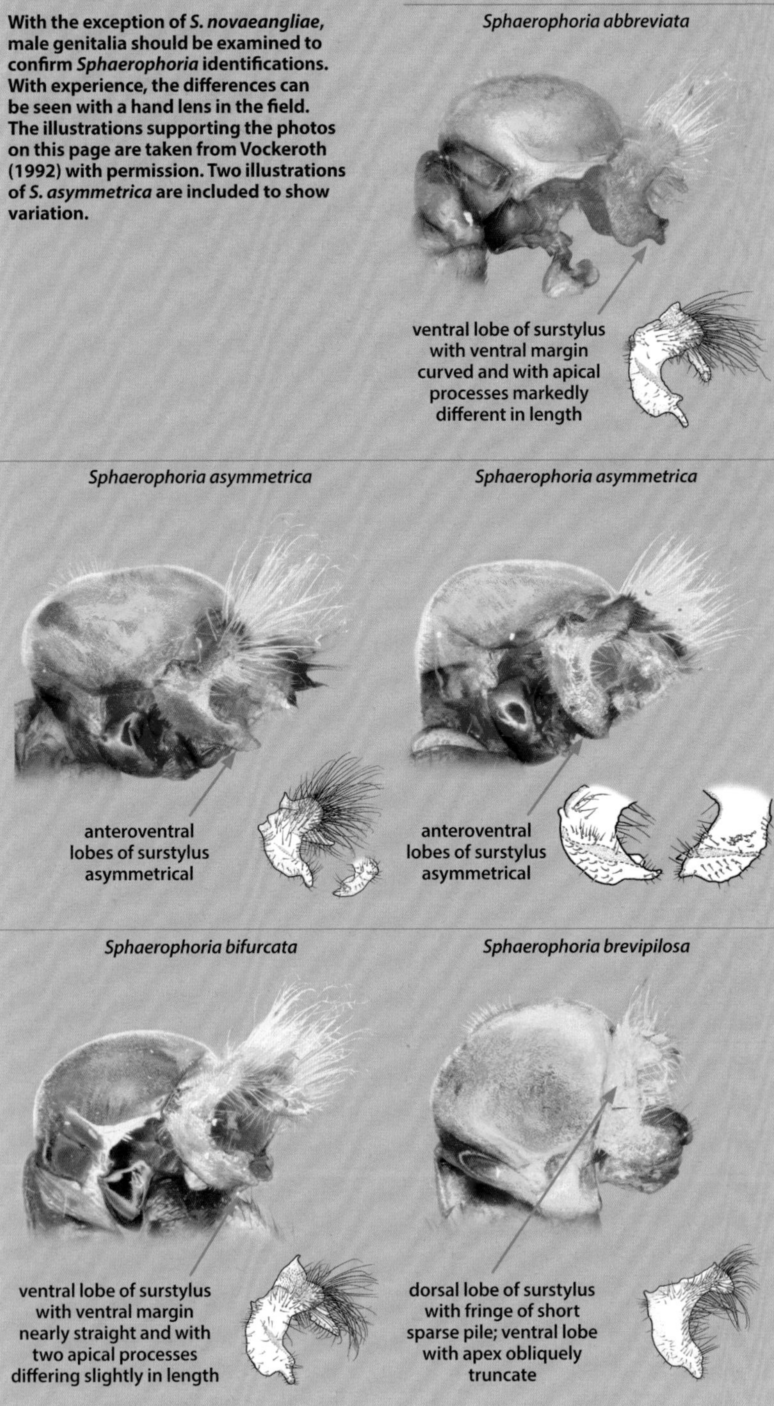
**With the exception of *S. novaeangliae*, male genitalia should be examined to confirm *Sphaerophoria* identifications. With experience, the differences can be seen with a hand lens in the field. The illustrations supporting the photos on this page are taken from Vockeroth (1992) with permission. Two illustrations of *S. asymmetrica* are included to show variation.**
*Sphaerophoria abbreviata*
ventral lobe of surstylus with ventral margin curved and with apical processes markedly different in length
*Sphaerophoria asymmetrica*
anteroventral lobes of surstylus asymmetrical
*Sphaerophoria asymmetrica*
anteroventral lobes of surstylus asymmetrical
*Sphaerophoria bifurcata*
ventral lobe of surstylus with ventral margin nearly straight and with two apical processes differing slightly in length
*Sphaerophoria brevipilosa*
dorsal lobe of surstylus with fringe of short sparse pile; ventral lobe with apex obliquely truncate

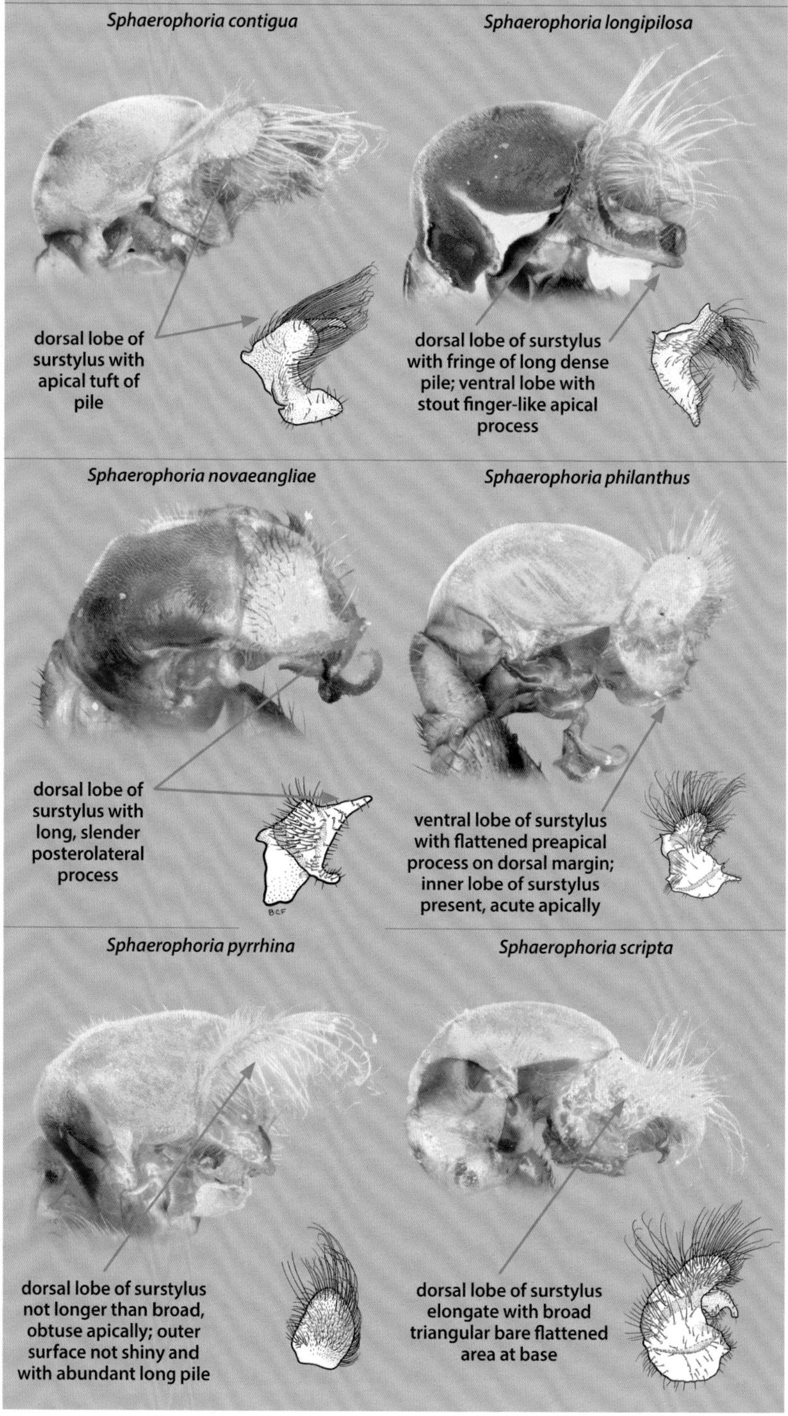
Sphaerophoria contigua
dorsal lobe of surstylus with apical tuft of pile
Sphaerophoria longipilosa
dorsal lobe of surstylus with fringe of long dense pile; ventral lobe with stout finger-like apical process
Sphaerophoria novaeangliae
dorsal lobe of surstylus with long, slender posterolateral process
BCF
Sphaerophoria philanthus
ventral lobe of surstylus with flattened preapical process on dorsal margin; inner lobe of surstylus present, acute apically
Sphaerophoria pyrrhina
dorsal lobe of surstylus not longer than broad, obtuse apically; outer surface not shiny and with abundant long pile
Sphaerophoria scripta
dorsal lobe of surstylus elongate with broad triangular bare flattened area at base

# *Xanthogramma*

This predominantly Old World genus contains 21 species. Only one occurs in North America. Although many adults have been found, its larval biology is unknown. Larvae of other *Xanthogramma* species live in ant nests where they are aphidophagous, feeding on root aphids tended by ants of the genus *Lasius*. Adults are slender, brightly colored flies often found low to the ground.

## *Xanthogramma flavipes*

American Harlequin

**SIZE RANGE:** 7.3–12.3 mm
**IDENTIFICATION:** Easily recognized by the vivid yellow stripes on the scutum, yellow markings on the thoracic pleura, and the yellow-banded abdomen. The abdomen varies from entirely black to mostly orange between the yellow bands. **ABUNDANCE:** Uncommon. **FLIGHT TIMES:** Early May to early September. **NOTES:** Adults are found in forest and field habitats. Larvae likely feed on aphids associated with ants. DNA suggests that *X. flavipes* may be a complex of two sympatric species.

# *Meliscaeva*

Only one of the 25 species of *Meliscaeva* occurs in North America. They occur in all regions except for Australasia and the Neotropics but are most diverse in southeast Asia. These are slender, medium-sized flies with a moderately pollinose scutum and a yellow-banded abdomen. As with most syrphines, the larvae are predators of aphids.

## *Meliscaeva cinctella* Common Thintail

**SIZE RANGE:** 8.0–10.8 mm
**IDENTIFICATION:** This species has a pilose anterior anepisternum, bare eye, and parallel-sided abdomen with broad, entire, yellow bands. There is no pile on the posteromedial apical angle of the metacoxa. Some specimens have lateral yellow stripes on the scutum. **ABUNDANCE:** Common. **FLIGHT TIMES:** Early May to late October (from mid-March in the south). **NOTES:** Adults are known from forest, meadow, bog, marsh, and tundra habitats and are also synanthropic. Flowers visited include *Myosotis* and *Solidago*. Larvae feed on aphids and psyllids of trees and shrubs. This species is Holarctic.

## *Xanthogramma flavipes*

yellow stripes on scutum

scutellum yellow posteriorly

yellow markings

## *Meliscaeva cinctella*

parallel-sided abdomen with broad, entire yellow bands

anterior anepisternum pilose

# *Meligramma*

There are three species of *Meligramma* in the Nearctic, two of which are found in our region. They have slender, parallel-sided abdomens with yellow markings or bands. Markings on tergite 2 are usually small and triangular. They have a narrow face (narrower than the width of the eye), bare metasternum, an unmargined abdomen, bare eyes, and no tuft of pile on the posteromedial apical angle of the metacoxa. Some specimens have lateral yellow stripes on their scutum and often have two yellow markings anterior to the scutellum. For a key to species of northern North America, see Vockeroth (1992).

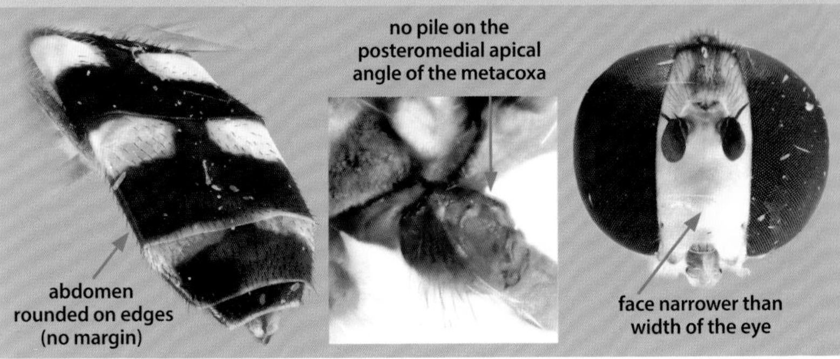


## *Meligramma guttata* Spotted Roundtail

**SIZE RANGE:** 7.6–9.6 mm

**IDENTIFICATION:** Some specimens have one or two round yellow spots on the scutum anterior to the scutellum. Tergites 3 and 4 each with pair of yellow markings. Males with entirely yellow frons, sometimes darkened on the upper ⅓. The female frons has a pollinose stripe along the eye margin at the level of the ocellar triangle. **ABUNDANCE:** Rare. **FLIGHT TIMES:** Early July to mid-August. **NOTES:** Adults are known from forest and tundra habitats and visit a variety of flowers. Larvae have been recorded feeding on arboreal aphids in the United Kingdom.

## *Meligramma triangulifera* Variable Roundtail

**SIZE RANGE:** 7.0–10.0 mm

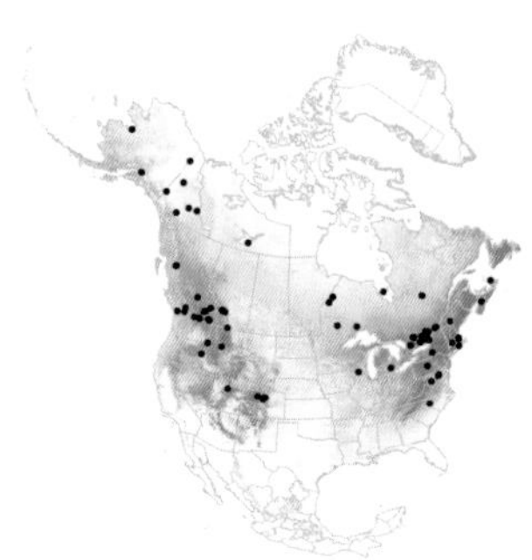

**IDENTIFICATION:** Tergites 3 and 4 with pair of yellow markings or single bands. Males with frons that is darkened at least on the midline, frequently entirely dark. Females with entirely shiny frons at the level of the ocellar triangle. **ABUNDANCE:** Uncommon. **FLIGHT TIMES:** Late April to early September (from late March in the west). **NOTES:** Adults are known from forest and bog habitats. They have been recorded from a variety of tree flowers and Apiaceae as well as other herbaceous plants. Larvae feed on aphids of trees and shrubs.

## *Meligramma guttata*

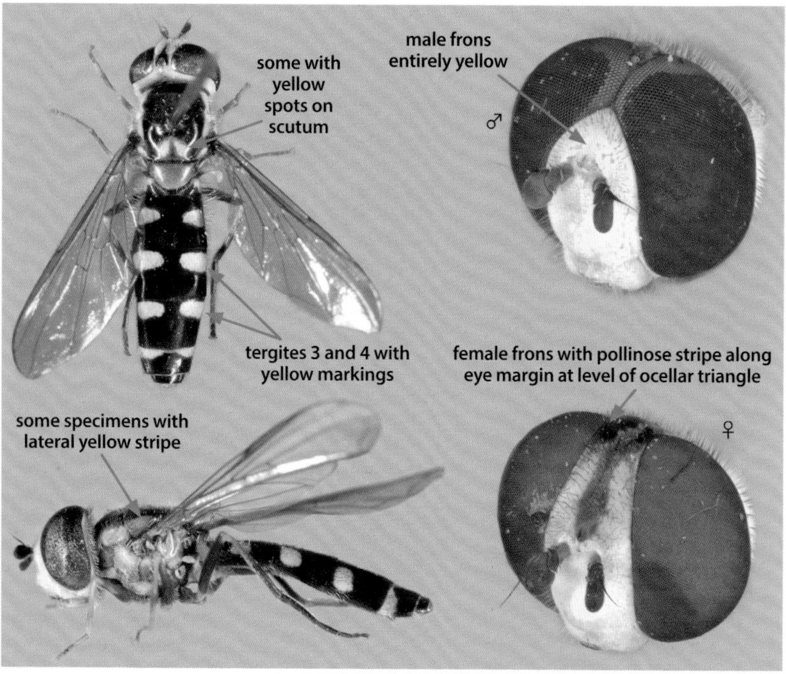

some with yellow spots on scutum

male frons entirely yellow

♂

tergites 3 and 4 with yellow markings

female frons with pollinose stripe along eye margin at level of ocellar triangle

some specimens with lateral yellow stripe

♀

## *Meligramma triangulifera*

♀

scutum all black

male frons darkened at least on midline, may be all black

♂

tergites 3 and 4 with yellow markings or single band

female frons shiny at level of ocellar triangle

♀

# Melangyna

There are seven species of *Melangyna* in the Nearctic, five of which are in our region. This genus has an unmargined abdomen, which is parallel sided to slightly oval. The eye is pilose or bare, the metasternum is bare, and they have a tuft of pile on the posteromedial apical angle of the metacoxa. Adults are flower visitors and larvae are aphidophagous. For a key to species of northern North America, see Vockeroth (1992).

## *Melangyna umbellatarum*

### Bare-winged Halfband

**SIZE RANGE:** 7.2–12.0 mm
**IDENTIFICATION:** Wing bare basally in cells c and bm (*M. fisherii* and *M. lasiophthalma* are our only other species with bare areas on the wing). Eyes bare. **ABUNDANCE:** Common. **FLIGHT TIMES:** Mid-May to mid-September (to late September in the west). **NOTES:** Adults are known from forest habitats, often near streams. They have been recorded visiting Apiaceae, *Euphorbia*, and Rosaceae flowers. Larvae feed on aphids found on Apiaceae, *Betula*, and *Rumex*. This species is Holarctic.

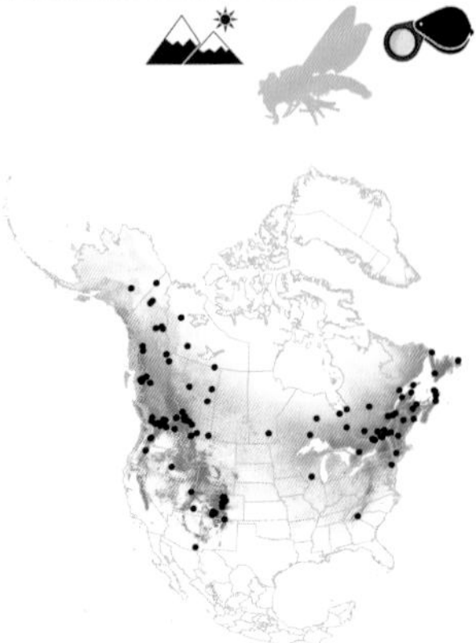

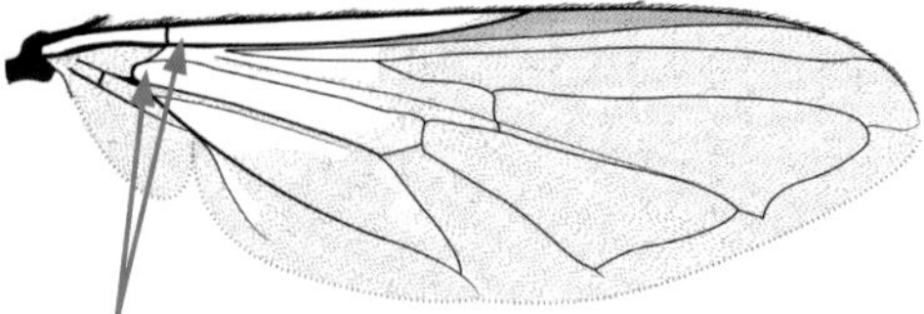

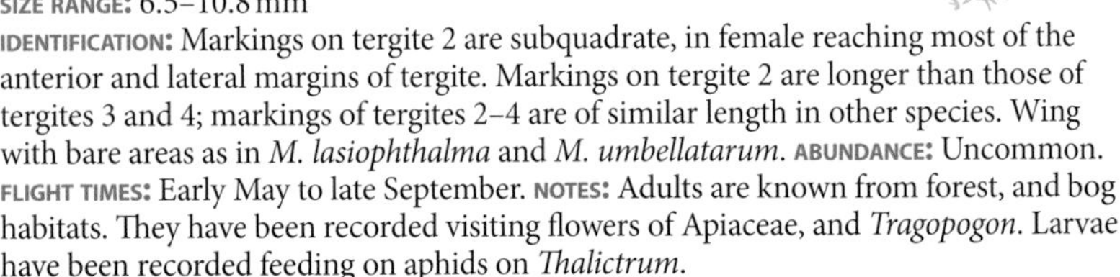

wing bare basally on cells c and bm in *M. umbellaturum, M. fisherii,* and *M. lasiophthalma*

## *Melangyna fisherii* Large-spotted Halfband

**SIZE RANGE:** 6.5–10.8 mm
**IDENTIFICATION:** Markings on tergite 2 are subquadrate, in female reaching most of the anterior and lateral margins of tergite. Markings on tergite 2 are longer than those of tergites 3 and 4; markings of tergites 2–4 are of similar length in other species. Wing with bare areas as in *M. lasiophthalma* and *M. umbellatarum*. **ABUNDANCE:** Uncommon. **FLIGHT TIMES:** Early May to late September. **NOTES:** Adults are known from forest, and bog habitats. They have been recorded visiting flowers of Apiaceae, and *Tragopogon*. Larvae have been recorded feeding on aphids on *Thalictrum*.

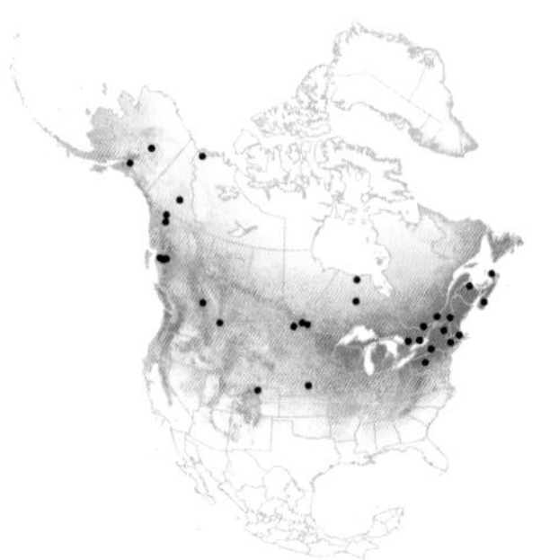

# Melangyna

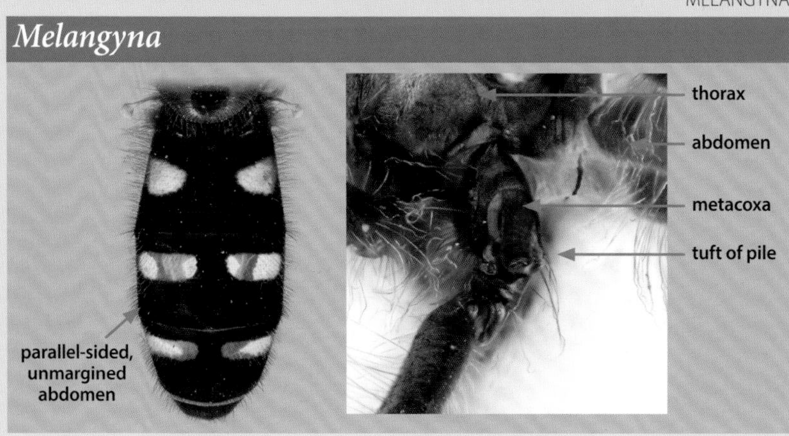


## Melangyna umbellatarum

## Melangyna fisherii

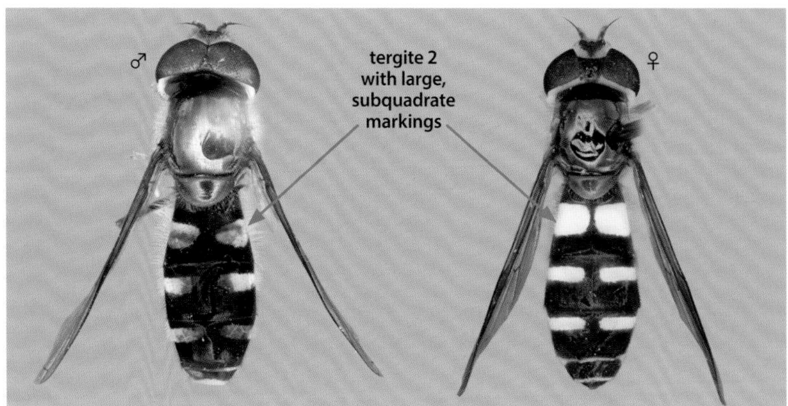

## *Melangyna lasiophthalma* Hair-eyed Halfband

**SIZE RANGE:** 6.1–10.9 mm
**IDENTIFICATION:** Eye typically pilose, but sometimes bare in females. Our other species have bare eyes. Females have a narrow pollinose band on the frons, which is typically separated medially. This band is broad in other species. Wing with bare areas as in *M. fisherii* and *M. umbellatarum*. **ABUNDANCE:** Common. **FLIGHT TIMES:** Early April to mid-August (late February to late November in the west). **NOTES:** This species is Holarctic. Adults are found in meadow, marsh, fen, and tundra habitats near forests. They have been recorded from *Amelanchier*, *Anemone*, *Euthamia*, *Salix*, *Solidago*, and *Tragopogon*. Larvae feed on arboreal aphids and adelgids.

## *Melangyna labiatarum* Bare-eyed Halfband

**SIZE RANGE:** 7.0–10.6 mm
**IDENTIFICATION:** Wing densely microtrichose. Markings on tergite 2 about equal in size to those of tergites 3 and 4, eyes bare. **ABUNDANCE:** Uncommon. **FLIGHT TIMES:** Late May to mid-September (to early October in the west). **NOTES:** Adults are known from forest, meadow, tundra, and alpine habitats. They have been recorded from Apiaceae, *Euthamia*, and *Solidago*. Larvae feed on aphids that have been recorded on *Euonymus europaeus* and *Heracleum sphondylium* in the United Kingdom. This species is Holarctic.

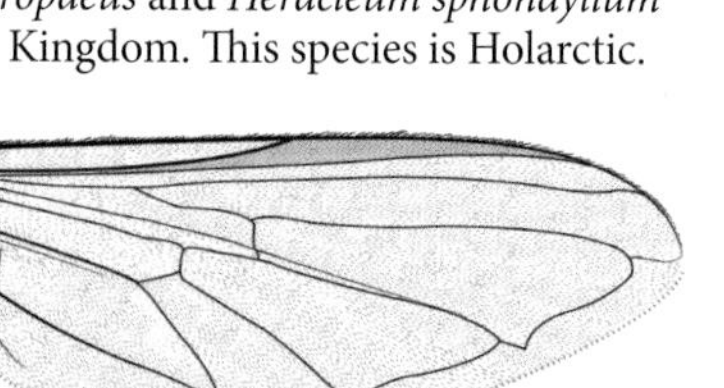

**wing entirely microtrichose in *M. labiatarum* and *M. arctica***

# *Melangyna lasiophthalma*

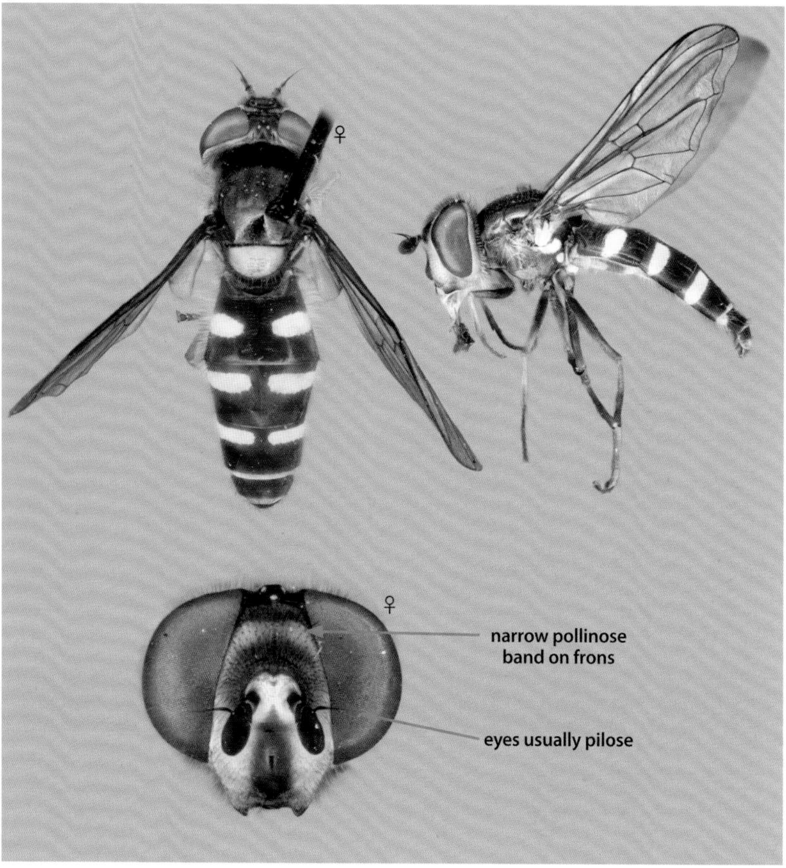

♀

♀

narrow pollinose band on frons

eyes usually pilose

# *Melangyna labiatarum*

## *Melangyna arctica* Pollinose Halfband

**SIZE RANGE:** 6.0–8.7 mm
**IDENTIFICATION:** Face yellow with broad black medial stripe to mostly dark, entirely covered in grayish pollinosity except sometimes facial tubercle bare. Wing densely microtrichose. Sternites also covered in gray pollinosity. **ABUNDANCE:** Common. **FLIGHT TIMES:** Early June to late August. **NOTES:** Adults are often found above the treeline in meadow, marsh, tundra and alpine habitats, also in forests. Larvae feed on aphids and in the United Kingdom are known to feed on arboreal aphids on *Alnus*. The biology of North American specimens is unknown.

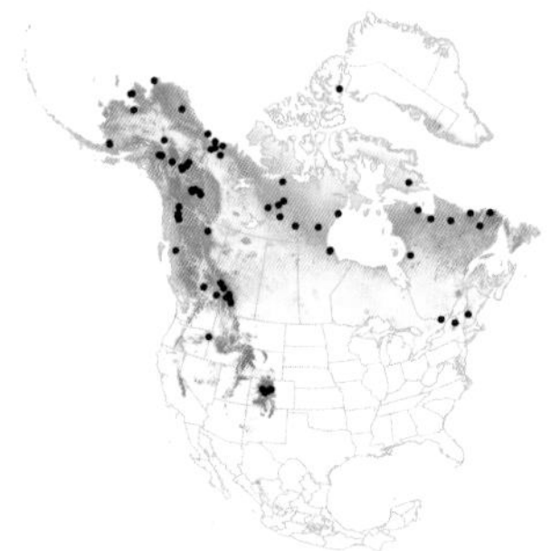

# *Dasysyrphus*

There are 45 recognized species of *Dasysyrphus*, occurring mainly in the Holarctic region. The Nearctic is home to 13 species, seven of which occur in the area covered by the guide. These black and yellow syrphines are easily confused with many other syrphine genera. The shape of the abdominal markings varies greatly in this genus. They are often curved, sometimes constricted medially, almost never meet medially, and may or may not meet the abdominal margin. The eyes are pilose and the wings are densely microtrichose, at least apically. Adults are often found on flowers in and around forests. Larvae are mottled to resemble bark and are often found on trees feeding on aphids and other soft-bodied insects. For a key to the species of the Nearctic, see Locke and Skevington (2013).

## *Melangyna arctica*

## *Dasysyrphus*

## *Dasysyrphus amalopis* Northern Conifer Fly

**SIZE RANGE:** 7.8–10.0 mm
**IDENTIFICATION:** This species is recognized by the yellow abdominal markings that are constricted medially, sometimes completely divided. Markings do not cross the abdominal margin. **ABUNDANCE:** Rare. **FLIGHT TIMES:** Late June to late July. **NOTES:** This is a northern species but was originally collected in the White Mountains, New Hampshire, presumably near the summit. Specimens have not been collected that far south since. Adults visit flowers and larvae are presumably aphid feeders like other *Dasysyrphus* species.

## *Dasysyrphus limatus*
Narrow-banded Conifer Fly

**SIZE RANGE:** 7.5–10.9 mm
**IDENTIFICATION:** This species is recognized by the small markings on tergite 2 and the straight, narrow markings on tergites 3 and 4, which cross the abdominal margin. **ABUNDANCE:** Uncommon. **FLIGHT TIMES:** Early May to late July (late April to mid-September in the west). **NOTES:** Adults are found in marsh, sand dune, boggy clearcut, and forest habitats. They have been recorded visiting *Heracleum* flowers. Larvae likely feed on aphids and other soft-bodied insects as other *Dasysyrphus* larvae do.

## *Dasysyrphus amalopis*

markings not crossing margin

markings constricted medially

## *Dasysyrphus limatus*

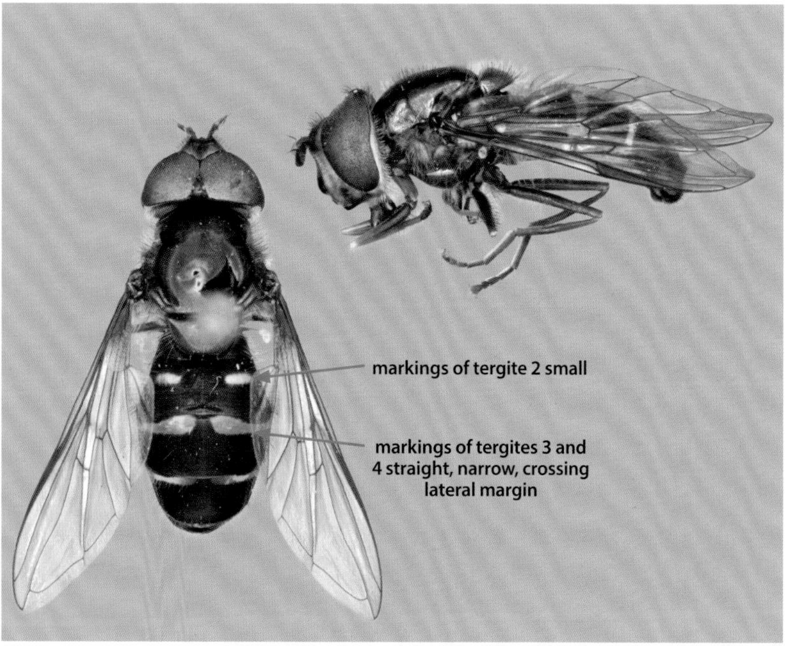

markings of tergite 2 small

markings of tergites 3 and 4 straight, narrow, crossing lateral margin

## *Dasysyrphus venustus*

### Transverse Conifer Fly

**SIZE RANGE:** 7.9–11.2 mm
**IDENTIFICATION:** This species is recognized by the large markings on tergite 2 and straight to slightly curved markings on tergites 3 and 4, which cross the abdominal margin. **ABUNDANCE:** Common. **FLIGHT TIMES:** Late April to mid-August. **NOTES:** Adults are found in forest, bog, and tundra habitats. They have been recorded from flowers of Apiaceae. Larvae are not known but likely feed on aphids and soft-bodied insects. The previous concept of *D. venustus* included the *D. intrudens* complex (below), which has now been split off. Old records in the literature containing this name must be closely scrutinized, as there is a chance that they are referring to a species in the *intrudens* complex.

## *Dasysyrphus intrudens*

### Confusing Conifer Fly

**SIZE RANGE:** 7.0–11.7 mm
**IDENTIFICATION:** This species, *D. venustus*, and the distinctive *D. limatus* are our only *Dasysyrphus* species with markings that cross the abdominal margin. The markings on tergite 2 are large and those on tergites 3 and 4 are variable. They are always curved and constricted medially, sometimes divided in ½. **ABUNDANCE:** Common. **FLIGHT TIMES:** Early May to early September (early January to mid-November in the west). **NOTES:** Adults are found in forest, meadow, bog, sandy, and tidal flat habitats. They have been recorded from flowers of Aceraceae, Apiaceae, Asteraceae, Betulaceae, Caprifoliaceae, Caryophyllaceae, Cornaceae, Ericaceae, Geraniaceae, Liliaceae, Onagraceae, Pinaceae, Poaceae, Polemoniaceae, Portulacaceae, Rosaceae, and Saxifragaceae. Larvae are not known but likely feed on aphids and soft-bodied insects. *Dasysyrphus intrudens* is certainly a complex of species, but the boundaries between those species are not clearly defined using morphology and DNA barcoding (mitochondrial gene COI). Rapidly evolving markers like ITS may be needed to help sort out species concepts. For now it is treated as a species complex.

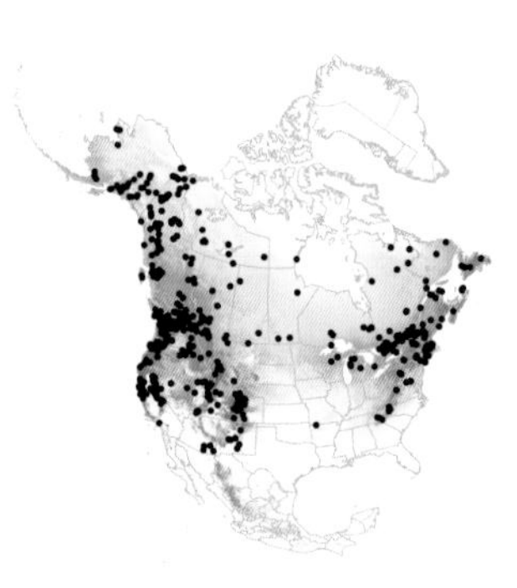

## *Dasysyrphus venustus*

markings on tergites 3 and 4 straight to slightly curved, crossing abdominal margin

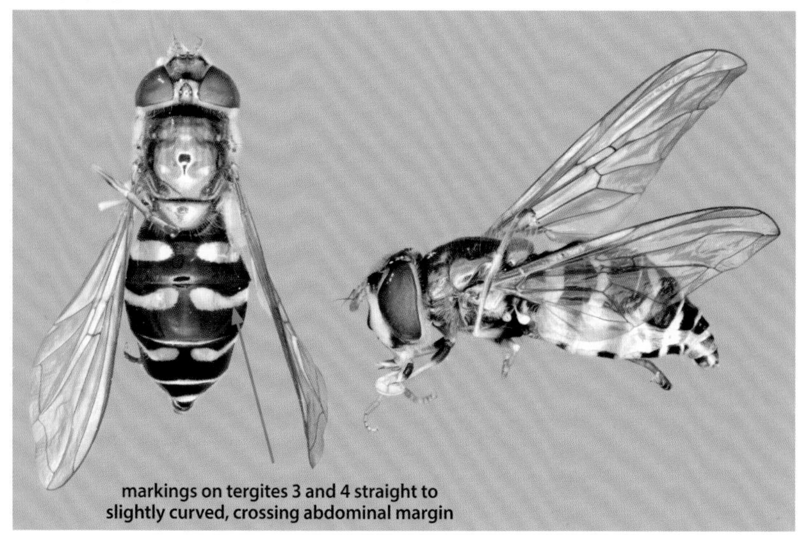

## *Dasysyrphus intrudens*

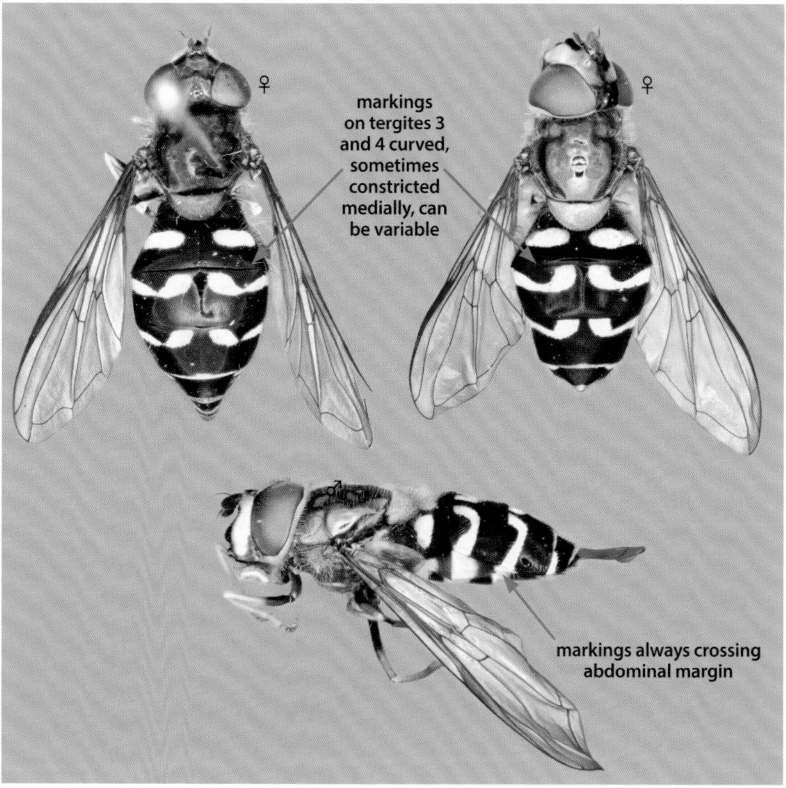
♀
♀
markings on tergites 3 and 4 curved, sometimes constricted medially, can be variable
♂
markings always crossing abdominal margin

## *Dasysyrphus laticaudus* Boreal Conifer Fly

**SIZE RANGE:** 5.0–8.2 mm

**IDENTIFICATION:** This species is recognized by markings on tergite 2 that extend to or toward the anterolateral edge. In females the markings reach the edge of the abdomen and in males the markings extend in a point toward the edge. Markings on tergites 3 and 4 are slightly curved and uniform in thickness, not crossing the abdominal margin. **ABUNDANCE:** Common. **FLIGHT TIMES:** Mid-April to late July. **NOTES:** Adults are found in forest habitats throughout the boreal region. Larvae are likely to feed on aphids.

## *Dasysyrphus nigricornis* Dusky Conifer Fly

**SIZE RANGE:** 5.5–7.7 mm

**IDENTIFICATION:** This species appears darker than others. Markings on tergite 2 of male are reduced to small lateral spots. In females they are larger but narrower. Markings on tergites 3 and 4 are slightly oblique and narrower medially, not crossing the abdominal margin, and narrower in females. **ABUNDANCE:** Uncommon. **FLIGHT TIMES:** Late May to late July. **NOTES:** This species is Holarctic. In Europe adults are known to visit *Caltha*, *Ranunculus*, *Rhododendron*, *Salix*, and *Taraxacum*. Larvae likely feed on aphids.

## *Dasysyrphus pinastri* Black-spotted Conifer Fly

**SIZE RANGE:** 8.4–10.2 mm

**IDENTIFICATION:** This species is similar to the two above in having slightly arcuate markings on tergites 3 and 4 that do not cross the abdominal margin. The black oval marking on sternite 2 is diagnostic (banded in other species). **ABUNDANCE:** Rare (common in Europe). **FLIGHT TIMES:** August (from March in Europe). **NOTES:** This species is mainly Palearctic, but found in Greenland in the Nearctic. Adults are flower visitors and larvae feed on aphids.

## *Dasysyrphus laticaudus*

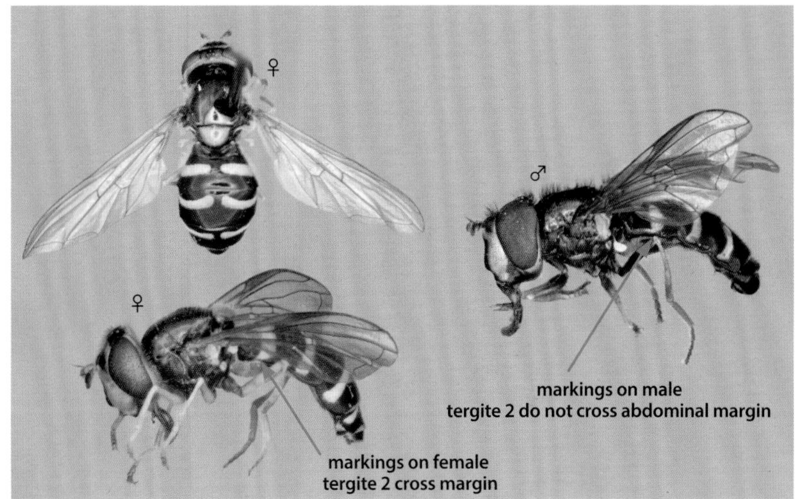


## *Dasysyrphus nigricornis*

## *Dasysyrphus pinastri*

# Scaeva

Twenty species of *Scaeva* can be found throughout the Palearctic, Nearctic, Neotropical, and Indomalayan regions. Two species have been recorded in the Nearctic. *Scaeva affinis* (in literature often referred to as *S. pyrastri*, but this is a Palearctic species) is found throughout the west with a few records known from the eastern Nearctic. The other species, *S. selenitica*, has been introduced to North Carolina from the Palearctic, but may not be established. *Scaeva* species are characterized by having tergites with paired, curved, yellow markings, a pilose eye, extensively bare wings (sparsely microtrichose on apical third), and wing vein $R_{4+5}$ curving up into cell $r_{2+3}$. Larvae are known to feed on arboreal and/or ground-layer aphids, with some species feeding on aphids on a wide variety of plants. *Scaeva pyrastri* in Europe and *S. affinis* are often used in biological control. *Scaeva* species are known to be migratory in Europe.

## *Scaeva affinis*

### White-bowed Smoothwing

**SIZE RANGE:** 11.1–15.7 mm

**IDENTIFICATION:** Since the only *Scaeva* species likely to be encountered in the Nearctic is *S. affinis*, the generic characters outlined above diagnose it. In the event that *S. selenitica* becomes established and spreads, the two species can be distinguished by the angle of the bands on tergites 3 and 4. Bands are oblique in *S. affinis* and parallel to the anterior margin of the tergites in *S. selenitica*. **ABUNDANCE:** Very rare in the east (common in the west). **FLIGHT TIMES:** Year round in the west. **NOTES:** Adults are known from forest, field, prairie and tundra habitats. They have been recorded visiting *Androsace*, *Hackelia*, *Leucanthemum*, *Pedicularis*, *Picea*, *Rosa*, *Solidago*, and *Symphoricarpos*. Larvae feed on aphids on *Acer negundo*, *Baccharis pilularis*, *Beta vulgaris*, *Brassica oleracea*, *Capsicum*, *Chamerion angustifolium*, *Erodium*, *Juglans*, *Lactuca serriola*, *Malus*, *Medicago sativa*, *Pinus ponderosa*, *Pisum sativum*, *Prunus persica*, *Rosa*, *Setaria viridis*, *Spinacia*, *Spiraea prunifolia*, *Symphoricarpos albus*, *Trifolium*, and *Vicia*. *Scaeva affinis* was considered synonymous with *S. pyrastri* of the Palearctic; however, DNA evidence indicates that the Nearctic taxon is a distinct species.

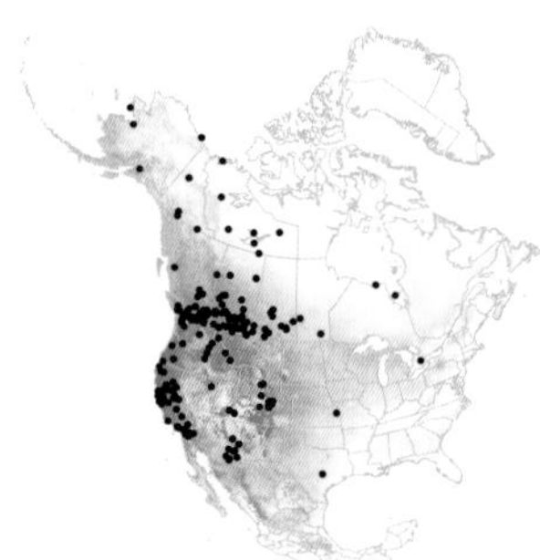

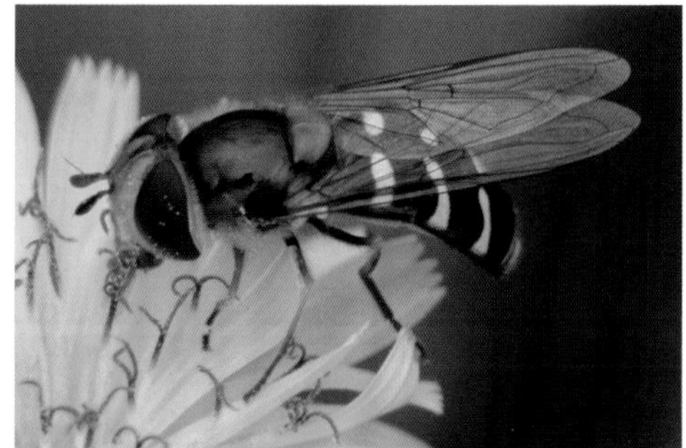

# *Scaeva affinis*

# *Lapposyrphus*

*Lapposyrphus* has previously been recognized as a subgenus of *Metasyrphus* and *Eupeodes*. Diagnostic features of this genus include $R_{4+5}$ curving up into cell $r_{2+3}$, bare eyes, and wings that are densely microtrichose on at least the apical third. There are two recognized species of *Lapposyrphus* worldwide. Only *L. lapponicus* is found in the area covered by this guide. The other species, *L. aberrantis*, is found in western North America only and has complete yellow bands on tergites 3 and 4.

## *Lapposyrphus lapponicus*

**Common Loopwing Aphideater**

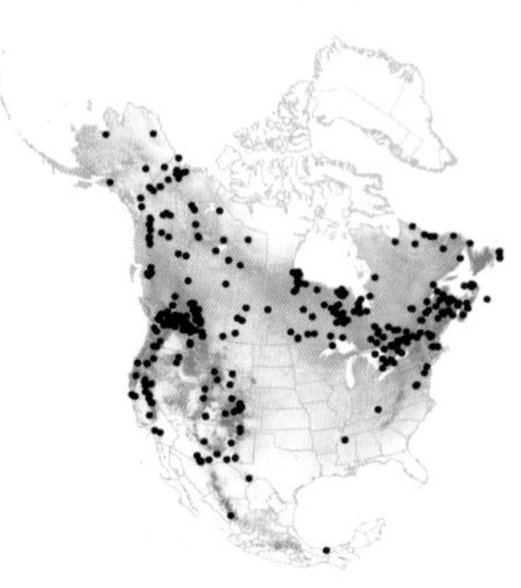

**SIZE RANGE:** 8.3–14.1 mm
**IDENTIFICATION:** Tergites 3 and 4 usually each with a pair of curved, yellow markings, but sometimes entirely black. The western species, *L. aberrantis*, has undivided yellow bands on tergites 3 and 4.
**ABUNDANCE:** Common. **FLIGHT TIMES:** Mid-April to mid-September (early January to late November in the west). **NOTES:** This species is Holarctic. Adults can be found mainly in forest habitats but have been recorded in fields, meadows, bogs, marshes, taiga, and tundra. Flowers visited include Apiaceae, Asteraceae, Caprifoliaceae, Ericaceae, Euphorbiaceae, Grossulariaceae, Hydrophyllaceae, Oleaceae, Onagraceae, Papaveraceae, Pinaceae, Primulaceae, Ranunculaceae, Rosaceae, and Salicaceae. Larvae feed on aphids and adelgids found on *Abies*, *Cedrus*, *Euonymus*, *Fagus*, *Gleditsia triacanthos*, *Larix*, *Malus*, *Picea*, *Pinus*, *Prenanthes purpurea*, and *Rhododendron*. This species is migratory.

# *Lapposyrphus lapponicus*

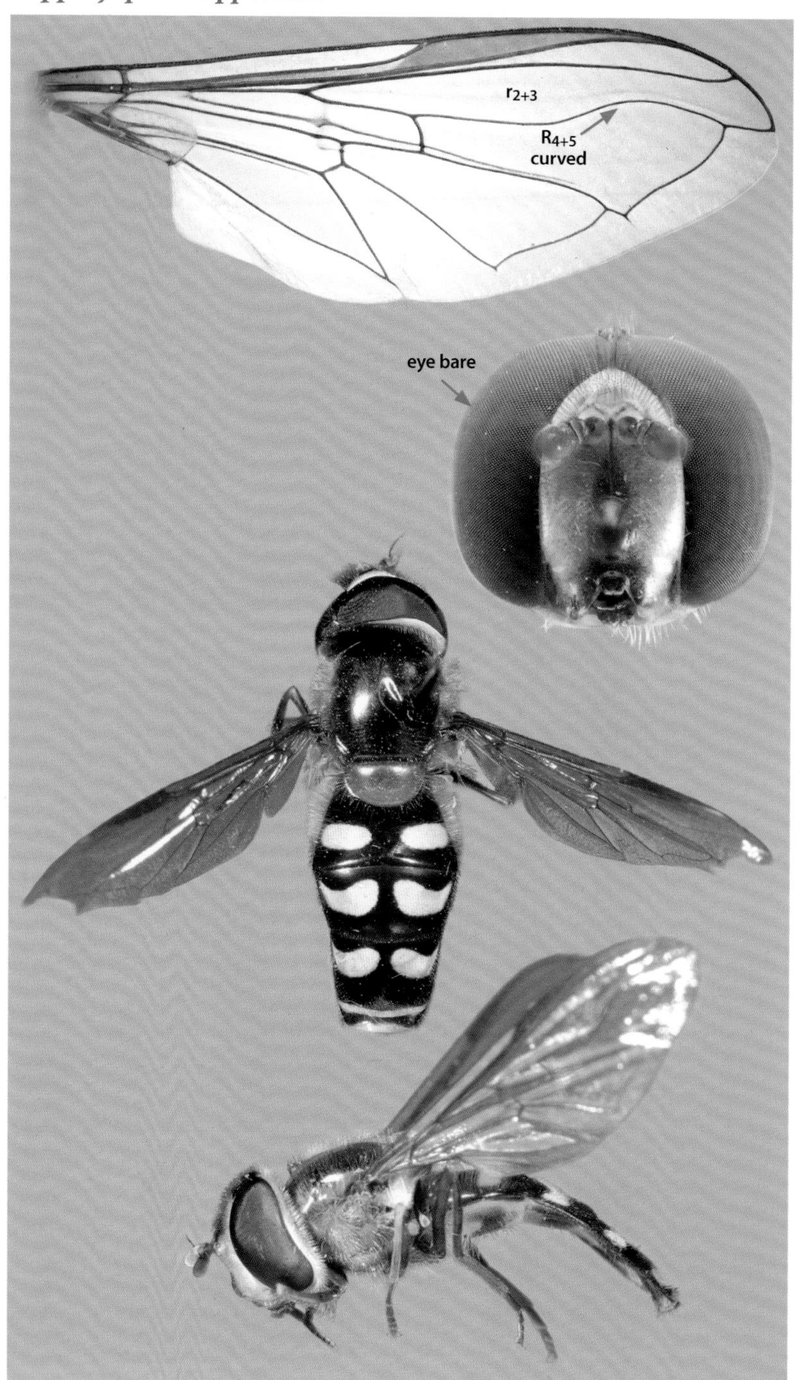

# Eupeodes

*Eupeodes* species are recognized by their pilose metasternum and margined abdomen. They have either yellow markings or bands on tergites 3 and 4. Adults are flower visitors and can be found in a wide variety of habitats. Larvae are often found feeding on arboreal aphids; however, some species are more generalist feeders and also feed on ground-layer aphids found on many species of shrubs and herbaceous plants. Many, if not all, species are multivoltine, and migratory behavior has been documented in one Old World species. There are 90 species in this predominantly Holarctic genus. Twenty-four species are known from the Nearctic, 11 in our region. Species can be difficult to tell apart and more scrutiny needs to be given to the species concepts of both the Nearctic and Palearctic species. For a key to species of northern North America, see Vockeroth (1992).

## *Eupeodes volucris* Large-tailed Aphideater

**SIZE RANGE:** 6.3–9.8 mm
**IDENTIFICATION:** Male genitalia visible dorsally as cylindrical protrusion from apex of abdomen. Tergites 3 and 4 with a pair of curved markings, wing extensively bare, and female sternites 4 and 5 yellow orange. **ABUNDANCE:** Common in the west; rare in the east. **FLIGHT TIMES:** Early May to mid-August (early January to late November in the west). **NOTES:** Adults are found in forest, prairie, meadow, field, tundra, fen, marsh, badland, and tidal flat habitats. They have been recorded from flowers of *Acer*, Fabaceae, *Hieracium*, *Leucanthemum*, *Penstemon*, and *Viburnum*. Larvae feed on aphids and Lepidoptera larvae found on Amaranthaceae, Anacardiaceae, Apiaceae, Asteraceae, Brassicaceae, Caprifoliaceae, Celastraceae, Cornaceae, Cucurbitaceae, Fabaceae, Fagaceae, Grossulariaceae, Juglandaceae, Pinaceae, Poaceae, Polygonaceae, Rosaceae, Rutaceae, Salicaceae, Sapindaceae, Solanaceae, Styracaceae, Typhaceae, and Ulmaceae. There are multiple generations per year.

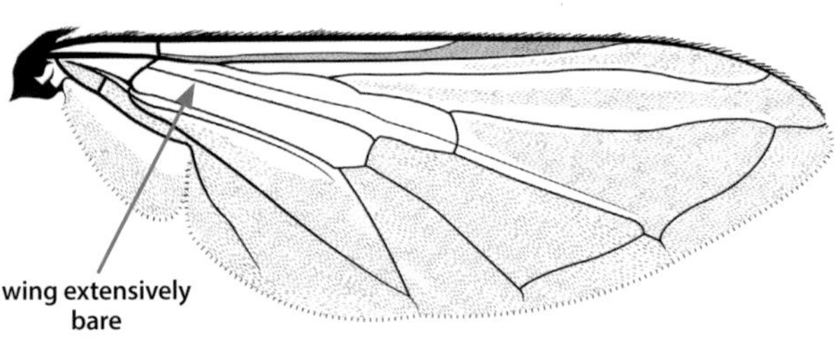

# Eupeodes

mesoleg
metasternum pilose
metaleg
abdomen margined

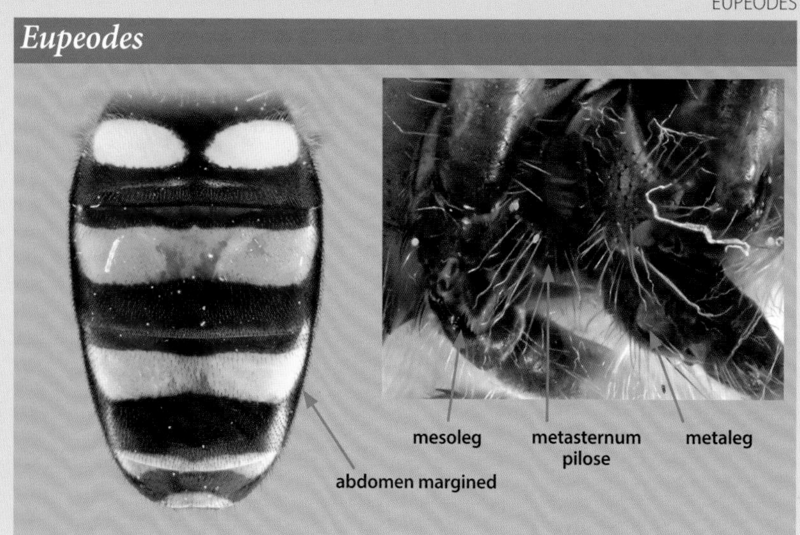

## Eupeodes volucris

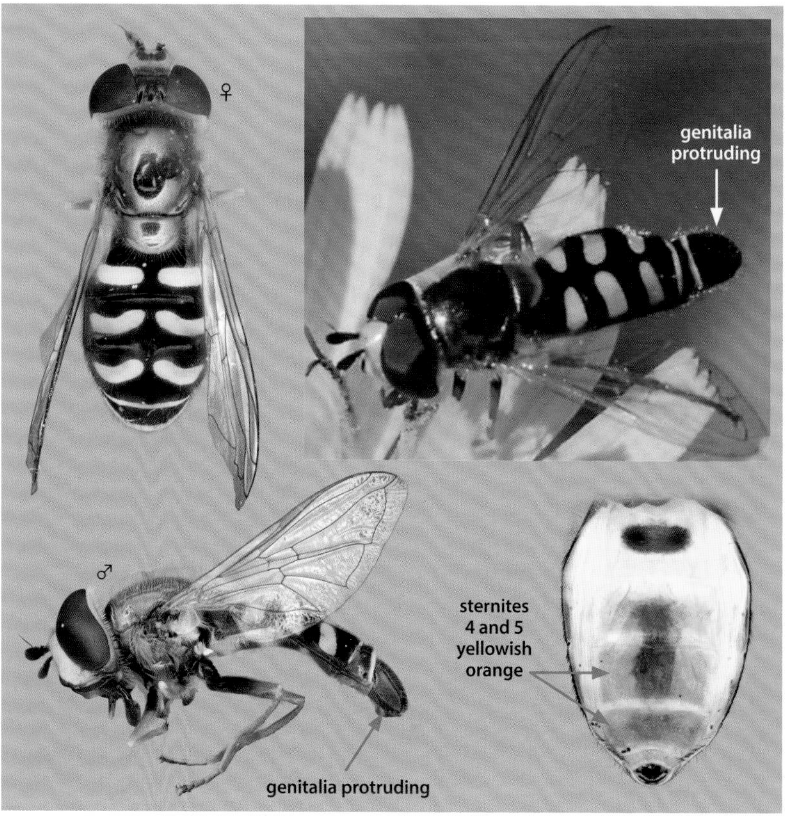
♀
genitalia protruding
♂
genitalia protruding
sternites 4 and 5 yellowish orange

## *Eupeodes flukei* Fluke's Aphideater

**SIZE RANGE:** 7.3–11.6 mm

**IDENTIFICATION:** This is one of two species in this region with tergite 5 entirely yellow orange (see also *Eupeodes abiskoensis*). Unlike in *E. abiskoensis*, the alula in *E. flukei* is always extensively bare anteriorly. Tergites 3 and 4 typically with pair of curved markings, sometimes meeting medially. **ABUNDANCE:** Rare. **FLIGHT TIMES:** Early May to early August (to early September in the west). **NOTES:** A northerly/alpine species that gets into northern Ontario. They have been found in scrub habitat and at the treeline on mountains. Larvae are unknown but are likely aphidophagous, like other *Eupeodes* larvae.

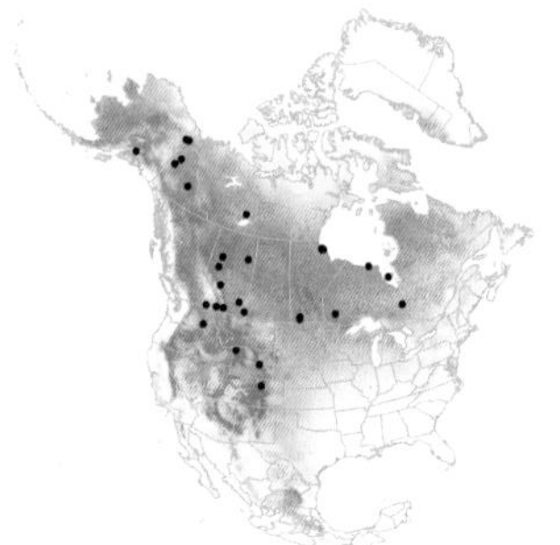

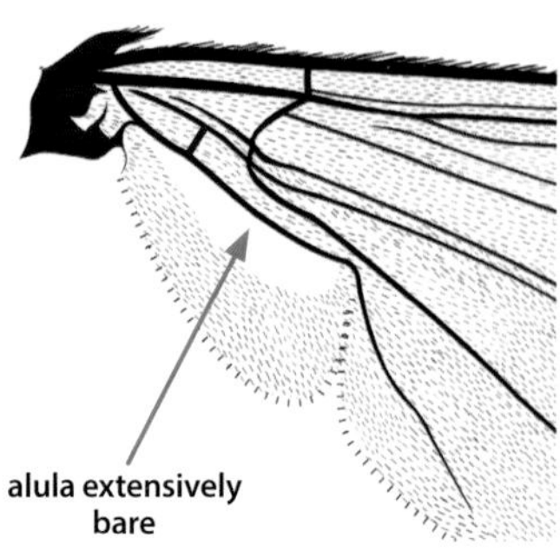


## *Eupeodes abiskoensis* Abisko Aphideater

**SIZE RANGE:** 7.6–9.9 mm

**IDENTIFICATION:** This species is one of the two *Eupeodes* species in this region with tergite 5 entirely or almost entirely yellow (see *E. flukei*). Unlike *E. flukei*, *E. abiskoensis* sometimes has a dark medial spot on tergite 5 and the alula is always entirely microtrichose. **ABUNDANCE:** Rare. **FLIGHT TIMES:** Mid-June to mid-August. **NOTES:** This is a circumpolar, Arctic species. Adults have been recorded in subalpine and alpine heathland habitats, as well as open coniferous and mixed forests. Flowers visited include *Bartsia alpina*, *Pinguicula*, and *Salix*. Larvae are unknown but are likely aphidophagous like other *Eupeodes* larvae.

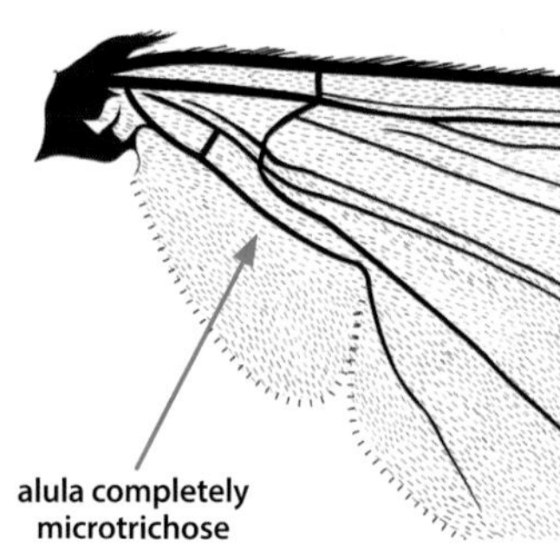

# *Eupeodes flukei*

tergite 5 entirely yellow orange

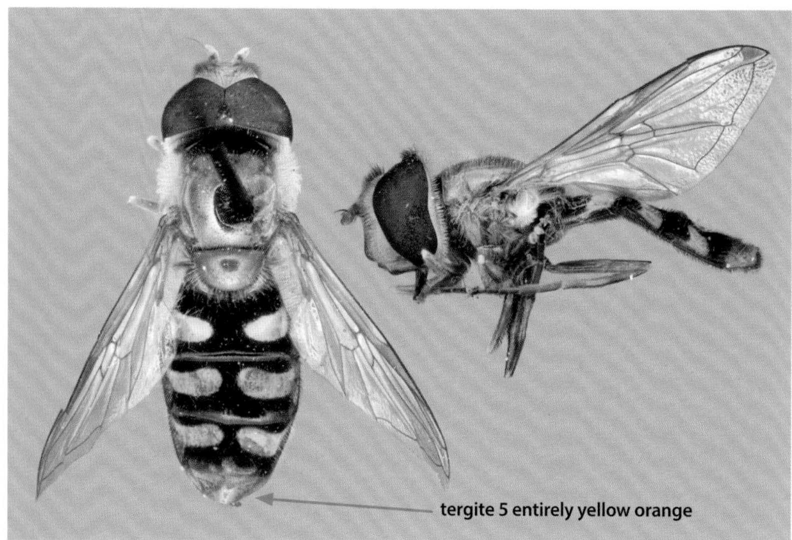

# *Eupeodes abiskoensis*

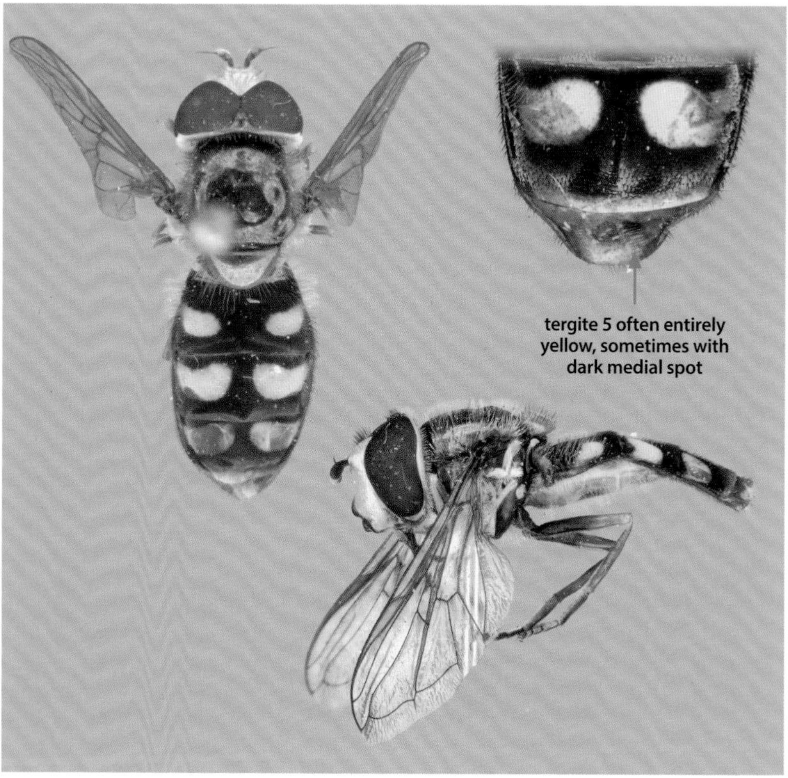
tergite 5 often entirely yellow, sometimes with dark medial spot

## *Eupeodes nigroventris* Black Aphideater

**SIZE RANGE:** 9.0–11.4 mm
**IDENTIFICATION:** Abdomen with reduced pairs of markings or entirely black. Face yellow with black oral margin and black medial stripe extending to tubercle only. **ABUNDANCE:** Very rare. **FLIGHT TIMES:** Early July to early August. **NOTES:** An Arctic species, also found in Iceland. Adults visit flowers. Larvae have been collected in Greenland from *Salix* where they were likely feeding on aphids or coccids. Böcher *et al.* (2015) applied the name *Eupeodes rufipunctatus* to this species. The type for *E. nigroventris* is from Greenland while the type locality for *E. rufipunctatus* is Lillooet, British Columbia. The latter is outside of the range of this species and Vockeroth (1986a) considered the type female specimen of *E. rufipunctatus* to simply be a melanic morph, possibly of *E. luniger*. We follow Vockeroth and apply the name based on the Greenland type here.

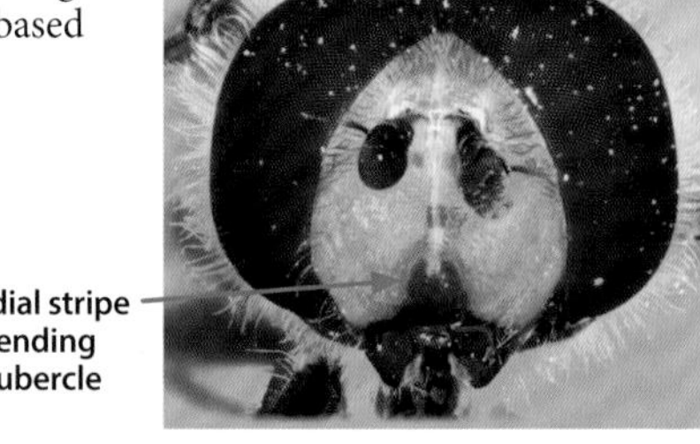


## *Eupeodes perplexus* Bare-winged Aphideater

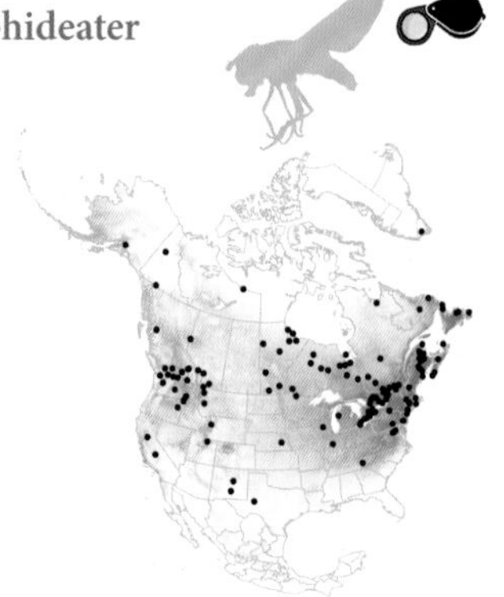

**SIZE RANGE:** 9.7–13.0 mm
**IDENTIFICATION:** Wing extensively bare on basal ½. Tergites 3 and 4 with pair of curved markings. Male genitalia small, not projecting beyond the end of the abdomen. Females with black band on sternites 4 and 5. **ABUNDANCE:** Common. **FLIGHT TIMES:** Late April to late October (from late March in the west). **NOTES:** Adults live in forest, sand dune, and riparian habitats. They are known to visit flowers of *Lactuca*, *Medicago*, *Packera*, and *Trifolium*. Larvae feed on aphids and psyllids on *Bidens*, *Celtis*, *Chrysanthemum*, *Lactuca*, *Populus*, *Rubus*, and *Verbesina*.

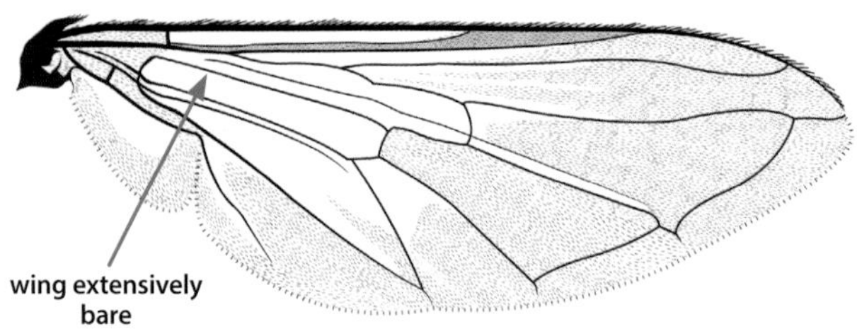

## Eupeodes nigroventris

abdomen black or with reduced markings

## Eupeodes perplexus

male genitalia not projecting beyond end of abdomen

sternites with black bands

♀

## *Eupeodes curtus* Comma-spot Aphideater

**SIZE RANGE:** 7.9–11.5 mm
**IDENTIFICATION:** Tergites 3 and 4 with pairs of yellow, concave markings. Wing not extensively bare as in *E. perplexus*. Alula entirely trichose. **ABUNDANCE:** Common. **FLIGHT TIMES:** Late May to early August (to late September in the west). **NOTES:** This northern Holarctic species can be found in Iceland, Norway, Sweden, Finland, and Russia as well. Adults are found in forest, field, open ground taiga, tundra, and marsh habitats. They have been recorded on *Caltha*, *Leucanthemum*, and *Taraxacum*. The larvae are not known but they probably feed on aphids, a typical habit for *Eupeodes*. Böcher *et al.* (2015) apply the name *E. punctifer* to this species. It is likely a synonym but types need to be checked to confirm this.

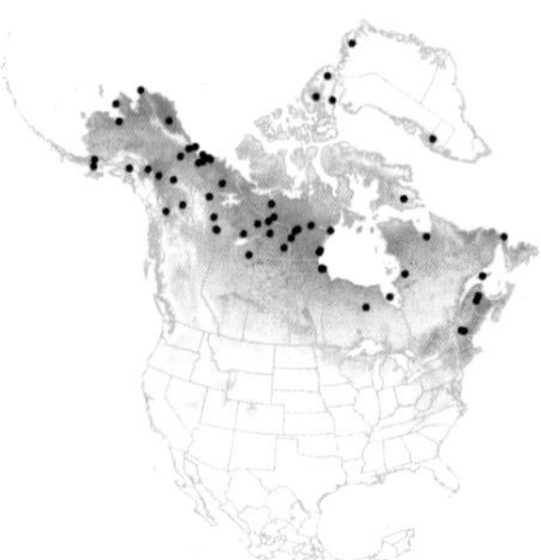

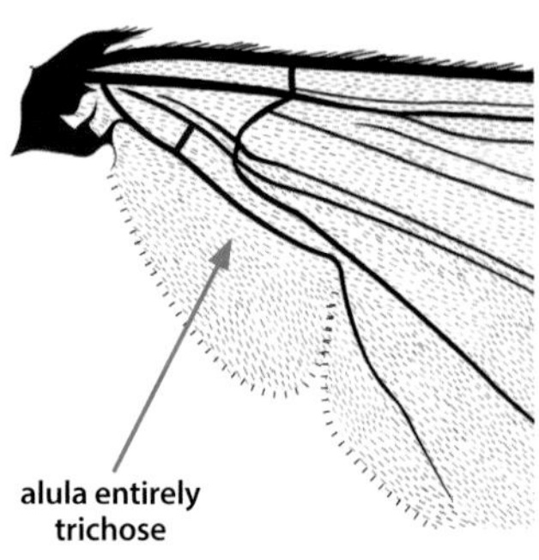


## *Eupeodes luniger* Black-tailed Aphideater

**SIZE RANGE:** 7.8–11.5 mm
**IDENTIFICATION:** Tergites 3 and 4 with pair of curved markings. Wing not extensively bare as in *E. perplexus*. Alula is bare anteriorly. Males with basiphallus with one short, arcuate tooth and one longer arcuate tooth. **ABUNDANCE:** Common. **FLIGHT TIMES:** Mid-March to early October. **NOTES:** Adults are found in forest, grassland, open ground, fen, bog, agricultural, and horticultural habitats. Flowers visited include Apiaceae, *Calluna*, *Fragaria*, *Leontodon*, *Malus*, *Polygonum*, *Prunus*, *Ranunculus*, *Rosa*, *Senecio*, and *Taraxacum*. Larvae feed on aphids on a wide variety of plants, from herbaceous plants to shrubs to trees and agricultural crops. Böcher *et al.* (2015) incorrectly apply the name *E. vockerothi* to this species.

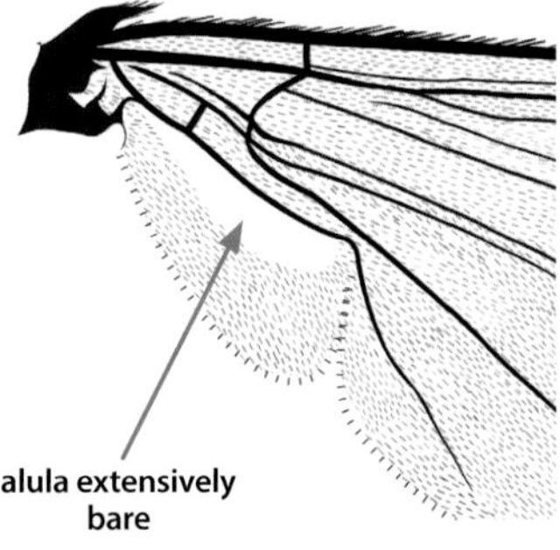

# *Eupeodes curtus*

anterior margin of markings concave

# *Eupeodes luniger*

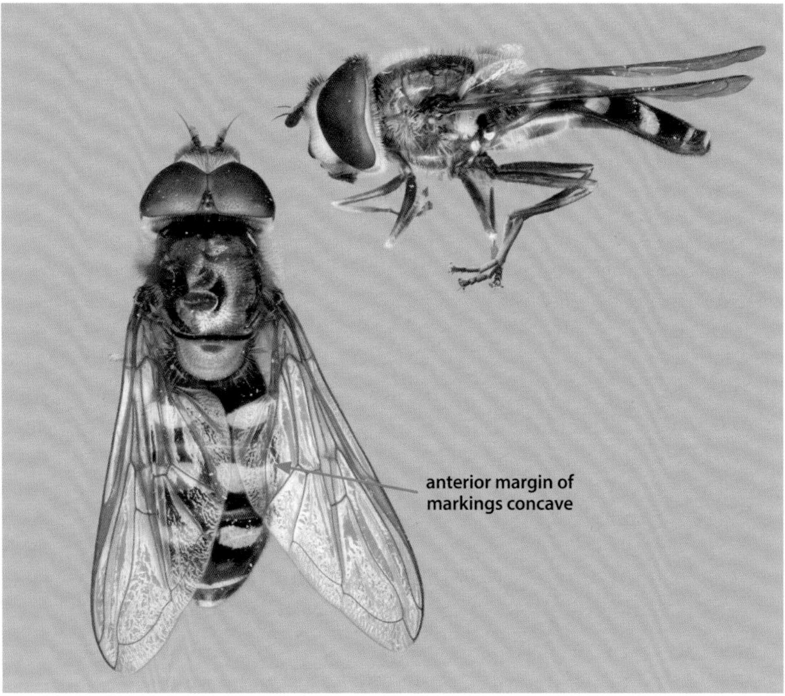

anterior margin of markings concave

## *Eupeodes latifasciatus* Variable Aphideater

**SIZE RANGE:** 7.0–10.1 mm

**IDENTIFICATION:** Yellow bands of tergites 3 and 4 entire, constricted or divided medially, anterior margins usually straight. Alula entirely trichose. Sternites 2–4 typically with black bands that do not reach the sternite edge. **ABUNDANCE:** Common. **FLIGHT TIMES:** Mid-April to mid-September (to late October in the west). **NOTES:** This widespread species occurs throughout the Palearctic and Nearctic regions; it has also been found in India. Adults are associated with wet habitats and have been found in forests, meadows, and fens in open areas, among low-growing vegetation or on moist soil. They have been recorded visiting flowers of Apiaceae, *Caltha*, *Circaea*, *Convolvulus*, *Euphorbia*, *Prunus*, *Ranunculus*, *Salix*, *Solidago*, *Taraxacum*, *Tussilago*, and *Ulex*. Larvae are known to feed on root aphids.

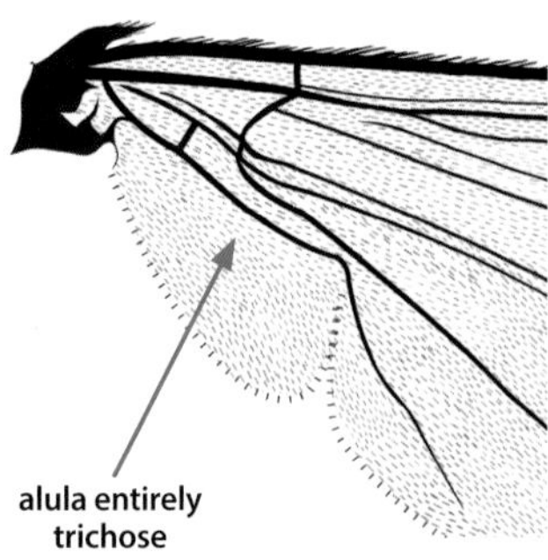


## *Eupeodes confertus* Black-bellied Aphideater

**SIZE RANGE:** 8.7–10.3 mm

**IDENTIFICATION:** Tergites 3 and 4 with single yellow bands. Alula entirely trichose. Sternites 2–4 with broad black bands that reach the sternite edge. **ABUNDANCE:** Very rare. **FLIGHT TIMES:** Late April to early September. **NOTES:** Originally described as a subspecies of *E. nitens*, a Palearctic species, this species has not been treated in later works and needs to be reviewed. The larvae have been recorded feeding on the aphid *Plocamaphis flocculosa* on *Salix*.

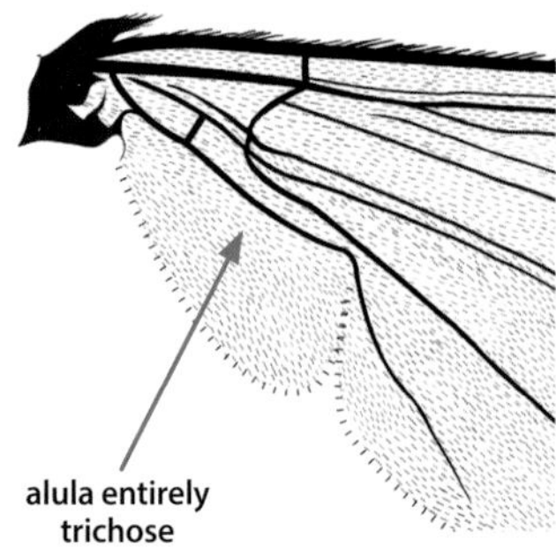

## *Eupeodes latifasciatus*

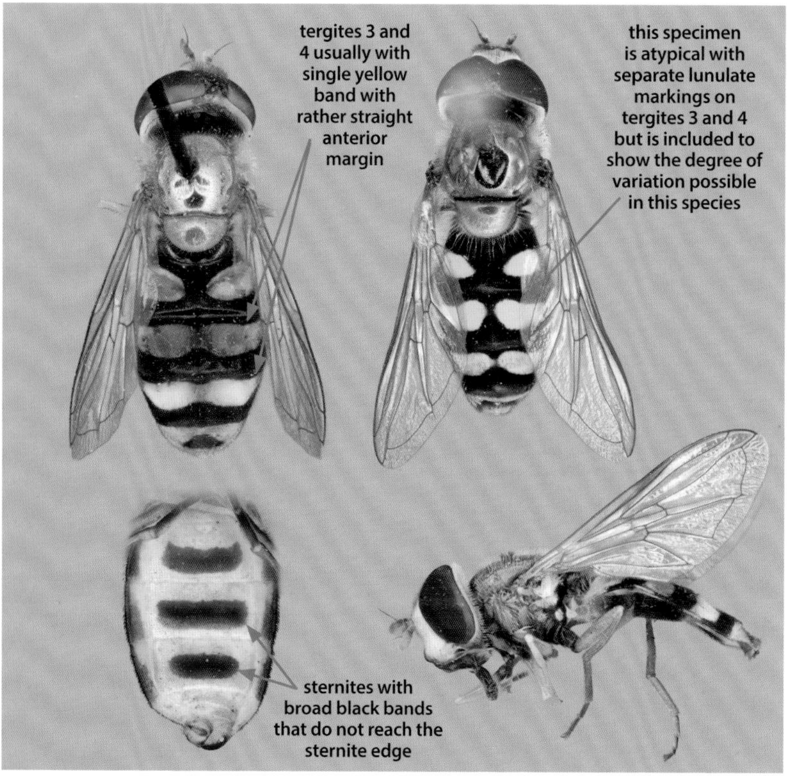

tergites 3 and 4 usually with single yellow band with rather straight anterior margin
this specimen is atypical with separate lunulate markings on tergites 3 and 4 but is included to show the degree of variation possible in this species
sternites with broad black bands that do not reach the sternite edge

## *Eupeodes confertus*

yellow bands
sternites with broad black bands that reach the sternite edge

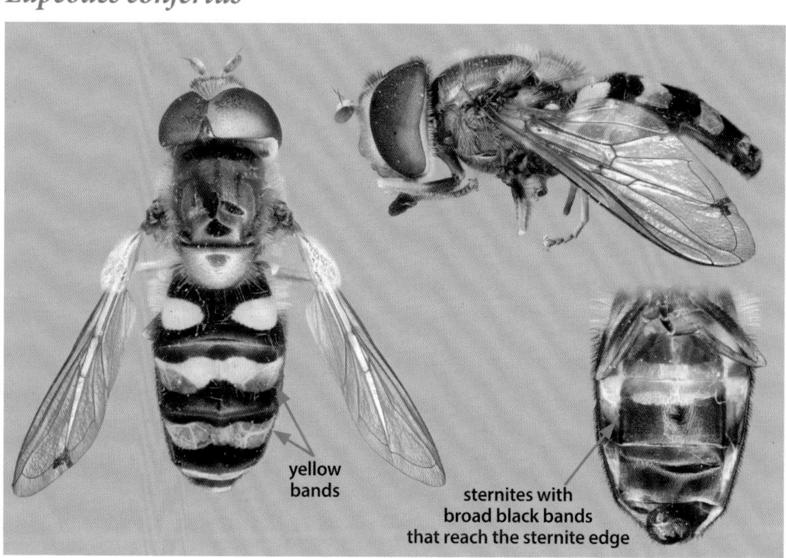

## *Eupeodes americanus* Long-tailed Aphideater

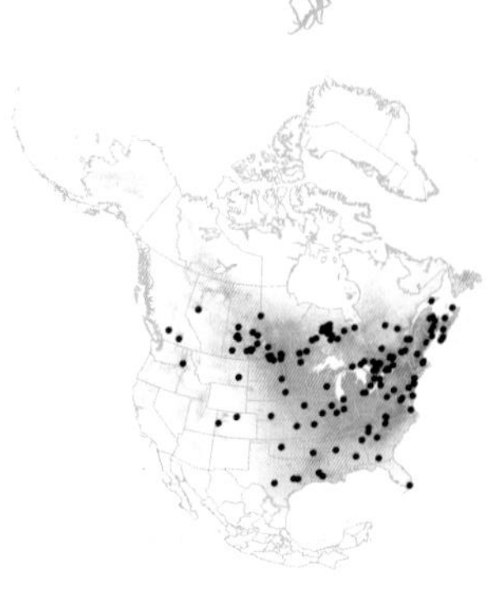

**SIZE RANGE:** 7.0–11.0 mm
**IDENTIFICATION:** Tergites with yellow bands that do not meet the abdominal margin and are generally not constricted medially. The alula is bare anteriorly. Males have surstyli that are elongate and twisted. Females are indistinguishable from female *E. pomus*. **ABUNDANCE:** Common. **FLIGHT TIMES:** Mid-April to mid-November (from mid-March in the south). **NOTES:** Adults have been found in forests, fields, plantations, open areas, gardens, and sand dunes. Flowers visited include *Chrysanthemum*, *Eupatorium*, *Medicago*, *Sonchus*, *Spiraea*, and *Symphyotrichum*. Larvae have been recorded feeding on aphids and adelgids from a wide variety of plants, including *Chrysanthemum*, *Oenothera*, *Phragmites*, *Rumex*, *Sambucus*, and *Trifolium*, as well as many crops and tree species.

## *Eupeodes pomus* Short-tailed Aphideater

**SIZE RANGE:** 6.8–12.0 mm
**IDENTIFICATION:** Tergites with yellow bands that do not meet the abdominal margin. Alula bare anteriorly. Surstylus short. Females indistinguishable from female *E. americanus*. **ABUNDANCE:** Uncommon. **FLIGHT TIMES:** Mid-May to late October (early March to late November in the south). **NOTES:** Adults are found in forests and open habitats. They have been recorded from *Fagopyrum esculentum*. Larvae have been recorded feeding on aphids from *Acer negundo*, *Brassica*, *Malus pumila*, *Oenothera*, *Phragmites*, *Populus grandidentata*, *Rhus*, *Rumex crispus*, *Salix nigra*, *Triticum*, and *Verbesina*.

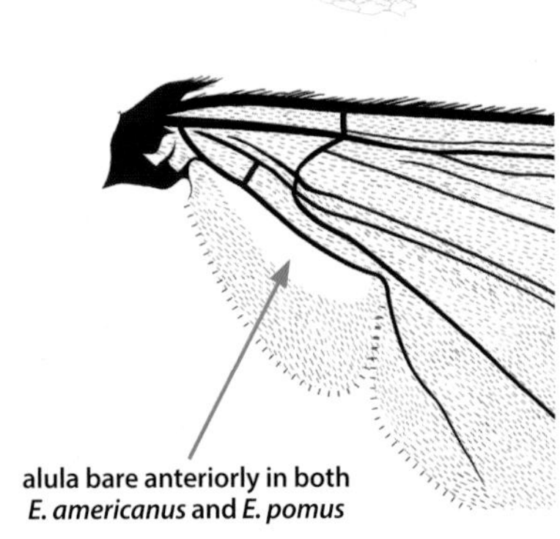

# *Eupeodes americanus* and *pomus*

*E. americanus* surstyli elongate and twisted

*E. pomus* surstyli short

# Epistrophe

*Epistrophe* species are yellow and black–banded syrphines. Species in this genus can have the metasternum pilose, sparsely pilose, or bare. The abdomen is weakly margined and the katepisternal pile patches are narrowly joined posteriorly. There are 50 species found throughout the Nearctic, Palearctic and Indomalayan regions. Five are found in the Nearctic and they are all found in our region. A sixth species, *E. ochrostoma*, is Palearctic with questionable Nearctic records. Adults are generalist flower visitors. Most of the larvae of the North American species are not known but other larvae in this genus are green and flattened aphid predators. Evidence suggests that members of the genus *Epistrophe* are univoltine, requiring a larval diapause. For a key to species of northern North America, see Vockeroth (1992).

## *Epistrophe metcalfi* Black-margined Smoothtail

**SIZE RANGE:** 11.0–12.3 mm

**IDENTIFICATION:** This species is easily identifiable by the yellow abdominal bands on tergites 2–4 that do not reach the margin, causing the entire abdominal margin to be black. **ABUNDANCE:** Very rare. **FLIGHT TIMES:** Mid-May to early August (from late March in the south). **NOTES:** The most recent specimen examined was collected in 1948. It is likely the larvae feed on aphids as other *Epistrophe* larvae do.

# *Epistrophe*

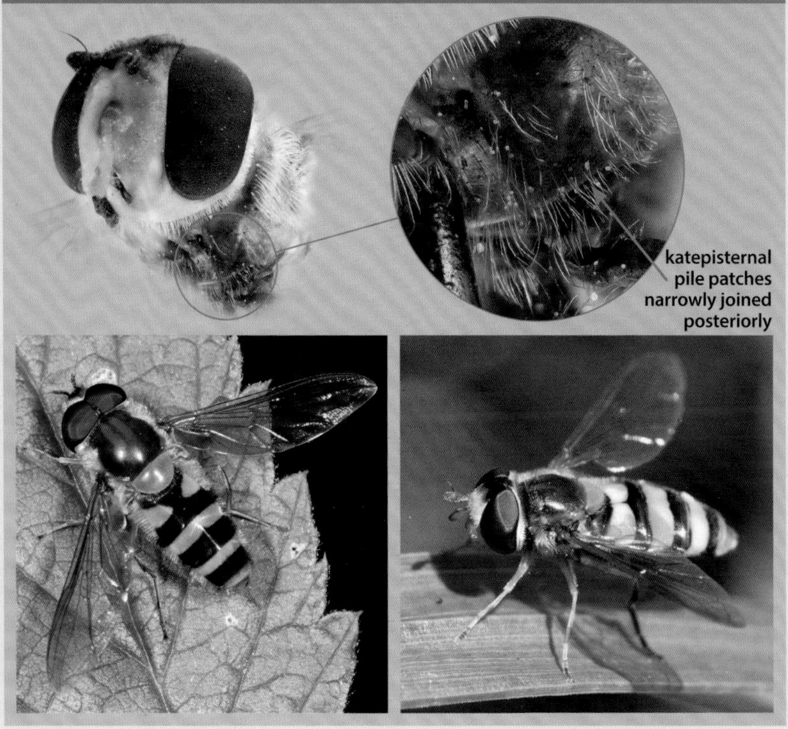
katepisternal
pile patches
narrowly joined
posteriorly

## *Epistrophe metcalfi*

abdomen black margined
as yellow bands do not
reach the edge

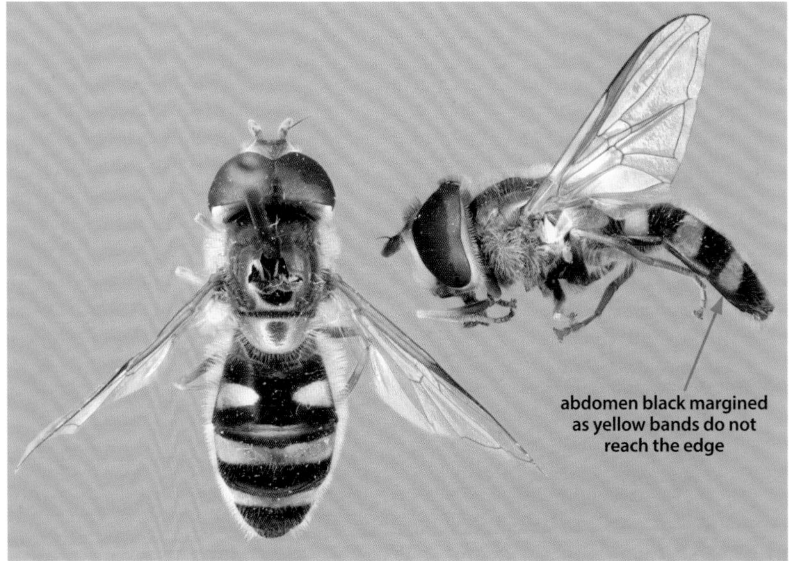

## *Epistrophe grossulariae*

### Black-horned Smoothtail

**SIZE RANGE:** 10.4–15.0 mm
**IDENTIFICATION:** This species is easily identifiable by the black antennae. Other species have antennae at least partially yellow. The long black isosceles triangle above the antennae (no yellow directly above antennae) is also diagnostic, and this is our only *Epistrophe* species with a pilose metasternum. **ABUNDANCE:** Common. **FLIGHT TIMES:** Early May to early October. **NOTES:** This is a Holarctic species. Adults are found in forests, meadows, fields, fens, and along streams. Flowers visited include *Anaphalis*, Apiaceae, *Centaurea*, *Cirsium*, *Eutrochium*, *Filipendula*, *Geranium*, *Knautia*, *Rhododendron*, *Rubus*, *Sambucus*, *Solidago*, *Succisa*, and *Valeriana*. Larvae are known to feed on arboreal aphids and aphids on *Brassica* and *Triticum* crops. DNA studies suggest that this is a complex of two species, both found in our region.

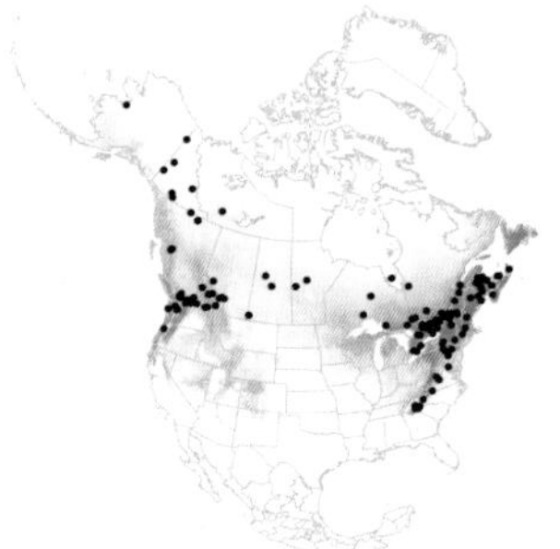

## *Epistrophe nitidicollis*

### Straight-banded Smoothtail

**SIZE RANGE:** 9.3–13.0 mm
**IDENTIFICATION:** This species has yellow antennae and wing cell bm is bare on the anterior ½. The black coxae distinguish *E. nitidicollis* from *E. xanthostoma*. The bands on tergites 3 and 4 are typically straight posteriorly. **ABUNDANCE:** Uncommon. **FLIGHT TIMES:** Late April to late July. **NOTES:** Adults are found in forest, bog, and fen habitats. They have been recorded visiting flowers of *Euphorbia*, *Malus*, and *Salix*. Larvae are known to feed on aphids on Adoxaceae, Amaranthaceae, Asteraceae, Caprifoliaceae, Celastraceae, Fabaceae, Grossulariaceae, Poaceae, Polygonaceae, and Rosaceae. DNA studies suggest that this is a complex of three species, two of which are found in the northeast.

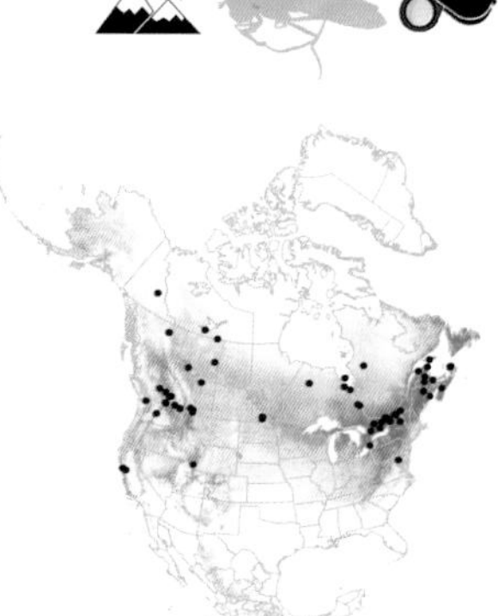

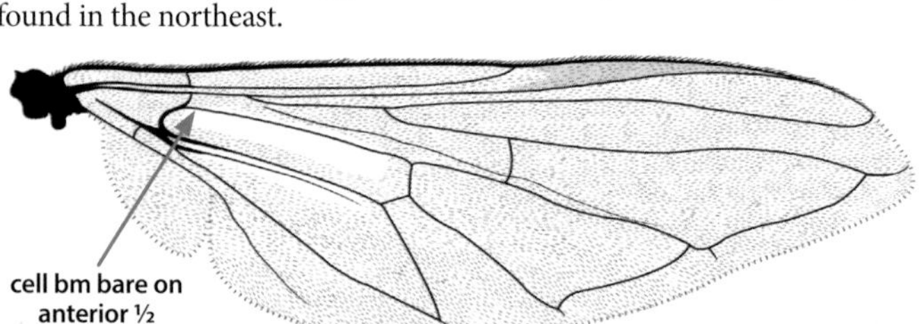

# *Epistrophe grossulariae*

antennae black
yellow bands
cross margin

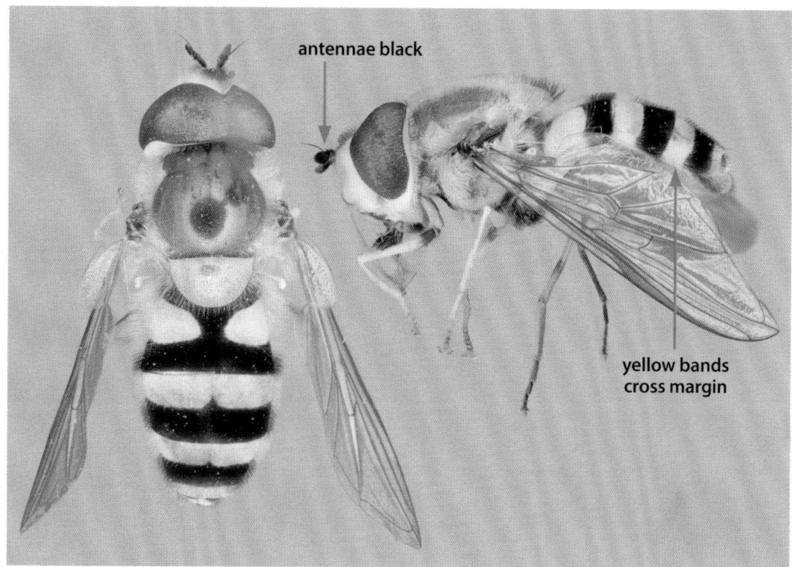

# *Epistrophe nitidicollis*

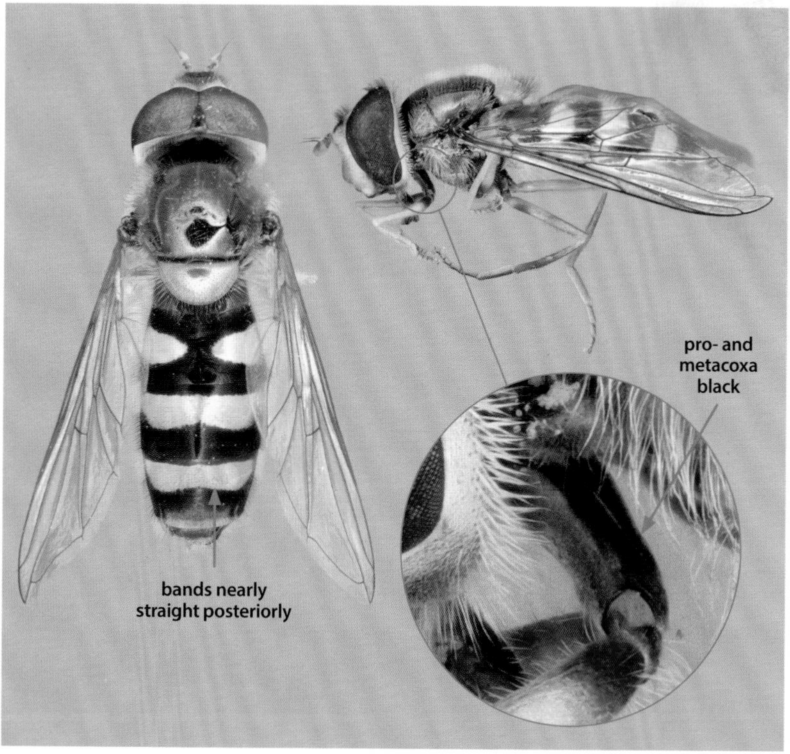
pro- and
metacoxa
black
bands nearly
straight posteriorly

# *Epistrophe xanthostoma*

## Emarginate Smoothtail

**SIZE RANGE:** 9.9–12.7 mm
**IDENTIFICATION:** This species has yellow antennae and wing cell bm is bare on the anterior ½. Yellow pro- and metacoxae distinguish *E. xanthostoma* from *E. nitidicollis*. The bands on tergites 3 and 4 are deeply emarginate posteriorly. **ABUNDANCE:** Uncommon. **FLIGHT TIMES:** Early May to late September. **NOTES:** Adults are found in forest and fen habitats. The only floral record is from *Eutrochium*. Larvae are unknown but likely feed on aphids.

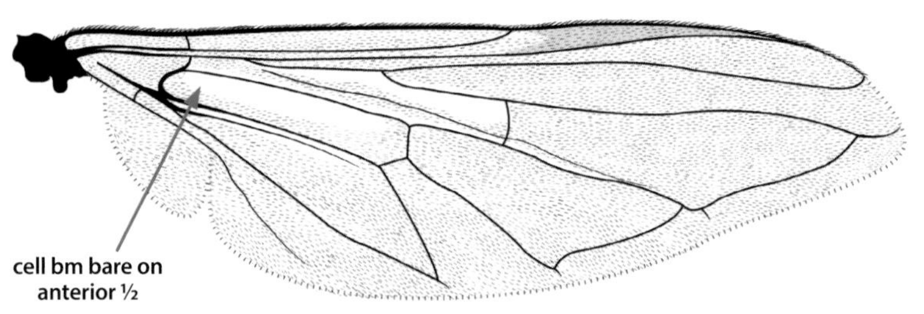


# *Epistrophe terminalis* Hairy-winged Smoothtail

**SIZE RANGE:** 9.0–12.0 mm
**IDENTIFICATION:** This species has yellow antennae, entirely microtrichose wings, and a bare metasternum. **ABUNDANCE:** Rare. **FLIGHT TIMES:** Early May to early July. **NOTES:** Adults are flower visitors. Larvae are unknown but likely feed on aphids as other *Epistrophe* larvae do.

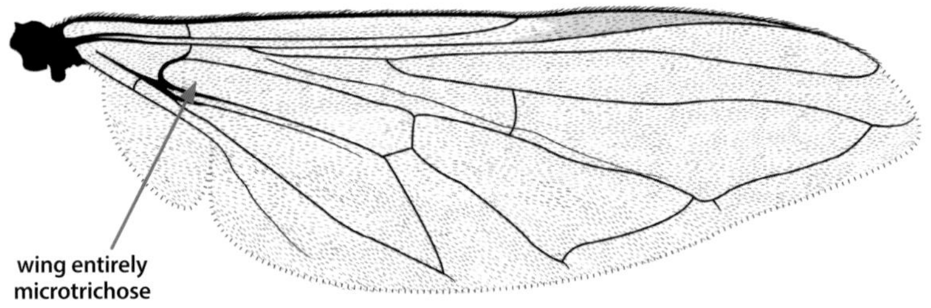

# *Epistrophe xanthostoma*

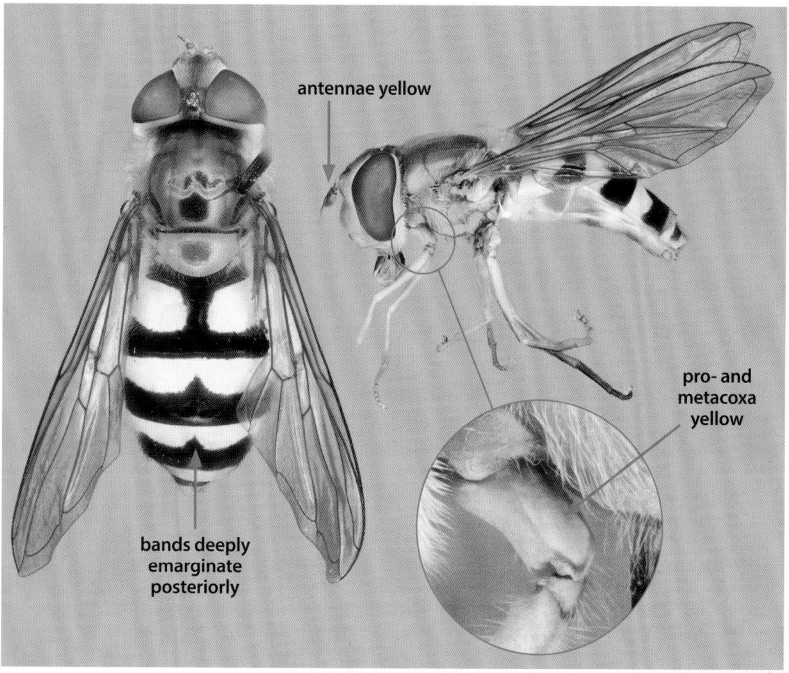

antennae yellow

pro- and metacoxa yellow

bands deeply emarginate posteriorly

# *Epistrophe terminalis*

antennae yellow

metasternum bare on all *Epistrophe* except *E. grossulariae*

# Epistrophella

*Epistrophella* species are typically slender, black and yellow flies with markings on tergite 3 either separated or joined medially and markings on tergite 4 typically separated medially. The scutum has obscure to distinct lateral stripes. The metasternum is bare and the katepisternum has broadly separated dorsal and ventral pile patches. There are two species of the genus worldwide, *E. emarginata* in the Nearctic and *E. euchroma* in the Palearctic, occurring from the United Kingdom to Russia. Adults visit flowers. Larvae are aphidophagous, feeding on aphids on a variety of plants from small herbaceous plants to trees. For a description of the one Nearctic species in this genus, see Vockeroth (1992).

## *Epistrophella emarginata*

### Slender Smoothtail

**SIZE RANGE:** 7.8–11.7 mm
**IDENTIFICATION:** Diagnosed by the generic characters given above. **ABUNDANCE:** Common. **FLIGHT TIMES:** Early May to early October (from early April in the south). **NOTES:** Adults can be found in hardwood forest, meadow, fen, and shore habitats. They have been recorded visiting *Ageratina*, *Angelica*, *Asclepias*, *Salix*, and *Symphyotrichum*. Larvae have been recorded feeding on a wide variety of aphids on plants such as *Chrysanthemum*, *Cornus*, *Hypericum calycinum*, *Lactuca*, *Populus*, *Prunus*, *Rubus*, *Salix*, *Solidago*, and *Verbesina helianthoides*. It has been reported that this species requires a winter diapause, so it is likely to be univoltine. *Epistrophella emarginata* is a highly variable species, with ten junior synonyms having been proposed over the years. Specimens collected at one collecting event show variation in abdominal pattern, color of frons and legs, and the distinctness of the lateral yellow stripes on the scutum. Despite this, DNA studies suggest that this is a complex of two or three species.

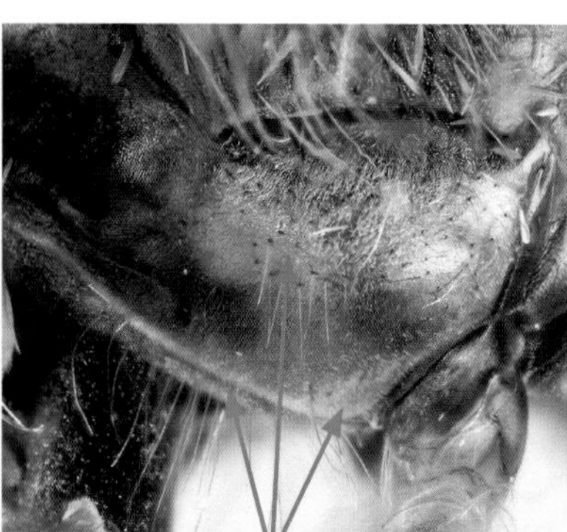

katepisternal pile patches broadly separated

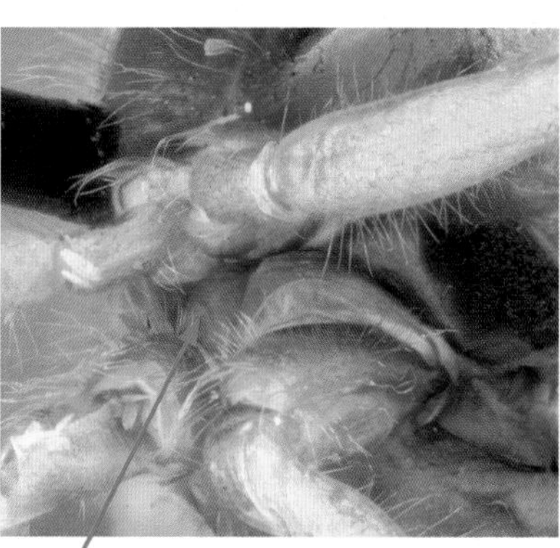
metasternum bare

# *Epistrophella emarginata*

# *Syrphus*

There are eight species of *Syrphus* in our region, 12 through the Nearctic, and 68 worldwide. *Syrphus* have the generic syrphine look, with a black and yellow abdominal pattern, but they differ from similar syrphines in having pile on the dorsal surface of the lower calypter (not to be confused with the pile on the edge of the calypter). This character is difficult to see without a hand lens or microscope, but it is unique to this genus. Adults of some species are ubiquitous and can be found on flowers of a wide variety of plants. Larvae feed on a variety of ground-layer and arboreal aphids. Several species are multivoltine. For a key to species of northern North America, see Vockeroth (1992).

## *Syrphus torvus* Hairy-eyed Flower Fly

**SIZE RANGE:** 8.6–13.3 mm
**IDENTIFICATION:** This is the only *Syrphus* species with pilose eyes. It has yellow bands on tergites 3 and 4 and wing cell bm is densely microtrichose.
**ABUNDANCE:** Common. **FLIGHT TIMES:** Mid-March to mid-October (to mid-November in the west).
**NOTES:** Adults can be found in a wide variety of habitats and have been noted on flowers of Asteraceae, *Caulophyllum*, *Ceanothus*, *Cicuta*, *Cornus*, *Rhododendron*, and *Rubus*. Larvae feed on aphids on many types of plants, including herbaceous plants, shrubs, and trees.

## *Syrphus sexmaculatus* Six-spotted Flower Fly

**SIZE RANGE:** 8.9–11.0 mm
**IDENTIFICATION:** This species is recognized by the yellow abdominal markings that are divided medially and the anteriorly bare wing cell bm.
**ABUNDANCE:** Rare. **FLIGHT TIMES:** Early July to late August (from early June in the west).
**NOTES:** This Holarctic species is also known from northern Europe and northern Russia. It can be found in forest/open ground, taiga, tundra, and subalpine habitats. Flowers visited include *Geranium*, *Leucanthemum*, *Matricaria*, *Polygonum*, *Ranunculus*, and *Taraxacum*. Larvae feed on aphids.

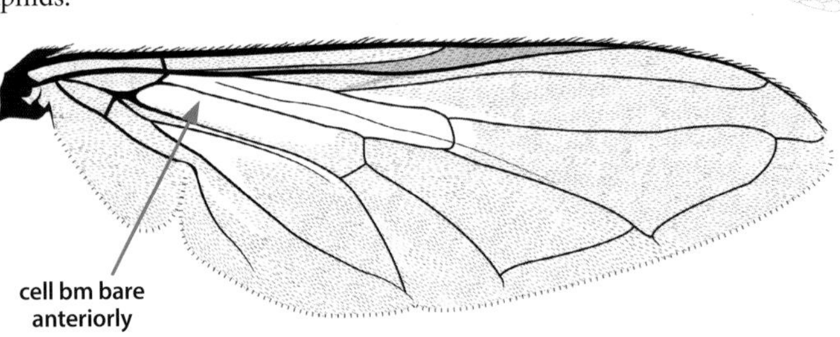

# Syrphus

pile present on dorsal surface of lower calypter (do not confuse this with the long pile on the edge of the calypter found in most genera)

## Syrphus torvus

## Syrphus sexmaculatus

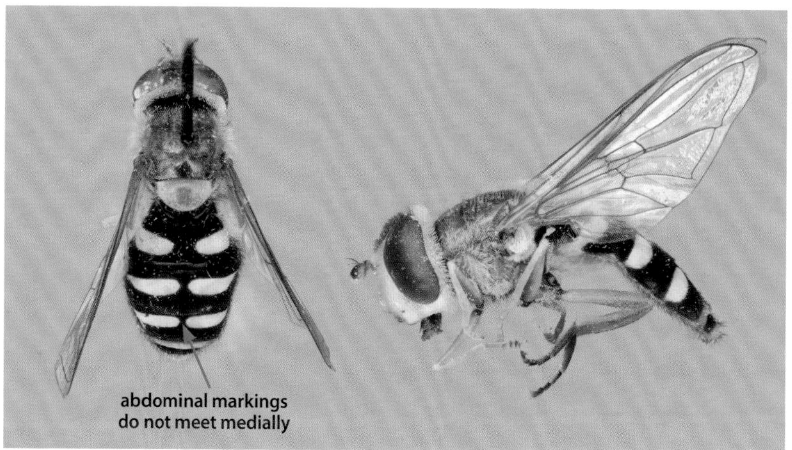

## *Syrphus attenuatus* Yellow-margined Flower Fly

**SIZE RANGE:** 8.1–12.8 mm

**IDENTIFICATION:** This species is recognized by the yellow abdominal markings that are divided medially, a densely microtrichose wing cell bm, and continuously yellow lateral margins of the tergites. **ABUNDANCE:** Rare in the east, uncommon in the west. **FLIGHT TIMES:** Early June to late August (as early as late April in the west). **NOTES:** Found in forest, fen, tundra, and subalpine habitats. Visits Apiaceae, *Carex*, *Papaver*, *Salix*, and *Taraxacum*. Larvae are known to feed on adelgids and likely aphids.

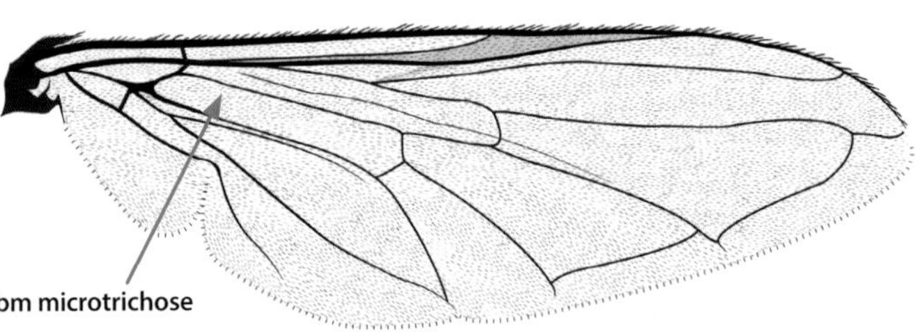


## *Syrphus ribesii* Common Flower Fly

**SIZE RANGE:** 8.1–13.3 mm

**IDENTIFICATION:** *Syrphus ribesii* has complete yellow bands on tergites 3 and 4, a densely microtrichose wing cell bm, bare eyes, and an alternating black and yellow abdominal margin. **ABUNDANCE:** Abundant. **FLIGHT TIMES:** Late April to mid-October (late March to mid-November in the west). **NOTES:** Both larvae and adults can be found on a wide variety of plants, from herbaceous plants to shrubs to trees. Larvae feed on aphids, adelgids, and other soft-bodied insects. They are found in many habitats from forests to agricultural land to gardens.

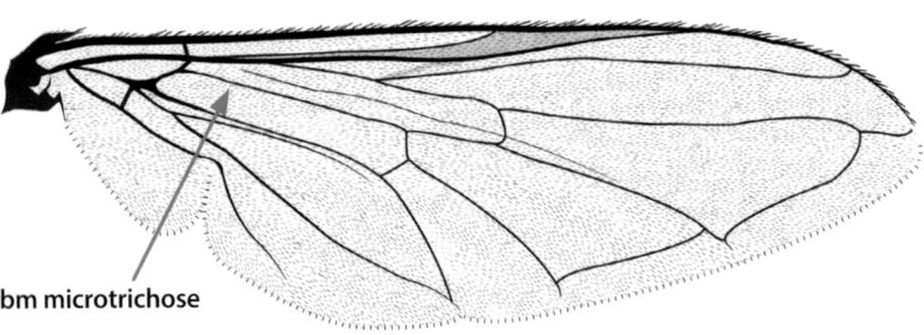

# *Syrphus attenuatus*

abdominal markings do not meet medially on tergites 3 and 4

abdominal margin continuously yellow

# *Syrphus ribesii*

yellow bands on tergites 3 and 4 connected medially

yellow bands cross abdominal margin

## *Syrphus opinator* Black-margined Flower Fly

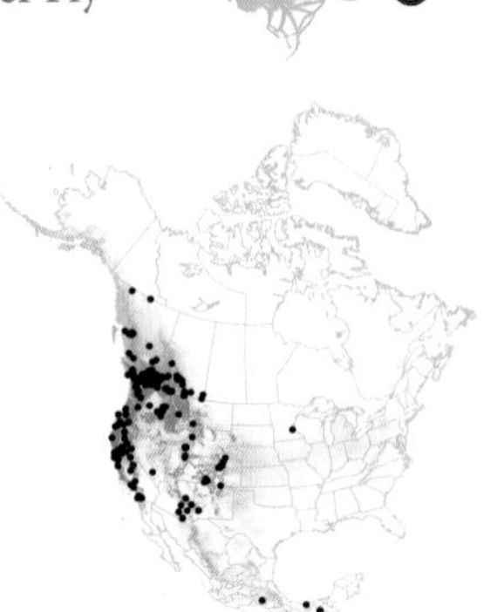

**SIZE RANGE:** 8.8–11.7 mm
**IDENTIFICATION:** This species is recognized by the yellow abdominal bands that do not reach the abdominal margin, creating an entirely black margin. Wing cell bm is bare anteriorly.
**ABUNDANCE:** Vagrant. **FLIGHT TIMES:** Late summer vagrant to the east (early January to late November in the west). **NOTES:** Adults are found in forest, meadow, and bog habitats as well as agricultural lands. Larvae feed on many adelgid and aphid species and have been collected from cultivated crops of *Capsicum annuum*, Fabaceae, and *Prunus*.

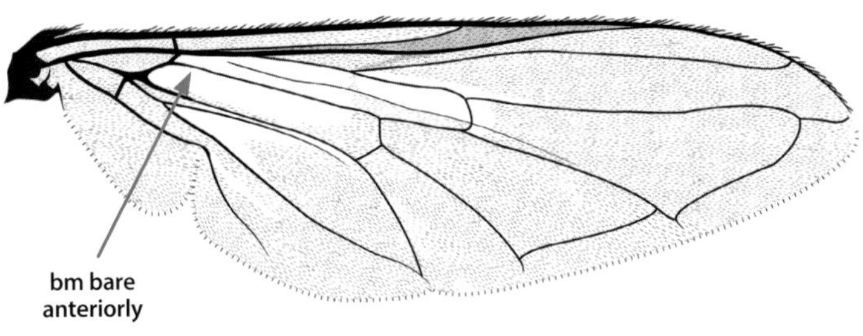


## *Syrphus vitripennis* Black-legged Flower Fly

**SIZE RANGE:** 7.6–11.4 mm
**IDENTIFICATION:** Wing cell bm is bare anteriorly and tergites 3 and 4 have yellow bands. The yellow bands narrowly meet the abdominal margin.
**ABUNDANCE:** Common. **FLIGHT TIMES:** Early April to early October (to late October in the west). **NOTES:** Adults live in a variety of habitats, from forests to agricultural lands. Adults are recorded from Apiaceae, *Caltha*, *Carex*, Rosaceae, and *Taraxacum*. Larvae feed on many species of aphids and other soft-bodied insects. They associate with many herbaceous plants and trees as well as many agricultural crops.

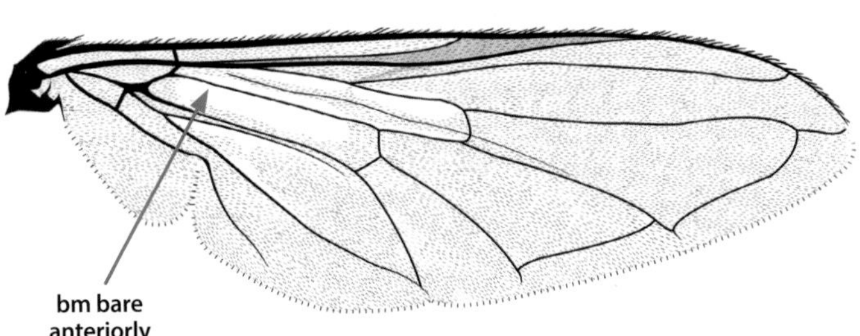

## *Syrphus opinator*

abdominal margin narrowly black

## *Syrphus vitripennis*

yellow bands narrowly meet abdominal margin

## *Syrphus knabi* Eastern Flower Fly

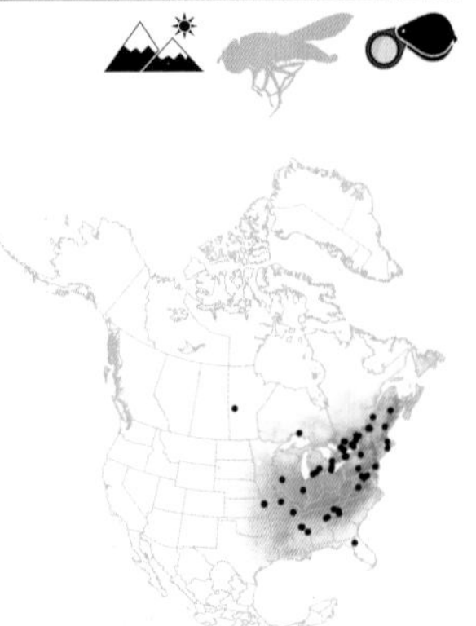

**SIZE RANGE:** 7.2–12.9 mm
**IDENTIFICATION:** Wing cell bm is bare anteriorly and tergites 3 and 4 have yellow bands that broadly meet the abdominal margin. The metafemur has a preapical dark band and the basitarsus of the metaleg is orange, contrasting with the other tarsomeres. **ABUNDANCE:** Uncommon. **FLIGHT TIMES:** Mid-May to mid-October (from mid-March in the south). **NOTES:** Adults are found in forests. Larvae have been recorded feeding on aphids on deciduous trees, *Brassica*, and *Thuja occidentalis*.

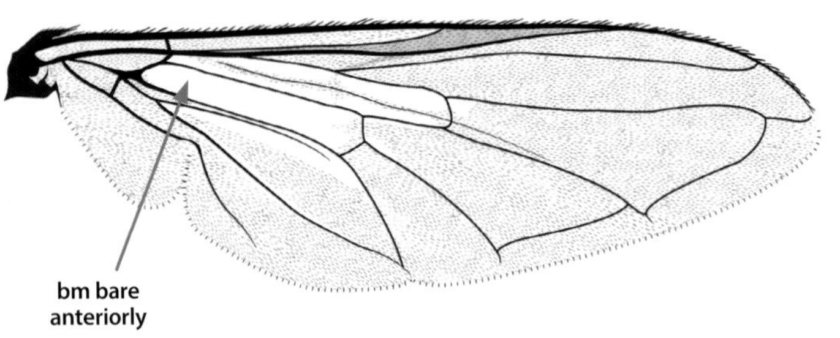


## *Syrphus rectus* Yellow-legged Flower Fly

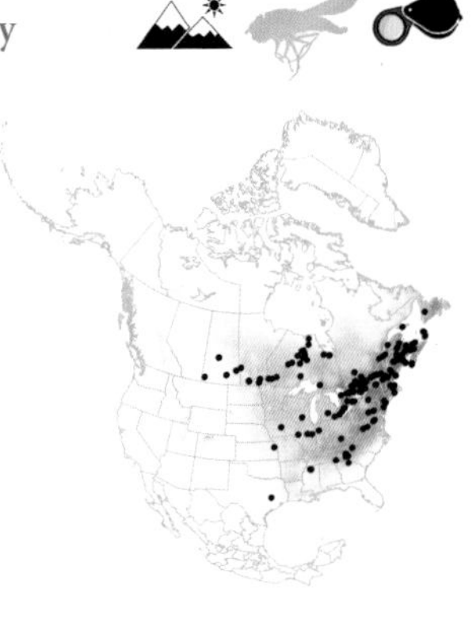

**SIZE RANGE:** 7.2–10.3 mm
**IDENTIFICATION:** Wing cell bm bare anteriorly; tergites 3 and 4 with yellow bands that broadly meet the abdominal margin; male metafemur dark on basal ⅔, female metafemur mostly yellow, sometimes darkened apically; basitarsus of metaleg dark or light but similar in color to following tarsomeres. **ABUNDANCE:** Common. **FLIGHT TIMES:** Late April to mid-November (from early April in the south). **NOTES:** Adults can be found in forest, bog, marsh, and grassy habitats. The single floral record is from *Symphyotrichum*. Larvae feed on aphids from herbaceous plants to trees.

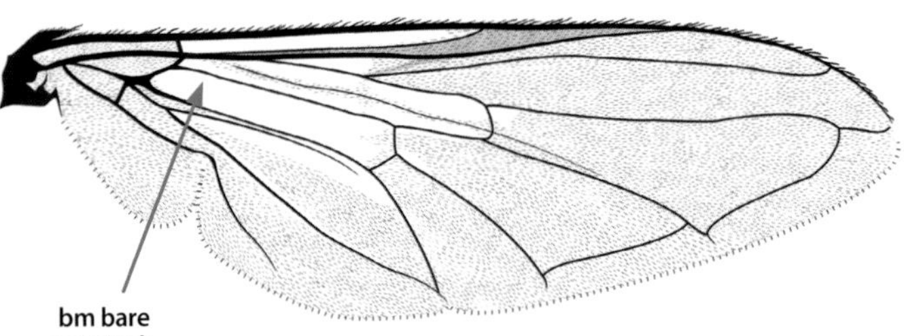

## *Syrphus knabi*

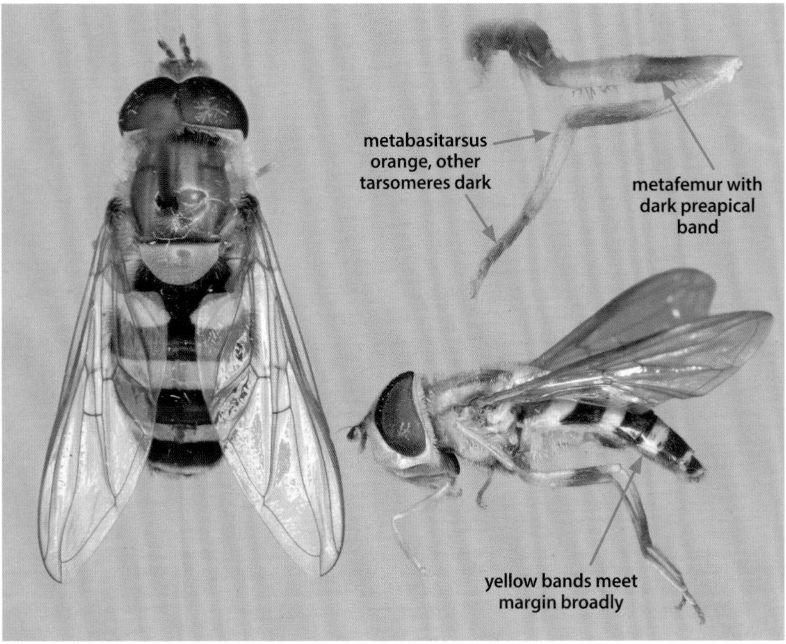

metabasitarsus orange, other tarsomeres dark

metafemur with dark preapical band

yellow bands meet margin broadly

## *Syrphus rectus*

♂

♀

male metafemur dark on basal ⅔

metatarsus uniform in color

yellow bands meet margin broadly

# Parasyrphus

*Parasyrphus* species are recognized by the pilose anterior anepisternum and the tuft of pile on the posteromedial apical angle of the metacoxa. They otherwise closely resemble *Syrphus* species. The 30 species are mainly Holarctic but the genus includes a few Indomalayan representatives. Eight of the 11 Nearctic species are treated here. Adults visit a variety of flowers and are often found around trees. Most known larvae feed on tree aphids and adelgids but some feed on chrysomelid beetle larvae. For a key to species of northern North America, see Vockeroth (1992). The undescribed species mentioned by Vockeroth (1992) was described as *P. vockerothi* by Thompson (2012).

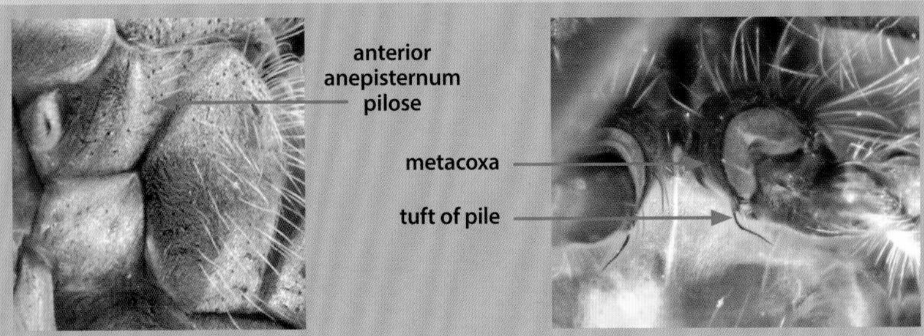


## *Parasyrphus groenlandica* Arctic Bristleside

**SIZE RANGE:** 7.2–7.9 mm

**IDENTIFICATION:** Eye pilose, face yellow, without dark medial stripe, males with a large, swollen frons, creating an angle between the eyes of ~130°. Males with pairs of yellow markings on tergites; female tergite markings are small to absent. **ABUNDANCE:** Common. **FLIGHT TIMES:** Mid-June to early August. **NOTES:** This species has been collected farther north than any other flower fly, at 82.5°N in Alert, Nunavut. It has been recorded feeding on *Salix* pollen. Larvae likely feed on aphids and other soft-bodied insects on High Arctic shrubs.

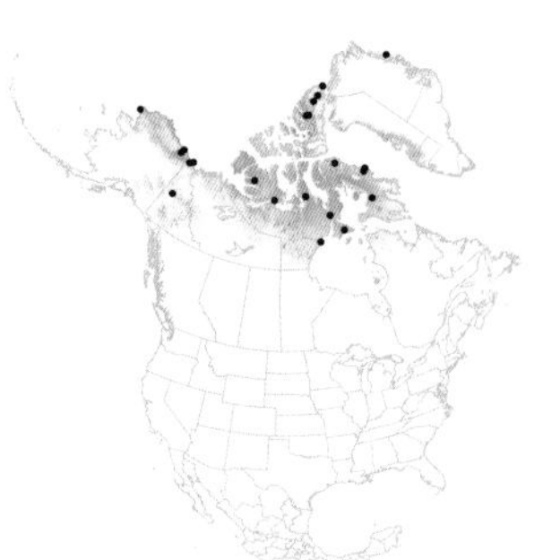

## *Parasyrphus tarsatus* Holarctic Bristleside

**SIZE RANGE:** 6.9–11.0 mm

**IDENTIFICATION:** Eye pilose, face yellow, without dark medial stripe, males with a small frons, creating an angle between the eyes of at most 100°. Males with pairs of yellow markings on tergites; female tergite markings vary from large to small to absent. Only females with large abdominal markings may be differentiated from female *P. groenlandica*. **ABUNDANCE:** Common. **FLIGHT TIMES:** Late May to early September. **NOTES:** Found in forest, tundra, meadow, and alpine habitats. Adults and larvae have been found on *Salix*. Larvae have been recorded feeding on aphids on both *Betula* and *Salix*.

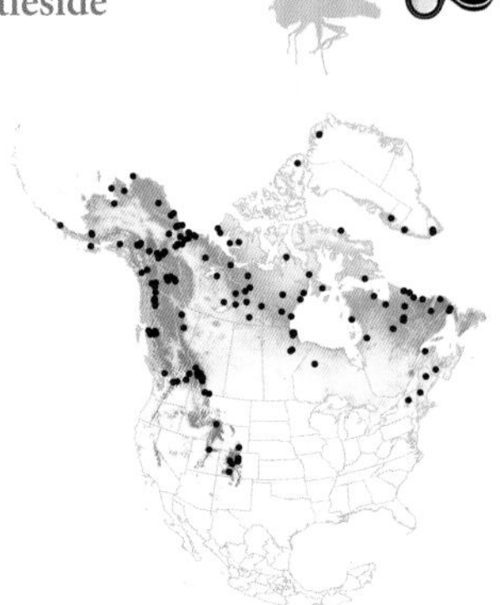

# Parasyrphus groenlandica

# Parasyrphus tarsatus

## *Parasyrphus nigritarsis* Yellow-faced Bristleside

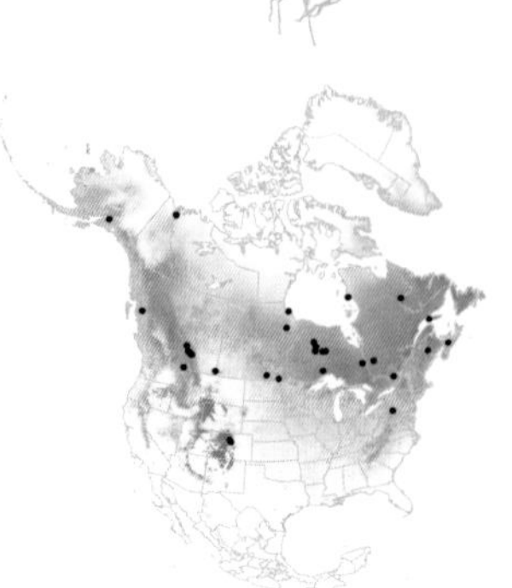

**SIZE RANGE:** 9.0–11.3 mm
**IDENTIFICATION:** Face yellow, without dark medial stripe, and tergites with single yellow band. Eyes sparsely pilose. All other species with yellow bands on the tergites have a black medial facial stripe. **ABUNDANCE:** Uncommon. **FLIGHT TIMES:** Late May to mid-July (mid-May to late July in the west). **NOTES:** This is a Holarctic species found through northern Europe, Russia, and Japan. They are often found in forest and wetland habitats. Adults have been recorded visiting *Anemone*, *Potentilla*, *Prunus*, *Ranunculus*, *Rhododendron*, *Rubus*, and *Salix*. Larvae are known to feed on chrysomelid beetle larvae in trees and shrubs.

## *Parasyrphus genualis* Common Bristleside

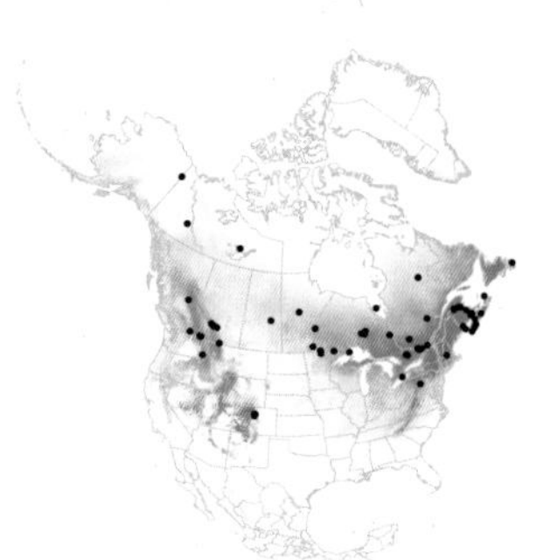

**SIZE RANGE:** 5.9–8.9 mm
**IDENTIFICATION:** Face yellow, with dark medial stripe, and abdominal tergites 3 and 4 with single yellow bands. Eyes bare. Wings partly bare, cell bm at least bare medially on basal ⅓. **ABUNDANCE:** Uncommon. **FLIGHT TIMES:** Early May to late August. **NOTES:** Adults can be found in forest, meadow, fen, and marsh habitats. They have been recorded on Apiaceae, *Malus*, *Prunus*, *Sambucus*, and *Solidago*. Larvae have been recorded feeding on adelgids on *Abies* and likely feed on tree aphids as well.

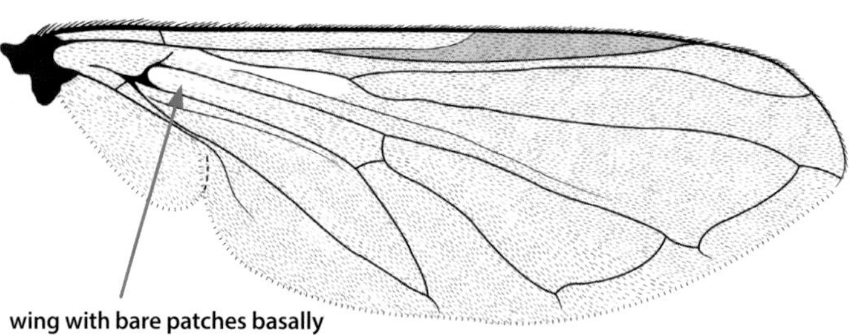

## *Parasyrphus nigritarsis*

single bands on tergites 3 and 4

face yellow

## *Parasyrphus genualis*

single bands on tergites 3 and 4

black medial stripe

## *Parasyrphus insolitus* Broad-striped Bristleside

**SIZE RANGE:** 5.6–8.1 mm
**IDENTIFICATION:** Face yellow, with dark medial stripe, and abdominal tergites 3 and 4 with single yellow bands. Eyes bare. Wing densely microtrichose. Pro- and mesotarsi yellow. Males with broad, black facial stripe (4/7 width of face). Female sternites yellow, without black markings/bands. **ABUNDANCE:** Rare. **FLIGHT TIMES:** Late June in Newfoundland (from mid-March to late July in the west). **NOTES:** This species is known from western North America and Labrador. The reason for this disjunct distribution is unclear. Little is known about this species. It is likely the larvae feed on tree/shrub aphids like other species in this genus.

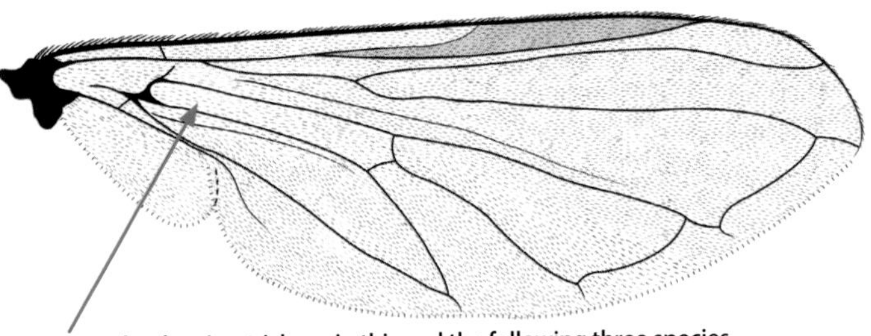


## *Parasyrphus relictus* Boreal Bristleside

**SIZE RANGE:** 5.7–11.1 mm
**IDENTIFICATION:** Face yellow, with narrow, dark medial stripe. Eyes bare. Wings densely microtrichose. Tergites with single yellow band. Sternites have obscure or indistinct black triangular or longitudinal markings and are sometimes entirely yellow in females. Males with frons broadly pollinose along eye margin (as shown for *P. vockerothi*), with anterior ⅓ of frons shiny. Female pro- and mesotarsi brown. **ABUNDANCE:** Rare in the east, common in the west. **FLIGHT TIMES:** Early June to late August (from mid-February to mid-October in the west). **NOTES:** This is a Holarctic species also occurring in Europe. Adults are found in scrub and marsh habitats. They visit flowers of Apiaceae, Asteraceae, *Rhododendron*, and *Sambucus*. Larvae have been recorded feeding on aphids and adelgids from trees. There is some doubt as to whether North American and European specimens are conspecific.

## *Parasyrphus insolitus*

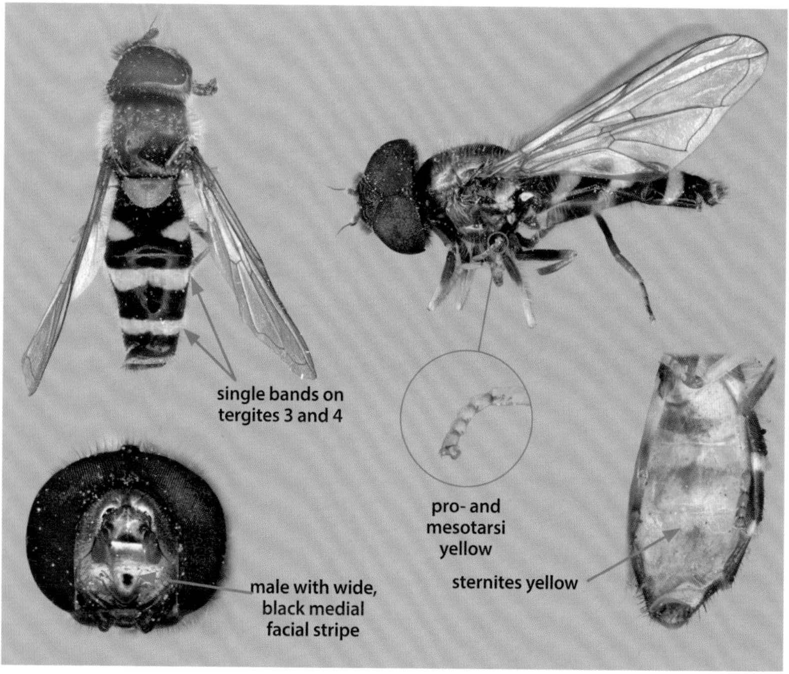

single bands on tergites 3 and 4

pro- and mesotarsi yellow

male with wide, black medial facial stripe

sternites yellow

## *Parasyrphus relictus*

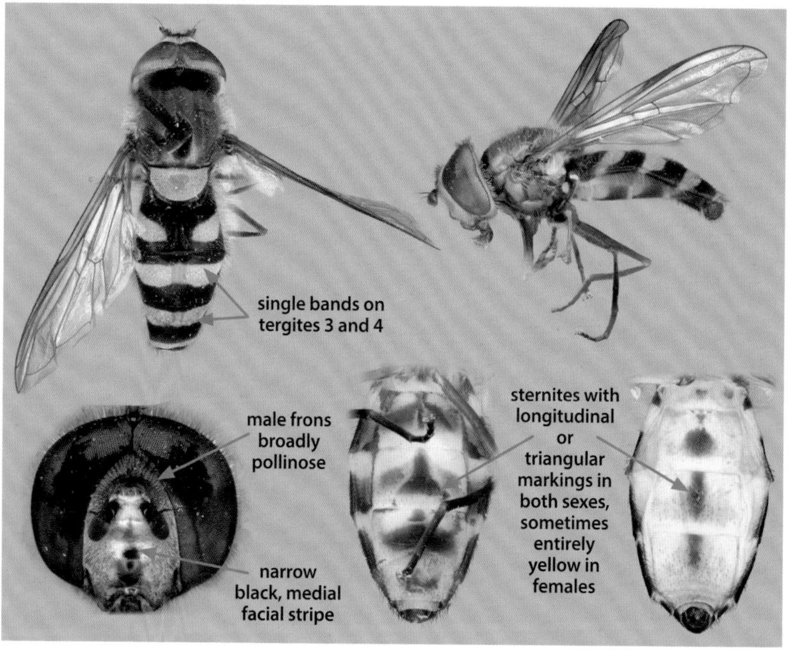

single bands on tergites 3 and 4

male frons broadly pollinose

narrow black, medial facial stripe

sternites with longitudinal or triangular markings in both sexes, sometimes entirely yellow in females

## *Parasyrphus semiinterruptus*

### Emarginate Bristleside

**SIZE RANGE:** 6.2–8.4 mm
**IDENTIFICATION:** Face yellow, with narrow, dark medial stripe. Eyes bare or with short sparse pile. Wings densely microtrichose. Tergites with single yellow band. Males with frons narrowly but distinctly pollinose along eye margin, with anterior 4/5 of frons shiny. Females, which have black bands on their sternites, are indistinguishable from females of *P. vockerothi*. **ABUNDANCE:** Rare. **FLIGHT TIMES:** Mid-April to late June. **NOTES:** Adults have been found in hardwood, mixed, and spruce forests as well as a rural garden; some have been collected from flowers of *Caltha*. Larvae presumably feed on tree aphids like other species in this genus.

## *Parasyrphus vockerothi* Vockeroth's Bristleside

**SIZE RANGE:** 6.3–8.2 mm
**IDENTIFICATION:** Face yellow, with narrow, dark medial stripe. Eyes bare. Wings densely microtrichose. Tergites with single yellow band. Sternites with black bands. Males with frons broadly pollinose along eye margin, with anterior ⅓ of frons shiny. Females are indistinguishable from female *P. semiinterruptus*. **ABUNDANCE:** Common. **FLIGHT TIMES:** Late April to early September. **NOTES:** Found in forest and bog habitats. Found on flowers of Apiaceae, *Caltha*, *Chamaedaphne*, and *Malus*. There is one record of the larvae feeding on aphids on *Abies*, which agrees with many other species in this genus being tree-aphid feeders.

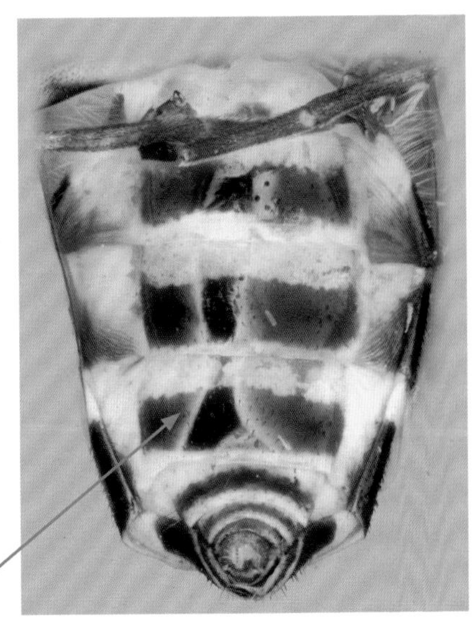

sternites of both sexes with bands

# *Parasyrphus semiinterruptus*

face with narrow black medial stripe

single emarginate bands on tergites 3 and 4

male frons narrowly pollinose

# *Parasyrphus vockerothi*

face with narrow black medial stripe

single bands on tergites 3 and 4

male frons broadly pollinose

# *Megasyrphus*

There are two species of *Megasyrphus* in the Nearctic (six species worldwide), one of which is found in the region covered here. The other, *Megasyrphus catalina*, is found in Arizona and New Mexico. *Megasyrphus* species can be recognized by their black and yellow abdominal pattern, sinuous $R_{4+5}$ vein, and the visible tracheal system, especially in live specimens. They can be confused with *Didea*, but *Megasyrphus* have a black and yellow margin on abdominal tergites 3 and 4, while the abdominal margin is all black in *Didea*. *Megasyrphus* was long considered a subgenus of *Eriozona*, so the species can sometimes be found in collections under that name. Adults are flower visitors. Larvae have been found feeding on arboreal aphids on *Pinus* and *Picea*; they have also been found on *Hordeum* and have been lab-reared on aphids from *Vicia faba*. The European species *M. erratica* is univoltine.

## *Megasyrphus laxus* Black-legged Gossamer

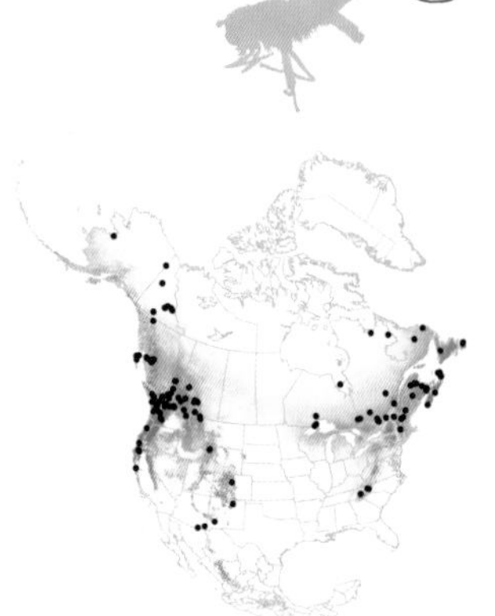

**SIZE RANGE:** 10.2–15.6 mm
**IDENTIFICATION:** This species can be recognized by the slightly sinuous $R_{4+5}$ vein and the alternating black and yellow abdominal margin. It also has sparsely pilose eyes and a black medial stripe on the yellow face. **ABUNDANCE:** Uncommon. **FLIGHT TIMES:** Late May to mid-September (early April to late October in the west). **NOTES:** Adults can be found in forest and bog habitats. They have been collected visiting flowers of Apiaceae and Asteraceae. Larvae are unknown but larvae of other species in this genus are known to feed on arboreal aphids.

color of abdominal bands varies from greenish to yellow in life, as in *Didea*

abdominal margin alternating yellow and black

# *Megasyrphus*

sinuous R4+5

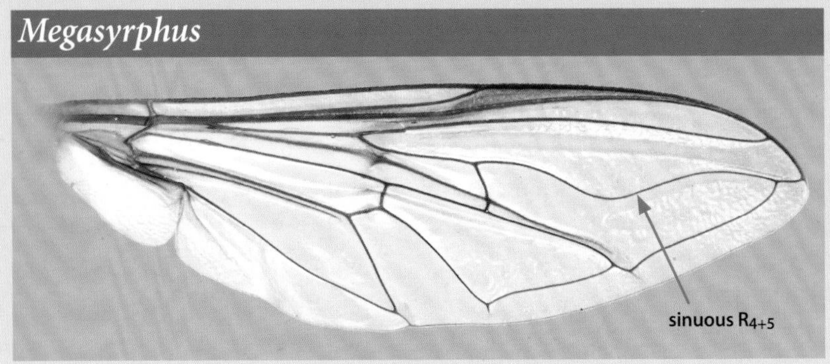

## *Megasyrphus laxus*

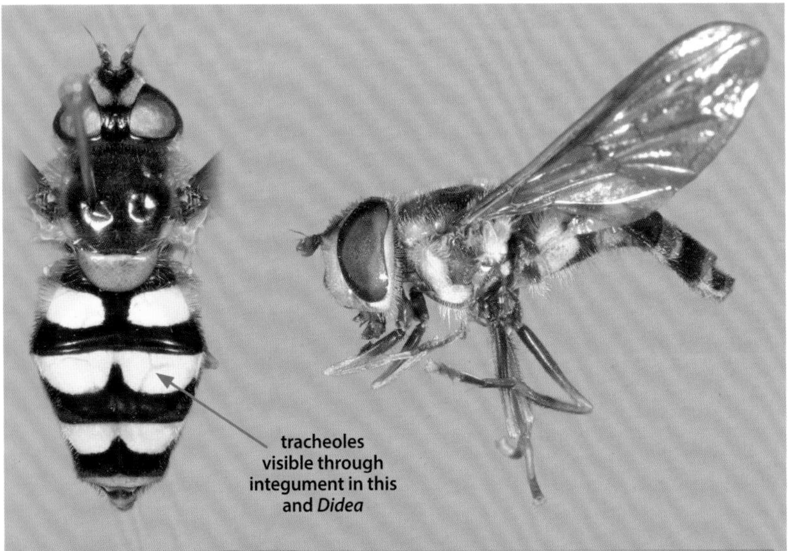

tracheoles visible through integument in this and Didea

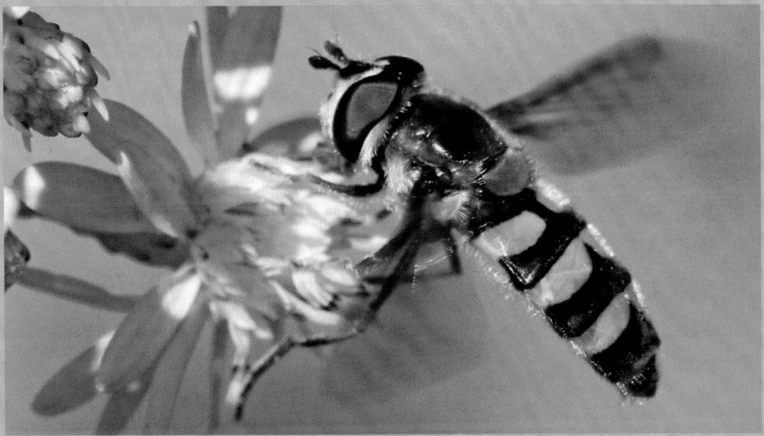

# *Didea*

There are two species of *Didea* in the Nearctic (seven worldwide) and they both occur in our region. *Didea* species can be recognized by their black and yellow abdominal pattern, sinuous $R_{4+5}$ vein, and the visible tracheal system, especially in live specimens. They can be confused with *Megasyrphus*, but *Didea* have a black margin on abdominal tergites 3 and 4, while the margin is black and yellow in *Megasyrphus*. Larvae feed on arboreal aphids. For a key to species of northern North America, see Vockeroth (1992).

## *Didea fuscipes* Undivided Lucent

**SIZE RANGE:** 9.7–15.0 mm
**IDENTIFICATION:** Tergite 4 with a complete yellow band. Tergite 5 of females and typical males with yellow markings. **ABUNDANCE:** Uncommon.
**FLIGHT TIMES:** Late April to late October (early February to early November in the west).
**NOTES:** Adults have been found in forest and bog habitats. They have been collected from flowers of *Foeniculum vulgare* and from tree and shrub leaves. Larvae feed on arboreal aphids found on *Abies*, *Platanus*, and *Tilia*.

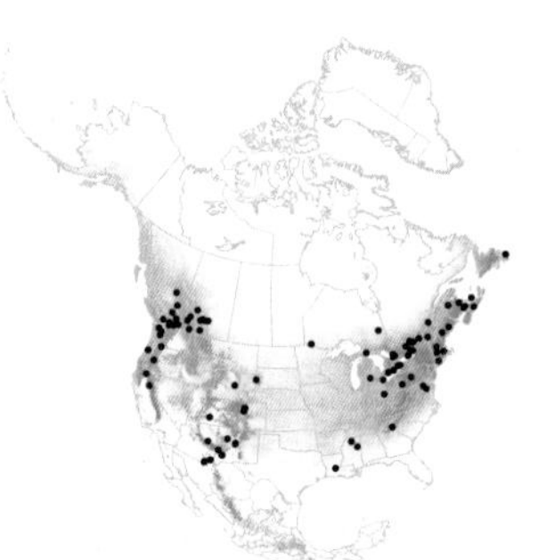

## *Didea alneti* Triangular Lucent

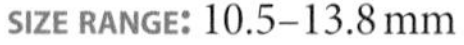

**SIZE RANGE:** 10.5–13.8 mm

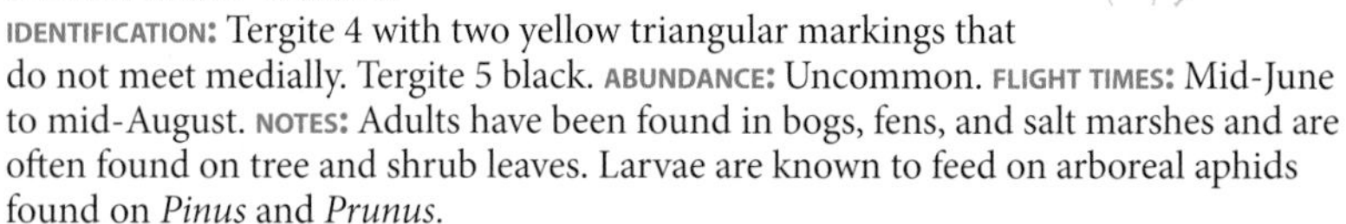

**IDENTIFICATION:** Tergite 4 with two yellow triangular markings that do not meet medially. Tergite 5 black. **ABUNDANCE:** Uncommon. **FLIGHT TIMES:** Mid-June to mid-August. **NOTES:** Adults have been found in bogs, fens, and salt marshes and are often found on tree and shrub leaves. Larvae are known to feed on arboreal aphids found on *Pinus* and *Prunus*.

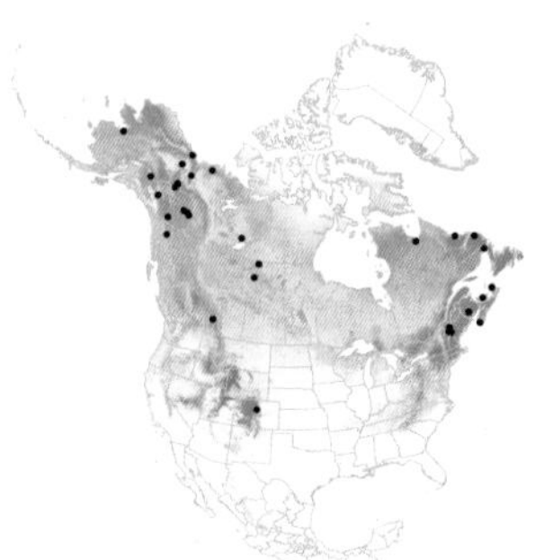

# Didea

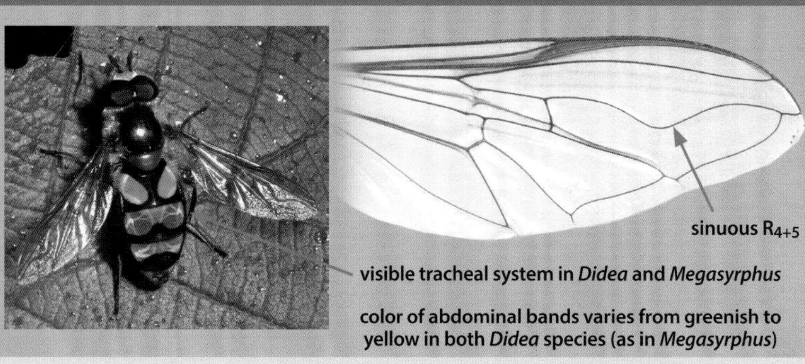


## Didea fuscipes

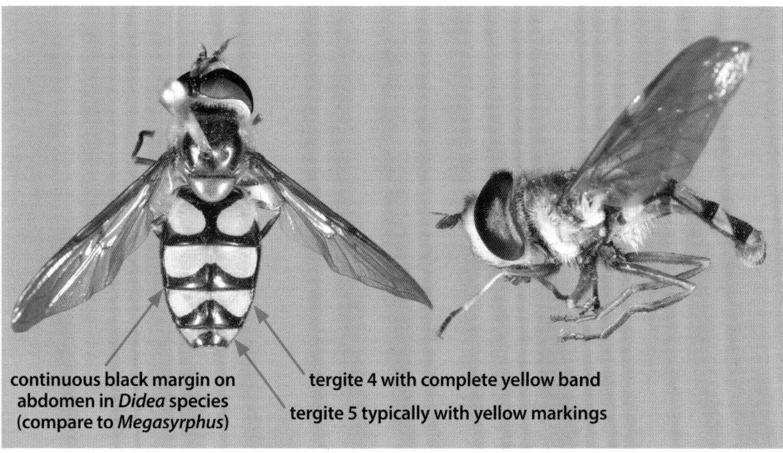


## Didea alneti

# Leucozona

Three of the 14 species of *Leucozona* occur in North America, two in our region. *Leucozona* larvae are predators of aphids. For a key to species of northern North America, see Vockeroth (1992).

## *Leucozona (Ischyrosyrphus) xylotoides*

Eastern Hoary

**SIZE RANGE:** 10.0–11.8 mm
**IDENTIFICATION:** This species is recognized by the presence of large subquadrate yellow-gray markings on tergite 2 that are much larger than markings on tergite 3. The eye is pilose and the wing is unbanded. The only species likely to be confused with this is *Melangyna fisherii*, but it has a bare eye. **ABUNDANCE:** Rare. **FLIGHT TIMES:** Late May to mid-July. **NOTES:** Flowers visited include *Daucus*, *Rubus*, and *Spiraea*. Larvae are unknown for this species but European species in this subgenus have been recorded feeding on aphids.

## *Leucozona (Leucozona) americana*

American Whitebelt

**SIZE RANGE:** 9.8–13.0 mm
**IDENTIFICATION:** This species is recognized by the distinct brown band extending from the anterior margin of the wing to crossvein r-m, and the yellow gray color of tergite 2 and the base of tergite 3 (the rest of the abdomen is black). **ABUNDANCE:** Uncommon. **FLIGHT TIMES:** Early June to late August. **NOTES:** Adults have been found in woodland, meadow, swamp, fen, bog, and tundra habitats. Flowers visited include *Cornus* and *Melilotus*. Larvae are presumed to feed on aphids like their European counterparts. This species was previously included under the European species *L. lucorum*; however, genetic evidence shows they are distinct.

## *Leucozona (Ischyrosyrphus) xylotoides*

## *Leucozona (Leucozona) americana*

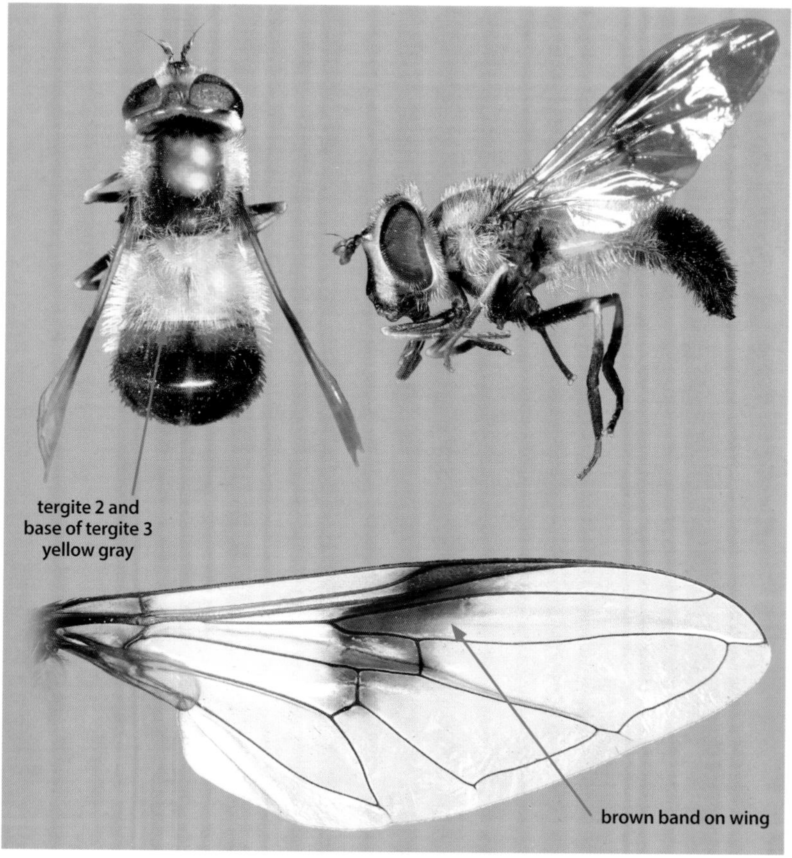

tergite 2 and base of tergite 3 yellow gray

brown band on wing

# Doros

*Doros* is a Holarctic genus of four species, only one of which occurs in North America. This distinctive but uncommonly encountered species is most often found at hilltop mating sites. Larval biology of the genus is unknown; however, due to its similarity to *Xanthogramma* it is thought that *Doros* may be associated with ants. It is suspected that they feed on root aphids associated with ant nests. Females of the European species have been observed searching for oviposition sites at the bases of trees, where pupae have been found. The European species have been observed visiting Apiaceae, *Chrysanthemum*, *Filipendula*, and *Rubus*. For a key to the genus and a species diagnosis, see Vockeroth (1992) or Miranda *et al.* (2013).

## *Doros aequalis* Canadian Potterfly

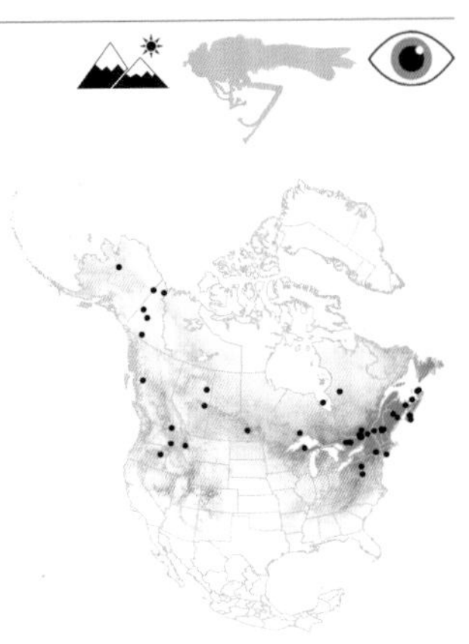

**SIZE RANGE:** 11.5–15.5 mm
**IDENTIFICATION:** This species can be recognized by the cylindrical black body with the characteristic yellow banding pattern and the parallel-sided, dorsally convex abdomen. The scutellum is brown and the lateral margins of the scutum are yellow. The anterior portion of the wing is darkened.
**ABUNDANCE:** Uncommon. **FLIGHT TIMES:** Mid-May to early August (one outlier in mid-October from Quebec). **NOTES:** Adults have been collected from forest and bog habitats. Flowers visited include *Maianthemum*, *Rosa*, and *Thalictrum*. *Doros* species are excellent mimics of potter wasps (such as *Ancistrocerus*).

# *Doros aequalis*

# *Chrysotoxum*

There are 101 world-recognized species of *Chrysotoxum*, 10 in the Nearctic. *Chrysotoxum* species are easily recognized by their long antennae (the flagellum and often the pedicel are elongate) and the projecting corners of the abdominal tergites. The abdomen is also broad with distinctive banding. *Chrysotoxum* species concepts have long been debated and Sommaggio and Skevington are currently revising the Nearctic taxa. Even with subtle genitalic characters, some specimens may not be identifiable without resorting to DNA. The four eastern species are fortunately easier to identify than the western ones and almost all individuals should be identifiable using this book. Keys by Curran (1924), Shannon (1926a), and Vockeroth (1992) are now out of date. Larval habits are unclear; larvae may be associated with ants and/or root aphids.

## *Chrysotoxum pubescens*

### Yellow-throated Meadow Fly

**SIZE RANGE:** 11.5–13.8 mm
**IDENTIFICATION:** This is our only *Chrysotoxum* species with a bright yellow proepimeron. Unlike in all of our other species, the antenna is short, with the scape shorter than the pedicel and the flagellum about length of the scape plus the pedicel.
**ABUNDANCE:** Common. **FLIGHT TIMES:** Early April to late September (to late October in the south).
**NOTES:** Adults have been recorded from flowers of *Oxypolis* and *Solidago*.

## *Chrysotoxum flavifrons*

### Blackshield Meadow Fly

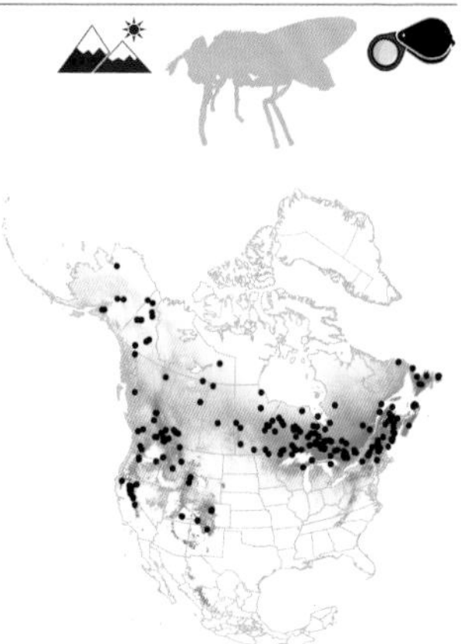

**SIZE RANGE:** 11.2–16.5 mm
**IDENTIFICATION:** This is the only eastern *Chrysotoxum* species with long, erect pile on the anterior anepisterum (five or more bristles). Flagellum longer than scape plus pedicel. Proepimeron black.
**ABUNDANCE:** Common. **FLIGHT TIMES:** Late May to mid-September. **NOTES:** Most specimens have been documented in conifer-dominated forests (spruce or pine). Flowers visited include *Cornus*, *Diervilla*, and *Linnaea*. This is the most likely *Chrysotoxum* species to encounter on hilltops in the east.

## *Chrysotoxum pubescens*

proepimeron yellow

posterior corners of the tergites project in all of our *Chrysotoxum* species

flagellum about length of scape plus pedicel

pedicel

scape shorter than pedicel

## *Chrysotoxum flavifrons*

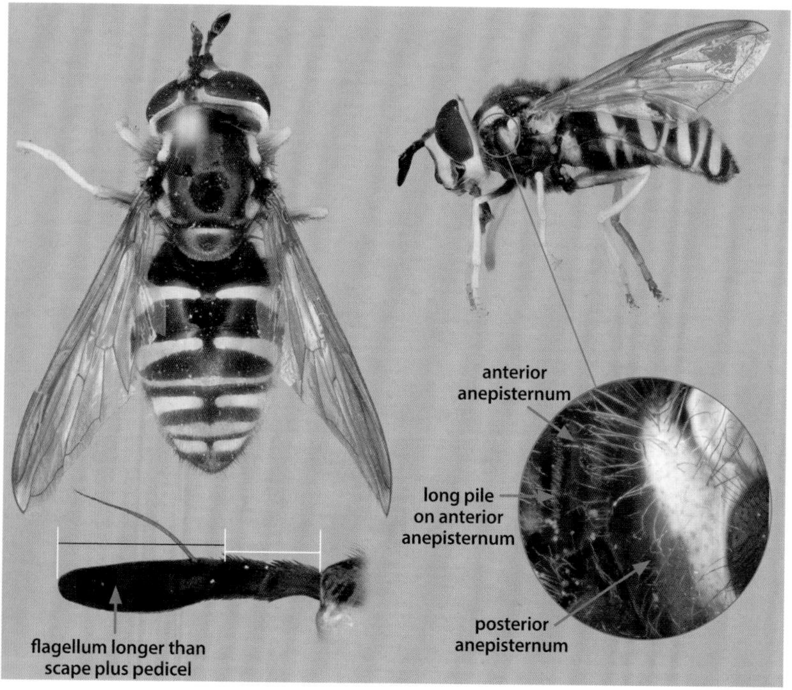

anterior anepisternum

long pile on anterior anepisternum

posterior anepisternum

flagellum longer than scape plus pedicel

# *Chrysotoxum plumeum*

## Broad-banded Meadow Fly

**SIZE RANGE:** 8.1–12.5 mm
**IDENTIFICATION:** Tergites 2 and 3 mostly yellow pilose, with only a few black pili along the margin. Abdominal bands broader than in *C. derivatum*, typically about ¼ the width of tergite. Metafemur usually completely yellow, with at most a narrow ring of black near base. Scape longer than pedicel. Proepimeron black. Anterior anepisterum usually bare, rarely with up to three bristles. **ABUNDANCE:** Common. **FLIGHT TIMES:** Late March to early October. **NOTES:** Specimens have been collected in many different forest types from hardwood to coniferous, as well as bogs and meadows. The only floral records are from *Solidago*. Only a few specimens have been captured hilltopping. This was considered a synonym of *C. derivatum* until our reconsideration of the limits and definition of that species.

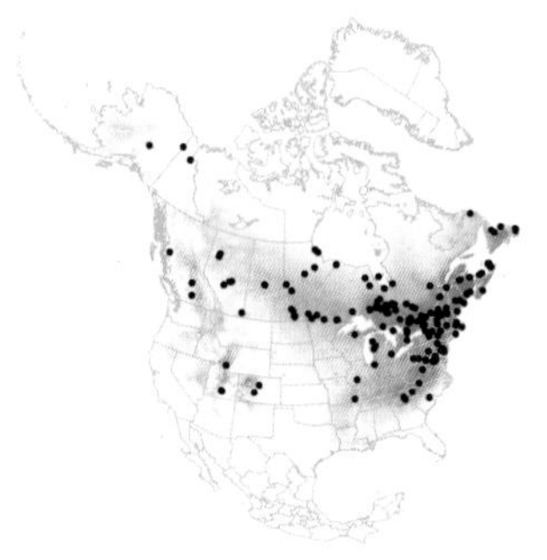

# *Chrysotoxum derivatum*

## Narrow-banded Meadow Fly

**SIZE RANGE:** 6.9–11.3 mm
**IDENTIFICATION:** Tergites 2 and 3 with extensive black pile. Abdominal bands narrow. Metafemur usually black on basal ⅓ or more. Scape longer than pedicel. Proepimeron black. Anterior anepisterum usually bare, rarely with up to three bristles. **ABUNDANCE:** Rare. **FLIGHT TIMES:** Mid-May to early October. **NOTES:** This appears to be a widely distributed but rare boreal species. All known specimens appear dark, with narrow abdominal bands and long black pile. Some western individuals can be confused with other species and in these cases only DNA can be used to confirm the identification. This was formerly the most common species of *Chrysotoxum* recognized in collections but we now consider it to be rare based on our revision of the species concept.

## *Chrysotoxum plumeum*

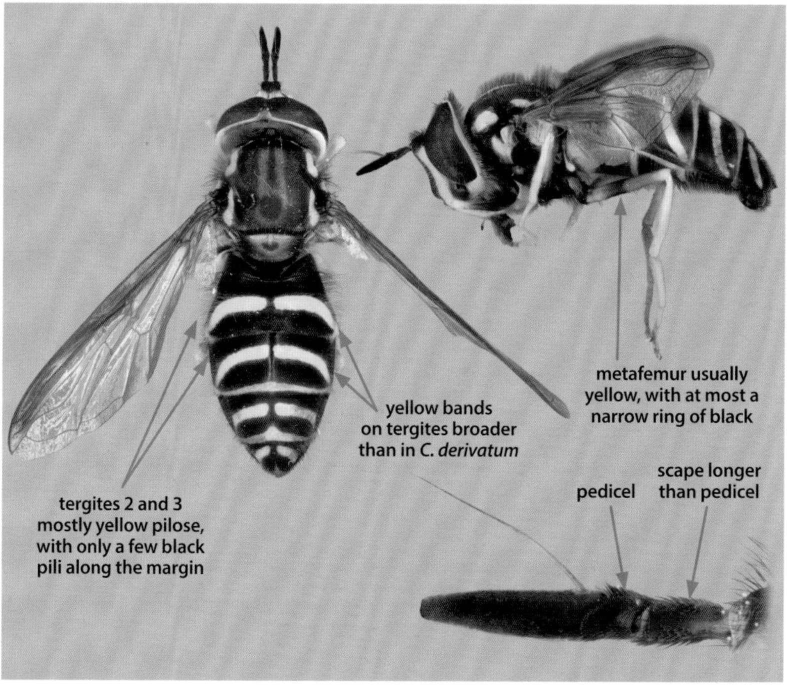

metafemur usually yellow, with at most a narrow ring of black

yellow bands on tergites broader than in *C. derivatum*

tergites 2 and 3 mostly yellow pilose, with only a few black pili along the margin

pedicel

scape longer than pedicel

## *Chrysotoxum derivatum*

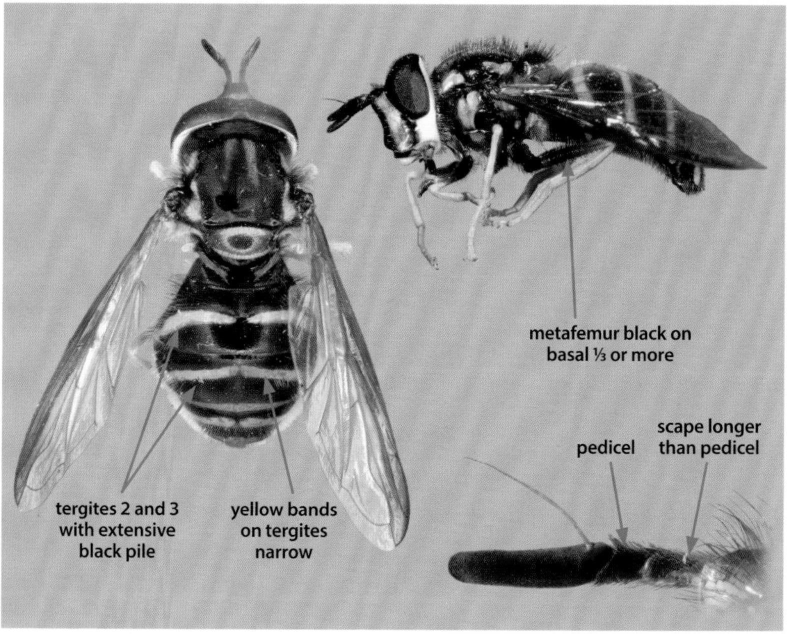

metafemur black on basal ⅓ or more

pedicel

scape longer than pedicel

tergites 2 and 3 with extensive black pile

yellow bands on tergites narrow

# Morphology

Most of the terms used in this book are labeled in the following drawings. The glossary, following the illustrations, provides definitions of these characters along with a few others that are not illustrated. Abbreviations: S - sternite; ta - tarsomere; T - tergite.

## Body (dorsal)

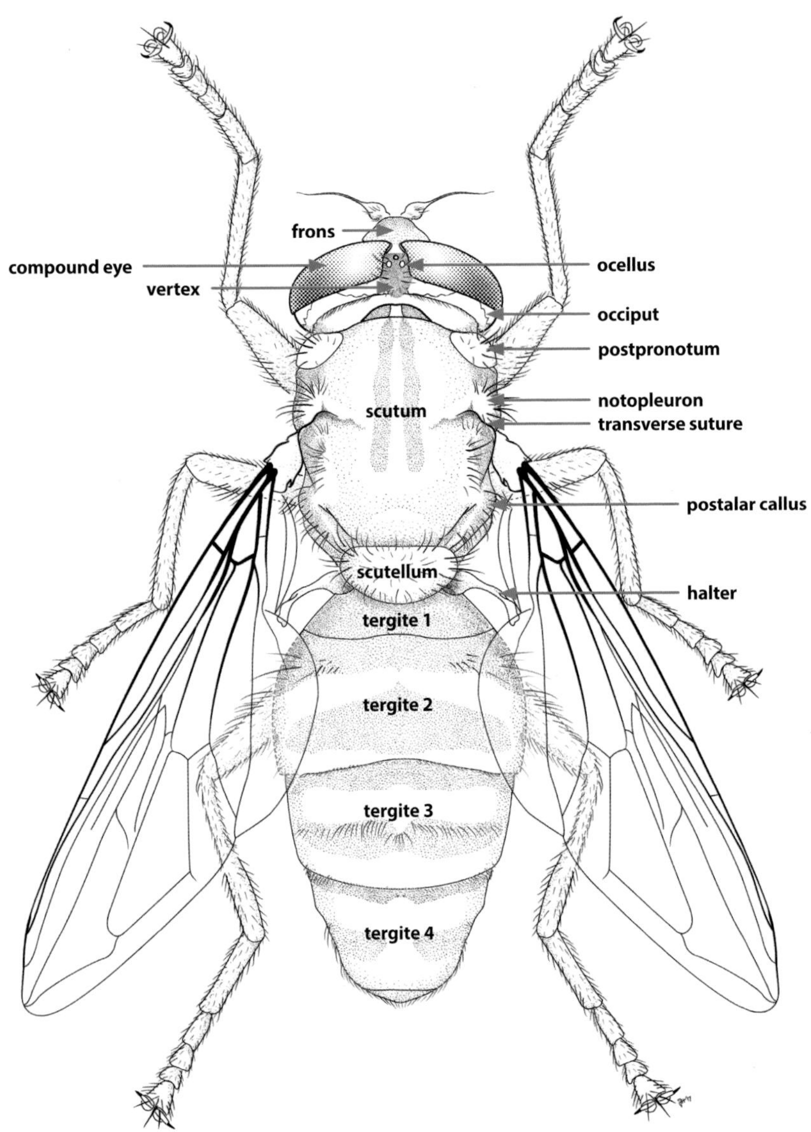

## Body (lateral)

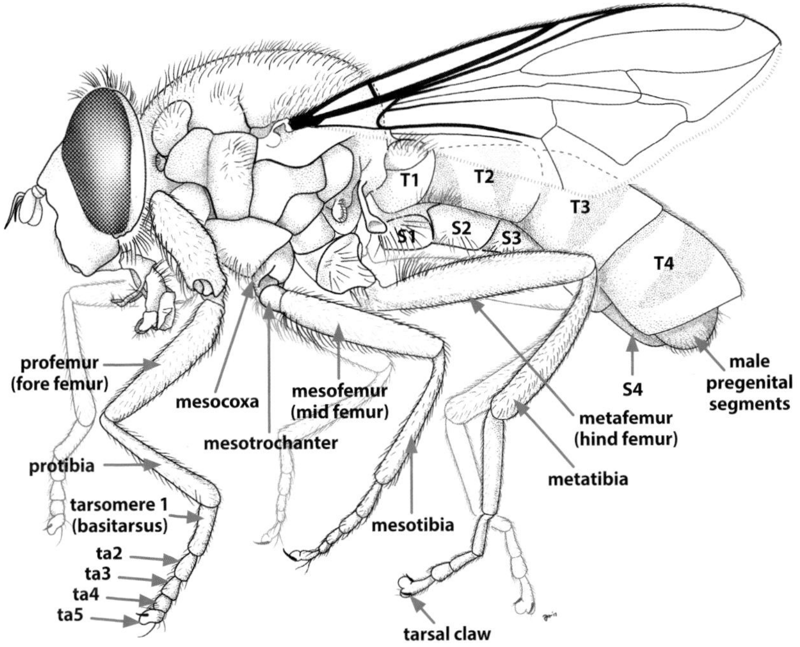

T1
T2
T3
T4
S1
S2
S3
S4
male pregenital segments
profemur (fore femur)
mesocoxa
mesofemur (mid femur)
mesotrochanter
metafemur (hind femur)
metatibia
protibia
tarsomere 1 (basitarsus)
mesotibia
ta2
ta3
ta4
ta5
tarsal claw

## Head and thorax (lateral)

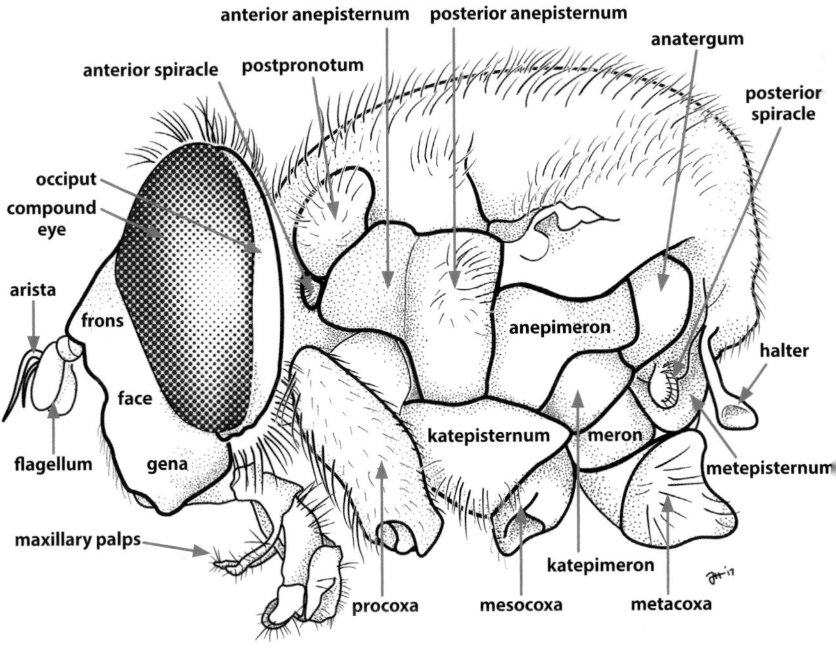

anterior anepisternum
posterior anepisternum
anatergum
anterior spiracle
postpronotum
posterior spiracle
occiput
compound eye
arista
frons
anepimeron
halter
face
katepisternum
meron
flagellum
gena
metepisternum
maxillary palps
katepimeron
procoxa
mesocoxa
metacoxa

## Wing

**Wing vein morphology has been recently revised based on homology across the Diptera (Wootton and Enos 1989; Saigusa 2006; Cumming and Wood 2017). To our knowledge, this is the first time that syrphid wing venation has been interpreted using this new system. Note: cells are lowercase, veins start with an upper-case letter, and crossveins are hyphenated.**

bc
h
C
Sc
R1
stigma
R2+3
R4+5
c
sc
r1
br
r-m
r2+3
bm
spurious vein
alula
M1
r4+5
bm-m
cua
m-cu
dm
R4+5+M1
cup
m4
A1
CuP
CuA
M4
dm-m
M2
CuA + CuP

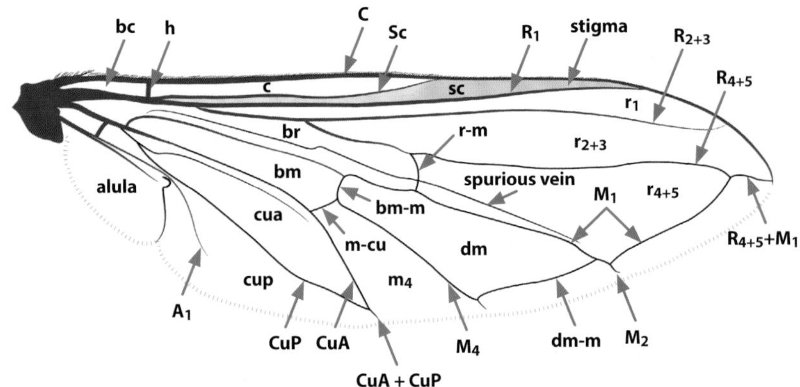

## Male genitalia

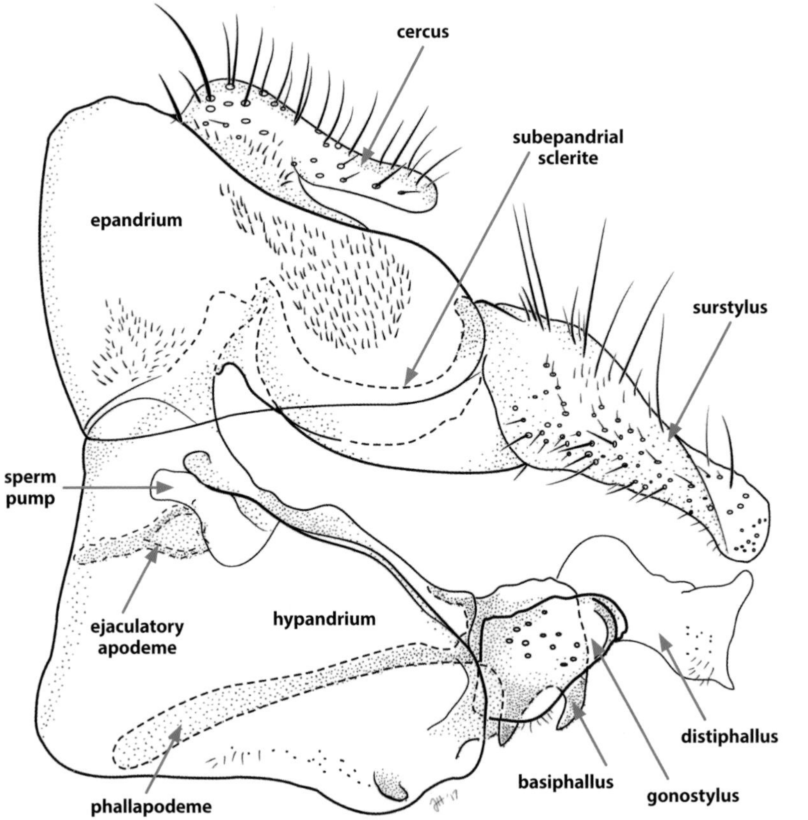

cercus
subepandrial
sclerite
epandrium
surstylus
sperm
pump
ejaculatory
apodeme
hypandrium
distiphallus
basiphallus
gonostylus
phallapodeme

## Head (anterior)

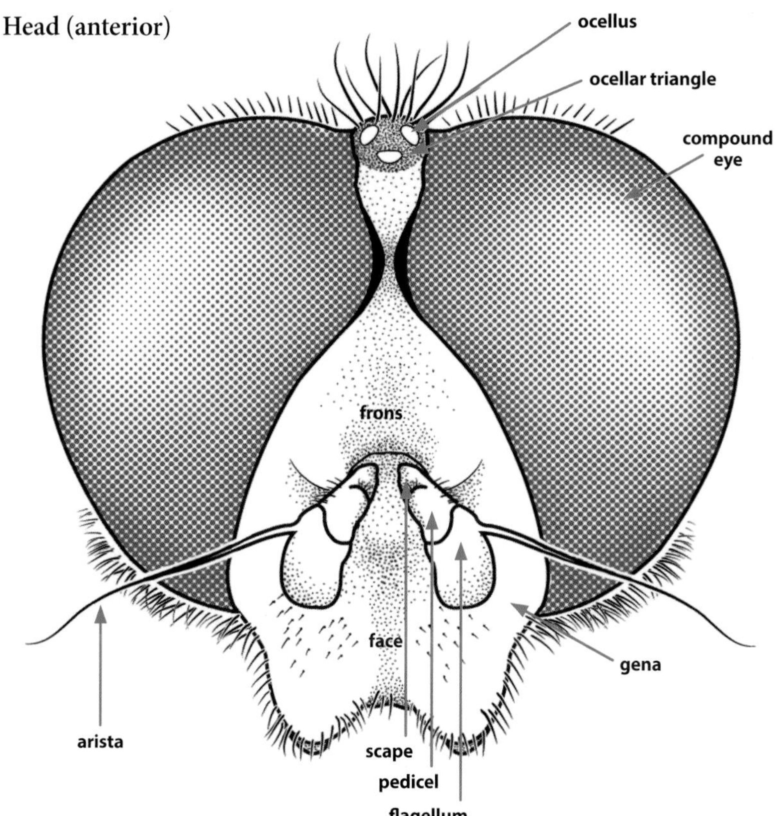

# Glossary of Taxonomic Terms

This glossary has been borrowed with some adaptation from Thompson (1999).

**abdomen** The posterior division of the insect body.

**abdominal margin** (adj. margined) This refers to a slightly flattened edge around the abdomen (edge appears crimped) and is often referred to as a premarginal sulcus or as margined or marginate.

**alula** A broad lobe on the posterior base of the wing. The size, shape, and vestiture of the alula are useful species characters.

**anal lobe** The area of the wing posterior to cell cua is frequently reduced in species with petiolate abdomens. The presence or absence of microtrichia on various areas of the anal lobe is a useful species character.

**anepimeron** (formerly pteropleuron). The medial plate of the mesothoracic pleuron, dorsal to the katepisternum and ventral to the wing base. Three distinct areas are recognized for taxonomic purposes: anterior, posterior, and dorsomedial portions. The anterior portion is always pilose, the posterior is occasionally pilose, and the dorsomedial is pilose only in some eristaline genera (e.g., *Eristalinus*).

**anepisternum** (formerly mesopleuron). The anterodorsal plate of the mesothoracic pleuron. In Syrphidae, the anepisternum is differentiated into a flattened anterior portion and a convex posterior portion. The pilosity of these areas varies, forming a critical generic character.

**antenna** A multisegmented sensory organ situated anteriorly on the head between the frons dorsally and the face ventrally. The shape of the antenna is one of the most widely used characters for recognition of genera; within genera the shape is usually constant, but antennal color does vary. The antenna consists of four major parts: the scape, pedicel, flagellum (formerly basoflagellomere, first flagellomere; now often called postpedicel), and arista.

**anterior** (adv. anteriorly). Adjective (adverb) meaning before, in front, toward the head end. Opposite of posterior. See orientation.

**apical** (adv. apically). Adjective (adverb) meaning toward the apex (cf. "the base" under basal). See orientation.

**appendix** A supplementary extension of a crossvein or vein.

**arcuate** Adjective for a marking or band that is slightly curved. The adjective lunulate is used for more deeply curved markings. See markings.

**arista** The apical flagellomeres (aristomeres) are usually reduced to a bristle-like structure, termed the arista. In most syrphids, the second and third flagellomeres are very small and only the fourth is greatly elongate; together they are termed the arista. The usual number of flagellomeres of the arista in syrphids is three. The arista varies in its insertion point from dorsobasal to apical and may be bare, pectinate, or plumose. See antenna.

**bare** The condition of lacking pile, or when referring to the surface of the wing, lacking microtrichia. See vestiture.

**basal** (adv. basally). Adjective (adverb) meaning toward the base. See orientation.

**basicosta** (also known as humeral plate). A small scale-like structure anterior to the base of vein C.

**basitarsus** The first tarsomere (closest to the body).

**bristle** Same as seta. Inserts into socket in the cuticle.

**calcar** (pl. calcaria, calcars, adj. calcarate). An articulating elongate extension of the exoskeleton. In Syrphidae, this term is usually restricted to an apicoventral extension of the metafemur, as in *Spilomyia*, or the apical extensions of the scutellum, as in *Microdon*.

**calypter** (formerly squama) (pl. calypteres). The membrane connecting the thorax to the posterior base of the wing forms two lobes called the calypteres. When the wing is folded, the calypteres are as well. The lobe attached to the thorax is referred to as the lower calypter as it is lower in position; the other lobe is the dorsal calypter.

**carina** (pl. carinae, adj. carinate). A sharp, raised, ridge-like extension of the exoskeleton. In Syrphidae, this is often used to refer to either a laterally compressed face as in *Tropidia*, or to abdominal sternites with a noticeable central ridge, as in *Neocnemodon*.

**cells** The cells of the wing are labeled above and are referred to by the name of the vein that forms the anterior margin of the cell.

**cercus** (pl. cerci). Plate lateral to the anus.

**costal section** The costal vein can be viewed as a series of sections defined by the various veins and crossveins that terminate in the costa. So, the first costal section is the area between the base of the costa and crossvein h, the second between h and the end of vein SC, the third between the end of vein SC and vein $R_1$, the fourth between vein $R_1$ and $R_{2+3}$, and the fifth between vein $R_{2+3}$ and vein $R_{4+5}$.

**coxa** (pl. coxae). The basal segment of the leg.

**crossvein** A short vein connecting long veins and referred to by the names of the two veins connected (lowercase), separated by a hyphen, such as crossvein r-m.

**dichoptic** The condition when the eyes are separated. Females are always dichoptic; males may be dichoptic. The alternative condition is holoptic.

**distal** Same as apical.

**dorsal** (adv. dorsally, dorsad). Adjective (adverb) meaning of or belonging to the upper surface. Opposite of ventral. See orientation.

**ejaculatory apodeme** Median sclerotized structure of the sperm pump in the male genitalia to which the muscles are attached. In shape, ranges from a simple short rod to a triangular plate to an umbrella-shaped piece.

**emarginate** Notched.

**epandrium** The dorsal sclerite (ninth tergum) of the male genitalia to which the surstyli are attached.

**epaulet** See tegula.

**eye** The term used for the large compound eye that occupies most of the side of the head. The small simple eyes on the vertex are referred to as ocelli.

**eye contiguity** In males, when the eyes are holoptic, the line of juncture is termed the eye contiguity. The length of this contiguity varies significantly in some genera and is a useful species character.

**face** The anteroventromedial area of the head, bordered laterally by the eyes and genae, dorsally by the antennae and frons, and ventrally by the subcranial cavity where the mouthparts are situated. Sometimes there is a distinctly demarcated region between the face and eyes, this area being the paraface. The shape of the facial region is one of the most important characters for the higher classification of syrphids.

**facet** The cornea of the individual ommatidia that make up the compound eye.

**fascia** (band) (pl. fasciae, adj. fasciate). A transverse (across the body) line. On the wing, fascia cross the narrow axis.

**femur** (pl. femora). The third segment of the leg.

**flagellomere** A division of the flagellum. In Syrphidae and most higher flies, the first flagellomere is greatly enlarged (flagellum) and the apical flagellomeres are reduced to a bristle-like structure, called the arista.

**flagellum** (basoflagellomere, first flagellomere). The third antennal segment. The shape is used extensively as a generic and specific character. This segment is now often known as the postpedicel, but there are ongoing issues with segmental homology. Because of this, we have chosen to use the older term flagellum. See antenna.

**flange** See lamina.

**frons** (front). The anterodorsomedial area of the head, bounded laterally by the eyes, dorsally by the vertex, and ventrally by the antennae. The anterior edge is usually differentiated and called the lunule.

**front** See frons.

**frontal lunule** See lunule.

**frontal prominence** (frontal tubercle, antennifer in *Ceriana* and relatives). In some syrphids, the region around the bases of the antennae is produced anteriorly; the result is termed the frontal prominence.

**frontal triangle** In males, when the eyes are holoptic, the frons forms this triangular area. The shape of this triangle varies greatly and is used as a species character.

**frontal tubercle** See frontal prominence.

**gena** (pl. genae). The lateral area of the head ventral to the eyes, anterior to the occiput, and posterior to the face. The term is used as equivalent to the cheek.

**genitalia** (terminalia). The reproductive structures.

**hair** (adj. hairy). See pile. The term is used colloquially and avoided scientifically, as the term hair is technically restricted to mammals.

**halter** (pl. halteres). The modified hind wing that consists of a base, pedicel (stem), and capitulum (head). Used like a gyroscope for monitoring body rotation in flight.

**head** The anterior division of the insect body.

**holoptic** The condition when the eyes are joined dorsally. Males are usually holoptic. The alternative condition is dichoptic.

**humeral crossvein** (crossvein h). A crossvein between the costal and subcostal veins at the base of the wing.

**humerus** See postpronotum.

**hyaline** Transparent. Used to describe a wing, hyaline means without a dark pattern. The term does not apply to the stigma, which may be dark.

**hypandrium** The ventral sclerite of the male genitalia.

**infuscate** Darkened with a brownish tinge. In reference to a completely darkened wing (compare to hyaline).

**katepimeron** (barette). The posteroventral plate of the mesothoracic pleuron, dorsal to the meron. Frequently the demarcation between the katepimeron and meron is weak. The katepimeron is occasionally pilose.

**katepisternum** (sternopleuron). The anteroventral plate of the mesothoracic pleuron.

**lamina** (plate or flange) (pl. laminae, adj. laminate). A large, thin extension of the exoskeleton.

**lateral** (adv. laterally, laterad). Of the side, away from the center. Opposite of medial. See orientation.

**leg** The ventrolateral locomotion organs.

**length** See size.

**lunulate** An adjective for markings that are crescent shaped. Less strongly curved markings are described as arcuate. See markings.

**lunule** The area of different texture on the anterior edge of the frons. This area is usually shiny, bare, and frequently differently colored than the rest of the frons.

**macula** (pl. maculae). Any kind of marking that is not a band or stripe. The term is usually used with adjectives to describe the mark, such as arcuate, lunulate, rectangular, punctate, or triangular.

**margin** See abdominal margin.

**markings** (pattern). Flower flies usually display intricate, bright color patterns, made up of various kinds of markings. These markings are of three basic types: fasciae, vittae, and maculae (bands, stripes, and markings). Various adjectives, such as arcuate, lunulate, or punctate, are used to more precisely define these markings.

**medial** Adjective meaning of the middle, toward the center. Opposite of lateral. See orientation.

**meron** The posteroventral plate of the mesothoracic pleuron ventral to the katepimeron. The demarcation between these two plates is usually indistinct. The meron is bare although the posterior margin anterior to the posterior spiracle may be pilose (see posterior spiracular fringe).

**meso-** Refers to components of the middle section of the thorax that includes the middle legs.

**mesonotum** The term for the dorsal part of the midthoracic segment and includes both the scutum and scutellum. Frequently incorrectly used for just the scutum.

**meta-** Refers to components of the posterior section of the thorax that includes the hind legs.

**metasternum** The area anterior to the metacoxa and ventral to the meron. The presence or absence of pile on the metasternum and the degree of development of the metasternum are important characters for the higher classification of syrphids.

**metathoracic pleuron** The lateral portions of the reduced third thoracic segment. Only the metepimeron and metepisternum provide taxonomic characters.

**metathoracic spiracular pile patch** See posterior spiracular fringe.

**metepimeron** The posterior portion of the metathoracic pleuron.

**metepisternum** The anterior portion of the metathoracic pleuron.

**microtrichium** (pl. microtrichia; adj. microtrichose). Small pile-like extension of exoskeleton (not socketed). See vestiture.

**notopleuron** The anterolateral region of the scutum bounded anteriorly by the postpronotum and posteriorly by the transverse suture.

**occiput** The posterior area of the head, limited anteriorly by the eyes, dorsally by the vertex, and ventrally by the genae. In some groups, the occiput may be uniformly swollen or dorsally swollen. In other groups, the nature of the vestiture on the lateral portions of the occiput is an important species group character.

**ocellar triangle** A region on the vertex defined by the ocelli. The ocellar triangle is a subregion of the vertical triangle, not synonymous with it.

**ocellus** (pl. ocelli). One of the three simple eyes present dorsally between the larger compound eyes.

**orientation** A fly or any object can be divided by three planes, the horizontal, sagittal, and transverse, each plane being 90 degrees to the other. Numerous terms define orientation, but only four pairs of terms are necessary: anterior–posterior, apical–basal, dorsal–ventral, and lateral–medial.

**pedicel** The second segment of the antenna, situated between the scape and flagellum.

**petiole** (adj. petiolate). A petiole is a stem and petiolate is an adjective meaning having a stem. In syrphid taxonomy, these terms are used in reference to various cells of the wing and to the shape of the abdomen.

**phallus** (aedeagus). The male copulatory organ, the median structure of the male genitalia, surrounded by the hypandrium. In Syrphidae, the phallus articulates with the hypandrium dorsally and ventrally, and laterally with the gonostylus.

**pile** (pilis [singular], pili [pl.], adj. pilose). The latin noun for hair, pilis the singular and pilose the adjective. Pilose is the condition of having hairs. See vestiture.

**plate** See lamina.

**pollinose** (pollen [singular]). The condition of being covered with opaque material that may appear like fine powder or dust. The material is made up of microtrichia. In syrphids, the term pollinose is used generally for any body area covered by microtrichia (except the wing). See vestiture.

**postalar callus** The posterolateral region of the mesoscutum separated from the scutum proper by a broad shallow furrow.

**postalar pile tuft** This is situated on the postalar ridge anterior to the postalar callus and posterior to the wing base.

**posterior** (adv. posteriorly). Adjective (adverb) meaning behind, in back, away from the head end. Opposite of anterior. See orientation.

**posterior spiracular fringe** (metathoracic spiracular pile patch). A row or patch of long pile anterior to and/or ventral of the posterior spiracle.

**postmetacoxal bridge** A sclerotized band that extends the metathoracic pleura from one side to the other. The bridge occurs in microdontines, some syrphines, and eristalines. Except in the microdontines, this bridge tends to be correlated with petiolate abdomens and enlarged metafemora.

**postpronotum** (humerus). A distinct plate on the anterolateral corner of the mesonotum.

**premarginal sulcus** A shallow groove medial to the margin of a sclerite. In flower flies, premarginal sulci are commonly found on the scutellum and abdomen. The presence of the sulcus is also referred to by stating that the structure is "emarginate or margined."

**presutural area** (callus, depression). See notopleuron.

**pro-** Refers to components of the anterior section of the thorax that includes the front legs.

**pronotum** The dorsal part of the most anterior thoracic segment. Only the posterior portion is distinct in flies (see postpronotum). Sometimes synonymous with postpronotum.

**proximal** See basal.

**pruinose** See pollinose. See vestiture.

**pubescent** An adjective referring to having very long, dense microtrichia, which appear like velvet. Sometimes a sclerite may appear to be pilose because the pubescence is long, but closer inspection should reveal the lack of alveoli (sockets, pits) at the base of the microtrichia. See vestiture.

**pulvillus** (pl. pulvilli). The apicolateral pad on the apex of the tarsus.

**punctate** Adjective used to modify macula to refer to a round spot. Rarely the adjective is used to refer to the integument or to pollinose areas; in these cases it means that the integument or the pollinosity has small, round pits.

**punctum** (pl. puncta). This refers to a small, round pit or hole in the integument.

**scale** A broadly flattened pilis (hair).

**scape** The basal or first segment of the antenna.

**sclerite** A component plate of the exoskeleton.

**scutellum** The posterior part of the notum. In flies, scutellum is accepted to apply to the mesoscutellum only. The shape, pile, and color of the scutellum offer valuable species characters. The apicoventral margin of the scutellum frequently has pile and such pile is referred to as the "ventral scutellar fringe." The dorsoapical margin of the scutellum may appear to have a rim, which is created by a premarginal sulcus or furrow.

**scutum** The large anterior portion of the notum. In flies, scutum is accepted to apply to the mesoscutum.

**seta** (pl. setae, adj. setate) (bristles). A long, thick macrotrichium. See vestiture.

**setula** (pl. setulae, adj. setulate). A short, thick macrotrichium. See vestiture.

**size** Syrphids vary intraspecifically in size based on their larval food consumption (especially the predaceous syrphines). Body length is measured here from the front of the face (not including antennae) to the tip of the straightened abdomen.

**spina** (pl. spinae, adj. spinate, spinose). A nonarticulating, elongate extension of the exoskeleton.

**spine** Same as a seta, usually described as "short spinose seta." The term usually is used for a seta found apicoventrally on the metafemur of eristaline syrphids, as in the genus *Xylota*. Also, used for elongate, nonarticulating extensions of the exoskeleton.

**spiracle** In Diptera, there are only two thoracic spiracles, the meso- and metathoracic spiracles, which for convenience are referred to as the anterior and posterior spiracles. The thoracic spiracles are mainly used as landmarks to identify the various parts of the pleuron.

**spot** See macula.

**spur** See calcar. Sometimes spur may be used in reference to wing venation. Spur in this sense is a short supplementary extension to a vein or crossvein, which is here referred to as an appendix.

**spurious vein** The sclerotization of the convexity between the radial and medial field of the wing appears like a vein and is termed the spurious vein. This is found in most flower flies in our region and rarely in other flies (such as Conopidae).

**sternite** See sternum.

**sternum** (pl. sterna) (sternite). The ventral plate of an abdominal segment.

**stigma** Pigmented region of the subcostal wing cell.

**stigmal (stigmatical) crossvein** A thickening of the wing membrane in the stigmal area that appears like a crossvein.

**stripe** See vitta.

**style** A term for the arista when the arista is thick (stout) and apical (terminal) in position. Seen in *Ceriana*.

**sulcus** (pl. sulci, adj. sulcate). A groove or shallow depression of purely functional origin. In flower flies, the scutellum and abdomen frequently have premarginal sulci (margins).

**surstylus** (pl. surstyli). The clasping organs of the male genitalia.

**synanthrope** (adj. synanthropic). Organisms that live near, and benefit from, an association with humans and the artificial habitats that humans create around them.

**syntergosternite 8**. The fusion of tergite 8 and sternite 8.

**tarsomere** (pl. tarsomeres). A part of the tarsus.

**tarsus** (pl. tarsi). The fifth part of the leg.

**tegula** (also epaulet). A large scale-like structure at the anterior base of the wing, the color of which is a useful species character among the eristalines.

**tentorium** An invagination of the exoskeleton that supports muscle attachments in the head. Where the exoskeleton is invaginated, pits are formed. These pits are either oval or elongate.

**tergite** See tergum.

**tergum** (pl. terga) (tergite). The dorsal plate of an abdominal segment.

**terminal** See apical.

**terminalia** See genitalia.

**thorax** The second (middle) division of the insect body.

**tibia** (pl. tibiae). The fourth part of the leg.

**tomentose** An adjective used to refer to very thick and opaque pile found in the genus *Meromacrus*.

**tooth** (pl. teeth) (dens [dentis]). A large, nonarticulating extension of the exoskeleton that is as as wide or wider than high.

**transverse suture** A transverse sulcus that runs across the mesonotum just anterior to the bases of the wings.

**trochanter** The small second part of the leg, connecting the coxa to the femur.

**tubercle** (adj. tuberculate). The face is frequently produced anteromedially into a distinct swelling, termed the tubercle. The presence or absence of a tubercle is frequently used as a generic character and the shape as a species character. Tubercle is also used for any small, distinct, rounded protuberance elsewhere, such as on the metafemur.

**unmarginate** As in "abdomen unmarginate." Not margined. See premarginal sulcus.

**veins** The names for the wing veins follow the "Wootton and Enos" system as interpreted and presented by Cumming and Wood (2017).

**ventral** (adv. ventrally, ventrad). Adjective (adverb) meaning of or belonging to the lower surface. Opposite of dorsal. See orientation.

**ventral scutellar fringe** The ventral surface of the scutellum may have long pile, which is referred to as the ventral scutellar fringe or scutellum with ventral fringe.

**vertex** The most dorsal portion of the head, bounded by the frons anteriorly, the eyes laterally, and the occiput posteriorly. The ocellar triangle is within the vertex. The vertex and associated pollen and pile are frequently of different color than found on the frons.

**vertical triangle** In males when the eyes are holoptic, the vertex forms a triangular area, termed the vertical triangle. The shape of this triangle can vary greatly.

**vestiture** The integument of flies is covered with two basic types of vestiture: microtrichia and macrotrichia. While there are only these two fundamental types, the form of each varies greatly, giving rise to a long, complex history of conflicting terminology. Microtrichia have been referred to as tomentum, pubescence, pruinescence, pollinosity, etc. Macrotrichia include setae (bristles), pile (hair), and setulae (spines). Macrotrichia differ from microtrichia as they are connected to nerves and are surrounded at the base by a membranous alveolus (socket). Here, macrotrichia are divided into three kinds based on thickness and length: setae are long and thick; setulae are short and thick; and pile is used for long or short, thin macrotrichia. Pile and pili are plural, pilis singular, and pilose, the adjective. Sometimes adjectives are used to described specialized pile, such as tomentose pile for the specialized thick, opaque pile found in the genus *Meromacrus*. Only two terms are used for microtrichia (villi) even though microtrichia can vary greatly in shape. Microtrichia (usually as the adjective, microtrichose) is used for the vestiture of the wing membrane. Pollen (usually as the adjective, pollinose) is used for the microtrichia on the body.

**villis** (pl. villi). Same as wing microtrichium. See vestiture.

**vitta** (pl. vittae, adj. vittate) An anterior to posterior (longitudinal) line (stripe). See markings.

**wing** The large dorsolateral membranous flight organ.

# Plant Names Used in Guide

| Scientific Name | Common Name |
|---|---|
| *Abies* | fir |
| *Acer* | maple |
| Aceraceae | maple family |
| *Acer negundo* | boxelder |
| *Acer spicatum* | mountain maple |
| *Achillea* | yarrow |
| *Actaea* | bugbane/baneberry |
| Adoxaceae | moschatel family |
| *Aegopodium* | goutweed |
| *Ageratina* | snakeroot |
| *Ajuga* | bugle |
| *Alisma* | water plantain |
| *Alliaria* | garlic mustard and relatives |
| *Allium* | onion |
| *Alnus* | alder |
| Amaranthaceae | amaranth family |
| *Amaryllis* | amaryllis |
| *Amelanchier* | serviceberry |
| Anacardiaceae | cashew/sumac family |
| *Anaphalis* | pearly everlasting |
| *Androsace* | rockjasmine |
| *Anemone* | anemone |
| *Angelica* | angelica |
| Apiaceae | carrot family/ umbellifers |
| *Apium* | celery and relatives |
| *Apocynum* | dogbane |
| *Arabis* | rockcress |
| *Aralia* | spikenard |
| *Arctium* | burdock |
| *Aronia* | chokeberry |
| *Aruncus* | aruncus |
| *Asclepias* | milkweed |
| *Asparagus* | asparagus |
| Asteraceae | aster family |
| *Avena* | oat |
| *Baccharis* | baccharis |
| *Baccharis pilularis* | coyotebrush |
| *Barbarea* | winter cress |
| *Bartsia alpina* | velvetbells |
| *Begonia* | begonia |
| *Berberis* | barberry |
| *Beta vulgaris* | common beet |
| *Betula* | birch |
| Betulaceae | birch family |
| *Bidens* | beggarticks |
| *Blephilia* | pagoda-plant |
| *Brassica* | mustard |
| Brassicaceae | mustard family/ crucifers |
| *Brassica oleracea* | cabbage |
| *Bromus* | brome |
| *Cakile* | searocket |
| *Calluna* | heather |
| *Caltha* | marsh marigold |
| *Camassia* | camas |
| *Campanula* | bellflower |
| Caprifoliaceae | honeysuckle family |
| *Capsicum* | pepper |
| *Capsicum annuum* | bell/cayenne pepper |
| *Cardamine* | bittercress |
| *Carex* | sedge |
| *Carum* | caraway |
| *Carya* | hickory |
| Caryophyllaceae | pink family |
| *Castanea* | chestnut |
| *Caulophyllum* | cohosh |
| *Ceanothus* | New Jersey tea |
| *Cedrus* | cedar |
| Celastraceae | bittersweet family |
| *Celtis* | hackberry |
| *Centaurea* | knapweed |
| *Cephalanthus* | buttonbush |
| *Chamaedaphne* | leatherleaf |
| *Chamerion angustifolium* | fireweed |
| *Chrysanthemum* | daisy |
| *Chrysothamnus* | rabbitbrush |
| *Cichorium* | chicory |
| *Cicuta* | water hemlock |
| *Circaea* | enchanter's nightshade |
| *Cirsium* | thistle |
| *Citrus* | citrus |
| *Clematis* | leather flower |
| *Clintonia* | bluebead |
| *Conium* | poison hemlock |
| Convolvulaceae | morning-glory family |
| *Convolvulus* | bindweed, morning glory |
| *Corema* | corema |
| *Coreopsis* | tickseed |
| *Coreopsis rosea* | pink tickseed |
| Cornaceae | dogwood family |
| *Cornus* | dogwood |
| *Corylus* | hazelnut |
| *Crataegus* | hawthorn |
| *Cryptotaenia* | honewort |
| Cucurbitaceae | gourd family/cucurbits |
| Cupressaceae | cypress family |
| *Cynara scolymus* | globe artichoke |
| Cyperaceae | sedge family |
| *Daucus* | carrot |
| *Diervilla* | bush honeysuckle |
| *Digitaria* | crabgrass |
| *Doellingeria* | tall flat-topped asters |
| *Dryas* | mountain-avens |
| *Echium* | viper's bugloss |
| *Elymus* | wildrye |
| *Epilobium* | willowherb |
| *Equisetum* | horsetail |
| Ericaceae | heath/heather family |
| *Erigeron* | fleabane |
| *Eriophorum* | cottongrass |

| Scientific Name | Common Name |
|---|---|
| *Erodium* | stork's bill |
| *Eschscholzia* | California poppy |
| *Euonymus* | euonymus |
| *Euonymus europaeus* | European spindletree |
| *Eupatorium* | thoroughwort/boneset |
| *Euphorbia* | spurge |
| Euphorbiaceae | spurge family |
| *Euthamia* | goldentop |
| *Eutrochium* | joe pye weed |
| Fabaceae | bean family/legumes |
| Fagaceae | beech family |
| *Fagopyrum esculentum* | buckwheat |
| *Fagus* | beech |
| *Farfugium* | leopard plant |
| *Filipendula* | queen |
| *Foeniculum vulgare* | sweet fennel |
| *Fragaria* | strawberry |
| *Fraxinus* | ash |
| *Galax* | galax |
| *Galearis* | galearis |
| *Galium* | bedstraw |
| Geraniaceae | geranium family |
| *Geranium* | geranium |
| *Geum* | avens |
| *Gleditsia triacanthos* | honeylocust |
| *Glyceria* | mannagrass |
| *Gnaphalium* | cudweed |
| *Gossypium* | cotton |
| *Grindelia* | gumweed |
| Grossulariaceae | currant family |
| *Gymnosperma* | gymnosperma |
| *Hackelia* | stickseed |
| *Hedyotis* | starviolet |
| *Helianthemum* | frostweed |
| *Helianthus* | sunflower |
| *Heracleum* | cowparsnip |
| *Hibiscus laevis* | halberd-leaved rose mallow |
| *Hieracium* | hawkweed |
| *Hippeastrum* | hippeastrum |
| *Hordeum vulgare* | common barley |
| *Houstonia* | bluet |
| *Hyacinthus* | hyacinths |
| *Hydrangea* | hydrangea |
| Hydrophyllaceae | waterleaf family |
| *Hydrophyllum* | waterleaf |
| *Hypericum calycinum* | Aaron's beard |
| *Hypochaeris* | cat's ear |
| *Ilex* | holly |
| *Impatiens* | touch-me-not |
| *Ipomoea* | morning-glory |
| *Iris* | iris |
| Juglandaceae | walnut family |
| *Juglans* | walnut |
| *Juglans nigra* | black walnut |
| *Juncus* | rush |
| *Juniperus virginiana* | eastern redcedar |
| *Kalmia* | laurel |
| *Knautia* | scabiosa |
| *Lactuca* | lettuce |

| Scientific Name | Common Name |
|---|---|
| *Lactuca serriola* | prickly lettuce |
| Lamiaceae | mint family |
| *Larix* | larch/tamarack |
| *Leontodon* | hawkbit |
| *Leucanthemum* | daisy |
| *Licania* | licania |
| *Ligusticum* | licorice-root |
| Liliaceae | lily family |
| *Lindera* | spicebush |
| *Linnaea* | twinflower |
| *Liriodendron* | tuliptree |
| *Lobularia* | lobularia |
| *Ludwigia* | primrose-willow |
| *Luzula* | woodrush |
| *Lycopus* | waterhorehound |
| *Lysimachia* | yellow loosestrife |
| *Lythrum* | loosestrife |
| *Maianthemum* | mayflower |
| *Malus* | apple |
| *Malus pumila* | paradise apple |
| Malvaceae | mallow family |
| *Matricaria* | mayweed |
| *Medicago* | alfalfa |
| *Medicago sativa* | alfalfa |
| *Melilotus* | sweetclover |
| *Mentha* | mint |
| *Menyanthes* | buckbean |
| *Miconia* | johnnyberry |
| *Myosotis* | forget-me-not |
| *Narcissus* | daffodil |
| *Nuphar* | pond-lily |
| *Nymphaea* | waterlily |
| *Nyssa* | tupelo |
| *Oenothera* | evening primrose |
| Oleaceae | olive family |
| Onagraceae | willowherb/evening primrose family |
| *Oryzopsis* | ricegrass |
| *Osmorhiza* | sweetroot |
| *Oxypolis* | cowbane |
| *Packera* | ragwort |
| *Papaver* | poppy |
| Papaveraceae | poppy family |
| *Parnassia* | grass of Parnassus |
| *Pastinaca* | parsnip |
| *Patrinia* | patrinia |
| *Pedicularis* | lousewort |
| *Penstemon* | beardtongue |
| *Perideridia* | yampah |
| *Persicaria* | smartweed |
| *Phalaris* | canarygrass |
| *Photinia* | chokeberry |
| *Phragmites* | reed |
| *Phyla nodiflora* | turkey tangle fogfruit |
| *Physocarpus* | ninebark |
| *Picea* | spruce |
| *Pilosella* | hawkweed |
| *Pimpinella* | burnet saxifrage |
| Pinaceae | pine family |
| *Pinguicula* | butterwort |

| Scientific Name | Common Name |
|---|---|
| *Pinus* | pine |
| *Pinus ponderosa* | ponderosa pine |
| *Pisum sativum* | garden pea |
| *Plantago* | plantain |
| *Platanus* | sycamore |
| Poaceae | grass family |
| Polemoniaceae | Jacob's-ladder/phlox family |
| Polygonaceae | knotweed/buckwheat family |
| *Polygonum* | knotweed |
| *Pontederia* | pickerelweed |
| *Populus* | poplar |
| *Populus balsamifera* | balsam poplar |
| *Populus grandidentata* | bigtooth aspen |
| Portulacaceae | purslane family |
| *Potentilla* | cinquefoil |
| *Prenanthes* | rattlesnake-root |
| *Primula* | primrose |
| Primulaceae | primrose family |
| *Prosopis* | mesquite |
| *Prunus* | plum, cherry |
| *Prunus avium* | sweet cherry |
| *Prunus persica* | peach |
| *Pycnanthemum* | mountainmint |
| *Pyrrocoma* | goldenweed |
| *Pyrus* | pear |
| *Quercus* | oak |
| *Quercus montana* | chestnut oak |
| Ranunculaceae | buttercup family |
| *Ranunculus* | buttercup |
| *Rhexia* | meadow-beauty |
| *Rhododendron* | rhododendron, Labrador tea |
| *Rhus* | sumac |
| *Rhus typhina* | staghorn sumac |
| *Ribes* | currant |
| *Rosa* | rose |
| *Rosa multiflora* | multiflora rose |
| Rosaceae | rose family |
| *Rubus* | blackberry, raspberry |
| *Rudbeckia* | coneflower |
| *Rumex* | dock |
| *Rumex crispus* | curly dock |
| Rutaceae | rue/citrus family |
| *Sagittaria* | arrowhead |
| Salicaceae | willow family |
| *Salix* | willow |
| *Salix nigra* | black willow |
| *Sambucus* | elderberry |
| *Sanicula* | sanicle |
| Sapindaceae | soapberry family |
| *Saxifraga* | saxifrage |
| Saxifragaceae | saxifrage family |
| *Schoenoplectus* | bulrush |
| *Scirpus* | bulrush |
| *Secale cereale* | cereal rye |
| *Sedum* | stonecrop |
| *Senecio* | ragwort |

| Scientific Name | Common Name |
|---|---|
| *Serjania* | serjania |
| *Seseli* | seseli |
| *Setaria viridis* | green bristlegrass |
| *Silene* | catchfly |
| Solanaceae | potato/nightshade family |
| *Solidago* | goldenrod |
| *Sonchus* | sowthistle |
| *Sorbus* | mountain ash |
| *Spartina* | cordgrass |
| *Spermacoce* | false buttonweed |
| *Spinacia* | spinach |
| *Spiraea* | spirea |
| *Spiraea prunifolia* | bridalwreath spirea |
| *Stachys* | hedgenettle |
| *Stellaria* | starwort, chickweed |
| Styracaceae | storax/silverbell family |
| *Succisa* | succisa |
| *Symphoricarpos* | snowberry |
| *Symphyotrichum* | aster |
| *Symplocos* | sweetleaf |
| *Tanacetum* | tansy |
| *Taraxacum* | dandelion |
| *Taxodium* | bald cypress |
| *Thalictrum* | meadow-rue |
| *Thlaspi* | pennycress |
| *Thuja occidentalis* | arborvitae |
| *Tiarella* | foamflower |
| *Tilia* | basswood |
| *Tragopogon* | goatsbeard |
| *Trifolium* | clover |
| *Trillium* | trillium |
| *Tripleurospermum* | mayweed |
| *Triticum* | wheat |
| *Triticum aestivum* | common wheat |
| *Tsuga* | hemlock |
| *Tussilago* | coltsfoot |
| *Typha* | cattail, bulrush |
| Typhaceae | bur-reed/cattail family |
| *Ulex* | gorse |
| Ulmaceae | elm family |
| *Ulmus* | elm |
| *Urtica* | nettle |
| *Vaccinium* | blueberry, cranberry |
| *Valeriana* | valerian |
| *Valerianella* | cornsalad |
| *Veratrum* | false hellebore |
| *Verbascum* | mullein |
| Verbenaceae | verbena famly |
| *Verbesina* | crownbeard |
| *Verbesina helianthoides* | gravelweed |
| *Viburnum* | viburnum |
| *Vicia* | vetch |
| *Viola* | violet |
| Vitaceae | grape family |
| *Washingtonia* | fan palm |
| *Wyethia* | mule-ears |
| *Zea mays* | corn |
| *Zizia* | alexanders |

# Checklist

The following list includes all species covered by the field guide. Synonyms are shown indented below the valid species. Authors and publication years for all species are provided for clarity. If you discover a name being used in the literature for a species, this will make cross-referencing easy. All nomenclatural changes in the guide are also summarized here **in bold** using the following abbreviations: **N. Comb.** (new combination), **N. Syn.** (new synonym), and **Res. Sp.** (resurrected species). These changes are described in the Introduction.

MICRODONTINAE

☐ ***Laetodon laetus*** (Loew 1864)
    ***scitulus*** (Williston 1887)
☐ ***Microdon (Chymophila) fulgens*** Wiedemann 1830
    ***englossoides*** (Gray 1832)
    ***euglossoides*** Gray 1832
    ***splendens*** (Macquart 1834)
☐ ***Microdon (Dimeraspis) abditus*** Thompson 1981
☐ ***Microdon (Dimeraspis) fuscipennis*** (Macquart 1834)
    ***agapenor*** Walker 1849
    ***pachystylum*** Williston 1887
☐ ***Microdon (Dimeraspis) globosus*** (Fabricius 1805)
    ***podagra*** (Newman 1838)
    ***marmoratus*** Bigot 1884
    ***albipilis*** Curran 1925
    ***conflictus*** Curran 1925
    ***pseudoglobsus*** Curran 1925
    ***hutchingsi*** Curran 1927
☐ ***Microdon (Microdon) abstrusus*** Thompson 1981
☐ ***Microdon (Microdon) albicomatus*** Novak 1977
☐ ***Microdon (Microdon) aurulentus*** (Fabricius 1805)
☐ ***Microdon (Microdon) cothurnatus*** Bigot 1884
    ***cockerelli*** Jones 1922
☐ ***Microdon craigheadii*** Walton 1912
☐ ***Microdon (Microdon) manitobensis*** Curran 1924
☐ ***Microdon (Microdon) megalogaster*** Snow 1892
    ***bombiformis*** Townsend 1895
☐ ***Microdon (Microdon) ocellaris*** Curran 1924
☐ ***Microdon (Microdon) ruficrus*** Williston 1887
    ***basicornis*** Curran 1925
    ***champlaini*** Curran 1925
☐ ***Microdon (Microdon) tristis*** Loew 1864
    ***robusta*** Telford 1939
☐ ***Microdon (Microdon)*** undescribed species 17-1
☐ ***Mixogaster breviventris*** Kahl 1897
☐ ***Mixogaster johnsoni*** Hull 1941
☐ ***Mixogaster*** undescribed species 1
☐ ***Omegasyrphus coarctatus*** (Loew 1864)
    ***baliopterus*** (Loew 1872)
☐ ***Serichlamys rufipes*** (Macquart 1842)
    ***limbus*** (Williston 1887)

ERISTALINAE

☐ ***Anasimyia anausis*** (Walker 1849) **Res. Sp., N. Comb.**
    ***hamatus*** (Loew 1863) **N. Comb.**
☐ ***Anasimyia bilinearis*** (Williston 1887) **N. Comb.**
    ***perfidiosus*** (Hunter 1897) **N. Comb.**
☐ ***Anasimyia chrysostoma*** (Wiedemann 1830) **N. Comb.**
    ***relicta*** (Curran & Fluke 1926) **N. Syn., N. Comb.**
    ***orion*** (Hull 1943) **N. Comb.**
☐ ***Anasimyia distincta*** (Williston 1887) **N. Comb.**
☐ ***Anasimyia grisescens*** (Hull 1943) **N. Comb.**
☐ ***Anasimyia*** undescribed species 1
☐ ***Anasimyia*** undescribed species 2
☐ ***Arctosyrphus willingii*** (Smith 1912) **N. Comb.**
    ***nitidulus*** Frey 1918 **N. Comb.**
    ***canadensis*** (Curran 1922) **N. Comb.**
☐ ***Blera analis*** (Macquart 1842)
☐ ***Blera armillata*** (Osten Sacken 1875)
    ***hunteri*** (Curran 1925)
    ***pacifica*** (Curran 1953)
☐ ***Blera badia*** (Walker 1849)
    ***intersistens*** (Walker 1849)
☐ ***Blera confusa*** Johnson 1913
☐ ***Blera nigra*** (Williston 1887)
☐ ***Blera notata*** (Wiedemann 1830)
    ***profusus*** (Walker 1849)
    ***pallipes*** (Bigot 1884)
☐ ***Blera pictipes*** (Bigot 1884)
☐ ***Blera umbratilis*** (Williston 1887)
☐ ***Brachyopa daeckei*** Johnson 1917
    ***nigricauda*** Curran 1922
☐ ***Brachyopa diversa*** Johnson 1917
☐ ***Brachyopa flavescens*** Shannon 1915
☐ ***Brachyopa notata*** Osten Sacken 1875
☐ ***Brachyopa perplexa*** Curran 1922
☐ ***Brachyopa*** undescribed species 17-5
☐ ***Brachyopa*** undescribed species 78-2
☐ ***Brachyopa vacua*** Osten Sacken 1875
    ***racua*** Osten Sacken 1875
☐ ***Brachypalpus (Brachypalpus) oarus*** (Walker 1849)
    ***frontosus*** Loew 1872
    ***margaritus*** Hull 1946
☐ ***Brachypalpus (Crioprora) femoratus*** (Williston 1882)

☐ ***Callicera erratica*** (Walker 1849)
  ***johnsoni*** Hunter 1896
  ***auripila*** Metcalf 1916
☐ ***Ceriana (Ceriana) abbreviata*** Loew 1864
  ***proxima*** (Curran 1925)
☐ ***Ceriana (Sphiximorpha) willistoni*** (Kahl 1897)
  ***ontarioensis*** (Curran 1921)
☐ ***Chalcosyrphus (Chalcosyrphus) aristatus*** (Johnson 1929)
☐ ***Chalcosyrphus (Chalcosyrphus) depressus*** (Johnson 1929)
  ***atra*** (Curran 1925)
  ***carri*** (Curran 1941)
☐ ***Chalcosyrphus (Xylotomima) anomalus*** (Shannon 1925)
☐ ***Chalcosyrphus (Xylotomima) anthreas*** (Walker 1849)
  ***facialis*** (Coquillett 1910)
☐ ***Chalcosyrphus (Xylotomima) chalybeus*** (Wiedemann 1830)
  ***violascens*** (Megerle 1803)
  ***purpurea*** (Walker 1849)
☐ ***Chalcosyrphus (Xylotomima) curvarius*** (Curran 1941)
☐ ***Chalcosyrphus (Xylotomima) inarmatus*** (Hunter 1897)
  ***apicaudus*** (Curran 1922)
  ***banksi*** (Hull 1945)
☐ ***Chalcosyrphus (Xylotomima) libo*** (Walker 1849)
  ***marginalis*** (Williston 1887)
☐ ***Chalcosyrphus (Xylotomima) metallicus*** (Wiedemann 1830)
  ***dascon*** (Walker 1849)
  ***subtropicus*** (Curran 1925)
  ***althaea*** (Hull 1943)
  ***astarte*** (Hull 1943)
  ***primavera*** (Hull 1944)
☐ ***Chalcosyrphus (Xylotomima) metallifer*** (Bigot 1884)
  ***rileyi*** (Williston 1887)
☐ ***Chalcosyrphus (Xylotomima) nemorum*** (Fabricius 1805)
  ***baton*** (Walker 1849)
  ***fraudulosa*** (Loew 1864)
  ***americana*** (Shannon 1926)
  ***arsenjevi*** (Violovitsh 1980)
☐ ***Chalcosyrphus (Xylotomima) ontario*** (Curran 1941)
☐ ***Chalcosyrphus (Xylotomima) piger*** (Fabricius 1794)
  ***haematodes*** (Fabricius 1805)
  ***crassipes*** (Wahlberg 1839)
  ***nigra*** (Walker 1849)
  ***pini*** (Perris 1870)
  ***rubbiginigaster*** (Bigot 1884)
☐ ***Chalcosyrphus (Xylotomima) plesius*** (Curran 1925)
☐ ***Chalcosyrphus (Xylotomima) sacawajeae*** (Shannon 1926)
  ***sacajawaeae*** (Shannon 1926)
  ***sacajaweai*** (Shannon 1926)
☐ ***Chalcosyrphus (Xylotomima) vecors*** (Osten Sacken 1875)
☐ ***Cheilosia albitarsis*** (Meigen 1822) **Res. Sp.**
  ***hiawatha*** (Shannon 1922) **N. Syn.**
☐ ***Cheilosia bigelowi*** (Curran 1926)
☐ ***Cheilosia capillata*** (Loew 1863)
☐ ***Cheilosia comosa*** (Loew 1863)
  ***tarda*** (Snow 1895)
  ***laevifrons*** (Jones 1907)
  ***brevichaeta*** (Shannon 1922)
  ***caltha*** (Shannon 1922) **N. Syn.**
  ***sensua*** (Curran 1922) **N. Syn.**
☐ ***Cheilosia cynoprosopa*** Hull & Fluke 1950
  ***cineralis*** Hull & Fluke 1950
☐ ***Cheilosia hunteri*** (Curran 1922)
  ***nigrofasciata*** (Curran 1926) **N. Syn.**
  ***callichroma*** Hull & Fluke 1950
  ***nuda*** Hull & Fluke 1950
  ***truncata*** Hull & Fluke 1950
☐ ***Cheilosia laevis*** (Bigot 1884)
☐ ***Cheilosia lasiophthalma*** Williston 1882
  ***nigroapicata*** (Curran 1926) **N. Syn.**
  ***browni*** Curran 1931 **N. Syn.**
☐ ***Cheilosia latrans*** (Walker 1849)
  ***aesyctes*** (Walker 1849)
  ***tristis*** (Loew 1863)
  ***longipilosa*** (Wehr 1924)
☐ ***Cheilosia leucoparea*** (Loew 1863)
☐ ***Cheilosia orilliaensis*** (Curran 1922)
  ***consentiens*** (Curran 1926) **N. Syn.**
☐ ***Cheilosia pallipes*** (Loew 1863)
  ***flavissima*** (Becker 1894)
☐ ***Cheilosia pontiaca*** (Shannon 1922)
☐ ***Cheilosia prima*** (Hunter 1896)
  ***ontario*** (Curran 1922)
  ***slossonae*** (Shannon 1922)
  ***slossoni*** (Shannon 1922)
☐ ***Cheilosia primoveris*** (Shannon 1915)
☐ ***Cheilosia rita*** (Curran 1922)
  ***sialia*** (Shannon 1922)
  ***alpinensis*** Fluke & Hull 1947
  ***argentipila*** Fluke & Hull 1947
☐ ***Cheilosia shannoni*** (Curran 1923)
  ***similis*** (Shannon 1916)
☐ ***Cheilosia*** species 17-2
☐ ***Cheilosia subchalybea*** (Curran 1923)
☐ ***Cheilosia swannanoa*** Brimley 1925
☐ ***Cheilosia*** undescribed species 17-1
☐ ***Cheilosia*** undescribed species 17-3
☐ ***Cheilosia*** undescribed species 76-1
☐ ***Cheilosia wisconsinensis*** Fluke & Hull 1947
☐ ***Cheilosia yukonensis*** (Shannon 1922)
  ***varipila*** Fluke & Hull 1946
  ***variseta*** Fluke & Hull 1946
☐ ***Chrysogaster antitheus*** Walker 1849
  ***nigripes*** Loew 1863
  ***ustulata*** (Loew 1869)
  ***greeni*** Shannon 1916

*texana* Shannon 1916
*ontario* Curran 1925
☐ ***Chrysogaster inflatifrons*** Shannon 1916
☐ ***Chrysosyrphus latus*** (Loew 1863)
*versipellis* (Williston 1887) **N. Syn.**
*ithaca* (Shannon 1925)
*bigelowi* (Curran 1926)
*canadensis* (Curran 1933)
☐ ***Copestylum barei*** (Curran 1925)
☐ ***Copestylum sexmaculatum*** (Palisot de Beauvois 1819)
*pallidum* (Macquart 1842)
*sexpunctata* (Loew 1861)
*quadripunctatum* Doesburg 1962
☐ ***Copestylum vesicularium*** (Curran 1947)
☐ ***Copestylum vittatum*** Thompson 1976
*fasciata* (Macquart 1842)
*americana* (Goot 1964)
☐ ***Criorhina nigriventris*** Walton 1911
*intermedia* Johnson 1917
*maritima* Curran 1924
*aurata* Curran 1925
*mystaceae* Curran 1925 **N. Syn.**
☐ ***Criorhina verbosa*** (Walker 1849)
☐ ***Criorhina villosa*** (Bigot 1879) **N. Comb.**
☐ ***Cynorhinella longinasus*** Shannon 1924
☐ ***Eristalinus (Lathyropthalmus) aeneus*** (Scopoli 1763)
*punctata* (Muller 1764)
*leucocephala* (Gmelin 1790)
*macrophthalma* (Preyssler 1791)
*cuprovittatus* (Wiedemann 1830)
*taphicus* (Wiedemann 1830)
*stygius* (Newman 1835)
*sincerus* (Harris 1841)
*aenescens* (Macquart 1842)
*sincerus* (Walker 1849)
*auricalcicus* (Rondani 1865)
*concolor* (Philippi 1865)
*nigrolineatus* (Herve-Bazin 1923)
☐ ***Eristalis anthophorina*** (Fallén 1817)
*bastardii* Macquart 1842
*nitidiventris* (Zetterstedt 1843)
*nebulosus* Walker 1849
*semimetallicus* Macquart 1850
*everes* Walker 1852
*montanus* Williston 1882
*occidentalis* Osburn 1907
*toyohare* Matsumura 1911
*toyoharensis* Matsumura 1916
*mellisoids* Hull 1925
*mellissoides* Hull 1925
*perplexus* Hull 1925
*luleoensis* (Kanervo 1934)
☐ ***Eristalis arbustorum*** (Linnaeus 1758)
*nemorum* (Linnaeus 1758)
*horticola* (De Geer 1776)
*lyra* (Harris 1776)
*parralleli* (Harris 1776)
*tricincta* (Muller 1776)
*deflagratus* (Preyssler 1793)
*succinctus* (Panzer 1804)
*sachalinensis* Matsumura 1916
*bulgarica* Szilady 1934
*polonica* Szilady 1934
*strandi* Duda 1940
*distincta* (Shiraki 1968)
☐ ***Eristalis brousii*** Williston 1882
☐ ***Eristalis cryptarum*** (Fabricius 1794)
*nubilipennis* Curtis 1832
*compactus* Walker 1849
*atriceps* Loew 1866
☐ ***Eristalis dimidiata*** Wiedemann 1830
*niger* Macquart 1834
*chalybaeus* Macquart 1842
*chalybeus* Macquart 1842
*lherminierii* Macquart 1842
*haesitans* Walker 1849
*inflexus* Walker 1849
*incisuralis* Macquart 1850
☐ ***Eristalis flavipes*** Walker 1849
*americana* (Swederus 1787)
*melanostomus* Loew 1866
*rufipilis* Hull 1925
*americanus* Goot 1964
☐ ***Eristalis fraterculus*** (Zetterstedt 1838)
*pilosa* Loew 1866
*lapponicus* Schirmer 1913
*vallei* (Kanervo 1934)
☐ ***Eristalis gomojunovae*** Violovitsh 1977
☐ ***Eristalis hirta*** Loew 1866
*temporalis* Thomson 1869
*alpha* Hull 1925
*beta* Hull 1925
☐ ***Eristalis interrupta*** (Poda 1761)
*fuscus* (Scopoli 1763)
*lineolae* (Harris 1776)
*obfuscata* (Gmelin 1790)
*sylvarum* Meigen 1838
*inornatus* Loew 1866
*carelica* Kanervo 1938
☐ ***Eristalis obscura*** Loew 1866
*beltrami* Telford 1949
*vandykei* Nayar & Cole 1969
☐ ***Eristalis oestracea*** (Linnaeus 1758)
*apiformis* (Fallén 1817)
*oestriformis* (Walker 1849)
☐ ***Eristalis rupium*** Fabricius 1805
*piceus* (Fallén 1817)
*nitidus* Wehr 1924
*hybrida* Kanervo 1938
*infuscata* Kanervo 1938
*nigrofasciata* Kanervo 1938
*nigrotarsata* Kanervo 1938
☐ ***Eristalis saxorum*** Wiedemann 1830
*pervagus* Walker 1849
☐ ***Eristalis stipator*** Osten Sacken 1877
*latifrons* Loew 1866
*maculipennis* Townsend 1897
*californicus* Nayar & Cole 1969

☐ ***Eristalis tenax*** (Linnaeus 1758)
 ***vulgaris*** (Scopoli 1763)
 ***porcina*** (De Geer 1776)
 ***apiformis*** (Geoffroy 1785)
 ***campestris*** Meigen 1822
 ***hortorum*** Meigen 1822
 ***sylvaticus*** Meigen 1822
 ***vulpinus*** Meigen 1822
 ***cognatus*** Wiedemann 1824
 ***sinensis*** Wiedemann 1824
 ***columbica*** Macquart 1855
 ***ventralis*** Thomson 1869
 ***alpinus*** Strobl 1893
 ***claripes*** Abreu 1924
☐ ***Eristalis transversa*** Wiedemann 1830
 ***vittata*** Macquart 1834
 ***philadelphicus*** Macquart 1842
 ***pumilus*** Macquart 1842
 ***calomera*** (Bigot 1880)
 ***zonatus*** Bigot 1880
 ***calomera*** (Bigot 1884)
☐ ***Eumerus funeralis*** Meigen 1822
 ***tuberculatus*** Rondani 1857
 ***robii*** (Jones 1917)
 ***victorianus*** Paramonov 1957
☐ ***Eumerus narcissi*** Smith 1928
☐ ***Eumerus strigatus*** (Fallén 1817)
 ***lineata*** (Geoffroy 1785)
 ***acanthodes*** (Rossi 1794)
 ***spinipes*** (Rossi 1794)
 ***grandicornis*** Meigen 1822
 ***lunulatus*** Meigen 1822
 ***planifrons*** Meigen 1822
 ***aeneus*** Macquart 1829
 ***melanopus*** Rondani 1857
 ***rufitarsis*** Strobl 1906
☐ ***Eurimyia stipata*** (Walker 1849) **Res. Sp., N. Comb.**
 ***conostomus*** (Williston 1887) **N. Syn., N. Comb.**
☐ ***Ferdinandea buccata*** (Loew 1863)
 ***dives*** (Osten Sacken 1877) **N. Syn.**
 ***nigripes*** (Osten Sacken 1877) **N. Syn.**
☐ ***Ferdinandea croesus*** (Osten Sacken 1877)
 ***midas*** Hull 1942
☐ ***Hadromyia aepalius*** (Walker 1849)
 ***sorosis*** (Williston 1887)
☐ ***Hammerschmidtia rufa*** (Williston 1882) **Res. Sp., N. Comb.**
☐ ***Hammerschmidtia*** undescribed species 1
☐ ***Helophilus bottnicus*** Wahlberg 1844
 ***stricklandi*** Curran 1927
 ***alaskensis*** Fluke 1949
☐ ***Helophilus fasciatus*** Walker 1849
 ***similis*** Macquart 1842
 ***decisus*** (Walker 1849)
 ***impositus*** (Walker 1860)
 ***susurrans*** Jaennicke 1867
☐ ***Helophilus groenlandicus*** (Fabricius 1780)
 ***bilineatus*** Curtis 1835
 ***arcticus*** Zetterstedt 1838
 ***notabilis*** Macquart 1842
 ***latro*** Walker 1849
☐ ***Helophilus hybridus*** Loew 1846
 ***scotiae*** Macquart 1847
 ***henricii*** Schnabl 1880
 ***latitarsis*** Hunter 1897
☐ ***Helophilus lapponicus*** Wahlberg 1844
 ***borealis*** Staeger 1845
 ***glacialis*** Loew 1846
 ***androclus*** (Walker 1849)
 ***frater*** (Walker 1849)
 ***chalepus*** (Walker 1852)
 ***dychei*** Williston 1897
 ***bruesi*** Graenicher 1910
☐ ***Helophilus latifrons*** Loew 1863
☐ ***Helophilus obscurus*** Loew 1863
☐ ***Hiatomyia cyanescens*** (Loew 1863)
 ***plumata*** (Loew 1863)
☐ ***Lejota aerea*** (Loew 1872)
☐ ***Lejota cyanea*** (Smith 1912)
 ***calcitrans*** (Curran 1922)
 ***bigelowi*** (Curran 1924)
☐ ***Mallota bautias*** (Walker 1849)
 ***bipartitus*** (Walker 1849)
 ***facialis*** Hunter 1896
 ***flavoterminata*** Jones 1917
☐ ***Mallota illinoensis*** Robertson 1901 **Res. Sp.**
 ***palmerae*** Jones 1917 **N. Syn.**
☐ ***Mallota mississipensis*** Hull 1946 **Res. Sp.**
☐ ***Mallota posticata*** (Fabricius 1805)
 ***coactus*** (Wiedemann 1830)
 ***balanus*** (Walker 1849)
 ***separata*** Hull 1945
☐ ***Merodon equestris*** (Fabricius 1794)
 ***bombyliformis*** (Geoffroy 1785)
 ***flavicans*** (Fabricius 1794)
 ***ferrugineus*** (Fabricius 1805)
 ***narcissi*** (Fabricius 1805)
 ***constans*** Wiedemann 1822
 ***nobilis*** Meigen 1822
 ***transversalis*** Wiedemann 1822
 ***validus*** Wiedemann 1822
 ***bulborum*** Rondani 1845
 ***tuberculatus*** Rondani 1845
 ***nigrithorax*** Bezzi 1900
☐ ***Meromacrus acutus*** (Fabricius 1805)
 ***cruciger*** (Wiedemann 1830)
 ***pictus*** (Macquart 1846)
 ***parvus*** Hull 1942
☐ ***Milesia virginiensis*** (Drury 1773)
 ***trifasciatus*** (Hausmann 1799)
 ***ornata*** Fabricius 1805
 ***limbipennis*** Macquart 1850
 ***fulvifrons*** (Bigot 1884)
☐ ***Myolepta nigra*** (Loew 1872)
 ***tuberans*** (Williston 1887)
☐ ***Myolepta pretiosa*** (Hull 1923)
☐ ***Myolepta strigilata*** (Loew 1872)
☐ ***Myolepta varipes*** (Loew 1870)

☐ ***Neoascia (Neoascia) metallica*** (Williston 1882)
 ***nasuta*** (Bigot 1884)
 ***quadrinotata*** (Bigot 1884)
 ***minuta*** Curran 1925
☐ ***Neoascia (Neoascia) tenur*** (Harris 1780)
 ***quadriguttata*** (Gravenhorst 1807)
 ***quadripunctata*** (Meigen 1822)
 ***bifasciata*** (Zetterstedt 1838)
 ***lapponica*** Kanervo 1934
 ***splendida*** Kanervo 1934
☐ ***Neoascia (Neoasciella) globosa*** (Walker 1849)
 ***distincta*** Williston 1887 **N. Syn.**
☐ ***Neoascia (Neoasciella) subchalybea*** Curran 1925
 ***petsamoensis*** Kanervo 1934
☐ ***Neoascia (Neoasciella)*** undescribed species 1
☐ ***Neoascia (Neoasciella)*** undescribed species 17-1
☐ ***Orthonevra anniae*** (Sedman 1966)
☐ ***Orthonevra nitida*** (Wiedemann 1830)
 ***hieroglyphica*** (Bigot 1859)
☐ ***Orthonevra pictipennis*** (Loew 1863)
 ***aeneus*** (Walker 1849)
☐ ***Orthonevra pulchella*** (Williston 1887)
☐ ***Orthonevra robusta*** (Shannon 1916)
☐ ***Orthonevra*** undescribed species 1
☐ ***Orthonevra weemsi*** (Sedman 1966)
☐ ***Palpada albifrons*** (Wiedemann 1830)
 ***albiceps*** (Macquart 1842)
 ***seniculus*** (Loew 1866)
☐ ***Palpada pusilla*** (Macquart 1842)
 ***annulipes*** (Macquart 1842)
 ***surinamensis*** (Macquart 1842)
 ***tricolor*** (Jaennicke 1867)
☐ ***Palpada*** undescribed species 1
☐ ***Palpada vinetorum*** (Fabricius 1799)
 ***surinamensis*** (De Geer 1776)
 ***trifasciatus*** (Say 1829)
 ***decora*** (Perty 1833)
 ***uvarum*** (Walker 1849)
 ***croceipes*** (Bigot 1880)
 ***soulougensis*** (Bigot 1880)
 ***hirtipes*** (Bigot 1883)
 ***trilimbatus*** (Giglio-Tos 1892)
☐ ***Parhelophilus brooksi*** Curran 1927
 ***pollinaria*** (Fluke 1939)
☐ ***Parhelophilus divisus*** (Loew 1863)
☐ ***Parhelophilus flavifacies*** (Bigot 1884)
 ***anniae*** (Brimley 1923)
☐ ***Parhelophilus integer*** (Loew 1863)
☐ ***Parhelophilus laetus*** (Loew 1863)
 ***aureopilis*** (Townsend 1895)
☐ ***Parhelophilus obsoletus*** (Loew 1863)
☐ ***Parhelophilus porcus*** (Walker 1849)
☐ ***Parhelophilus rex*** Curran & Fluke 1926
☐ ***Pelecocera pergandei*** (Williston 1884)
☐ ***Polydontomyia curvipes*** (Wiedemann 1830) **N. Comb.**
 ***albiceps*** (Macquart 1846) **N. Comb.**
 ***morosus*** (Walker 1849) **N. Comb.**
 ***bicolor*** (Macquart 1850) **N. Comb.**
 ***bicolor*** (Walker 1852) **N. Comb.**
☐ ***Psilota buccata*** (Macquart 1842)
☐ ***Psilota flavidipennis*** Macquart 1855
☐ ***Psilota*** undescribed species 17-1
☐ ***Pterallastes thoracicus*** Loew 1863
☐ ***Rhingia nasica*** Say 1823
☐ ***Sericomyia arctica*** Schirmer 1913
 ***obscurior*** (Stackelberg 1927)
☐ ***Sericomyia bifasciata*** Williston 1887
☐ ***Sericomyia carolinensis*** (Metcalf 1917)
☐ ***Sericomyia chrysotoxoides*** Macquart 1842
 ***limbipennis*** Macquart 1847
 ***filia*** Walker 1849
☐ ***Sericomyia jakutica*** (Stackelberg 1927)
☐ ***Sericomyia lata*** (Coquillett 1907)
☐ ***Sericomyia militaris*** Walker 1849
 ***calcarata*** Curran 1923
☐ ***Sericomyia nigra*** Portschinsky 1873
 ***intermedia*** (Ringdahl 1922)
☐ ***Sericomyia sexfasciata*** Walker 1849
☐ ***Sericomyia slossonae*** Curran 1934
☐ ***Sericomyia transversa*** (Osburn 1926)
☐ ***Sericomyia vockerothi*** Skevington 2012
☐ ***Somula decora*** Macquart 1847
☐ ***Somula mississippiensis*** Hull 1922
 ***marivirginiae*** Brimley 1923
☐ ***Sphecomyia vittata*** (Wiedemann 1830)
 ***ornatus*** (Wiedemann 1830)
☐ ***Sphegina (Asiosphegina) biannulata*** Malloch 1922
☐ ***Sphegina (Asiosphegina) campanulata*** Robertson 1901
☐ ***Sphegina (Asiosphegina) petiolata*** Coquillett 1910
☐ ***Sphegina (Asiosphegina) rufiventris*** Loew 1863
☐ ***Sphegina (Sphegina) appalachiensis*** Coovert 1977
☐ ***Sphegina (Sphegina) brachygaster*** Hull 1935
 ***perplexa*** Hull 1935
 ***brimleyi*** Shannon 1940
☐ ***Sphegina (Sphegina) flavimana*** Malloch 1922
☐ ***Sphegina (Sphegina) flavomaculata*** Malloch 1922
 ***notata*** Hull 1935
☐ ***Sphegina (Sphegina) keeniana*** Williston 1887
☐ ***Sphegina (Sphegina) lobata*** Loew 1863
 ***monticola*** Malloch 1922
☐ ***Sphegina (Sphegina) lobulifera*** Malloch 1922
☐ ***Spilomyia alcimus*** (Walker 1849)
 ***hamifera*** Loew 1864
 ***texana*** Johnson 1921
☐ ***Spilomyia fusca*** Loew 1864
☐ ***Spilomyia longicornis*** Loew 1872
 ***banksi*** Nayar & Cole 1968
☐ ***Spilomyia sayi*** (Goot 1964)
 ***quadrifasciatus*** (Say 1824)
☐ ***Syritta flaviventris*** Macquart 1842
 ***nigricornis*** Macquart 1842
 ***spinigera*** Loew 1848
 ***albifacies*** Bigot 1859

*aculeipes* Schiner 1868
*armipes* Thomson 1869
*spinigerella* Thomson 1869
*cortesi* (Marnef 1967)

☐ ***Syritta pipiens*** (Linnaeus 1758)
*proxima* (Say 1824)
*obscuripes* Strobl 1899
*albicincta* Abreu 1924
*flavicans* Szilady 1940
*vicina* Szilady 1940
*tenofemorus* (Dzhafarova 1974)

☐ ***Temnostoma alternans*** Loew 1864
☐ ***Temnostoma balyras*** (Walker 1849)
☐ ***Temnostoma barberi*** Curran 1939
*acra* Curran 1939

☐ ***Temnostoma daochus*** (Walker 1849)
*pictulum* Williston 1887
*greenei* Shannon 1939

☐ ***Temnostoma excentrica*** (Harris 1841) **Res. Sp.**
*aequalis* Loew 1864 **N. Syn.**

☐ ***Temnostoma obscurum*** Loew 1864
☐ ***Temnostoma trifasciatum*** Robertson 1901
☐ ***Temnostoma venustum*** Williston 1887
*nipigonensis* Curran 1923

☐ ***Teuchocnemis bacuntius*** (Walker 1849)
☐ ***Teuchocnemis lituratus*** (Loew 1863)
☐ ***Tropidia albistylum*** Macquart 1847
☐ ***Tropidia calcarata*** Williston 1887
☐ ***Tropidia mamillata*** Loew 1861
☐ ***Tropidia quadrata*** (Say 1824)
☐ ***Volucella arctica*** Johnson 1916 **Res. Sp.**
☐ ***Volucella evecta*** Walker 1852 **Res. Sp.**
*sanguinea* Williston 1887 **N. Syn.**
*americana* Johnson 1916 **N. Syn.**

☐ ***Volucella facialis*** Williston 1882 **Res. Sp.**
*lateralis* Johnson 1916 **N. Syn.**
*rufomaculata* Jones 1917 **N. Syn.**

☐ ***Xylota (Ameroxylota) flukei*** (Curran 1941)
☐ ***Xylota (Xylota) angustiventris*** Loew 1866
*elongata* Williston 1887

☐ ***Xylota (Xylota) annulifera*** Bigot 1884
☐ ***Xylota (Xylota) bicolor*** Loew 1864
☐ ***Xylota (Xylota) confusa*** Shannon 1926
*bigelowi* (Curran 1941)

☐ ***Xylota (Xylota) ejuncida*** Say 1824
*viridaenea* Shannon 1926

☐ ***Xylota (Xylota) flavifrons*** Walker 1849
*communis* Walker 1849
*obscura* Loew 1866
*arcticus* Curran 1941

☐ ***Xylota (Xylota) flavitibia*** Bigot 1884
☐ ***Xylota (Xylota) hinei*** (Curran 1941)
☐ ***Xylota (Xylota) naknek*** Shannon 1926
*atlantica* Shannon 1926
*mixtus* (Curran 1941)

☐ ***Xylota (Xylota) ouelleti*** (Curran 1941)
☐ ***Xylota (Xylota) quadrimaculata*** Loew 1866
*artemita* Hull 1943

☐ ***Xylota (Xylota) segnis*** (Linnaeus 1758)
*maritima* (Scopoli 1763)
*fucatus* (Harris 1780)
*contracta* (Geoffroy 1785)
*melanochrysa* (Gmelin 1790)

☐ ***Xylota (Xylota) subfasciata*** Loew 1866
*notha* Williston 1887

☐ ***Xylota (Xylota) tuberculata*** (Curran 1941)
☐ ***Xylota (Xylota)*** undescribed species 78-1
☐ ***Xylota (Xylota)*** undescribed species 78-3

## PIPIZINAE

☐ ***Heringia canadensis*** Curran 1921
☐ ***Heringia intensica*** Curran 1921
☐ ***Heringia salax*** (Loew 1866)
*radicum* (Walsh & Riley 1869)
*pistica* (Williston 1887)

☐ ***Neocnemodon calcarata*** (Loew 1866)
☐ ***Neocnemodon carinata*** (Curran 1921)
☐ ***Neocnemodon cevelata*** (Curran 1921)
☐ ***Neocnemodon coxalis*** (Curran 1921)
☐ ***Neocnemodon elongata*** (Curran 1921)
☐ ***Neocnemodon intermedia*** (Curran 1921)
☐ ***Neocnemodon longiseta*** (Curran 1921)
☐ ***Neocnemodon myerma*** (Curran 1921)
☐ ***Neocnemodon ontarioensis*** (Curran 1921)
☐ ***Neocnemodon pisticoides*** (Williston 1887)
☐ ***Neocnemodon rita*** (Curran 1921)
☐ ***Neocnemodon squamulae*** (Curran 1921)
☐ ***Neocnemodon trochanterata*** (Malloch 1918)
☐ ***Neocnemodon unicolor*** (Curran 1921)
☐ ***Neocnemodon venteris*** (Curran 1921)
☐ ***Pipiza atrata*** Curran 1922
☐ ***Pipiza cribenni*** Coovert 1996
☐ ***Pipiza femoralis*** Loew 1866
*albipilosa* Williston 1887
*binotatus* (Johnson 1925)

☐ ***Pipiza macrofemoralis*** Curran 1921
☐ ***Pipiza nigripilosa*** Williston 1887
*tricolor* Curran 1921

☐ ***Pipiza puella*** Williston 1887
*severnensis* Curran 1921
*nigrotibiata* Curran 1924

☐ ***Pipiza quadrimaculata*** (Panzer 1804)
*quadriguttata* Macquart 1829
*maculata* Zetterstedt 1859
*bipunctata* Strobl 1898
*immaculata* Strobl 1898
*insolata* Violovitsh 1985

☐ ***Trichopsomyia apisaon*** Walker 1849
*transatlanticus* (Walker 1849)
*modestus* (Loew 1863)
*nigribarba* (Loew 1866)
*pulchella* (Williston 1887)

☐ ***Trichopsomyia banksi*** (Curran 1921)
☐ ***Trichopsomyia pubescens*** (Loew 1863)
☐ ***Trichopsomyia recedens*** (Walker 1852)
*fraudulenta* (Loew 1866)

☐ ***Trichopsomyia*** undescribed species 1

**SYRPHINAE**

☐ ***Allograpta exotica*** (Wiedemann 1830)
  *quadrigemina* (Thomson 1869)
  *fracta* Osten Sacken 1877
  *bilineata* Enderlein 1938
  *duplofasciata* Enderlein 1938
  *flavibuca* Enderlein 1938
  *interrupta* Enderlein 1938
  *skottsbergi* Enderlein 1938
  *tucumana* Enderlein 1938
☐ ***Allograpta obliqua*** (Say 1823)
  *securiferus* (Macquart 1842)
  *baccides* (Walker 1849)
  *dimemsus* (Walker 1852)
  *signatus* (Wulp 1867)
  *dejongi* Doesburg 1958
☐ ***Baccha cognata*** Loew 1863 **Res. Sp.**
☐ ***Chrysotoxum derivatum*** Walker 1849
  *columbianum* Curran 1927
  *minor* Curran 1927
☐ ***Chrysotoxum flavifrons*** Macquart 1842
  *occidentale* Curran 1924
☐ ***Chrysotoxum plumeum*** Johnson 1924 **Res. Sp.**
  *perplexum* Johnson 1924 **N. Syn.**
☐ ***Chrysotoxum pubescens*** Loew 1864
  *cuneatum* Wehr 1924
  *currani* Wehr 1924
  *luteopilosum* Curran 1924
☐ ***Dasysyrphus amalopis*** (Osten Sacken 1875)
☐ ***Dasysyrphus intrudens*** (Osten Sacken 1877)
  *disgregus* (Snow 1895)
  *laticaudatus* (Curran 1925)
  *osburni* (Curran 1925)
☐ ***Dasysyrphus laticaudus*** (Curran 1925)
☐ ***Dasysyrphus limatus*** (Hine 1922)
☐ ***Dasysyrphus nigricornis*** (Verrall 1873)
  *obscura* (Zetterstedt 1838)
  *nigrolimbatus* (Duda 1940)
  *obscuraticeps* (Frey 1950)
☐ ***Dasysyrphus pinastri*** (De Geer 1776)
☐ ***Dasysyrphus venustus*** (Meigen 1822)
  *arcuata* (Fallén 1817)
  *implicatus* (Meigen 1822)
  *lunulatus* (Meigen 1822)
  *solitaria* (Zetterstedt 1838)
  *berberidis* (Loew 1840)
  *hilaris* (Zetterstedt 1843)
  *reflectipennis* (Curran 1921)
  *pauxilloides* (Petch & Maltais 1932)
  *atricornis* (Szilady 1940)
☐ ***Didea alneti*** (Fallén 1817)
  *pellucidulus* (Wiedemann 1822)
  *japonica* Matsumura 1917
  *sachalinensis* Matsumura 1917
☐ ***Didea fuscipes*** Loew 1863
  *pacifica* Lovett 1919
☐ ***Doros aequalis*** Loew 1863
☐ ***Epistrophe grossulariae*** (Meigen 1822)
  *formosus* (Harris 1780)
  *lesueurii* (Macquart 1842)
  *conjugens* Walker 1852
  *melanis* (Curran 1922)
☐ ***Epistrophe metcalfi*** (Fluke 1933)
☐ ***Epistrophe nitidicollis*** (Meigen 1822)
  *protritus* (Osten Sacken 1877)
  *hunteri* (Curran 1925)
☐ ***Epistrophe terminalis*** (Curran 1925)
  *submarginalis* (Curran 1925)
☐ ***Epistrophe xanthostoma*** (Williston 1887)
☐ ***Epistrophella emarginata*** (Say 1823) **N. Comb.**
  *felix* (Osten Sacken 1875) **N. Comb.**
  *disjunctus* (Williston 1882) **N. Comb.**
  *divisa* (Williston 1882) **N. Comb.**
  *maculifrons* (Bigot 1884) **N. Comb.**
  *disjectus* (Williston 1887) **N. Comb.**
  *aenea* (Jones 1907) **N. Comb.**
  *invigorus* (Curran 1921) **N. Comb.**
  *fragila* (Fluke 1922) **N. Comb.**
  *infuscatus* (Fluke 1931) **N. Comb.**
  *weborgi* (Fluke 1931) **N. Comb.**
☐ ***Eupeodes abiskoensis*** (Dusek & Laska 1973)
☐ ***Eupeodes americanus*** (Wiedemann 1830)
  *medius* (Jones 1917)
  *wiedemanni* (Johnson 1919)
  *canadensis* (Curran 1926)
  *lebanoensis* (Fluke 1930)
  *flavipes* (Enderlein 1938)
☐ ***Eupeodes confertus*** (Fluke 1952)
☐ ***Eupeodes curtus*** (Hine 1922)
☐ ***Eupeodes flukei*** (Jones 1917)
  *palliventris* (Curran 1925)
☐ ***Eupeodes latifasciatus*** (Macquart 1829)
  *affinis* (Loew 1840)
  *abbreviata* (Zetterstedt 1849)
  *affinis* (Palma 1864)
  *submaculatus* (Frey 1918)
  *pallifrons* (Curran 1925)
  *depressus* (Fluke 1933)
  *chillcotti* (Fluke 1952)
☐ ***Eupeodes luniger*** (Meigen 1822)
  *maricolor* (Enderlein 1938)
  *azureus* (Szilady 1940)
  *transcendens* (Szilady 1940)
  *astutus* (Fluke 1952)
  *vockerothi* (Fluke 1952)
☐ ***Eupeodes nigroventris*** (Fluke 1933)
☐ ***Eupeodes perplexus*** (Osburn 1910)
  *meadii* (Jones 1917)
☐ ***Eupeodes pomus*** (Curran 1921)
  *vinelandi* (Curran 1921)
☐ ***Eupeodes volucris*** Osten Sacken 1877
  *perpallidus* (Bigot 1884)
  *braggii* Jones 1917
  *weldoni* Jones 1917
  *longipenis* (Enderlein 1938)
☐ ***Lapposyrphus lapponicus*** (Zetterstedt 1838) **N. Comb.**
  *agnon* (Walker 1849) **N. Comb.**
  *alcidice* (Walker 1849) **N. Comb.**
  *arcucinctus* (Walker 1849) **N. Comb.**

*bipunctatus* (Girschner 1884) **N. Comb.**
*komabensis* (Matsumura 1917) **N. Comb.**
*marginatus* (Jones 1917) **N. Comb.**
*mediaconstrictus* (Fluke 1930) **N. Comb.**
*sibericus* (Kanervo 1938) **N. Comb.**

☐ ***Leucozona (Ischyrosyrphus) xylotoides*** (Johnson 1916)

☐ ***Leucozona (Leucozona) americana*** Curran 1923 **Res. Sp.**

☐ ***Megasyrphus laxus*** (Osten Sacken 1875) **N. Comb.**
*daphne* (Hull 1925) **N. Comb.**
*syrphoides* (Hull 1925) **N. Comb.**
*nigrocomus* (Hull 1943) **N. Comb.**

☐ ***Melangyna arctica*** (Zetterstedt 1838)
*geniculatus* (Macquart 1842)
*glacialis* (Johnson 1898)
*gracilis* (Coquillett 1900)
*atrogenatus* (Kanervo 1934)
*melanatus* (Kanervo 1934)
*nasutus* (Kanervo 1934)
*nigrofaciatus* (Kanervo 1934)
*pseudomaculatus* (Kanervo 1934)
*coquilletti* Sedman 1965

☐ ***Melangyna fisherii*** (Walton 1911)
*diversipunctatus* (Curran 1925)

☐ ***Melangyna labiatarum*** (Verrall 1901)

☐ ***Melangyna lasiophthalma*** (Zetterstedt 1843)
*sexquadratus* (Walker 1849)
*mentalis* (Williston 1887)
*constrictus* (Matsumura 1917)
*elongatus* (Matsumura 1917)
*saghalinensis* (Matsumura 1917)
*nikkoensis* (Matsumura 1918)
*yezoensis* (Matsumura 1918)
*vittifacies* (Curran 1923)
*abruptus* (Curran 1924)
*columbiae* (Curran 1925)
*garretti* (Curran 1925)
*flavosignatus* (Hull 1930)

☐ ***Melangyna umbellatarum*** (Fabricius 1794)
*amoenus* (Loew 1840)
*pullula* (Snow 1895)
*cherokeensis* (Jones 1917)
*albipunctatus* (Curran 1925)
*nudifrons* (Curran 1925)
*remotus* (Curran 1925)

☐ ***Melanostoma mellinum*** (Linnaeus 1758)
*facultas* (Harris 1780)
*mellarius* (Meigen 1822)
*melliturgus* (Meigen 1822)
*minutus* (Macquart 1829)
*unicolor* (Macquart 1829)
*laevigatus* (Meigen 1838)
*concolor* (Walker 1851)
*parva* (Williston 1882)
*bicruciata* Bigot 1884
*cruciata* Bigot 1884
*ochripes* Bigot 1884
*pachytarse* Bigot 1884
*pruinosa* Bigot 1884
*angustatum* Williston 1887
*bellum* Giglio-Tos 1892
*nigricornis* Strobl 1893
*montivagum* Johnson 1916
*inornatum* Matsumura 1919
*interruptum* Matsumura 1919
*ochiaianum* Matsumura 1919
*ogasawarae* Matsumura 1919
*sachalinense* Matsumura 1919
*fallax* Curran 1923
*pallitarsis* Curran 1926
*melanderi* Curran 1930
*angustatoides* Kanervo 1934
*melanatus* Kanervo 1934
*obscuripes* Kanervo 1934
*deficiens* Szilady 1940
*dilatatum* Szilady 1940

☐ ***Meligramma guttata*** (Fallén 1817) **N. Comb.**
*flavifrons* (Verrall 1873) **N. Comb.**
*habilis* (Snow 1895) **N. Comb.**
*interrupta* (Jones 1917) **N. Comb.**
*savtshenkoi* (Violovitsh 1965) **N. Comb.**
*sajanica* (Violovitsh 1975) **N. Comb.**

☐ ***Meligramma triangulifera*** (Zetterstedt 1843) **N. Comb.**
*tenuis* (Osburn 1908) **N. Comb.**
*oronoensis* (Metcalf 1917) **N. Comb.**
*nielseni* (Violovitsh 1982) **N. Comb.**

☐ ***Meliscaeva cinctella*** (Zetterstedt 1843)
*libatrix* (Scopoli 1763)
*diversipes* (Macquart 1850)
*formosana* (Shiraki 1930)

☐ ***Ocyptamus fascipennis*** (Wiedemann 1830)
*aurinota* (Walker 1849)

☐ ***Ocyptamus fuscipennis*** (Say 1823)
*fiscipennis* (Say 1823)
*fascipennis* Macquart 1834
*amissas* (Walker 1849)
*peas* (Walker 1849)
*radaca* (Walker 1849)
*lugens* (Loew 1863)
*longiventris* Loew 1866
*fenestratus* Bigot 1885

☐ ***Paragus (Pandasyopthalmus) haemorrhous*** Meigen 1822
*sigillatus* Curtis 1836
*trianguliferus* Zetterstedt 1838
*substitutus* Loew 1858
*dimidiatus* Loew 1863
*auricaudatus* Bigot 1884
*ogasawarae* Matsumura 1916
*pallipes* Matsumura 1916
*tamagawanus* Matsumura 1916
*coreanus* Shiraki 1930

☐ ***Paragus (Paragus) angustifrons*** Loew 1863

☐ ***Paragus (Paragus) angustistylus*** Vockeroth 1986

☐ ***Paragus (Paragus) bispinosus*** Vockeroth 1986

☐ ***Paragus (Paragus) cooverti*** Vockeroth 1986
☐ ***Paragus (Paragus) variabilis*** Vockeroth 1986
  ***vockerothi*** Vujic, Simic & Radenkovic 1999
☐ ***Parasyrphus genualis*** (Williston 1887)
  ***johnsoni*** (Curran 1924)
☐ ***Parasyrphus groenlandica*** (Nielsen 1910)
  ***interruptus*** (Malloch 1919)
  ***mallochi*** (Curran 1923)
  ***latifrons*** (Ringdahl 1928)
  ***bulbosus*** (Fluke 1954)
☐ ***Parasyrphus insolitus*** (Osburn 1908)
☐ ***Parasyrphus nigritarsis*** (Zetterstedt 1843)
  ***nigrigena*** (Matsumura 1917)
  ***imperialis*** (Curran 1925)
  ***helvetica*** (Sack 1938)
☐ ***Parasyrphus relictus*** (Zetterstedt 1838)
  ***vittigera*** (Zetterstedt 1843)
  ***quinquelimbatus*** (Bigot 1884)
  ***rectoides*** (Curran 1921)
☐ ***Parasyrphus semiinterruptus*** (Fluke 1935)
☐ ***Parasyrphus tarsatus*** (Zetterstedt 1838)
  ***adolescens*** (Walker 1849)
  ***dryadis*** (Holmgren 1869)
  ***contumax*** (Osten Sacken 1875)
  ***sodalis*** (Williston 1887)
  ***bryantii*** (Johnson 1898)
  ***nigropilosa*** (Curran 1927)
  ***monachus*** (Hull 1930)
  ***distinctus*** (Kanervo 1934)
  ***immaculatus*** (Kanervo 1934)
  ***scutellatus*** (Kanervo 1934)
  ***evanescens*** (Enderlein 1938)
  ***extrema*** (Enderlein 1938)
  ***flavifacies*** (Enderlein 1938)
  ***lanata*** (Enderlein 1938)
  ***violaceiventris*** (Enderlein 1938)
☐ ***Parasyrphus vockerothi*** Thompson 2012
☐ ***Pelecinobaccha costata*** (Say 1829)
  ***costalis*** (Wiedemann 1830)
  ***tarchetius*** (Walker 1849)
☐ ***Platycheirus aeratus*** (Coquillett 1900)
  ***occidentalis*** Curran 1927
  ***angustitarsis*** Kanervo 1934
  ***pauper*** Hull 1944
☐ ***Platycheirus albimanus*** (Fabricius 1781)
  ***cyanea*** (Muller 1764)
☐ ***Platycheirus amplus*** (Curran 1927)
☐ ***Platycheirus angustatus*** (Zetterstedt 1843)
  ***major*** (Szilady 1940)
☐ ***Platycheirus chilosia*** (Curran 1922)
  ***carinata*** (Curran 1927)
☐ ***Platycheirus clypeatus*** (Meigen 1822)
  ***dilatatus*** (Macquart 1834)
  ***alpinus*** Strobl 1893
☐ ***Platycheirus coerulescens*** (Williston 1887)
☐ ***Platycheirus confusus*** (Curran 1925)
☐ ***Platycheirus discimanus*** (Loew 1871)
☐ ***Platycheirus flabella*** Hull 1944
☐ ***Platycheirus granditarsis*** (Forster 1771)
  ***confusus*** (Harris 1780)
  ***ocymi*** (Fabricius 1794)
  ***lobatus*** (Meigen 1822)
  ***apicauda*** (Curran 1925)
  ***lindrothi*** (Ringdahl 1930)
  ***digitalis*** (Fluke 1939)
☐ ***Platycheirus groenlandicus*** Curran 1927
  ***monticolus*** (Nielsen 1972)
  ***boreomonatus*** Nielsen 1981
☐ ***Platycheirus hyperboreus*** (Staeger 1845)
  ***erraticus*** Curran 1927
  ***chirosphena*** Hull 1944
☐ ***Platycheirus immarginatus*** (Zetterstedt 1849)
  ***navus*** (Harris 1780)
  ***palmulosus*** Snow 1895
  ***felix*** Curran 1931
☐ ***Platycheirus inversus*** Ide 1926
☐ ***Platycheirus jaerensis*** (Nielsen 1971)
☐ ***Platycheirus kelloggi*** (Snow 1895)
  ***johnsoni*** (Jones 1917)
☐ ***Platycheirus latitarsis*** Vockeroth 1990
☐ ***Platycheirus lundbecki*** (Collin 1931)
  ***fjellbergi*** Nielsen 1974
☐ ***Platycheirus luteipennis*** (Curran 1925)
  ***atra*** (Curran 1925)
  ***agens*** (Curran 1931)
☐ ***Platycheirus modestus*** Ide 1926
☐ ***Platycheirus naso*** (Walker 1849)
  ***rostrata*** (Zetterstedt 1838)
  ***holarcticus*** Vockeroth 1990
☐ ***Platycheirus nearcticus*** Vockeroth 1990
☐ ***Platycheirus neoperpallidus*** (Young 2016)
☐ ***Platycheirus nielseni*** Vockeroth 1990
☐ ***Platycheirus nigrofemoratus*** (Kanervo 1934)
☐ ***Platycheirus nodosus*** Curran 1923
☐ ***Platycheirus normae*** Fluke 1939
☐ ***Platycheirus obscurus*** (Say 1824)
  ***rostrata*** (Bigot 1884)
  ***rufipes*** (Bigot 1884)
  ***ontario*** (Davidson 1922)
  ***nitidiventris*** (Curran 1931)
☐ ***Platycheirus orarius*** Vockeroth 1990
☐ ***Platycheirus parmatus*** Rondani 1857
  ***ovalis*** Becker 1921
  ***bigelowi*** Curran 1927
☐ ***Platycheirus perpallidus*** Verrall 1901
  ***paramushiricus*** Mutin 1998
☐ ***Platycheirus pictipes*** (Bigot 1884)
  ***rufipes*** (Williston 1882)
  ***concinnus*** (Snow 1895)
  ***willistoni*** (Goot 1964)
  ***rufimaculatus*** Vockeroth 1990
☐ ***Platycheirus podagratus*** (Zetterstedt 1838)
☐ ***Platycheirus quadratus*** (Say 1823)
  ***fuscuanipennis*** (Macquart 1855)
☐ ***Platycheirus rosarum*** (Fabricius 1787)
  ***duplicata*** (Fluke 1922)
☐ ***Platycheirus scamboides*** Curran 1927
☐ ***Platycheirus scambus*** (Staeger 1843)
  ***chaetopodus*** Williston 1887

☐ ***Platycheirus scutatus*** (Meigen 1822)
   ***quadratus*** (Macquart 1829)
   ***sexnotatus*** (Meigen 1838)
   ***pygmaeus*** (Frey 1907)
☐ ***Platycheirus striatus*** Vockeroth 1990
☐ ***Platycheirus thompsoni*** Vockeroth 1990
☐ ***Platycheirus thylax*** Hull 1944
☐ ***Platycheirus trichopus*** (Thomson 1869)
☐ ***Platycheirus varipes*** Curran 1923
☐ ***Pseudodoros clavatus*** (Fabricius 1794)
   ***fusciventris*** (Wiedemann 1830)
   ***scutellaris*** (Walker 1836)
   ***babista*** (Walker 1849)
   ***varia*** (Walker 1849)
   ***facialis*** (Thomson 1869)
   ***quadrimaculata*** (Ashmead 1880)
   ***bacchoides*** (Bigot 1884)
   ***scutellata*** (Williston 1886)
☐ ***Scaeva affinis*** Say 1823 **N. Comb.**
☐ ***Sphaerophoria abbreviata*** Zetterstedt 1849
☐ ***Sphaerophoria asymmetrica*** Knutson 1972
☐ ***Sphaerophoria bifurcata*** Knutson 1972
☐ ***Sphaerophoria brevipilosa*** Knutson 1972
☐ ***Sphaerophoria contigua*** Macquart 1847
   ***cylindricus*** (Say 1824)
   ***fulvicauda*** Bigot 1884
   ***nigricauda*** Metcalf 1913
☐ ***Sphaerophoria longipilosa*** Knutson 1972
☐ ***Sphaerophoria novaeangliae*** Johnson 1916
☐ ***Sphaerophoria philanthus*** (Meigen 1822)
   ***nigricoxa*** Zetterstedt 1843
   ***dubia*** Zetterstedt 1849
   ***nigritarsi*** Fluke 1930
   ***robusta*** Curran 1930
   ***sarmatica*** Bankowska 1964
☐ ***Sphaerophoria pyrrhina*** Bigot 1884
   ***guttulata*** Hull 1942
☐ ***Sphaerophoria scripta*** (Linnaeus 1758)
   ***invisito*** (Harris 1780)
   ***strigata*** Staeger 1845
   ***scutellata*** Portevin 1909
   ***violacea*** Abreu 1924
   ***brunettii*** Joseph 1967
☐ ***Syrphus attenuatus*** Hine 1922
   ***pilisquamus*** Ringdahl 1928
   ***hinei*** Fluke 1933
☐ ***Syrphus knabi*** Shannon 1916
☐ ***Syrphus opinator*** Osten Sacken 1877
☐ ***Syrphus rectus*** Osten Sacken 1875
   ***transversalis*** Curran 1921
   ***bretoletensis*** Goeldlin 1996
☐ ***Syrphus ribesii*** (Linnaeus 1758)
   ***vacua*** (Scopoli 1763)
   ***blandus*** (Harris 1780)
   ***concava*** (Say 1823)
   ***philadelphicus*** Macquart 1842
   ***vittafrons*** Shannon 1916
   ***japonicus*** Matsumura 1917
   ***jezoensis*** Matsumura 1917
   ***moiwanus*** Matsumura 1917
   ***similis*** Jones 1917
   ***yamahanensis*** Matsumura 1917
   ***maculifer*** Matsumura 1918
   ***teshikaganus*** Matsumura 1918
   ***bigelowi*** Curran 1924
   ***interruptus*** Ringdahl 1930
   ***brevicinctus*** Kanervo 1938
   ***nigrigena*** Enderlein 1938
   ***jonesii*** Fluke 1949
   ***autumnalis*** Fluke 1954
   ***himalayanus*** Nayar 1968
   ***beringi*** Violovitsh 1975
☐ ***Syrphus sexmaculatus*** (Zetterstedt 1838)
   ***tshekanovskyi*** Kuznetzov 1987
☐ ***Syrphus torvus*** Osten Sacken 1875
   ***discretus*** Szilady 1940
☐ ***Syrphus vitripennis*** Meigen 1822
   ***topiarius*** Meigen 1822
   ***confinis*** (Zetterstedt 1838)
   ***akakurensis*** Matsumura 1917
   ***chujenjianus*** Matsumura 1917
   ***tsukisappensis*** Matsumura 1917
   ***agitatus*** Matsumura 1918
   ***campestris*** Matsumura 1918
   ***candidus*** Matsumura 1918
   ***conspicuus*** Matsumura 1918
   ***dubius*** Matsumura 1918
   ***kitakawae*** Matsumura 1918
   ***kuccharensis*** Matsumura 1918
   ***kushirensis*** Matsumura 1918
   ***okadensis*** Matsumura 1918
   ***palliventralis*** Matsumura 1918
   ***shibechensis*** Matsumura 1918
   ***tenuis*** Matsumura 1918
   ***velox*** Matsumura 1918
   ***strandi*** Duda 1940
☐ ***Toxomerus boscii*** Macquart 1842
   ***gurges*** (Walker 1852)
   ***arethusa*** (Hull 1945)
☐ ***Toxomerus corbis*** (Walker 1852)
   ***coalescens*** (Walker 1852)
   ***quintius*** (Walker 1852)
   ***planiventris*** (Loew 1866)
☐ ***Toxomerus geminatus*** (Say 1823)
   ***interrogans*** (Walker 1852)
   ***privernus*** (Walker 1852)
   ***notatus*** Macquart 1855
☐ ***Toxomerus jussiaeae*** Vige 1939
☐ ***Toxomerus marginatus*** (Say 1823)
   ***limbiventris*** (Thomson 1869)
   ***circumdata*** (Bigot 1884)
   ***comma*** (Giglio-Tos 1892)
☐ ***Toxomerus politus*** (Say 1823)
   ***anchoratus*** (Macquart 1842)
   ***cingulatulus*** (Macquart 1850)
   ***hecticus*** (Jaennicke 1867)
☐ ***Toxomerus verticalis*** (Curran 1927)
   ***mitis*** (Curran 1930)
   ***rhodope*** (Hull 1951)
☐ ***Xanthogramma flavipes*** (Loew 1863)

# Photo Credits

Photos not taken by the authors are acknowledged here. Most of these come via the tremendous web resource BugGuide. A virtual army of taxonomists and parataxonomists are constantly at work curating BugGuide images. Whenever we found a photo for a species we had not photographed in the field, or a photo that was simply better than ours, we wrote to the photographer and asked permission to use it. In order to keep the cost of the book down, none of the book's authors or photographers received payment. We appreciate the gift that these people have given to the naturalist community of allowing the use of their images in this book. These photographs were often obtained at considerable personal expense, and many required hours of effort to be able to achieve the depth, lighting, and natural poses illustrated.

**John Acorn**: *Sericomyia vockerothi*, Hinton, Alberta, 8 June 2008, Page 146

**Yurika Alexander**: *Copestylum vittatum*, Atco, New Jersey, 18 October 2011, Page 125; *Palpada* undescribed species 1, Atco, New Jersey, 24 October 2012, Page 120

**Gilles Arbour**: *Helophilus fasciatus*, Réserve naturelle du Bois-des-Patriotes, St-Denis sur Richelieu, Quebec, 14 August 2017, Page 60; *Spilomyia fusca*, Réserve naturelle du Bois-des-Patriotes, St-Denis sur Richelieu, Quebec, 6 July 2017, Page 133; *Syrphus* sp., Réserve naturelle du Bois-des-Patriotes, St-Denis sur Richelieu, Quebec, 30 August 2016, Page 16; *Syrphus torvus*, Réserve naturelle du Bois-des-Patriotes, St-Denis sur Richelieu, Quebec, 5 September 2017, Page 443; *Temnostoma obscurum*, Monteregie County, Quebec, 18 June 2014, Page 142; *Toxomerus marginatus*, Réserve naturelle du Bois-des-Patriotes, St-Denis sur Richelieu, Quebec, 6 July 2017, Page 380

**Tom Bentley**: *Copestylum vittatum*, Indiana Dunes National Lakeshore, Indiana, 2 June 2012, Page 124; *Epistrophe xanthostoma*, Woods Dam No 4 East Forest Preserve, Illinois, 16 August 2008, Page 435; *Lapposyrphus lapponicus*, San Isabel National Forest, Monarch Park Campground, Colorado, 5 July 2009, Page 420; *Ocyptamus cylindricus* group, Santa Ana National Wildlife Refuge, Texas, 24 November 2007, Page 377; *Ocyptamus fascipennis*, Hiawatha National Forest, Grand Island National Recreation Area, Michigan, 6 September 2009, Page 376; *Pseudoscaeva diversifasciatus*, Red Rock Canyon State Park, Colorado, 28 March 2015, Page 377; *Syritta flaviventris*, Pena Blanca Lake, Arizona, 20 October 2007, Page 207; *Teuchocnemis lituratus*, Reed-Turner Woodland Nature Preserve, Illinois, 14 May 2012, Page 166

**Betsy Betros**: *Sphegina campanulata*, Overland Park Arboretum, Kansas, 20 April 2016, Page 222

**Philip Bjork**: *Mallota illinoensis*, Rapid City, South Dakota, 22 May 2015, Pages 33, 96, 97

**Rick Bohler**: *Microdon fulgens*, Jacksonville, Florida, 16 May 2015, Page 42

**Ashley Bradford**: *Myolepta varipes*, Alexandria, Virginia, 27 April 2011, Page 245

**Karen Campbell**: *Copestylum barei*, Lehigh County, Pennsylvania, 17 July 2014, Page 127

**Dave Cheung**: *Temnostoma balyras*, Guelph, Ontario, date unknown, Page 140; *Toxomerus marginatus*, Guelph, Ontario, 3 July 2006, Page 381

**Patrick Coin**: *Callicera erratica*, Fews Ford, Eno River State Park, North Carolina, 6 April 2007, Pages 12, 129; *Microdon megalogaster*, Little River Regional Park, North Carolina, 27 May 2005, Page 42; *Teuchocnemis bacuntius*, Parkwood, North Carolina, 31 March 2012, Page 166

**Stephen Cresswell**: *Helophilus hybridus*, Banff, Alberta, 15 June 2013, Page 62; *Xylota* sp., Near Ruraldale, West Virginia, 16 February 2006, Page 13

**Peter Cristofono**: *Brachyopa vacua*, Salem, Massachusetts, 12 May 2015, Page 216

**Rob Curtis**: *Chalcosyrphus nemorum*, Reed Turner Woodland, Illinois, 26 September 2013, Page 177; *Eristalis interrupta*, Ninilchik, Alaska, 4 June 1999, Page 116; *Palpada vinetorum*, Willow Springs, Illinois, 18 October 2016, Page 121; *Xylota segnis*, Gaspé National Park, Quebec, 26 July 2015, Page 188

**Bill Dean**: *Brachypalpus femoratus*, Riding Mountain National Park, Manitoba, 14 June 2009, Page 90; *Eristalis dimidiata*, Sandy Lake, Manitoba, 19 June 2011, Page 113; *Microdon manitobensis*, Clear Lake, Riding Mountain National Park, Manitoba, 26 July 2012, Page 45

**Tony DiTerlizzi**: *Platycheirus hyperboreus* Copyright © 2005 Tony DiTerlizzi, Hampshire County, Massachusetts, 17 May 2005, Page 340

**Jason Dombroskie**: *Sericomyia vockerothi*, Blackfoot Lake, Alberta, 14 May 2006, Page 146

**Denis Doucet**: *Blera nigra*, Fundy National Park, New Brunswick, 13 June 2017, Page 160; *Chalcosyrphus anthreas*, Riverside-Albert, New Brunswick, 16 July 2015, Page 176; *Didea alneti*, Fundy National Park, New Brunswick, 7 July 2015, Page 460; *Eupeodes latifasciatus*, Fundy National Park, New Brunswick, 20 July 2016, Page 430; *Megasyrphus laxus*, Riverside-Albert, New Brunswick, 16 September 2014, Page 458; *Megasyrphus laxus*, Fundy National Park Headquarters, New Brunswick, 25 September 2014, Page 459; *Parhelophilus rex*, MacLaren Pond Trail, Fundy National Park, New Brunswick, 5 July 2015, Page 71; *Polydontomyia curvipes*, Fundy National Park, New Brunswick, 21 July 2015, Page 84; *Rhingia nasica*, Riverside-Albert, New Brunswick, 25 August 2015, Page 209; *Sericomyia bifasciata*, MacLaren Pond, Fundy National Park, New Brunswick; 5 July 2015, Page 150

**Bill DuPree**: *Laetodon laetus*, Atlanta, Georgia, 17 May 2006, Page 38; *Ocyptamus fuscipennis*, Atlanta, Georgia, 11 October 2005, Page 376

**Susan Ellis**: *Copestylum vesicularium*, Brandywine, Maryland, 29 June 2007, Page 126

**Mardon Erbland**: *Megasyrphus laxus*, St. Philips, Northeast Avalon, Newfoundland, 2 August 2009, Page 458

**Michele Esposito**: *Neocnemodon unicolor* holotype male, Copyright California Academy of Sciences, Page 295

**Tamra Feenstra**: *Polydontomyia curvipes*, Bear River Migratory Bird Refuge, Utah, 23 September 2012, Page 85

**Cara Flinn**: *Temnostoma daochus*, Hagarville, Arkansas, 2 May 2015, Page 137

**Leif Gabrielsen**: *Sericomyia nigra*, Øyslebø, Marnardal, Vest-Agder, Norway; 7 June 2014, Page 152

**Judy Gallagher**: *Copestylum sexmaculatum*, Okaloacoochee Slough State Forest, Florida, 12 December 2013, Page 125; *Mixogaster* sp., Long Pine Key, Everglades National Park, Florida, 10 January 2017, Page 55; *Chalcosyrphus metallifer*, Julie Metz Wetlands, Virginia, 22 February 2016, Page 181

**Greg K. Gerber**: *Syritta pipiens*, Harrisonburg, Virginia, 16 June 2011, Page 207

**Richard R. Greene**: *Platycheirus obscurus*, Montebello Open Space Preserve, California, 9 May 2010, Page 330

**Henri Goulet**: *Sericomyia lata*, Ottawa, Ontario, Page 148; *Spilomyia longicornis*, Ottawa, Ontario, Page 135

**Martin Hauser**: *Myolepta pretiosa*, Pages 244, 245; *Myolepta varipes*, Pages 244, 245

**Ron Hemberger**: *Eupeodes volucris*, Peters Canyon, Orange, California, 14 July 2009, Page 423; *Scaeva affinis*, Nix Nature Center, Laguna Beach, California, 19 March 2008, Page 419

**Karl W. Hillig**: *Brachyopa notata*, Ballston Lake, New York, 26 April 2010, Page 212; *Myolepta strigilata*, Ballston Lake, New York, 22 April 2010, Page 242

**Kurt Holmqvist**: *Helophilus groenlandicus*, Stavreviken, Sweden, 23 September 2009, Page 64

**Sam Houston**: *Ceriana willistoni*, Sand Springs, Oklahoma, 27 April 2008, Page 145

**Corey Husic**: *Microdon aurulentus*, Kunkletown, Pennsylvania, 28 May 2011, Page 44

**Stephen R. Johnson**: *Myolepta nigra*, Pella, Iowa, 15 June 2016, Page 242

**Thaddeus Charles Jones**: *Platycheirus kelloggi*, Beartooths, Wyoming, 20 July 2016, Page 356

**John Klymko**: *Blera armillata*, Nipisiguit Protected Natural Area, New Brunswick, 22 June 2017, Page 160; *Hammerschmidtia rufa*, Carleton, New Brunswick, 7 June 2011, Page 219; *Leucozona americana*, location and date unknown, Page 462; *Melangyna labiatarum*, Amherst, Nova Scotia, 13 September 2014, Page 408; *Microdon manitobensis*, Jacquet River Gorge Protected Natural Area, New Brunswick, 23 June 2010, Page 44

**Kendra Kocab**: *Mallota mississipensis*, Tomball, Texas, 27 February 2011, Page 96

**Monica Krancevic**: *Palpada albifrons*, Lake Jackson, Texas, 20 December 2015, Page 122; *Palpada pusilla*, Lake Jackson, Brazoria County, Texas, 4 August 2014, Page 123

**John Lampkin**: *Meromacrus acutus*, Hillsborough County, Florida, 26 March 2016, Back cover; *Palpada pusilla*, Rothenbach Park, Sarasota County, Florida, 16 September 2015, Page 122; *Toxomerus verticalis*, Long Key Natural Area, Florida, 24 February 2018, Page 383

**Cheryl Lavers**: *Chalcosyrphus metallicus*, Jonesboro, Arkansas, 26 June 2015, Page 178

**Tim Lethbridge**: *Palpada pusilla*, Lake Placid, Highlands County, Florida, 15 November 2009, Page 123

**Allen D. Levine**: *Anasimyia chrysostoma*, Bashakill Wildlife Reserve, New York, 17 August 2016, Page 80

**John van der Linden**: *Helophilus fasciatus* ovipositing, Iowa, 2 October 2011, Page 60

**Stephen Luk**: *Platycheirus scambus*, Preservation Park, Guelph, Ontario, 26 May 2012, Page 344

**Emil Lütken**: *Anasimyia lunulata*, Denmark, 1 May 2011, Page 78

**Kerry S. Matz**: *Platycheirus trichopus*, Salt Lake City, Utah, 16 May 2007, Page 332

**Robin McLeod**: *Eristalis dimidiata*, near Ailsa Craig, Ontario, 13 April 2004, Page 112

**Tom Murray**: *Anasimyia bilinearis*, Groton, Massachusetts, 2 May 2014, Page 76; *Brachyopa daeckei*, Groton, Massachusetts, 3 May 2010, Page 214; *Chalcosyrphus chalybeus*, Ripton, Vermont, 16 July 2014, Page 172; *Chalcosyrphus curvarius*, Topsham, Vermont, 22 May 2012, Page 172; *Chalcosyrphus inarmatus*, Plymouth, Massachusetts, 28 May 2005, Page 178; *Chalcosyrphus libo*, Topsham, Vermont, 22 June 2010, Page 175; *Cheilosia* sp., Lancaster, Massachusetts, 4 September 2005, Page 12; *Chrysosyrphus latus*, Groton, Massachusetts, 5 May 2012, Page 256; *Copestylum vesicularium*, Harvard, Massachusetts, 20 September 2005, Page 127; *Epistrophe grossulariae*, Dixville, New Hampshire, 29 August 2006, Page 436; *Helophilus obscurus*, Pittsburg, New Hampshire, 15 August 2005, Page 62; *Heringia salax*, Townsend, Massachusetts, 8 June 2014, Page 292; *Mallota bautias*, Woodford, Vermont, 19 July 2012, Page 94; *Microdon tristis*, Lunenburg, Massachusetts, 10 June 2007, Page 46; *Orthonevra pulchella*, Middlebury, Vermont, 22 July 2009, Pages 247, 248; *Pipiza quadrimaculata*, Albany, New Hampshire, 10 May 2005, Page 308; *Platycheirus granditarsis*, Woodford, Vermont, 25 June 2009, Page 332

**Steve Nanz**: *Mallota posticata*, Pymatuning Reservoir, Pennsylvania, 3 July 2007, Page 94; *Platycheirus quadratus*, Prospect Park, New York, 8 June 2008, Page 334

**Vazrick Nazari**: K.W. Neatby Building (home of the CNC), Ottawa, Ontario, Page 27

**Torry Nergart**: *Temnostoma trifasciatum*, Sapphire, North Carolina, 16 May 2015, Page 140

**Jake Paredes**: *Ocyptamus parvicornis*, Boca Raton, Florida, 10 December 2011, Page 377

**Sheryl Pollock**: *Criorhina villosa*, Riverbend Regional Park, Virginia, 23 February 2017, Page 101

**John Rosenfeld**: *Chalcosyrphus piger*, Allison Park, Pennsylvania, 2 September 2013, Page 174; *Tropidia albistylum*, Allison Park, Pennsylvania, 2 August 2013, Page 202

**T. Saigusa**: Dick Vockeroth on a field trip in Miyazaki Prefecture, Kyushu, Japan, September 2006, Page 6

**Phil Schappert**: *Toxomerus politus*, Roaches Pond, Spryfield, Nova Scotia, 18 September 2015, Page 383

**Harvey Schmidt**: *Teuchocnemis lituratus*, Creighton, Saskatchewan, 15 June 2012, Page 167

**Marie L. Schmidt**: *Anasimyia grisescens*, Delaware County, Pennsylvania, 14 July 2009, Page 82

**Nolie Schneider**: *Pipiza femoralis*, Ottawa, Ontario, 11 May 2009, Page 314

**Stephen Schueman**: *Mallota mississipensis*, Glen Burnie, Maryland, 15 April 2010, Page 97

**Angela Skevington**: Jeff and Alexander Skevington, Arizona, August 2007, Page 26

**Emily Stanley**: *Blera pictipes*, Glyndon, Maryland, 2 May 2013, Page 164

**Tam Stuart**: *Spilomyia alcimus*, Assunpink Wildlife Management Area, New Jersey, 10 July 2003, Pages 134, 135

**Thomas M. Stuart**: *Sphaerophoria* sp., Plainsboro, New Jersey, 16 May 2005, Page 391

**Stuart Tingley**: *Eristalis interrupta*, Murray Corner, New Brunswick, 23 May 2013, Page 117; *Helophilus lapponicus*, Grand Harbour, Grand Manan Island, New Brunswick, 19 May 2013, Page 64

# Bibliography

Citations made in text are marked with an asterisk; other important overview papers not cited are not marked as such.


*Bagatshanova, A. K. (1990) Fauna i ekologiya mukh-zhurchalok (Diptera, Syrphidae) Yakutii. *Yakutsk Nauchnye Tsentr*, 1–164.

Ball, S. & Morris, R. (2013) *Britain's Hoverflies: An Introduction to the Hoverflies of Britain.* Princeton, NJ: Princeton University Press. 296 pp.

Barkalov, A. V. S. & Ståhls, G. (1997) Revision of the Palearctic bare-eyed and black-legged species of the genus *Cheilosia* Meigen (Diptera, Syrphidae). *Acta Zoological Fennica, 208*, 1–74.

*Barkalov, A. V. & Mutin, V. A. (1991) Revision of hover-flies of the genus *Blera* Billberg, 1820 (Diptera, Syrphidae). 1. *Entomologicheskoe Obozrenie, 70*, 204–213.

Bartsch, H., Binkiewicz, K., Klintbjer, A., Råden, A. & Nasibov, K. (2009) *Nationalnyckeln till Sveriges flora och fauna. Tvåvingar: Blomflugor: Eristalinae & Microdontinae. Diptera: Syrphidae: Eristalinae & Microdontinae.* Uppsala: ArtDatabanken, SLU. 478 pp.

Bartsch, H., Binkiewicz, E., Råden, A. & Nasibov, E. (2009) *Nationalnyckeln till Sverigesflora och fauna. Tvåvingar: Blomflugor: Syrphinae. Diptera: Syrphidae: Syrphinae.* Uppsala: ArtDatabanken, SLU. 406 pp.

*Belliure, B. & Michaud, J. P. (2001) Biology and behavior of *Pseudodorus clavatus* (Diptera: Syrphidae), an important predator of citrus aphids. *Annals of the Entomological Society of America, 94*, 91–96.

*Böcher, J., Láska, P., Mazánek, L. & Nielsen, T. R. (2015) 17.16. Syrphidae (Hoverflies). *In:* J. Böcher, N. P. Kristensen, T. Pape & L. Vilhelmsen (Eds), *The Greenland Entomofauna*. Brill, Leiden, pp. 589–599.

Brown, B. V. (1993) A further chemical alternative to critical-point-drying for preparing small (or large) flies. *Fly Times, 11*, 10.

Chandler, A. E. (1968) A preliminary key to eggs of some of the commoner aphidophagous Syrphidae (Diptera) occurring in Britain. *Transactions of the Royal Entomological Society of London, 85*, 131–139.

*Coovert, G. A. (1996) A revision of the genus *Pipiza* Fallén (Diptera, Syrphidae) of America North of Mexico with notes on the placement of the tribe Pipizini. *Ohio Biological Survey, 11*, 1–68.

*Coovert, G. A. & Thompson, F. C. (1977) The *Sphegina* species of eastern North America (Diptera: Syrphidae). *Proceedings of the Biological Society of Washington, 90*, 536–552.

*Cumming, J. M. & Wood, D. M. (2017) Adult morphology and terminology. *In: Manual of Afrotropical Diptera.* South African National Biodiversity Institute, Pretoria, pp. 89–133.

*Curran, C. H. (1921) Revision of the *Pipiza* group of the family Syrphidae (flower-flies) from north of Mexico. *Proceedings of the California Academy of Sciences, 11*, 345–393.

*Curran, C. H. (1922) The syrphid genera *Hammerschmidtia* and *Brachyopa* in Canada. *Annals of the Entomological Society of America, 15*, 239–255.

*Curran, C. H. (1924) Synopsis of the genus *Chrysotoxum* with notes and descriptions of new species (Syrphidae, Diptera). *The Canadian Entomologist, 56*, 34–40.

*Curran, C. H. (1925a) Contribution to a monograph of the American Syrphidae from north of Mexico. *Bulletin of the University of Kansas – Science Bulletin, 15*, 1–283.

*Curran, C. H. (1925b) Revision of the genus *Neoascia* Williston (Diptera: Syrphidae). *Proceedings of the Entomological Society of Washington, 27*, 51–62.

*Curran, C. H. (1930) Synopsis of the American species of *Volucella* (Syrphidae; Diptera). *American Museum Novitates, 1027*, 1–7.

*Curran, C. H. (1932) New American Syrphidae (Diptera), with notes. *American Museum Novitates, 519*, 1–9.

Curran, C. H. (1939) The species of *Temnostoma* related to *Bombylans* Linné (Syrphidae, Diptera). *American Museum Novitates, 1040*, 1–3.

*Curran, C. H. (1940) Some new Neotropical Syrphidae (Diptera). *American Museum Novitates, 1086*, 1–14.

*Curran, C. H. (1941) New American Syrphidae. *Bulletin of the American Museum of Natural History, 78*, 243–304.

*Curran, C. H. (1953) Notes and descriptions of some Mydaidae and Syrphidae (Diptera). *American Museum Novitates, 1645*, 1–15.

*Curran, C. H. & Fluke, C. L. (1926) Revision of the Nearctic species of *Helophilus* and allied genera. *Transactions of the Wisconsin Academy of Sciences, Arts, and Letters, 22*, 207–281.

Dixon, T. J. (1960) Key to and descriptions of the third instar larvae of some species of Syrphidae (Dipt.) occurring in Britain. *Transactions of the Royal Entomological Society of London, 112*, 345–379.

*Fluke, C. L. & Hull, F. M. (1945) The *Cartosyrphus* flies of North America (Syrphidae). *Transactions of the Wisconsin Academy of Sciences, Arts, and Letters, 37*, 221–263.

*Fluke, C. L. & Hull, F. M. (1946) Syrphid flies of the genus *Cheilosia*, subgenus *Chilomyia* in North America (Part II). *Transactions of the Wisconsin Academy of Sciences, Arts, and Letters, 36*, 327–347.

*Fluke, C. L. & Weems, H. V. J. (1956) The Myoleptini of the Americas (Diptera, Syrphidae). *American Museum Novitates, 1758*, 1–23.

Gilbert, F. (2005a) The evolution of imperfect mimicry in hoverflies. *In:* M. D. E. Fellowes, G. J. Holloway & J. Rolff (Eds), *Insect Evolutionary Ecology*. Wallingford, United Kingdom, CABI, pp. 231–288.

Gilbert, F. (2005b) Syrphid aphidophagous predators in a food-web context. *European Journal of Entomology, 102*, 325–333.

Gittings, T., O'Halloran, J., Kelly, T. & Giller, P. S. (2006) The contribution of open spaces to the maintenance of hoverfly (Diptera, Syrphidae) biodiversity in Irish plantation forests. *Forest Ecology and Management, 237*, 290–300.

*Greene, C. T. (1923) The larva and pupa of *Microdon megalogaster* Snow (Diptera). *Proceedings of the Entomological Society of Washington, 25*, 140–141.

* Haarto, A. & Ståhls, G. (2014) When mtDNA COI is misleading: Congruent signal of ITS2 molecular marker and morphology for North European *Melanostoma* Schiner, 1860 (Diptera, Syrphidae). *ZooKeys, 431*, 93–134.

*Hartley, J. C. (1961) A taxonomic account of the larvae of some British Syrphidae. *Proceedings of the Zoological Society of London (Reprinted), 136*, 505–573.

Heiss, E. M. (1938) *A Classification of the Larvae and Puparia of the Syrphidae of Illinois Exclusive of Aquatic Forms* (Vol. 187). Urbana: University of Illinois Press. 142 pp.

*Hippa, H. (1978) Classification of Xylotini (Diptera, Syrphidae). *Acta Zoological Fennica, 156*, 1–153.

*Hull, F. M. (1942a) The flies of the genus *Meromacrus* (Syrphidae). *American Museum Novitates, 1200*, 1–11.

*Hull, F. M. (1942b) The genus *Ferdinandea* Rondani. *Journal of the Washington Academy of Sciences, 32*, 239–241.

*Hull, F. M. (1943) The Genus *Mesogramma*. *Entomologica Americana. A Journal of Entomology, 23 (New Series)*, 1–99.

Hull, F. M. (1945) A revisional study of the fossil Syrphidae. *Bulletin of the Museum of Comparative Zoology at Harvard College, 95*, 251–355.

Hull, F. M. (1949) The morphology and inter-relationship of the genera of syrphid flies, recent and fossil. *Transactions of the Zoological Society of London, 26*, 257–409.

*Hull, F. M. (1954) The genus *Mixogaster* (Diptera, Syrphidae). *American Museum Novitates*, 1652, 1–28.

*Hull, F. M. & Fluke Jr., C. L. (1950) The genus *Cheilosia* Meigen (Diptera, Syrphidae). The subgenera *Cheilosia* and *Hiatomyia*. *Bulletin of the American Museum of Natural History, 94*, 301–401.

*Jones, C. R. (1922) A contribution to our knowledge of the Syrphidae of Colorado. *Colorado Agricultural Experiment Station Bulletin*, 269, 72.

Kearns, C. A. (1992) Anthophilous fly distribution across an elevation gradient. *American Midland Naturalist, 127*, 172–182.

Kearns, C. A. (2001) North American dipteran pollinators: Assessing their value and conservation status. *Conservation Ecology, 5*, 5.

*Kerr, P. H., Fisher, E. M. & Buffington, M. (2008) Dome lighting for insect imaging under a microscope. *American Entomologist*, Winter, 198–200.

Kevan, P. (2002) Flowers, pollination, and the associated diversity of flies. *Biodiversity and Conservation, 3*, 16–18.

*Knutson, L. V. (1973) Taxonomic revision of the aphid-killing flies of the genus *Sphaerophoria* in the Western Hemisphere (Syrphidae). *Miscellaneous Publications of the Entomological Society of America, 9*, 2–50.

Krivosheina, N. P. (2004) The morphology of flies of the *Temnostoma apiforme* and *T. vespiforme* groups (Diptera, Syrphidae): Communication II. *Entomological Review, 84*, 100–117.

Kuznetzov, S. Y. (1988) Morphology of the eggs of hover-flies (Diptera, Syrphidae). *Entomologicheskoe Obozrenie, 67*, 741–753.

Larson, B. M., Kevan, P. G. & Inouye, D. W. (2001) Flies and flowers: Taxonomic diversity of anthophiles and pollinators. *The Canadian Entomologist, 133*, 439–466.

*Latta, R. & Cole, F. R. (1933) A comparative study of the species of *Eumerus* known as the lesser bulb flies. *Monthly Bulletin of the California Department of Agriculture, 22*, 142–152.

*Locke, M. M. & Skevington, J. H. (2013) Revision of Nearctic *Dasysyrphus* Enderlein (Diptera: Syrphidae). *Zootaxa, 3660*, 1–80.

*Lyneborg, L. & Barkemeyer, W. (2005) *The Genus* Syritta. *A World Revision of the Genus* Syritta *Le Peletier & Serville, 1828 (Diptera: Syrphidae)* (Vol. 15). Stenstrup: Apollo Books. 224 pp.

*MacGowan, I. (1994) Creating breeding sites for *Callicera rufa* Schummel (Dipt., Syrphidae) and a further host tree. *Dipterists Digest, 1*, 6–8.

*Maibach, A. & Detiefenau, P. G. (1994) Generic limits and taxonomic features of some genera belonging to the tribe of Chrysogasterini (Diptera, Syrphidae). 3. Description of immature stages of some West Palearctic species. *Revue Suisse de Zoologie, 101*, 369–411.

*Maibach, A. & Goeldlin de Tiefenau, P. (1993) Description et clé de détermination des stades immatures de plusieurs espèces du genre *Neoascia* Williston de la région paléarctique occidentale (Diptera, Syrphidae). *Bulletin de la Société Entomologique Suisse, 66*, 337–257.

Maier, C. T. & Waldbauer, G. P. (1979) Dual mate-seeking strategies in male syrphid flies (Diptera: Syrphidae). *Annals of the Entomological Society of America, 52*, 54–61.

*Martin, J. E. H. (1977) *Collecting, Preparing and Preserving Insects, Mites, and Spiders* (Vol. 1). Hull, Québec: Supply and Services Canada.

*Mengual, X., Kazerani, F., Talebi, A. A. & Gilasian, E. (2015) A revision of the genus *Pelecocera* Meigen with the description of the male of *Pelecocera persiana* Kuznetzov from Iran (Diptera: Syrphidae). *Zootaxa, 3947*, 99–108.

Mengual, X., Ruiz, C., Rojo, S., Ståhls, G. & Thompson, F. C. (2009) A conspectus of the flower fly genus *Allograpta* (Diptera: Syrphidae) with description of a new subgenus and species. *Zootaxa, 2214*, 1–28.

Mengual, X., Ståhls, G. & Rojo, S. (2008) First phylogeny of predatory flower flies (Diptera, Syrphidae, Syrphinae) using mitochondrial COI and nuclear 28S rRNA genes: Conflict and congruence with the concurrent tribal classification. *Cladistics, 24*, 543–562.

Mengual, X., Ståhls, G. & Rojo, S. (2008) Molecular phylogeny of *Allograpta* (Diptera, Syrphidae) reveals diversity of lineages and non-monophyly of phytophagous taxa. *Molecular Phylogenetics and Evolution, 49*, 715–727.

Mengual, X., Ståhls, G. & Rojo, S. (2012) Is the mega-diverse genus *Ocyptamus* (Diptera, Syrphidae) monophyletic? Evidence from molecular characters including the secondary structure of 28S rRNA. *Molecular Phylogenetics and Evolution, 62*, 191–205.

Mengual, X., Ståhls, G. & Rojo, S. (2015) Phylogenetic relationships and taxonomic ranking of pipizine flower flies (Diptera: Syrphidae) with implications for the evolution of aphidophagy. *Cladistics, 31(5)*, 491–508.

Mengual, X. & Thompson, F. C. (2011) Carmine cochineal killers: The flower fly genus *Eosalpingogaster* Hull (Diptera: Syrphidae) revised. *Systematic Entomology*, *36*, 713–731.

*Michaud, J. P. & Belliure, B. (2001) Impact of syrphid predation on production of migrants in colonies of the brown citrus aphid, *Toxoptera citricida* (Homoptera: Aphididae). *Biological Control*, *21*, 91–95.

*Miranda, G. F. G. (2011) An overview of the genus *Ocyptamus* Macquart, 1834, with a revision of the *Ocyptamus tristis* species group. PhD thesis. Department of Environmental Biology, University of Guelph, Guelph, 617 pp.

*Miranda, G. F. G. (2017a) Identification key for the genera of Syrphidae (Diptera) from the Brazilian Amazon and new taxon records. *Acta Amazonica*, *47*, 53–62.

*Miranda, G. F. G. (2017b) Revision of the *Hybobathus arx* and *Pelecinobaccha summa* species groups (Diptera: Syrphidae). *Zootaxa*, *4338*, 1–43.

*Miranda, G. F. G., Marshall, S. A. & Skevington, J. H. (2014) Revision of the genus *Pelecinobaccha* Shannon, description of *Relictanum* gen. nov., and redescription of *Atylobaccha flukiella* (Curran, 1941) (Diptera: Syrphidae). *Zootaxa*, *3819*, 1–154.

*Miranda, G. F. G., Skevington, J. H., Marshall, S. A. & Kelso, S. (2016) The genus *Ocyptamus* Macquart (Diptera: Syrphidae): A molecular phylogenetic analysis. *Arthropod Systematics and Phylogeny*, *74*, 161–176.

*Miranda, G. F. G., Young, A. D., Locke, M. M., Marshall, S. A., Skevington, J. H. & Thompson, F. C. (2013) Key to the genera of Nearctic Syrphidae. *Canadian Journal of Arthropod Identification*, *23*, 1–351.

Nielsen, T. R. & Svendsen, S. (2014) Hoverflies (Diptera, Syrphidae) in North Norway. Norwegian *Journal of Entomology*, *61*, 119–134.

Pape, T. & Thompson, F. C. (2013) Systema Dipterorum. The Biosystematic Database of World Diptera. Version 1.5. http://www.diptera.org/.

Penney, H. D., Hassall, C., Skevington, J. H., Abbott, K. R. & Sherratt, T. N. (2012) A comparative analysis of the evolution of imperfect mimicry. *Nature*, *483*, 461–466.

Penney, H. D., Hassall, C., Skevington, J. H., Lamborn, B. & Sherratt, T. N. (2014) The relationship between morphological and behavioral mimicry in hover flies (Diptera: Syrphidae). *The American Naturalist*, *183*, 281–289.

Rader, R., Bartomeus, I., Garibaldi, L. A., Garratt, M. P. D., Howlett, B. G., Winfree, R., et al. (2016) Non-bee insects are important contributors to global crop pollination. *Proceedings of the National Academy of Sciences*, *113*, 146–151.

*Reemer, M. & Ståhls, G. (2013a) Generic revision and species classification of the Microdontinae (Diptera, Syrphidae). *ZooKeys*, *288*, 1–213.

Reemer, M. & Ståhls, G. (2013b) Phylogenetic relationships of Microdontinae (Diptera: Syrphidae) based on molecular and morphological characters. *Systematic Entomology*, *38*, 661–688.

Roberts, M. J. (1970) The structure of the mouthparts of syrphid larvae (Diptera) in relation to feeding habits. *Acta Zoologica*, *51*, 43–65.

*Rojo, S., Gilbert, F. S., Marcos-Garcia, M. A., Nieto, J. M. & Mier, M. P. (2003) *A World Review of Predatory Hoverflies (Diptera, Syrphidae: Syrphinae) and Their Prey*. Alicante, Spain: CIBIO Ediciones. 319 pp.

Rotheray, G. & Gilbert, F. (1999) Phylogeny of Palearctic Syrphidae (Diptera): Evidence from larval stages. *Zoological Journal of the Linnean Society*, *127*, 1–112.

*Rotheray, G. & Gilbert, F. (2011) *The Natural History of Hoverflies*. Cardigan, UK: Forrest Text. 334 pp.

Rotheray, G. E. (1989) *Aphid Predators*. Exeter, UK: Pelagic Publishing.

Rotheray, G. E. (1993) *Color Guide to Hoverfly Larvae (Diptera, Syrphidae)* (Vol. 9). Sheffield, England: Derek Whiteley. 156 pp.

Rotheray, G. E. & Gilbert, F. S. (1989) The phylogeny and systematics of European predaceous Syrphidae (Diptera) based on larval and puparial stages. *Zoological Journal of the Linnaean Society*, *95*, 29–70.

*Saigusa, T. (2006) *Homology of Wing Venation of Diptera*. Fukuoka, Japan: Privately published. 26 pp.

*Sedman, Y. S. (1964) The *Chrysogaster (Orthonevra) bellula* group in North America. *Proceedings of the Entomological Society of Washington, 66*, 169–176.

*Sedman, Y. S. (1966) The *Chrysogaster (Orthonevra) pictipennis* group in North America. *Proceedings of the Entomological Society of Washington, 68*, 185–194.

*Shannon, R. C. (1916) Notes on some genera of Syrphidae with descriptions of new species. *Proceedings of the Entomological Society of Washington, 18*, 101–113.

*Shannon, R. C. (1926a) The Chrysotoxine Syrphid-flies. *Proceedings of the United States National Museum, 69*, 1–20.

*Shannon, R. C. (1926b) Review of the American Xylotine Syrphid-flies. *Proceedings of the United States National Museum, 69*, 1–52.

Shannon, R.C. (1939) *Temnostoma bombylans* and related species (Syrphidae. Diptera) *Proceedings of the Entomological Society of Washington, 41*, 215–224.

*Skevington, J. H. & Thompson, F. C. (2012) Review of New World *Sericomyia* (Diptera, Syrphidae), including description of a new species. *The Canadian Entomologist, 144*, 216–247.

Skevington, J. H. & Yeates, D. K. (2000) Phylogeny of the Syrphoidea (Diptera) inferred from mtDNA sequences and morphology with particular reference to classification of the Pipunculidae (Diptera). *Molecular Phylogenetics and Evolution, 16*, 212–224.

Sommaggio, D. (1999) Syrphidae: Can they be used as environmental bioindicators? *Agriculture, Ecosystems and Environment, 74*, 343–356.

*Speight, M. C. D. (2016) *Species Accounts of European Syrphidae (Diptera), 2016* (Vol. 93). Dublin: Syrph the Net Publications. 289 pp.

*Speight, M. C. D., Hauser, M. & Withers, P. (2013) *Eumerus narcissi* Smith (Diptera, Syrphidae), presence in Europe confirmed, with a redescription of the species. *Dipterists Digest, 20*, 17–32.

Ssymank, A., Kearns, C. A., Pape, T. & Thompson, F. C. (2008) Pollinating flies (Diptera): A major contribution to plant diversity and agricultural production. *Biodiversity, 9*, 86–89.

Ståhls, G., Hippa, H., Rotheray, G., Muona, J. & Gilbert, F. (2003) Phylogeny of Syrphidae (Diptera) inferred from combined analysis of molecular and morphological characters. *Systematic Entomology, 28*, 433–450.

Ståhls, G. & Nyblom, K. (2000) Phylogenetic analysis of the genus *Cheilosia* (Diptera, Syrphidae) using mitochondrial COI sequence data. *Molecular Phylogenetics and Evolution, 15*, 235–241.

Ståhls, G., Stuke, J.-H., Vujic, A., Doczkal, D. & Muona, J. (2004) Phylogenetic relationships of the genus *Cheilosia* and the tribe Rhingiini (Diptera, Syrphidae) based on morphological and molecular characters. *Cladistics, 20*, 105–122.

Ståhls, G., Vujić, A. & Milankov, V. (2008) *Cheilosia vernalis* (Diptera, Syrphidae) complex: Molecular and morphological variability. *Annales Zoologici Fennici, 45*, 149–159.

Stubbs, A. E. & Falk, S. (2002) *British Hoverflies: an Illustrated Identification Guide. 2nd*. British Entomological and Natural History Society. 469 pp.

*Telford, H. S. (1970) *Eristalis* (Diptera: Syrphidae) from America north of Mexico. *Annals of the Entomological Society of America, 63*, 1201–1210.

*Thompson, F. C. (1974) The genus *Pterallastes* Loew (Diptera: Syrphidae). *Journal of the New York Entomological Society, 82*, 15–29.

Thompson, F. C. (1975) Notes on the status and relationships of some genera in the tribe Milesiini (Diptera: Syrphidae). *Proceedings of the Entomological Society of Washington, 77*, 291–305.

*Thompson, F. C. (1980) The North American species of *Callicera* Panzer (Diptera: Syrphidae). *Proceedings of the Entomological Society of Washington, 82*, 195–211.

*Thompson, F. C. (1981) Revisionary notes on Nearctic *Microdon* flies (Diptera: Syrphidae). *Proceedings of the Entomological Society of Washington, 83*, 725–758.

*Thompson, F. C. (1997) *Spilomyia* flower flies of the New World (Diptera: Syrphidae). *Memoirs of the Entomological Society of Washington, 18*, 261–272.

*Thompson, F. C. (1999) A key to the genera of the flower flies (Diptera: Syrphidae) of the Neotropical region including descriptions of new genera and species and a glossary of taxonomic terms. *Contributions on Entomology, International, 3*, 319–378.

*Thompson, F. C. (2012) Fabulous flower flies for famous fly fanatics (Diptera: Syrphidae). A tribute to the dipterists of the Canadian National Collection. *The Canadian Entomologist, 144*, 1–16.

Thompson, F. C., Fee, F. D. & Berzark, L. G. (1990) Two immigrant synanthropic flower flies (Diptera: Syrphidae) new to North America. *Entomological News, 101*, 69–74.

Thompson, F. C., Rotheray, G. E. & Zumbado, M. A. (2010) Syrphidae (Flower Flies). *In:* B. V. Brown, A. Borkent, J. M. Cumming, D. M. Wood, N. E. Woodley & M. A. Zumbado (Eds), *Manual of Central American Diptera*. NRC Research Press, Ottawa, pp. 763–792.

van Veen, M. P. (2004) *Hoverflies of Northwest Europe: Identification Keys to the Syrphidae*. Utrecht, Netherlands: KNNV Publishing. 254 pp.

Vockeroth, J. R. (1969) A revision of the genera of the Syrphini (Diptera: Syrphidae). *Memoirs of the Entomological Society of Canada, 62*, 1–176.

Vockeroth, J. R. (1980) A review of the Nearctic species of *Melangyna* (*Meligramma*) Frey (Diptera: Syrphidae). *The Canadian Entomologist, 112*, 775–778.

Vockeroth, J. R. (1983) Nomenclatural notes on Nearctic Syrphinae, with descriptions of new species of *Syrphus* and keys to Nearctic species of *Didea*, *Epistrophe* s. str., and *Syrphus* (Diptera: Syrphidae). *The Canadian Entomologist, 115*, 175–182.

*Vockeroth, J. R. (1986a) Nomenclatural notes on Nearctic *Eupeodes* (including *Metasyrphus*) and *Dasysyrphus* (Diptera: Syrphidae). *The Canadian Entomologist, 118*, 199–204.

*Vockeroth, J. R. (1986b) Revision of the New World species of *Paragus* Latreille (Diptera: Syrphidae). *The Canadian Entomologist, 118*, 183–198.

*Vockeroth, J. R. (1992) *The Flower Flies of the Subfamily Syrphinae of Canada, Alaska and Greenland* (Vol. 18). Ottawa: Canada Communications Group – Publishing. 456 pp.

Vockeroth, J. R. & Thompson, F. C. (1987) Syrphidae. *In:* J. F. McAlpine, B. V. Peterson, G. E. Shewell, H. J. Teskey, J. R. Vockeroth & D. M. Wood (Eds), *Manual of Nearctic Diptera*. Canadian Government Publishing Center, Ottawa, pp. 713–743.

*Vujić, A., Ståhls, G., Ačanski, J., Bartsch, H., Bygebjerg, R. & Stefanović, A. (2013) Systematics of Pipizini and taxonomy of European *Pipiza* Fallén: molecular and morphological evidence (Diptera, Syrphidae). *Zoologica Scripta, 42*, 288–305.

Vujić, A., Ståhls, G., Rojo, S., Radenković, S. & Šimić, S. (2008) Systematics and phylogeny of the tribe Paragini (Diptera: Syrphidae) based on Molecular and morphological characters. *Zoological Journal of the Linnean Society, 152*, 507–536.

Webb, D. W. L. & Lisowski, E.A. (1985) The immature stages of *Orthonevra flukei* (Diptera: Syrphidae). *The Southwestern Naturalist, 30*, 312–315.

Weems Jr., H. V. (1953) Notes on collecting syrphid flies (Diptera: Syrphidae). *Florida Entomologist, 36*, 91–98.

Williston, S. W. (1887) Synopsis of the North American Syrphidae. *Bulletin of the United States National Museum, 31*, 1–335.

Williston, S. W. (1891) *Family Syrphidae* (Vol. 3). London. 63 pp.

Wirth, W. W., Sedman, Y. S. & Weems, H. V. J. (1965) Family Syrphidae. *In:* A. Stone, C. W. Sabrosky, W. W. Wirth, R. H. Foote & J. R. Coulson (Eds), *A Catalog of the Diptera of America North of Mexico*. United States Department of Agriculture, Washington, pp. 557–625.

*Wootton, R. J. & Ennos, A. R. (1989) The implications of function on the origin and homologies of the dipterous wing. *Systematic Entomology, 14*, 507–520.

Young, A. D., Lemmon, A. R., Skevington, J. H., Mengual, X., Ståhls, G., Reemer, M., et al. (2016) Anchored enrichment dataset for true flies (order Diptera) reveals insights into the phylogeny of flower flies (family Syrphidae). *BMC Evolutionary Biology, 16*, 1–13.

*Young, A. D., Marshall, S. A. & Skevington, J. H. (2016) Revision of Nearctic *Platycheirus* Lepeletier and Serville (Diptera: Syrphidae). *Zootaxa, 4082*, 1–317.

# Index

# About the Authors

**Jeff Skevington** is a research scientist at the Canadian National Collection of Insects, Arachnids and Nematodes (CNC) in Ottawa, Canada, where he studies the taxonomy and phylogenetics of Syrphidae and related flies. He did his MSc at the University of Guelph, Canada, with Steve Marshall and his PhD in Brisbane, Australia, with David Yeates.

**Michelle Locke** is a collection management technician at the CNC and previously worked at the American Museum of Natural History. Her work involves all groups of insects but she specializes in flower flies. She revised the taxonomy of *Dasysyrphus* as part of her MSc training at Carleton University with Jeff Skevington.

**Andrew Young** recently completed his PhD with Jeff Skevington at Carleton University. His PhD focused on the phylogenetics of Syrphidae and related flies (worldwide) and a revision of Australian *Psilota*. He did his MSc project on Nearctic *Platycheirus* at the University of Guelph with Steve Marshall and Jeff Skevington.

**Kevin Moran** is a doctoral student working with Jeff Skevington at Carleton University. His PhD focuses on the phylogenetics of Eristalinae and a revision of world *Criorhina* and related species. He completed an internship at the Smithsonian Institution in Washington, DC, with Torsten Dikow (studying *Temnostoma* and *Euscelidia* [Asilidae]).

**Bill Crins** is recently retired and spent much of his career working with the Parks and Protected Areas Program with the Ministry of Natural Resources in Peterborough, Ontario. His PhD from the University of Toronto was on sedge systematics, but he has a longtime passion for hover flies and has supported many student projects involving syrphids.

**Steve Marshall** is a professor at the University of Guelph, where he also did his graduate work. He is the world expert on Sphaeroceridae and Micropezidae and holds the C. P. Alexander Award that acknowledges him as the leading North American dipterist. He is well known for his books on flies, beetles and other insects and has an upcoming book on aquatic insects.